MW01012687

About Island Press

Since 1984, the nonprofit Island Press has been stimulating, shaping, and communicating the ideas that are essential for solving environmental problems worldwide. With more than 800 titles in print and some 40 new releases each year, we are the nation's leading publisher on environmental issues. We identify innovative thinkers and emerging trends in the environmental field. We work with world-renowned experts and authors to develop cross-disciplinary solutions to environmental challenges.

Island Press designs and implements coordinated book publication campaigns in order to communicate our critical messages in print, in person, and online using the latest technologies, programs, and the media. Our goal: to reach targeted audiences—scientists, policymakers, environmental advocates, the media, and concerned citizens—who can and will take action to protect the plants and animals that enrich our world, the ecosystems we need to survive, the water we drink, and the air we breathe.

Island Press gratefully acknowledges the support of its work by the Agua Fund, Inc., Annenberg Foundation, The Christensen Fund, The Nathan Cummings Foundation, The Geraldine R. Dodge Foundation, Doris Duke Charitable Foundation, The Educational Foundation of America, Betsy and Jesse Fink Foundation, The William and Flora Hewlett Foundation, The Kendeda Fund, The Andrew W. Mellon Foundation, The Curtis and Edith Munson Foundation, Oak Foundation, The Overbrook Foundation, the David and Lucile Packard Foundation, The Summit Fund of Washington, Trust for Architectural Easements, Wallace Global Fund, The Winslow Foundation, and other generous donors.

Ecosystem-Based Management for the Oceans

Ecosystem-Based Management for the Oceans

edited by

Karen McLeod and Heather Leslie

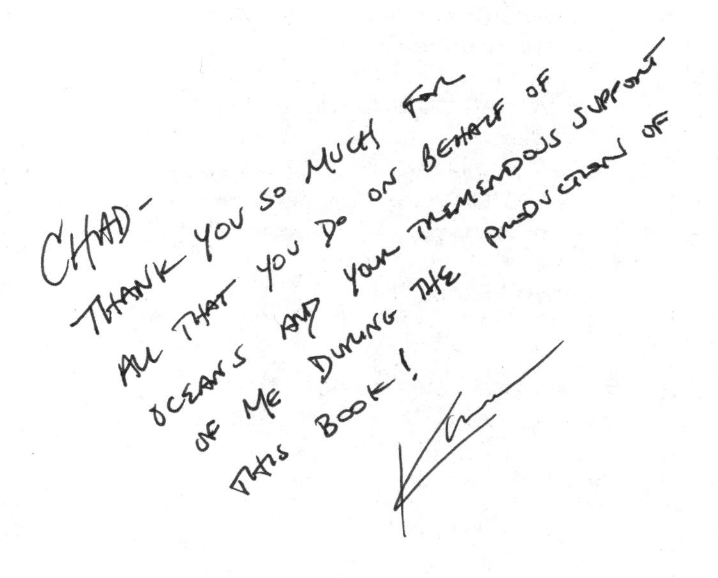

Washington • Covelo • London

Library of Congress Cataloging-in-Publication Data
McLeod, Karen, 1972-
Ecosystem-based management for the oceans / Karen McLeod and Heather Leslie.
p. cm.
ISBN-13: 978-1-59726-154-8 (cloth : alk. paper)
ISBN-10: 1-59726-154-8 (cloth : alk. paper)
ISBN-13: 978-1-59726-155-5 (pbk. : alk. paper)
ISBN-10: 1-59726-155-6 (pbk. : alk. paper) 1. Marine ecosystem management. I. Leslie, Heather, 1974- II. Title.
QH541.5.S3M374 2009
577.7—dc22 2008044994

Printed on recycled, acid-free paper
Manufactured in the United States of America

10 9 8 7 6 5 4 3 2 1

For Fiona and Isaac,
whose enthusiasm and sense of adventure inspire

Contents

Foreword
Lessons from the Ice Bear

Jane Lubchenco

I held my breath as the bear cub peered over its mother's back, sneaking a peek at our ship. Through my binoculars, the cub's nose and eyes were startlingly black against its white face. Keeping close watch, the mother was partially obscured behind a large boulder-shaped block of ice, one of many motley chunks ice-welded to the sparkling floe. After watching us watch her, she emerged from behind the ice barrier, sniffing the air, then climbed effortlessly up and down steep slabs of ice and leaped across the water to another frozen platform. The cub stayed close, revealing itself to be much larger than we initially thought, likely around 2 ½ years old. Not quite as adept as its mom, the cub slipped slightly on its jump—rump and hind legs splashing in the water before it regained control on the far side of the floe. The comical moment triggered general laughter, betraying our collective delight in witnessing such exquisite creatures in their home.

My colleagues and I were visiting the high Arctic of Svalbard to see ice-dependent ecosystems, including these "ice bears," as they are called in Norway. The scientific evidence that these species are at great risk as the global climate changes is compelling in and of itself. But witnessing the stark beauty of the place, observing different species' superb adaptations to this icy world, understanding the complex interconnections among the species, and seeing the precarious nature of the sea ice on which they all depend transformed that knowledge into deep concern. Here, at the top of the world, in the bounteous light of polar summer, the reality of climate change was tangible—nay, visceral. So too were the interconnections among species in the ecosystem and the need for responsible action and wise stewardship.

We saw, and felt, that polar bears are intimately connected to the rest of their ice-centric ecosystem. We spotted over a hundred ringed seals—the staple food of the ice bear—who not only mate, rest, molt, and pup on pack ice, but feed on ice-associated crustaceans and fish, mostly polar cod. We caught glimpses of the bears' alternate prey—swimming pods of the social harp seals out fishing. And we watched glaucous and ivory gulls hovering around a ringed seal kill, biding their time until the bear was sated and leftovers were available. What was less obvious was the hidden portion of the ecosystem—the minute plankton that are the base of this spectacular food web and the bottom-dwelling invertebrates that nourish fishes, seals, walruses, and more. Bears depend upon—and they link—all of these species, above, on, and under the ice. Although no native peoples currently live in Svalbard, Dorset and Thule artifacts and Inuit stories, skills, and art from elsewhere in the Arctic reflect the key role of bears in their lives. Ice bears symbolize the entire web of life in the Arctic.

The high-latitude-sea ecosystem we found so riveting provides a compelling case for why new approaches to ecosystems are so urgently needed and why this book is so timely. This Arctic Ocean is, in fact, a microcosm of all ocean ecosystems: rich in its beauty, bounty, and history but fragile in its susceptibility to unsustainable practices on land and in the oceans. Climate change is but one of many threats. The fact that the top nonhuman predator in the Arctic lives on the air side of the ice and is more easily seen by humans means that the stressed nature of this Arctic system is obvious to even casual observers and easily documented by photographers. In contrast, the threats to other ocean apex predators, such as sharks that live beneath the waves, and the disrupted nature of their ecosystems, remain mostly hidden.

Indeed, all ocean ecosystems are changing rapidly—from the top predators like bears to the base of the food web, the plankton. Ocean ecosystems are undergoing rapid disruption and depletion due to the combination of climate change, overfishing and destructive fishing gear, pollution, habitat destruction, and ocean acidification. Polar bears who now serve as the icon for climate change could well represent all apex ocean predators, everywhere: tunas, sharks, marlin, swordfish, orcas, sperm whales, and others. Although the plight of polar bears is more commonly recognized than that of these other charismatic species, they are all threatened because they and their ecosystems are under assault.

And people, too, are impacted by ocean degradation. Humans depend upon ocean ecosystems for a wealth of benefits—food, oxygen, protection of shores from storm damage, nutrient cycling, climate regulation, recreation, inspiration, cultural heritage, and religious value. All of these benefits are at risk as oceans are depleted and disrupted. Witness the demise of more and more fisheries, the vulnerability of homes and people on coastlines bereft of mangroves, the increasing blooms of toxic algae or nuisance jellyfish, and the conversion of spectacular coral reefs to weedy wastelands. As awareness of the seriousness of the threats grows, and understanding of the consequences dawns, calls for solutions become more intense. Discussions about how to address climate change abound today; calls for solutions to ocean degradation are just beginning.

I've heard encouraging calls for new approaches over the last decade, from governments, civil society, business, faith-based communities, and new alliances across traditional lines. For example, the Pew Oceans Commission and the US Commission on Ocean Policy issued compelling findings about US oceans in trouble and strong recommendations for vastly improved management, leadership, and scientific information. The European Union, Canada, and Australia have issued similar reports and passed new legislation in support of reform. From a moral perspective, the Forum on Religion and Ecology; six Religion, Science and the Environment symposia; the Earth Charter; and evangelical Christians' "creation care" all articulate a deep-rooted ethical responsibility for stewardship of creation. Big businesses such as Unilever, Wal-Mart, the Compass Group, and Wegmans are increasingly showing leadership in their pledges to sell only sustainably caught or farmed seafood. Their actions are driven by new awareness of the plight of oceans, by a sense of responsibility, and by market opportunities in response to consumer demand. Third-party seafood certifiers such as the Marine Stewardship Council; novel partnerships among fishermen, fish buyers, and fish consumers such as the Seafood Choices Alliance; or consumer information providers such as the Monterey Bay Aquarium, Environmental Defense Fund, and Blue Ocean Institute all reflect a growing interest in making fishing—one of the sectors with the biggest impact on ocean ecosystems—more sustainable. Numerous local, national, and international conservation organizations educate citizens about ocean disruption, its consequences, and solutions. And some, like Environmental Defense Fund, work directly with fishermen and policymakers to align economic and conservation incentives.

Scientific knowledge underpins all of the above actions, providing understanding to inform choices about oceans. Some of the new scientific information is sector-specific: How can fishing (or aquaculture, or coastal development, etc.) be more sustainable? But much of the emerging scientific information points to the need for more holistic approaches, managing across, not just within, sectors: What combination of activities is sustainable? How should we manage various activities in light of climate change? How can we manage for resilience of coupled natural–social ecosystems?

Likewise, the search for solutions is prompting more holistic approaches to ocean management. Traditional management focuses on individual sectors, activities, or problems such as fisheries, oil and gas exploration and extraction, coastal development, recreation, and endangered species. Historically, each of these was, and for the most part still is, considered and regulated separately. Doing so inevitably misses the key role of habitat, biodiversity, complexity, and connectedness of the pieces to the whole system. There is increasing recognition that good sectoral management is necessary and important, but it's not enough. From this realization has emerged a new approach. Ecosystem-based management, also called "the ecosystem approach," is beginning to consider the interdependencies, to integrate the collective activities, and to consider the cumulative impacts of the relevant activities on the ecosystem. Ecosystem-based management provides a mechanism for making decisions about those activities in light of the goal of maintaining (or restoring) the ecosystem in (to) a healthy, productive, and resilient state. In this new approach, the system is the focus. The system sustains the pieces. And, because the sectors are numerous and the drivers of the degradation are complex and multiple, solutions must be comprehensive.

The science and the art of this more holistic integration are the subjects of this book. The chapters herein provide an excellent introduction to the developing knowledge and practical experiences of ecosystem-based management of coastal and marine places. While there has been a growing awareness of the need for more holistic and integrated marine management practices in recent years (witness the Pew Oceans Commission, the US Commission on Ocean Policy, and the Millennium Ecosystem Assessment), there have been few specific guidelines for best practices. Drawing upon the expertise of more than forty of the world's experts in this area, this book synthesizes the science relevant to EBM to inform both policy and practice.

The focus on ecosystems should not be construed as the elevation of ecosystems over people, of nature over jobs, or of bears over progress. Rather, the focus on ecosystems reveals the ultimate dependence of people upon the benefits provided by ecosystems. Human well-being depends on healthy, productive, resilient ecosystems, and vice versa. The ecosystem approach is an expression of an emerging imperative to care for, understand, and use marine ecosystems in a fundamentally different fashion than current practice. It reflects the growing awareness that people not only are responsible for, but also benefit from good stewardship. It reflects the reality that using ecosystems inherently involves trade-offs and that choices based on a full understanding of those trade-offs will be better options.

Ecosystem-based management—including both policies and practices—is the immediate focus of this book, but the needs go far beyond just management. Management should reflect our values, goals, and knowledge and should be enabled by our institutions. Thus, we also need ecosystem-based understanding and education; ecosystem-based monitoring, research, and training; and ecosystem-based institutions. And, of course, these activities must be integrated or in close communication. Integrated research, monitoring, training, and outreach programs at the large-marine-ecosystem scale are urgently needed in every large marine ecosystem. Valuable lessons have been learned about how to organize and manage such programs, for example, by the Partnership for Interdisciplinary Studies of Coastal Oceans, or

PISCO, a consortium of scientists across academic institutions that is focused on the nearshore portion of the California Current large marine ecosystem off the coasts of Washington, Oregon, and California. New knowledge enables better management.

Make no mistake: The challenges inherent in transitioning to these ecosystem-based ways of thinking and acting are considerable. But, they are absolutely necessary. Humans can use land and ocean resources wisely. The keys to doing so are to (1) enable much broader awareness of the consequences of current actions and opportunities for alternatives; (2) acknowledge the ethical responsibilities that people have to one another and to all of life on Earth; (3) reject the current trajectory and choose a more sustainable path; (4) incorporate current knowledge into new practices, policies, and institutions; and (5) create new mechanisms to acquire, share, and use scientific information at the ecosystem scale to inform future decisions. The result will not be a naive attempt to return to some past state, nor a distorted view of current realities, nor an impossible dream, but viable options for the future.

The increasingly recognized plight of the ice bear and the Arctic foreshadows the growing awareness that business-as-usual is not sustainable, that current land-based and ocean-based practices are collectively disrupting and depleting every ocean ecosystem on Earth. This depletion and disruption threaten the future of humanity because they impair the delivery of key ecosystem services, that is, the benefits from oceans that sustain and enrich people's lives and livelihoods, as well as those of bears.

Polar bears are the iconic species of climate change, but they are much more. They serve to remind us not only of the need to reduce greenhouse gas emissions, but also of how connected people and other species are—even at the ends of the Earth. The Arctic is changing dramatically, warming at two to three times the rate of the rest of the planet. The melting of polar sea ice is indeed threatening not only polar bears, but the ring seals upon which they depend and, indeed, the entire web of life that is associated with ice: phytoplankton, zooplankton, benthic invertebrates, fishes, seals, walruses, seabirds, bears, whales, and native peoples. The fragility of the system reminds us of our global responsibility as stewards and our interest in maintaining the flow of services we need and want from all ecosystems. This book provides concrete assistance for understanding the complexity, the connections, and the resilience of ecosystems, in short, for becoming good stewards.

Preface
A Puget Sound Story

Anne D. Guerry

Our oceans and coasts are in jeopardy. Although they are vast and resilient, it is increasingly clear they are not limitless—the deleterious effects of overfishing, development, pollution, and other threats can be seen from the equator to the poles. Yet there are many reasons for hope. First, and perhaps foremost, people care deeply about the oceans. We live along coasts, walk along beaches, sail, surf, and dive, eat the ocean's bounty, and watch our children splash in the waves. Oceans support and flavor the human communities that surround them. Second, we have admitted that there is a problem with the current management of oceans and coasts and have begun to chart a way forward. There is broad consensus that oceans are in crisis and that new solutions are essential to restoring their health (POC 2003; USCOP 2004). In particular, the move toward more holistic ecosystem-based management has immense potential—and is already making strides toward ensuring the sustainability of marine systems and of the billions of human lives that are so intricately tied to them.

One place that is emblematic of the biological, social, and political challenges and opportunities facing our coasts and oceans is the majestic Puget Sound region of Washington State in the northwestern corner of the United States. It is a place of exquisite beauty and a rich and productive ecosystem, but all is not well beneath its waters. Fortunately, the region has a long history of human connection to the natural environment that is enabling its inhabitants to develop promising new approaches to ecosystem-based management.

Puget Sound is a dramatic sight. Nestled between the Olympic Mountains to the west and the Cascade Mountains to the east, the Puget Sound basin encompasses approximately 42,000 km^2, an area roughly the size of West Virginia. Surrounding the marine waters of the sound are dense urban areas, sprawling suburban neighborhoods, farms, oak woodlands, prairies, and the world's largest remaining tracts of temperate rain forest. Approximately ten thousand streams and rivers bring freshwater from the snowy crests of the mountains through the forests, fields, and cities to the marine waters of the sound.

The sound is, technically, a fjord—a system of flooded glacial valleys with steep banks. It averages about 130 m in depth but has troughs as deep as 280 m. The sound's 4,000 km of shoreline are composed of salt marshes, mudflats, beaches, bluffs, and rocky shores as well as jetties, piers, bulkheads, and other human-built structures. The Strait of Juan de Fuca, the sound's 150 km long pathway from the Pacific Ocean, creates tidal ranges as great as 4.6 m, yielding generous intertidal areas that are home to a dazzling array of organisms that live in this zone of not quite land, not quite sea.

Marine life throughout the sound is rich and varied. Primary productivity—the foundation of the food web, generated by phytoplankton, lush meadows of eelgrass, and over 625 species of seaweed—is relatively high compared with that in other temperate estuaries (Strickland 1983). Beneficiaries of the sound's productivity include two hundred species of fish, twenty-six marine mammal species, and thousands of invertebrate species (Sound Science 2007). Among the most notable of this cast of characters are some of the world's largest and most diverse runs of salmon and the iconic orcas.

The Puget Sound region also supports a burgeoning population of over 3.5 million people. Human residents of the basin reside in twelve counties and four cities with populations greater than a hundred thousand. Nineteen Native American tribes, eight with reservations and most with treaty-reserved rights for harvesting shellfish and finfish, also call the Puget Sound basin home.

The human population of this region depends on the sound for a wide array of ecosystem services (the benefits humans derive from natural systems), creating a diversity of (often competing) interests and stakeholders. Aquaculture and commercial and recreational fisheries are an important component of local economies and identities. Revenues from aquaculture and commercial (including tribal) harvest in Puget Sound averaged over $88 million from 2000 to 2007 (Plummer, unpub.). And people love to fish here for salmon—more than 50% of the salmon caught recreationally in Washington State are caught in the Puget Sound basin (PSP 2008).

The sound also provides for a broad range of less obvious but no less critical benefits. The sound's food webs support those species we like to eat, catch, and watch. Bivalves such as geoducks and clams pump water across their gills and clean it in the process. Eelgrass habitats sequester carbon and thereby help regulate the Earth's climate; lock up and degrade toxins in their sediments, keeping them from entering food webs; and—along with kelp beds and other complex nearshore habitats—stabilize shorelines and prevent erosion. These services are difficult to value directly but are nonetheless vital.

Puget Sound is also an essential marine transportation corridor, linking the Seattle region with the rest of the world and moving local people to work and play. Looking out from a bluff at any given time, one might see a ferry lumbering across the sound, a container ship arriving from China, a tugboat, a few sailboats, and a kayak. The Seattle–Tacoma port combined is the second largest US port for container traffic, over 25 million people ride ferries in Puget Sound in a given year, and the sound is a nexus for recreational and commercial boating (PSP 2008).

Opportunities for outdoor activities and the scenic beauty of the sound are important in numerous tangible and intangible ways. The sound plays a critical role in drawing tourists, with an estimated 80% of statewide revenues from tourism and travel generated in the Puget Sound region (PSP 2008). In addition, people simply want to live near the sound, stimulating the construction industry. For example, nearly 20% of all housing units in the Seattle–Everett metropolitan region were within 300 feet of bodies of water in 2004, and nearly 20% of new construction in the previous 4 years occurred within this zone (HUD 2005). One study estimated that a high-quality view of the sound can increase the market price of an otherwise comparable home by up to 60% (Benson et al. 1998). No less important are the more abstract benefits of educational value, cultural importance, and spiritual sustenance.

As humans have prospered from all these ecosystem benefits, Puget Sound has suffered. Water quality has been degraded, habitats have been destroyed, and the future of many species has been put at risk. All of the thirty-nine Puget Sound sites regularly monitored for fecal coliform, nitrogen, ammonium, dissolved oxygen, and stratification showed levels of at least some concern for at least one monitored parameter in 2005. In the same year, nearly a third of the sound's shellfish-growing areas had fecal coliform levels that triggered harvest restrictions (PSAT 2007). Two icons of the region, salmon and orcas, have high concentrations of toxics in their tissues. In fact, the state department of health has advised that people consume no more than two meals of Puget Sound chinook salmon per month. Puget Sound's southern resident orcas do not listen to these advisories. Their levels of PCBs (polychlorinated biphenyls) and PBDEs (polybrominated diphenyl ethers) are three to four times higher than their northern counterparts, and such contamination was one of the main reasons they have been listed under the Endangered Species Act. If orca blubber was regulated under the US Environmental Protection Agency's disposal requirements for PCB articles, disposal of stranded orca carcasses could get quite complicated. The blubber from one transient female orca that washed up dead on the beach in Puget Sound had concentrations of 1,300 ppm (Krahn et al. 2004); concentrations of 500 ppm trigger regulations for PCB hazardous waste disposal (EPA 2007).

Although it is easy to assume that human behavior on land is of marginal relevance to the marine environment, human actions are intricately linked to the sound. According to one monitoring program, one of the region's signature substances, caffeine, was found in 55% of the offshore water-column samples taken from around the sound; it was found in one sample taken from a depth of 195 m (King County 2001). Also emblematic of the intimate connection between human actions and the state of the sound, a University of Washington study even found a spike in vanilla and cinnamon in treated sewage headed for the sound during the holiday season (Keil 2008).

The cumulative impacts of innumerable local decisions are taking their toll on the sound. For example, a growing human population in the region increased impervious surfaces—rooftops, sidewalks, roads, and parking lots—in regional watersheds by 10% between 1991 and 2001. These increases lead to greater storm water runoff, decreased water quality, and degraded aquatic and terrestrial habitats. Concomitant decreases in forest cover also threaten habitat and water quality (PSAT 2007). Degraded habitat has, in turn, affected some of the sound's signal species. Sound-wide, chinook salmon are currently at 10% of historic numbers, with some basins currently supporting as little as 1% of the historic population size (SSPS 2007). The total population of wintering marine birds in northern Puget Sound decreased by nearly 30% from 1978 to 2004. And ten marine-dependent Puget Sound species are listed as endangered or threatened on federal or state endangered species lists (PSAT 2007).

Clearly, existing management of human activities that affect the sound is insufficient. Current management in the Puget Sound region, as in other regions, is based on a sector-by-sector model. Separate departments, agencies, or programs manage fisheries, aquaculture, water quality, transportation, and land use. Managing each sector in isolation ignores the inherent connections between them and effectively hides trade-offs and synergies between ecosystem services. The single-sector management approach focuses on each of the puzzle pieces, with no one responsible for putting those pieces together to see the whole.

Emblematic of the problem is the jurisdictional tangle affecting the life of a single Puget Sound chinook salmon. A chinook salmon born in a clear pool in the lower Snoqualmie River in the central section of the sound is affected by decisions made by a bewildering array of organizations, including the Snohomish Basin Salmon Recovery Forum; the Washington State Departments of Ecology, Fisheries and Wildlife, and Transportation; the US National Marine Fisheries Service; the Pacific Fisheries Management Council; the Tulalip tribal government; the Northwest Indian Fisheries Commission; and the Canadian Department of Fisheries and Oceans. As in other regions, the layers of jurisdiction in Puget Sound are complex, making coordination of regional management at once more difficult and more critical.

Given the current state of the sound and

projections for a human population increase of another 1.7 million residents by 2025, it is clear that some difficult choices are going to have to be made. Fortunately, people in the Puget Sound region are strongly connected to the natural world. They are known for their appreciation of their often wet, but always spectacular, natural environment. In a survey of citizens across the central Puget Sound region, 80% of respondents thought that their town was a good or excellent place to live, with natural environment/beauty (31%), climate (22%), and recreation/entertainment opportunities (18%) cited most frequently as best liked. Interestingly, 96% of all respondents agreed that it is important for cities and towns within the region to coordinate their planning activities, and 67% favored concentrating growth into already developed areas (Northwest Research Group 2004).

Indicative of the cultural capital for conservation in this region, Puget Sound has numerous examples of how communities are addressing multiple diffuse threats. One such example is the recent effort to restore Drayton Harbor in northern Puget Sound. After a long history of commercial, recreational, and tribal harvest of shellfish in the harbor, degradation of water quality forced the Washington State Department of Health to close the bay to harvest in 1999. In 2001, the Puget Sound Restoration Fund led a community-based effort to plant oyster seed in the harbor, publicize that effort, and then use the 3 years it would take to grow mature oysters to investigate and mitigate the sources of water pollution that would prevent the harvesting of those oysters. Progress toward cleaning up the bay was reported quarterly in the local newspaper, and after replacing cracked sewer lines, tightening leases regulating discharges in local marinas, monitoring storm water samples, studying circulation patterns, and ramping up public education, a portion of the bay was opened to shellfish harvest in 2004. Proceeds from the sale of oysters from the Drayton Harbor Community Oyster Farm were used to fund further restoration activities. Water quality has not yet been improved enough to ensure the viability of commercial oyster harvesting in Drayton Harbor, but efforts to address nonpoint source threats in the watershed are ongoing.

Moving from one bay and its challenges to the whole of the sound takes bold thinking, political sophistication, and a regional approach to coordination. Yet the people of Puget Sound are finding ways to tackle the problem. For example, recognizing the magnitude of the cumulative effects of innumerable local land use decisions on the sound, Washington State passed a Growth Management Act that, though imperfect, is a powerful tool for conservation and one of the country's most progressive attempts at comprehensive growth management. It mandates the regulation of development to protect the ecological functions of critical areas and has played a key role in slowing the state's per capita consumption of land since its promulgation in 1990. Similarly, the Shoreline Management Act calls for no net loss of ecological functions of shoreline areas. However, these and other current management tools, though wide-ranging, are insufficiently comprehensive and flexible to fully address the complex and interdependent threats facing Puget Sound.

National calls for comprehensive ecosystem-based management of marine systems have been heard in this region, illustrated most recently by the formation of the Puget Sound Partnership in 2007. This new state agency is charged by the governor and state legislature with restoring and protecting the entire Puget Sound ecosystem, from the crests of the Cascade and Olympic mountains to the depths of the sound. The partnership is not a regulatory entity—instead they are attempting to coordinate governmental and private resources at an ecosystem scale, set priorities with input from communities and scientists, and create an action plan for a healthy ecosystem and human population. In support of the partnership, the National Oceanic and Atmospheric Administration (NOAA) is spearheading several efforts that will contribute directly to ecosystem-based management of the sound. Two of these include conducting the first integrated ecosystem assessment of the Puget Sound ecosystem (see Ruckelshaus et al., chap. 12 of this volume) and an explicit examination of the ecosystem services provided by the sound and potential trade-offs among them.

Hallmarks of the partnership's progressive ecosystem-based approach include the development of a solid scientific foundation by a well-respected science panel, an action plan that is flexible and emphasizes adaptive management, and an inclusive and transparent process for decision making that involves those who are most affected by the outcomes. Additionally, the partnership integrates upland and marine ecosystem analysis and management and recognizes that Puget Sound ecosystem processes are intimately connected to Canada's Georgia Basin to the north. And finally, as exemplified by their charter, the partnership places humans squarely within the Puget Sound ecosystem; human health and well-being are central to the explicitly stated objectives of the ecosystem plan. This shift to an overarching perspective fundamentally changes the necessary analyses, the framework for identifying indicators, and the ways in which strategies are prioritized. The formation and funding of the Puget Sound Partnership were recently heralded by the national Joint Ocean Commission Initiative as an example of progress toward regional and state ocean governance reform.

In Puget Sound, as in many other marine systems around the world, the combination of population pressure, tight connections between actions on land and the condition of the marine ecosystem, and signs of degradation make the situation a difficult one. However, political and community will and empowerment, a tradition of community mobilization, and the bold new ideas embodied in ecosystem-based management—and fleshed out in this book—make the future of Puget Sound a realistically bright one.

Acknowledgments

I would like to thank Mary Ruckelshaus for encouragement, guidance, and review; Josh Lawler, Tim Dickinson, DuPont Guerry IV, Ann Seiter, and two anonymous referees for review; the National Research Council and NOAA for support; and the Puget Sound for inspiration.

References

Benson, E. D., J. L. Hansen, A. L. Schwarz, Jr., and G. T. Smersh. 1998. Pricing residential amenities: The value of a view. *Journal of Real Estate Finance and Economics* 16(1):55–73.

EPA (Environmental Protection Agency). 2007. Title 40, chap. 1, part 761. Polychlorinated biphenyls (PCBs) manufacturing, processing, storage, and disposal. 40CFR761.60. http://edocket.access.gpo.gov/cfr_2007/julqtr/40cfr761.60.htm.

HUD (US Department of Housing and Urban Development). 2005. American housing survey, Seattle–Everett metropolitan area 2004. US HUD and US Dept. of Commerce. Issued October 2005. http://www.census.gov/prod/2005pubs/h170-04-60.pdf.

Keil, R. 2008. Species in Puget Sound and in Seattle's sewage effluent. Online symposium. http://water.washington.edu/Outreach/Events/AnnualReview/2008AR/2008ARpresentations.html.

King County. 2001. Water quality status report for marine waters, 1999 and 2000. Seattle, WA: King County Dept. of Natural Resources.

Krahn, M. M., D. P. Herman, G. M. Ylitalo, C. A. Sloan, D. B. Burrows, R. C. Hobbs, B. A. Mahoney, G. K. Yanagida, J. Calambokidis, and S. E. Moore. 2004. Stratification of lipids, fatty acids and organochlorine contaminants in blubber of white whales and killer whales. *Journal of Cetacean Research and Management* 6:175–89.

Northwest Research Group. 2004. Vision 2020 scoping survey final report. Puget Sound Regional Council 2003. http://www.psrc.org/projects/vision/outreach/execsum.pdf.

Plummer, M. Unpublished. Analysis from PACFIN (Pacific Fisheries Information Network, Pacific States Marine Fisheries Commission, Portland, OR). http://www.psmfc.org/pacfin/overview.html.

POC (Pew Oceans Commission). 2003. *America's living ocean: Charting a course for sea change. A report to the nation.* Washington, DC: Pew Trusts.

PSAT (Puget Sound Action Team). 2007. State of the sound 2007. Office of the Governor, State of Washington. Publication no. PSAT 07-01. http://www.psat.wa.gov/Publications/state_sound07/sos.htm.

PSP (Puget Sound Partnership). 2008. Puget Sound facts. Olympia, WA: PSP Resource Center. http://www.psparchives.com/puget_sound/psfacts.htm.

Sound Science. 2007. Sound Science: Synthesizing ecological and socioeconomic information about the Puget Sound ecosystem. M. H. Ruckelshaus and M. M. McClure, coordinators; prepared in cooperation with the Sound Science collaborative team. Seattle, WA: National Oceanic and Atmospheric Administration (NMFS), Northwest Fisheries Science Center.

SSPS (Shared Strategy for Puget Sound). 2007. Puget Sound salmon recovery plan. Seattle, WA: SSPS. http://www.nwr.noaa.gov/Salmon-Recovery-Planning/Recovery-Domains/Puget-Sound/.

Strickland, R. M. 1983. *The fertile fjord: Plankton in Puget Sound.* Seattle: University of Washington Press.

USCOP (US Commission on Ocean Policy). 2004. *An ocean blueprint for the twenty-first century.* Final report. Washington, DC: USCOP. ISBN 0 9759462 0 X.

Acknowledgments

This book would not have been possible without the contributions of countless individuals, many of whom are named here, and some of whom will remain anonymous. We are particularly grateful to the talented pool of reviewers who gave valuable feedback on these chapters. This book would also not have been possible without logistical and administrative support from Currie Saray Dugas, Ginger Hopkins, Cindy Kent, and Lynn Rutter as well as the patience and perseverance of all of the staff at Island Press, especially Todd Baldwin, Emily Davis, and Sharis Simonian. Permission to use the cover photo taken in Sitka, Alaska was generously provided to us by Gary Bagley, an independent photographer who now resides in South Carolina.

We thank Jane Lubchenco for her encouragement and gracious sharing of ideas and opportunities that led to the development of this book. We also thank Andy Rosenberg for his role in bringing this book into being and for his strong leadership on this topic. In addition, we thank many other friends, colleagues, and mentors for their insights and challenges that shaped the ideas contained in this volume, especially Brian Baird, Mike Beck, Meg Caldwell, Lisa Campbell, Mark Carr, Sarah Carr, Billy Causey, Sarah Chasis, Elizabeth Chornesky, Chris Costello, Gretchen Daily, Jon Day, David Festa, Dave Fluharty, Mike Fogarty, Carl Folke, Helen Fox, Rod Fujita, Steve Gaines, Barry Gold, Emily Goodwin, Gary Griggs, Churchill Grimes, Lynne Zeitlin Hale, Ben Halpern, Dennis Heinemann, Burr Henemann, Terry Hughes, Glen Jamieson, David Keeley, Jim Kramer, Chris Krenz, John Largier, Judy Layzer, Kai Lee, Sarah Lester, Simon Levin, Amber Mace, Marc Mangel, Becky Mansfield, Kathryn Mengerink, Fiorenza Micheli, Steve Murawski, Elliott Norse, Mike Orbach, Julia Parrish, Ellen Pikitch, Kevin Ranker, Walt Reid, Steve Rumrill, Carl Safina, Jim Sanchirico, Maja Schlueter, Kristin Sherwood, Paul Siri, Louise Solliday, Will Stelle, Bob Steneck, Bill Sydeman, Heather Tallis, Mike Weber, Jim Wilson, and Amy Windrope.

We would also like to thank friends and colleagues at the Communication Partnership for Science and the Sea (COMPASS), past and present, for their support of this volume and enormous dedication to the communication of science to the wider world: Adina Abeles, Nancy Baron, Kasey Brown, Verna DeLauer, Chad English, Kirsten Grorud-Colvert, Kimberly Heiman, Sarah Lester, Dawn Martin, Pete McDougall, John Meyer, Liz Neeley, Steve Palumbi, Brooke Simler Smith, Vikki Spruill, Mike Sutton, and Matt Wright.

Special thanks to our husbands, J. Brock McLeod and Jeremy Rich, without whom this endeavor would not have been possible. We are incredibly grateful to Brock and Jeremy for their ongoing support, and for their generous parenting of Fiona and Isaac. We also thank our extended families and friends for their love and support.

This book was made possible through financial support from the David and Lucile Packard Foundation through a grant to Heather and to Simon Levin and through their support of COMPASS. Other support for Karen's position at COMPASS comes from the Gordon and Betty Moore Foundation, Meyer Memorial Trust, Resources Legacy Fund, and other charitable contributions. Additional support to Heather was provided by the Princeton Environmental Institute, Brown University's Center for Environmental Studies, and the Santa Fe Institute's Social Robustness Program through a grant from the James S. McDonnell Foundation.

Finally, we thank the multitude of contributors to this volume for their hard work, commitment, attention to detail, and most of all, passion. The strength of this book clearly lies in the diversity of perspectives, cultures, geographies, and personalities that make up this unique synthesis of the knowledge and practice of marine ecosystem-based management. We hope that you, our readers, are able to benefit and grow as scientists, practitioners, and citizens from this compilation, as much as we have during its creation.

Karen McLeod
Corvallis, Oregon

Heather Leslie
Providence, Rhode Island

PART 1
Setting the Stage

CHAPTER 1
Why Ecosystem-Based Management?

Karen L. McLeod and Heather M. Leslie

As illustrated by the preceding stories of ice bears and Puget Sound, ocean and coastal ecosystems around the globe are in trouble. Both the severity and scale of impacts to these systems—including those from climate change, biodiversity loss, overfishing, pollution, coastal development, habitat loss, and fragmentation—are increasing (MA 2005a, b), with no corner of the globe untouched (Halpern et al. 2008a). Acting in concert, these impacts decrease the ability of marine ecosystems to deliver vital ecosystem services to humankind, such as abundant seafood, clean water, renewable energy, and the protection of coastal areas from storm damage.

The unprecedented environmental challenges facing the oceans require us as scientists, practitioners, and citizens to embrace a broader vision than ever before of what we want to achieve through coastal and ocean management (UNEP 2006). This vision must encompass not only the long-term health of coasts and oceans, but also human well-being. Sustaining the long-term capacity of systems to deliver ecosystem services is the core goal of ecosystem-based management (EBM) for the oceans (Rosenberg and McLeod 2005). Moving forward with EBM requires synthesizing and applying knowledge from across the social and natural sciences as well as the humanities (Leslie and McLeod 2007) and raises numerous questions:

- How can we better account for the interactive and cumulative effects of the growing number of human activities affecting marine ecosystems?
- How well do we understand feedbacks between the social and ecological components of systems, and what are the broader implications of these linkages?
- In an increasingly dynamic world, how can management institutions respond more rapidly to changing, and often surprising, conditions?
- How can small-scale management decisions make a difference in light of large-scale change (especially climate change)?
- How can we better recognize that systems are approaching critical thresholds? In other words, how do we know how likely they are to shift to a fundamentally different state (e.g., from coral to algal dominance on tropical reefs) that will produce a radically different set of services?
- To what extent are these shifts reversible (especially over the scale of a human lifetime)? How can we identify and bolster attributes that decrease vulnerability to such shifts?

We tackle these and related questions in this book through synthesis of the science, policy, and practice of ecosystem-based management.

What Is Ecosystem-Based Management for the Oceans?

EBM is a new approach to managing the range of human activities that affect marine ecosystems, as called for by numerous national and international bodies (WSSD 2002; POC 2003; USCOP 2004). In 2005, more than two hundred academic scientists and policy experts from US institutions agreed by consensus on the following definition of EBM for the oceans: "Ecosystem-based management is an integrated approach to management that considers the entire ecosystem, including humans. The goal of ecosystem-based management is to maintain an ecosystem in a healthy, productive and resilient condition so that it can provide the services humans want and need. Ecosystem-based management differs from current approaches that usually focus on a single species, sector, activity or concern; it considers the cumulative impacts of different sectors" (McLeod et al. 2005). The following key elements are based on that definition:

1. *Connections* At its core, EBM is about acknowledging connections (Guerry 2005; Leslie and McLeod 2007), including first and foremost, the inextricable linkages between marine ecosystems and social systems. Human well-being is intimately connected to ecosystems through the delivery of ecosystem services across a range of scales. Cultures, economies, and institutions form and evolve in response to their local or regional ecosystem contexts. Human behavior, including the extent, intensity, and type of activity, affects natural systems. Humans interact with coasts and oceans as individuals (consumers, surfers, or fishers), as organizations (local fish markets or canneries), and as institutions (trade organizations, fishery management councils, or conservation organizations), each within a particular cultural context. These dynamic, linked systems of humans and nature are called "coupled social–ecological systems" (fig. 1.1).

   EBM is fundamentally a place-based approach, and coupled systems occur across a range of spatial scales from a local ecosystem, such as an individual estuary, to an entire large marine ecosystem, such as the California Current off the US west coast. Thus, there is no single "correct" scale at which to do EBM. Instead, it is an approach to be implemented over a range of scales, acknowledging the connections and leaky boundaries among scales.
2. *Cumulative impacts* EBM focuses on how individual actions affect the ecosystem services that flow from these coupled systems. In other words, what are the cumulative impacts of multiple activities, both within and among sectors, on the delivery of ecosystem services? Accounting for cumulative impacts also involves recognition of interactions with drivers of change that operate over smaller or larger scales than the scale of management (see Guichard and Peterson, chap. 5 of this volume, for more on cross-scale interactions). Inevitably, a comprehensive accounting of cumulative impacts requires that sectors ultimately work toward a common goal (as expanded upon by Rosenberg and Sandifer in chap. 2 of this volume).
3. *Multiple objectives* EBM focuses on the range of benefits that we receive from marine systems, rather than single ecosystem services. In a particular place, people want and expect to receive multiple services, which may include vibrant commercial and recreational fisheries, biodiversity conservation, renewable energy from wind or waves, coastal protection, diving, and sea kayaking. In order to fully implement EBM, we need to understand the connections among these services and the suite of factors affecting their production and delivery. This will allow us to make informed choices about trade-offs among the multiple objectives that are fundamental to a more

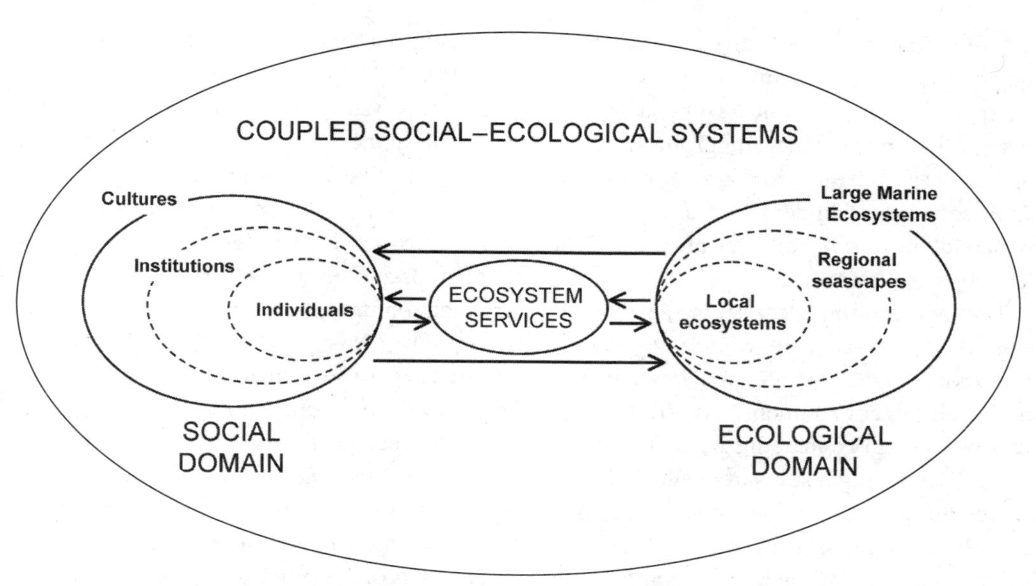

Figure 1.1 Dynamic human and ecological systems referred to as "coupled social–ecological systems." Interactions between the social and ecological domains occur over multiple geographic and organizational scales, and understanding connections across scales is critical to the long-term success of EBM efforts. While some domains may be relatively smaller in scale, such as individuals and institutions, they are not necessarily all nested. For example, cultures occur at geographic scales that are parallel to or larger than institutions. Ecosystem services represent a key connection between domains, and the flow of services is affected by both social and ecological factors.

comprehensive approach. Being explicit about trade-offs among multiple objectives (which often correspond to services) is a critical component of EBM. Importantly, decisions about trade-offs among services or sectors certainly occur under current management, but they are most often made implicitly, rather than explicitly (see Rosenberg and Sandifer, chap. 2 of this volume, for a detailed treatment of this topic).

Finally, it is important to note that the concept of ecosystem-based management is grounded in the idea that ultimately we are managing people's influences on ecosystems, not ecosystems themselves. Thus, this volume brings together knowledge of scientists, managers, and practitioners to advance *ecosystem-based management* of the human activities that affect these systems, rather than *ecosystem management*, which implies that we are managing systems, rather than people.

Conservation of Ecosystem Services

Although elements of an EBM approach have been applied for decades both on land and in the sea, what is innovative about the current EBM movement is its focus on conservation of the long-term potential of systems to sustain the delivery of a broad suite of ecosystem services (McLeod et al. 2005; Rosenberg and McLeod 2005). Thus, each sector of human activity (fisheries, coastal development, tourism, etc.) must consider how it affects ecosystem structure, functioning, and key processes, and all sectors must collectively work toward the common goal of maintaining these components of ecosystem health. This sharply contrasts with current management practices that tend to focus on the short-term provision of single services and under which individual sectors are often working at cross-purposes (POC 2003; USCOP 2004).

The Millennium Ecosystem Assessment assessed the status of the world's ecosystems, the services they produce, and how changes in the global environment are affecting human well-being, concluding that 60% of ecosystem services globally are degraded (MA 2005b). Supporting services, or the fundamental ecological processes that sustain ecosystem functioning (e.g., nutrient cycling and photosynthesis), underlie all other services that are directly used by people. Remaining services are categorized as (1) provisioning services, such as food or fiber; (2) regulating services, such as climate regulation or coastal protection; or (3) cultural services, or the nonmaterial benefits that are important to our well-being (table 1.1). Individual services may fall into multiple categories; the specific labels that we attach to them are less important than acknowledging the breadth of services that affect our well-being. In particular, we have historically focused management on maintaining the flow of provisioning services, to the detriment of regulating, cultural, and supporting services.

Importantly, the focus on ecosystem services, rather than ecosystem functioning per se, explicitly acknowledges social–ecological connections (see fig. 1.1). Specifically, it requires accounting for the social, ecological, and physical factors affecting the production of services, as well as those that affect their delivery. For example, the provision of local, healthy, wild salmon for food depends upon (1) salmonid populations that are robust to fishing pressure; (2) suitable habitat; (3) other ecosystem components such as availability of prey or nursery habitat; (4) water quality; (5) local fleet access, including harbors; and (6) local markets and restaurants (Halpern et al. 2008b).

The ecosystem service focus is not without its limitations, and it raises key moral and ethical questions. For one, this perspective is fundamentally anthropocentric, reflecting human values and experience. In many ways this is appropriate, given that EBM is focused on managing human behavior. Intrinsic values of ecosystems ("existence values") are captured as cultural services to humans, although these are arguably the most challenging to value from an economic perspective (see Wainger and Boyd, chap. 6 of this volume). Ultimately, the ecosystem services perspective privileges human well-being over the well-being of other species, which is at odds with an ecological worldview that does not grant special status to humans. Thus, the evolution of management practices may call for a coevolution of ethics, as discussed in detail by Moore and Russell (chap. 18 of this volume).

Building on the Legacy of EBM on Land

Management of human activities that influence coastal and marine ecosystems lags behind management of terrestrial areas for several reasons. First, our awareness and knowledge of the terrestrial realm is greater than that of the sea—it is our home, and consequently, more scientific and technical resources have been devoted to understanding the dynamics and drivers of change to terrestrial ecosystems as compared with marine systems. Moreover, the watery medium that dominates coastal and ocean ecosystems creates technical challenges to understanding ecosystem dynamics and how these dynamics influence the provision of ecosystem services. Finally, the institutions governing allocation and use of marine resources are quite different from those governing terrestrial resources; private property is much less common in the sea, and common pool or open

Table 1.1. The range of ecosystem services provided by different coastal and marine habitats

Ecosystem services	Coastal systems								Marine systems			
	Estuaries and marshes	Mangroves	Lagoons and salt ponds	Intertidal areas	Kelp	Rock and shell reefs	Seagrass	Coral reefs	Inner shelf	Outer shelves, edges, & slopes	Seamounts & midocean	Deep sea & central gyres
Provisioning services												
Food	x	x	x	x	x	x	x	x		x	x	x
Fiber, timber, fuel	x	x	x						x	x		x
Medicines	x	x	x		x			x	x			
Regulating services												
Biological regulation	x	x	x	x		x		x				
Water storage and retention	x		x									
Climate regulation	x	x	x	x		x	x	x	x			x
Human disease control	x	x	x	x		x	x	x				
Waste processing	x	x	x				x	x				
Flood/storm protection	x	x	x	x	x	x	x	x				
Erosion control	x	x	x				x	x				
Cultural services												
Culture and amenity	x	x	x	x	x	x	x	x				
Recreation	x	x	x	x	x			x				
Aesthetics	x	x	x	x				x				
Education and research	x	x	x	x	x	x	x	x	x	x	x	x
Supporting services												
Biochemical processes	x	x			x			x				
Nutrient cycling	x	x	x	x	x	x		x	x	x	x	x

Source: Modified from UNEP (2006) with permission.

Note: The Xs denote that the habitat provides a significant amount of the service.

access institutions are more the norm. For all of these reasons, we need to think about managing marine systems differently from their terrestrial counterparts. This means building on the experience of managers and conservation practitioners on land but not necessarily trying to duplicate their approaches or aims.

That being said, decades of ecosystem-based approaches to land conservation and management (particularly in forests) provide some critical insights regarding the challenges and opportunities of marine EBM. Many, if not most, of the 428,000 km^2 of publicly held lands in the United States (almost 1/3 of the US land area) are managed for multiple objectives, including timber harvest, consumptive and nonconsumptive recreation, energy development, and wildlife conservation. Scientists and practitioners face a similar set of challenges in the sea, which requires considering the cumulative and potentially synergistic impacts of multiple activities, and developing ways to effectively communicate and integrate this information in a policy environment. Many regulatory tools developed to aid terrestrial ecosystem-based approaches (e.g., zoning and protected areas, market-based incentives) also apply to coastal and marine systems. Moreover, the legal mechanisms that guide land management decisions, such as the National Environmental Policy Act and the Endangered Species Act, are often invoked in marine settings as well (see Searles Jones and Ganey, chap. 10 of this volume, for further discussion).

Resilience Science as a Conceptual Backbone for EBM

Another key feature of EBM is the ability to embrace change, both in the increasingly dynamic world around us and in ourselves. The world's people are more connected than ever before; the flow of information, commerce, and people themselves is unprecedented. These connections bring opportunities as well as challenges. Embracing change requires us to better understand what influences the responses of systems, both human and natural, to a range of disturbances. Will a system resist disturbance, rebound quickly, slowly degrade, or shift to a completely new state? Once a threshold is crossed, is it possible for a system to return to a preexisting state; in other words, is the change reversible? Thus, understanding *resilience*—the extent to which a system can maintain its structure, function, and identity in the face of disturbance—can enable us to better predict how systems will respond not only to a growing array of perturbations, but also to a spectrum of management alternatives.

Importantly, resilience is not always a positive attribute. In other words, our aim will not always be to maintain or enhance the resilience of an existing state. As Boesch and Goldman (chap. 15 of this volume) describe, poor water quality, depleted fisheries, and the considerable inputs of nutrients and sediments from upland areas of the watershed characterize the current state of the Chesapeake Bay system, one that stakeholders from throughout the region are working to change. Restoration of sea grass beds, oyster reefs, and other bay habitats, and changes in farming, waste management, and development practices, are among the strategies being used to "erode" the resilience of the current state of the system and to help shift it to a more desirable one characterized by abundant fisheries, clean water, and vibrant habitats.

Key characteristics of resilience science include recognizing (1) the close coupling between social and ecological systems, (2) the existence of multiple possible states and abrupt changes among them, and (3) the contributions to system resilience of diversity and interactions across scales of space, time, and organization (Leslie and Kinzig, chap. 4 of this volume). A central contribution of resilience science to EBM is recognition that systems are constantly changing in ways that cannot be fully predicted or controlled. Consequently, an adaptive management framework is recommended (Anderies et al. 2006; Leslie and Kinzig, chap. 4 of this volume; Guichard and Peterson, chap. 5 of this volume). Learning to sustain and enhance the ability of systems to cope with uncertainty and surprise is in sharp contrast with conventional management approaches that tend to focus on maximizing production of particular services by reducing variability and controlling changes in systems that are assumed to be stable (e.g., maximum sustainable yield of fisheries; see Anderies et al. 2006 for further discussion). Resilience science also includes a strong multiscale perspective that emphasizes interactions and cooperation across scales, rather than centralization or decentralization

(Adger et al. 2005; Kinzig et al. 2006). Such an approach is critical to the effective implementation of EBM, which requires both bottom-up and top-down approaches (see McLeod and Leslie, chap. 17 of this volume). As discussed above, there is no correct spatial scale at which EBM should be implemented. Thus, efforts that range from local-scale, community-based efforts like that in Port Orford, Oregon (Wedell et al. 2005), to national-level mandates such as Canada's Oceans Act (see Rosenberg et al., chap. 16 of this volume) are critical to advancing EBM. The challenge lies in determining how to create bridges for meaningful connections across scales. Despite increasing application of resilience frameworks to management in terrestrial and freshwater systems (Walker et al. 2006), the use of these concepts in the marine realm is in its infancy (but see Nyström et al. 2000; Hughes et al. 2003; Bellwood et al. 2004; Hughes et al. 2005).

A Guide to This Volume

To date, the science and management of marine ecosystems and coastal communities in North America—and many other parts of the world—have tended to focus on aspects of the ecological system *or* the social system, but rarely integrate them. Yet, solutions to what ails the oceans will not be found through an examination of either of these systems alone. This volume provides the first comprehensive synthesis of the considerable knowledge needed to implement EBM. It brings together key ecological and social theory and emphasizes cross-fertilization of knowledge among varied disciplines. From a more practical perspective, it draws on empirical examples and case studies, particularly but not exclusively from the coasts and oceans of North America, to offer lessons broadly applicable around the world.

Perhaps most importantly, it addresses the key challenges facing scientists and managers in implementing EBM for the oceans and suggests specific approaches for implementation that can be put into practice immediately (see Rosenberg and Sandifer, chap. 2 of this volume).

Conceptual Basis for EBM

After a general setting of the stage, part 2 of the book focuses on key concepts that underpin EBM. Shackeroff and colleagues (chap. 3) review current thinking on the status and drivers of change in the world's oceans and address whether the current state of knowledge is sufficient to address these issues from the perspective of coupled social–ecological systems. Next, Leslie and Kinzig (chap. 4) introduce us to resilience science, highlighting the characteristics of systems that contribute to ecological and social resilience and the implications of this body of knowledge for the implementation of EBM. Building on these ideas, Guichard and Peterson (chap. 5) highlight the role of cross-scale interactions (across both space and time) and discuss how these phenomena can be used to address problems of uncertainty and controllability. This section concludes with Wainger and Boyd's (chap. 6) exploration of the potential and realized contributions of economics to EBM, addressing the roles of economics in measuring the value of nature and capturing the quality, quantity, and reliability of ecosystem services.

Connecting Concepts to Practice

Part 3 draws on diverse expert knowledge and tools and explores how to effectively link these in order to implement marine EBM. Kaufman and colleagues (chap. 7) begin by highlighting the relationships between EBM and adaptive management and show how monitoring, research, and modeling can provide the information necessary to enable management programs to move toward resilience and sustainability. Next, Barbier (chap. 8) assesses economic approaches to the valuation of ecosystem services and explores their application to EBM through an examination of a case study from Thailand. Kliskey and colleagues (chap. 9) discuss the challenges and opportunities for incorporating local and traditional ecological knowledge into EBM. Searles Jones and Ganey (chap. 10) conclude this section with an overview of the legal landscape within which we are working to implement marine EBM in the United States.

The State of Marine EBM in Practice

Although there are few examples where EBM in coastal and marine areas has been fully implemented, part 4 highlights initiatives where key elements have been put into practice. Cases explored in chapters 11–15 include ongoing efforts in Morro Bay, California, USA (Wendt and colleagues); Puget Sound, Washington, USA (Ruckelshaus and colleagues); Gulf of California, Mexico (Ezcurra and colleagues); the Eastern Scotian Shelf, Canada (O'Boyle and Worcester); and Chesapeake Bay, USA (Boesch and Goldman). These efforts span a range of geographic locations; include bottom-up, stakeholder-driven processes and more top-down, government-mandated efforts; and vary in age from the nascent effort in Morro Bay to over three decades of experience with EBM in the Chesapeake. Individual case study chapters are followed by Rosenberg and colleagues' larger-scale examination of national-level implementation of EBM concepts around the world (Australia, Canada, European Union, New Zealand, and United States; chap. 16). This section concludes with a synthesis of lessons learned from these diverse case studies by the editors, focusing on the themes of scale, stakeholders, integration of science and management, and success.

Looking Ahead

The concluding section of this volume begins with a thought-provoking chapter from Moore and Russell, who urge us to consider the moral and ethical landscape within which we approach our interactions with the world. They show that new ethical foundations are possible (perhaps even necessary) and that EBM might offer the opportunity to think more carefully and creatively about the principles that guide us in moving forward. This is followed by a final synthetic chapter from the editors in which we highlight key concepts that have emerged from the book, discuss future research needs, and suggest strategies for moving forward with implementation over the near-term and longer time horizons.

Notably, there is no single correct path to EBM. The approach will be put into practice in many different places across a range of spatial scales, each with its own unique historical, ecological, and social context. The range of suitable strategies will also vary based on the types of management and governance already in place. In some locations, it might be best to start by laying a foundation for future action to build and nurture a constituency and initiate a collaborative process to identify overarching goals and objectives. Other places may be primed to establish cross-jurisdictional management goals, develop mechanisms to consider trade-offs among ecosystem services, or deliberate over the spectrum of feasible spatial management options (ranging from no-take marine reserves to multiple-use areas). Further along this spectrum, other locations may be ripe to move forward with implementation of comprehensive ocean zoning in which areas are designated for particular allowable uses in both space and time. Regardless of the starting point, EBM must ultimately include (1) a means for sectors to work toward a common goal, (2) a mix of strategies to allow for both protection and use, (3) long-term monitoring and research, and (4) adaptive frameworks to allow us to learn from management actions, test alternate approaches, and readjust as either knowledge or systems change. Thus, EBM relies not on prescription, but on adapting a set of approaches suited to a particular context. Our aim is to provide the bricks and mortar from which practitioners can build an EBM approach appropriate to their circumstances.

Acknowledgments

We thank Anne Guerry, Kai Lee, and Sarah Lester for their comments on an earlier version of this chapter, as well as the contributors of this volume and the many other colleagues with whom we have interacted regarding these ideas.

References

Adger, W. N., T. P. Hughes, C. Folke, S. R. Carpenter, and J. Rockström. 2005. Social–ecological resilience to coastal disasters. *Science* 309:1036–39.

Anderies, J. M., B. H. Walker, and A. P. Kinzig. 2006. Fifteen weddings and a funeral: Case studies and resilience-based management. *Ecology and Society* 11(1):21.

Bellwood, D. R., T. P. Hughes, C. Folke, and M. Nystrom. 2004. Confronting the coral reef crisis. *Nature* 429:827–33.

Halpern, B. S., S. Walbridge, K. A. Selkoe, C. V. Kappel, F. Micheli, C. D'Agrosa, J. F. Bruno, et al. 2008a. A global map of human impacts on marine ecosystems. *Science* 319:948–52.

Halpern, B. S., K. L. McLeod, A. A. Rosenberg, and L. B. Crowder. 2008b. Managing for cumulative impacts in ecosystem-based management through ocean zoning. *Ocean and Coastal Management* 51:203–11.

Hughes, T. P., A. H. Baird, D. R. Bellwood, M. Card, S. R. Connolly, C. Folke, R. Grosberg et al. 2003. Climate change, human impacts, and the resilience of coral reefs. *Science* 301:929–33.

Hughes, T. P., D. R. Bellwood, C. Folke, R. S. Steneck, and J. Wilson. 2005. New paradigms for supporting the resilience of marine ecosystems. *Trends in Ecology and Evolution* 20:380–86.

Kinzig, A. P., P. Ryan, M. Etienne, H. Allison, T. Elmqvist, and B. H. Walker. 2006. Resilience and regime shifts: Assessing cascading effects. *Ecology and Society* 11(1):20.

Leslie, H. M., and K. L. McLeod. 2007. Confronting the challenges of implementing marine ecosystem-based management. *Frontiers in Ecology and the Environment* 5(10):540–48.

MA (Millennium Ecosystem Assessment). 2005a. *Ecosystems and human well-being: Current state and trends.* Washington, DC: Island Press.

MA (Millennium Ecosystem Assessment). 2005b. *Ecosystems and human well-being: Synthesis.* Washington, DC: Island Press.

McLeod, K. L., J. Lubchenco, S. R. Palumbi, and A. A. Rosenberg. 2005. *Scientific consensus statement on marine ecosystem-based management.* The Communication Partnership for Science and the Sea (COMPASS). Signed by 221 academic scientists and policy experts with relevant expertise. http://www.compassonline.org/pdf_files/EBM_Consensus_Statement_v12.pdf.

Nyström, M., C. Folke, and F. Moberg. 2000. Coral reef disturbance and resilience in a human-dominated environment. *Trends in Ecology and Evolution* 15:413–17.

POC (Pew Oceans Commission). 2003. *America's living ocean: Charting a course for sea change. A report to the nation.* Washington, DC: Pew Trusts.

Rosenberg, A. A., and K. L. McLeod. 2005. Implementing ecosystem-based approaches to management for the conservation of ecosystem services. In Politics and socio-economics of ecosystem-based management of marine resources, ed. H. I. Browman and K. I. Stergiou. *Marine Ecology Progress Series* 300:270–74.

UNEP (United Nations Environment Programme). 2006. *Marine and coastal ecosystems and human well-being: A synthesis report based on the findings of the Millennium Ecosystem Assessment.* Nairobi: UNEP.

USCOP (US Commission on Ocean Policy). 2004. *An ocean blueprint for the twenty-first century.* Final report. Washington, DC: USCOP. ISBN 0 9759462 0 X.

Walker, B. H., J. M. Anderies, A. P. Kinzig, and P. Ryan. 2006. Exploring resilience in social–ecological systems through comparative studies and theory development: Introduction to the special issue. *Ecology and Society* 11(1):12.

Wedell, V., D. Revell, L. Anderson, and L. Cobb. 2005. Port Orford Ocean Resources Team: Partnering local and scientific knowledge with GIS to create a sustainable community in southern Oregon. In *Place Matters*, ed. D. J. Wright and A. J. Scholz. Corvallis: Oregon State University Press.

WSSD. 2002. *World Summit on Sustainable Development plan of implementation.* Johannesburg, South Africa: WSSD.

CHAPTER 2
What Do Managers Need?

Andrew A. Rosenberg and Paul A. Sandifer

From national and international perspectives, there have been clear calls for a move toward ecosystem-based approaches to management and numerous efforts to define it. However, many managers remain unclear about just what an ecosystem-based approach entails and are seeking tangible advice for moving forward with implementation.

An ecosystem approach to management is intended to directly address the long-term, sustainable delivery of ecosystem services and the resilience of marine ecosystems to perturbations. It encompasses both a process for the development of policy actions and a conceptual framework for the formulation of policy principles, goals, and objectives. Most importantly, the ecosystem-based management (EBM) concept helps guide the development of these goals for a given ecosystem, that is, conserving the delivery of the full range of services from a particular system.

Fortunately, EBM has now been implemented in several marine areas, including areas within the United States, Canada, Australia, New Zealand, and the European Union (Garcia et al. 2003; Lafolley et al. 2004; SAFMC 2004; Frid et al. 2005; NOAA 2006; Merrick et al. 2007; Rosenberg et al., chap. 16 of this volume). However, challenges remain, relating to both the scientific support structure for EBM and to taking management actions (e.g., Sandifer and Rosenberg 2005). There are institutional as well as technical issues to be resolved, as we discuss below. The strategies for attaining ecosystem goals—that is, the general management approach—and tactics or specific measures taken to realize these strategies are the essence of implementation and will be specific to a particular ecosystem and management setting. Here we describe the principles of EBM and the challenges to developing strategies and tactics for implementation of an ecosystem-based approach to policymaking.

EBM Principles for Managers

From a management perspective, five principles can guide the development of an ecosystem-based approach (see box 2.1): (1) setting goals that include the full range of ecosystem services, (2) determining the spatial scale for management planning, (3) integrating across sectors of human activity (e.g., transportation, fisheries, energy production, recreation), (4) accounting for cumulative impacts within and across sectors, and (5) making decisions under uncertainty. Note that principles addressing these areas should be accounted for in all management planning. Particular to an ecosystem-based approach is that the ecosystem (including humans) and its services become the organizing framework for management (see also Juda 1999; Shepherd 2004; McLeod et al. 2005; Leslie and McLeod 2007).

Setting Goals

Goal setting is a critical function for management planning (e.g., Clark 2002). The setting of goals shapes the effort to develop workable policy in practice. For example, a sole focus on the goal of obtaining a specific level of abundance of a single species or even a group of

Box 2.1. Principles of Ecosystem-Based Management

The essence of an ecosystem-based approach to management rests on five basic principles:

1. **Diverse ecosystem service provision**:EBM focuses on the ability of the ecosystem to continuously provide the services that support human well-being and includes recognition that humans are inherently part of the ecosystem. Such ecosystem services go beyond simple extractive uses, such as fisheries harvest, and can be categorized as having provisioning, supporting, regulating, and cultural roles for society (MA 2005; McLeod and Leslie, Chapter 1 of this volume).
2. **Importance of natural boundaries**:Management recognizes that natural boundaries are more relevant to the conservation of ecosystem services than artificial boundaries (e.g., between legal jurisdictions) and that these natural boundaries are highly porous.
3. **Integrated management**:Management strategies of the various sectors of human activity that potentially impact a particular marine ecosystem can affect one another and require some level of management integration if those impacts are to be adequately controlled.
4. **Accounting for cumulative impacts and necessary trade-offs among services**:Impacts of human activities on a given ecosystem are often cumulative across time and space, with each activity or impact contributing to overall ecosystem change and the cumulative effect determining how the ecosystem continues to function. Moreover, policy decisions are not likely to have the same effect on all services. Some explicit decisions that involve trade-offs in services among sectors must be made. If management is not integrated across the sectors of human activities, these trade-offs are often implicit or completely ignored, with potentially disastrous and usually unintended results.
5. **Making decisions under uncertainty**:Management decisions should be made in the context of a precautionary approach as elaborated by the US Commission on Ocean Policy (2004), that is, by relying on the best available information and management practices from the beginning, weighing decisions in light of the level of uncertainty of available information and the level of potential risks to the ecosystem, and including continued gathering and analysis of information, with periodic reassessment and modifications of permit conditions or other requirements.

species (e.g., marine mammals) may place great constraints on subsequent policy choices. Each statute—for example, from the United States, the Magnuson–Stevens Fishery Conservation and Management Act (MSA), the Marine Mammal Protection Act, or various wetlands protection laws—sets goals for that activity or sector, most often in isolation from others or as if the impacts of areas of management concern are of minor importance. This approach shapes overall ocean policy by isolating decisions among sectors. In an ecosystem-based approach, goal setting is based on the concept of sustaining the delivery of a full range of ecosystem services (McLeod et al. 2005; Palumbi et al. 2009) and ensuring that the ecosystem does not lose its inherent resilience, that is, its ability to absorb perturbations and recover (Walker and Salt 2006; Leslie and Kinzig, chap. 4 of this volume). Addressing these ecosystem-level goals cannot be achieved by each sector in isolation. It necessitates integration.

Establishing goals for conserving services and maintaining or enhancing resilience will be affected by the scale defined for each ecosystem. There is no one "correct" scale for management, and the boundaries between ecosystems will always be fuzzy, porous, and zones of conflict. Nonetheless, the intent of focusing management effort at the ecosystem scale—whatever that might be in a particular situation—is to facilitate the ability of policymakers to consider factors affecting the production of the full range of ecosystem services and to have a coherent basis for evaluating impacts on ecosystem resilience. Managers need to focus on more natural or scientifically delimited boundaries of ecosystems, such as major biogeographical features or the well-established limits of large marine ecosystems (LMEs) (Sherman and Skjoldal 2002; Fanning et al. 2007), rather than on the often artificial borders established by political jurisdictions (e.g., state borders, territorial sea lines, limits of the exclusive economic zone). As NOAA

(2004) concluded, "The chief goal of any delineation scheme is to develop a geographic frame of reference that is useful in describing biological populations, their interactions, and the effects of human activities that influence ecosystem outcomes."

Determining the Spatial Scale

In a sense, using ecosystem boundaries rather than jurisdictional boundaries should minimize the number of factors outside the boundaries of the management system that need to be considered. It is never possible for natural systems (particularly marine systems) to be entirely self-contained, and many factors that affect ecosystems cross arbitrarily defined boundaries. For example, climate change is likely an external driver that cannot be "managed" within the boundaries of any one marine ecosystem. However, action can still be taken within a given ecosystem to help mitigate and adapt to climate change (see Kliskey et al., chap. 9 of this volume; Leslie and Kinzig, chap. 4 of this volume). Also, market forces for services are likely to be affected by multiple ecosystems. However, it is better to be able to consider ecosystem functioning, and the impacts of human activities on resulting services, as completely as possible.

Clearly, there are effects of the choice of boundaries with the setting of specific ecosystem goals. If ecosystem boundaries are defined at a small scale, goal setting may be simplified, but attaining those goals will be more difficult because there will likely be more external factors. If boundaries are broad, then the dimensionality of the goals will increase substantially, simply because there are likely more "parts" to the ecosystem. Yet, policy could be more coherent across the sectors of human activity and might work better in attaining goals by being more inclusive of all the human impacts within the ecosystem. For example, managing just the estuary of a single watershed may be workable and allow clear goals to be set, but it may be very difficult to achieve ecosystem-level goals without the ability to manage interactions with drivers occurring at a larger scale. On the other hand, global action to address marine ecosystem protection may be effective in one sense, but setting goals at the global scale may be next to impossible. In practice, a balance must be found in setting ecosystem boundaries between these two extremes. Realistic boundaries for EBM, certainly at the regional scale, are generally going to be larger than the watershed of a single, small estuary but may be smaller than an LME. Delineating a finite number of regional ecosystems that can become the foci of management efforts will entail a compromise between our understanding of biological populations and oceanographic processes and the established structure of human institutions and political jurisdictions (NOAA 2004). For the United States, one approach worth considering (and that advocated by the US Commission on Ocean Policy [2004]) would be the ecoregions defined by the eight regional fishery management councils (fig. 2.1).

Integrating Across Sectors

Integrating management across sectors of human activity, such as linking up fisheries management actions with water quality management and coastal development, is key to an ecosystem approach. However, while critical, such integration is not independent from the other principles. It is possible to integrate management actions (and there is substantial literature on integrated management, e.g., Cairns and Crawford 1991; Cicin-Sain and Knecht 1998) without a focus on the goals of maintaining the ability to provide a full suite of ecosystem services or maintaining resilience. Integrated management without ecosystem goals might improve efficiency of management and streamline some regulations and controls, but it will not help to ensure ecosystem

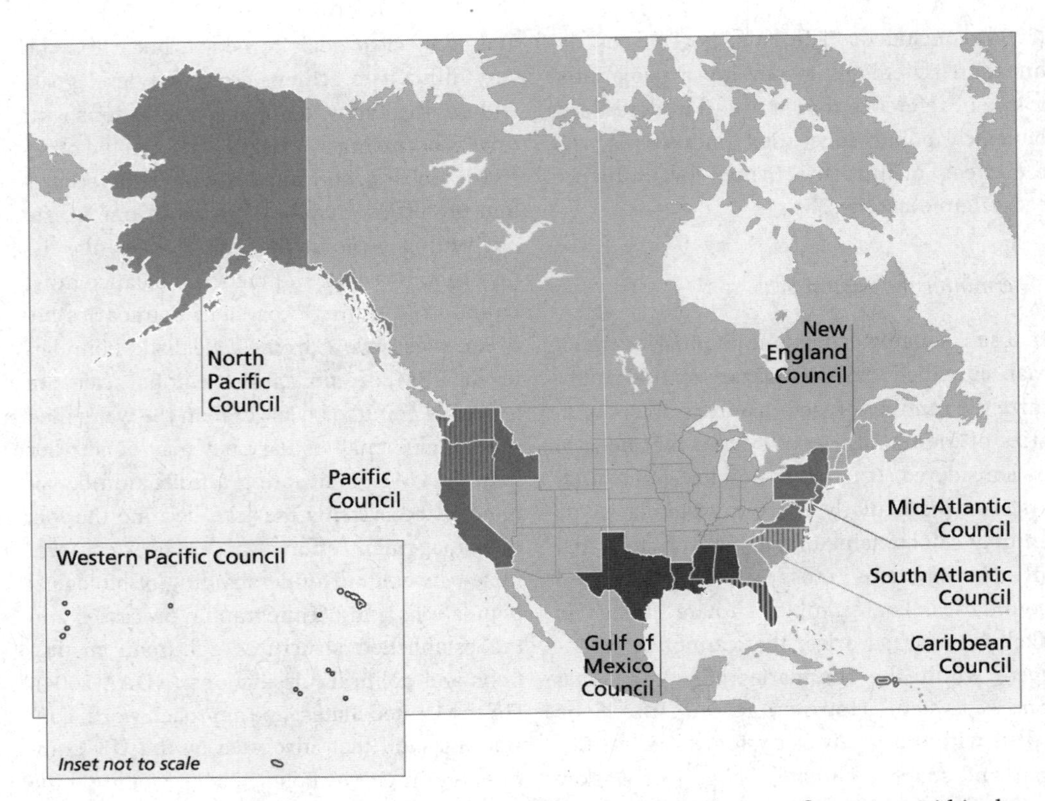

Figure 2.1 Eight regional fishery management councils manage the harvest of resources within the United States. Several states, illustrated with vertical lines, belong to more than one council. For example, Oregon and Washington are both members of the Pacific and the North Pacific councils. The US Commission on Ocean Policy (2004) advocated that these ecoregions be used to define the boundaries of regional marine ecosystems in US waters.

resilience. As discussed in chapter 1, it is the integration across sectors that enables policy to address conflicts and synergies and allows simplification of management to meet multiple objectives. Similarly, integrating the scientific advice for a given ecosystem will reveal policy options as well as new avenues for investigation beyond any one discipline's or sector's conventional view.

Accounting for Cumulative Impacts and Trade-offs among Services

Recognizing and accounting for the cumulative effects of human activities is difficult but, nonetheless, necessary for management to address EBM goals. A simple example is modification of wetlands. In a thousand-hectare area, perhaps filling or dredging a hectare or two may not affect the functioning of the wetland (such functions include sediment trapping, exporting of productivity, providing nursery and feeding areas for a variety of species, and supporting recreation and other cultural activities). But suppose two hundred permits are requested for one hectare each over time? At what point are the ecosystem services provided impaired? Should permits be issued on a first-come, first-served basis, or some other criterion? At what point do the minor impacts of each individual permit cascade into a major cumulative effect? Are the services provided by the filled or dredged areas more important than the natural wetland areas? To what limit? These questions relate directly to the principle of evaluating trade-offs between different services and user groups.

Under sector-by-sector management, trade-offs within a sector may be considered, but those among sectors are largely ignored and often remain unaccounted for. Suppose coastal development of harbor facilities is to be managed. What are the trade-offs between space for recreational vessels, commercial fishing vessels, and commercial transport vessels? What about working waterfront areas versus tourist areas, or natural coastal areas and developed areas? How should the decisions be made? In a sector-by-sector management system, fisheries managers might comment on harbor needs but might not have direct input into the management process for coastal planning. Recreational and commercial fishing interests each might push for as much space as possible, but other than political pressure, what is the means of resolving the trade-offs between these groups? What is the forum for considering overall need and for evaluating effects—both potentially beneficial and negative—on other sectors or industries? About the only process currently available in the United States is the environmental-permitting system operated by the state coastal zone management agencies and the US Army Corps of Engineers and related environmental impact analysis requirements. However, this system also typically deals with one permit at a time. Generally, no comprehensive, regionwide evaluations are conducted, and without some clear analysis of alternatives, political considerations may dominate. There are often transport plans, coastal zone management plans, fishery plans, and others, but the trade-offs among these sectors can only be evaluated if planning is integrated (box 2.2).

Box 2.2. Regional Ocean Councils

Both US ocean commission reports (POC 2003; USCOP 2004) included substantial consideration of regional ocean councils. Such councils or other regional coordinating mechanisms provide an opportunity for the cross-sectoral discussions, associated information gathering, and clearinghouse functions that must occur in order for an EBM approach to be implemented. It is important to link the efforts to create coordinating bodies with the development of science infrastructure in order to build a coherent system. Regional councils or other coordinating entities do not supplant the need for sectoral management plans, though hopefully they would bring new considerations to bear as well as create better management across the board. Recent interest at the state level, as expressed through the initiation of several interstate, ecoregional alliances, provides leadership examples that the federal government could emulate and opportunities to establish regional governance entities.

Making Decisions under Uncertainty

There is inevitable uncertainty in the status and dynamics of any ecosystem, our knowledge about that system, and the effects of potential management actions. Management must determine how to proceed in implementing policy with imperfect information and substantial uncertainty in almost all cases. Given the focus on conserving ecosystem services and resilience in an EBM approach, a precautionary approach to management (Rosenberg 2002; USCOP 2004; Rosenberg 2007) is clearly necessary; that is, to avoid irreversible changes, be more cautious when uncertainty about the impacts of an activity is greater, and evaluate potential impacts before allowing activities to move forward, rather than allowing them first and trying to recover later.

Challenges to Implementing Ecosystem-Based Management Policies

Implementing an ecosystem-based approach under the five principles outlined here is no simple task. From a manager's perspective, there are both incentives and disincentives to apply these principles. The clear incentive is that an ecosystem-based approach is more comprehensive, and theoretically it should result in more effective and successful management if success is to be measured principally in terms of long-term sustainability and the continued delivery of necessary and desired ecosystem services. Comprehensiveness is more effective because it does not ignore the interaction between management actions nor the cumulative nature of impacts, as often occurs with sector-by-sector management. It is unlikely in real systems that these interactions and cumulative effects are negligible and can be safely ignored (Halpern et al. 2008). EBM attempts to deal with these interactions and cumulative effects up front, where choices and trade-offs can be made consciously, as opposed to simply allowing them to happen with unknown and unconsidered consequences.

For example, water quality, coastal development, and fishery productivity are almost certainly related in most systems because the former two human impacts affect the productivity of fisheries habitat. While some efforts over the last 10 years have been made to include the consideration of fisheries habitat in the fishery management process (Rosenberg et al. 2000), these fall far short of integration, because neither the goals for each sector nor the impacts of each on the other are jointly addressed. Rather, the mandate to protect essential fish habitat relates mostly to potential impacts of fishing itself, along with weak provisions to consult with other sector managers on actions that might impact habitat. There is no consideration of trade-offs nor a clear focus on a broader range of ecosystem services. Clearly, managing ecosystem services such as essential fish habitat in a more integrated way will be very challenging. When the mandate was first created in the 1996 amendments to the MSA (NMFS 1996), several groups such as the National Association of Home Builders became concerned that home building would be impacted by fishery management planning (Rosenberg, personal observation). Bringing together more sectors brings a broader array of constituencies into the management arena.

If all the sectors are managed separately, there is no real opportunity to leverage actions across the ecosystem as a whole. This will only occur if an integrated or comprehensive management approach is taken where all managers involved consider the same things and, to the degree possible, use a common approach (Rosenberg and McLeod 2005; Palumbi et al. 2009). Consider, for example, the complex set of marine protected areas in the Gulf of Maine (fig. 2.2). Each area was established for a different purpose and usually for a single sector or part of a sector, such as a specific fishery. The areas do not complement or leverage one another, and the result is an overlapping set of requirements that have been somewhat effective in providing some ecosystem protections, such as leading to partial fish stock recovery (Rosenberg et al. 2006) or habitat protection (Collie et al. 2005; Grizzle et al. in press). But these successes are fragmentary and fall far short of EBM goals of maintaining ecosystem services as a whole, even though each may have had some success for a given particular resource. If the protected areas were designed to meet multiple purposes and rationalized to be a nonoverlapping set of protections, then they would also be more coherent to the affected community on the water.

Beyond just the creation of protected areas, if the goals for ecosystem services and resilience were considered together, it should be possible to simplify the rules to determine whether sets of activities should be prohibited

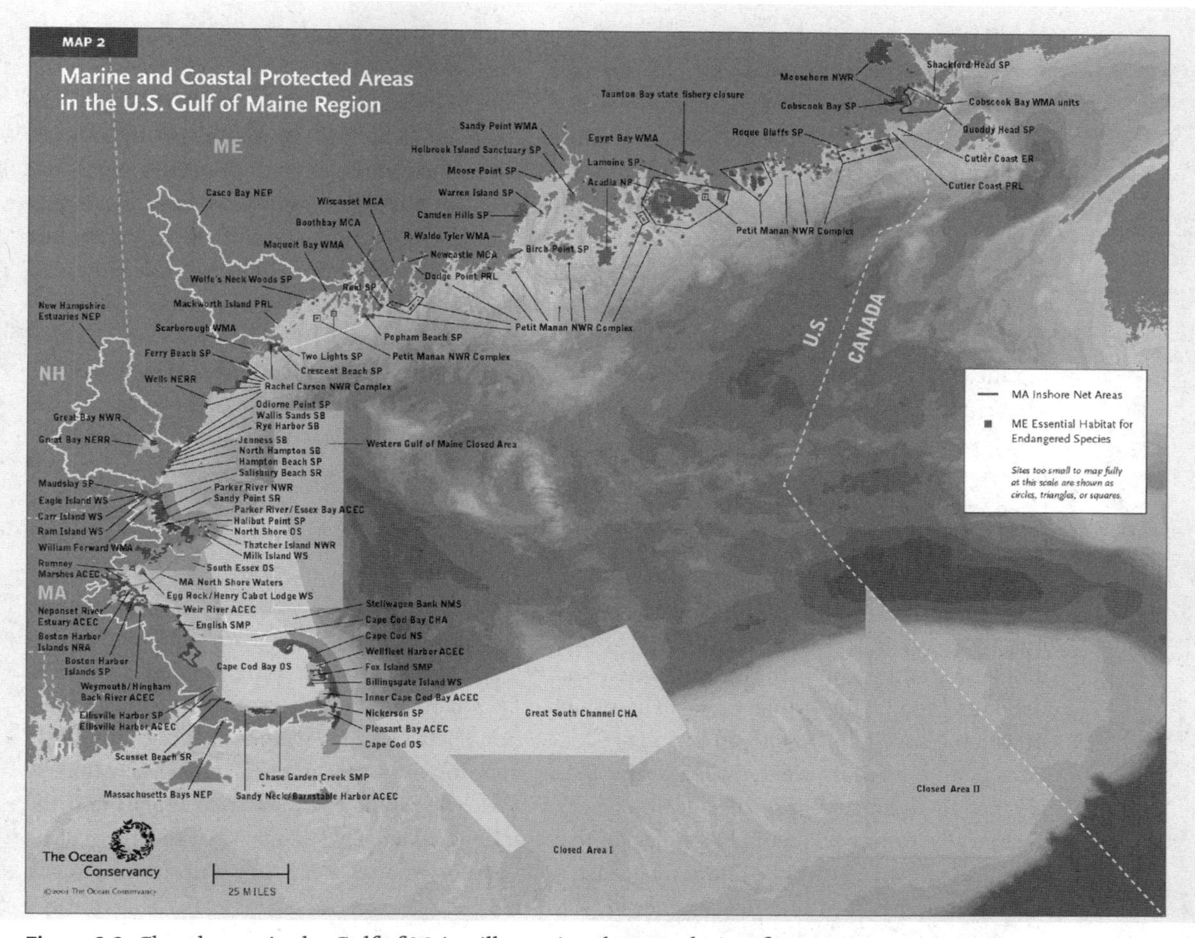

Figure 2.2 Closed areas in the Gulf of Maine illustrating the complexity of management on a sector-by-sector basis. Adapted from Recchia et al. 2001 and reprinted with the permission of The Ocean Conservancy.

or allowed within different areas in a system of contiguous zones, much as occurs on land. While many interest groups are wary of a zoning approach in the ocean, the results may ultimately be more coherent and simpler that the current fragmented approach (Crowder et al. 2006; Young et al. 2007).

Current sector-by-sector management has a set of procedures that is well known by those tasked to implement policy. The mandates are reasonably clear—for example, to achieve optimum fishery yields, or fishable and swimmable waters. Taking an ecosystem-based approach means changing those procedures. As with most human activities, change is difficult. As a result, managers are likely to be conflicted with respect to their desire to implement EBM. The goals of EBM are widely accepted by managers, but the changes needed to attain those goals mean working outside of the existing boundaries for sectoral management. Management implementation in each of the sectors of ocean activities, such as fisheries, water quality, habitat conservation, protection of species at risk, transportation, or energy development, is intensely controversial and already difficult under the current system. Getting managers to take on EBM, with its new controversies, broader constituencies, and more-complex decision making that involves trade-offs and interactions with other sectors, even if it may be the right thing to do, is a major challenge and will require that new mandates and management systems be developed carefully (USCOP 2004).

In statutory language, the goal of maintaining the full range of ecosystem services across sectors is addressed in only a general sense, as noted above. For example, the MSA calls for fisheries to be managed for optimum yield, defined as maximum sustainable yield as reduced by relevant social, economic, and ecological considerations. The 2007 reauthorization of the MSA strengthens the role of science in setting these limits and takes some steps toward developing ecosystem plans but still focuses on fisheries yield as the primary goal. The extent to which a specific plan should include reductions from maximum yield is a matter for fishery management councils to decide and justify. Technically, it may be justifiable to decide that only a very low yield will be allowed because it is more important to produce another service at a high level (e.g., allow only a low yield of forage species because of the desire to increase whale-watching opportunities near shore, or a low yield of an anadromous species because of hydropower or transportation needs). But is it reasonable to expect that a council set up to manage fisheries and composed of officials, representatives, and individuals knowledgeable about fisheries will decide to reduce fishery productivity to meet the needs of another sector when that other sector has only minor involvement in the fishery management process, and vice versa? To take another example, recreational and commercial fisheries both depend on port facilities in one way or another. But port facilities may impact natural barriers that provide storm protection, or water quality, or habitat conservation. Should we expect fishery managers to reduce fishery yields to reduce the demand for ports? While these examples are intended to be illustrative and not exact (each is far more complex than described here), the problem should be clear. If trade-offs are only considered within a sector, then policy choices will be made implicitly and not explicitly with regard to the suite of ecosystem services from any given system.

It is the governance structures at the federal, state, and tribal levels that must set the goals for policy and the approach to making decisions with respect to trade-offs among services. That is, legislative bodies must create the mandates, because each administrative agency at the various levels does not have a sufficiently broad purview to cover the spectrum of services. At the administrative or implementing level, specific coordinating mechanisms must be put in place to work through the interactions between management actions, and these coordinating mechanisms must be able to make decisions that have real impacts on the individual agency actions. That means that the statutory mandate has to set up means to create decision rules or give the authority to some entity to create those decision rules for how to resolve conflicts and interactions.

Then too, there are human resource and funding issues. Most managers and scientists see development of an ecosystem-based approach as something additional to do, not a different way of doing things. Natural resource agencies and science efforts are often understaffed and underfunded. Providing more funds and resources for regulatory efforts may not be the most popular political initiative in debates over spending priorities. But EBM will not happen without additional human resources and funding, even if the long-term prospect is a more coherent and efficient system that better maintains provision of critical ecosystem services.

Developing the Science for an Ecosystem-Based Approach to Management

Natural and social scientists certainly understand that marine ecosystems, including the roles of humans and their individual and collective actions within those systems, are highly interconnected (Shackeroff et al., chap. 3 of this volume). While the principles given above are not necessarily universally accepted in the scientific community, there is broad agreement on the interconnectedness and complexity of marine ecosystems and the importance of taking a more integrated approach to management (McLeod et al. 2005). So, does it follow that scientists want and are prepared to implement an ecosystem-based approach and are waiting for managers to act? And does it follow that managers want to implement EBM and are just waiting for scientists to provide a workable blueprint?

The role of science in policy and management is to provide advice on conservation limits, impacts, and projected changes in the ecosystem (including humans) under different management scenarios. A key area of advice should relate to ecosystem resilience, that is, the ability of the system to recover from both acute and chronic impacts (Leslie and Kinzig, chap. 4 of this volume). This is reflected in statements describing the precautionary approach, for example, those that emphasize the avoidance of irreversible changes. Science is not policy but, rather, informs policy and management. Scientists certainly want management to work better and thus have the same incentive as managers to support a more comprehensive approach to management, such as EBM, that accounts for interactions and cumulative impacts. But many of the same disincentives for managers apply to scientists as well. The current system of sectoral management requires scientific advice using generally known procedures and involves scientists from a limited number of disciplines and with specifically applicable expertise. The mandates are well known, and the constituencies scrutinizing the advice are well established. While scientific advice in almost all sectors can be highly controversial, it is reasonably predictable, though often limited in its influence on how management proceeds in practice. Still, it is critically important to recognize that questions raised by managers must fit within the existing scientific framework for a particular sector. Otherwise, those questions will remain unanswered.

Change is as difficult for scientists to accept as for managers; a new requirement to work outside of their "known world" and with other scientists and managers who have widely different backgrounds, expertise, and perceptions is likely to be intimidating or unacceptable for some. Additionally, in many scientific analyses, the first step is often to simplify the problem to make it as tractable as possible. Considering ecosystem interactions and cumulative impacts in management decisions is a major complication that makes analyses much more difficult. For example, procedures to develop fishery stock assessments and measure fishery performance with regard to accepted conservation standards are well known and routinely implemented (see Hilborn and Walters 1992; NRC 1998). Management is, nonetheless, difficult and controversial, and by extension, so is the scientific advice underpinning the management decisions. However, the inclusion of the cumulative impacts of habitat changes and fishery effects as well as trade-offs with, say, energy development, coastal land use, or water quality are much more complicated, and procedures for doing so are poorly defined to date. Under the time pressure to move forward with management, providing advice in an already contentious environment will become a key challenge for scientists when such interactions are included.

Developing a Scientific Advisory Structure for EBM

If a scientific advisory structure is to be developed for EBM, do we know enough about ecosystem interactions, cumulative effects, and trade-offs between services to provide useful advice to management? Clearly, detailed knowledge of ecosystem structure, functioning, and dynamics is incomplete and uncertain in virtually all cases. But we do know some of the key interactions, we have observed changes in numerous ecosystems, and we have sufficient information from other, similar ecosystems to advise management on likely impacts of human activities. We also know, to a great extent, the consequences of many types of human impacts and of the past and present management of those human activities. In addition, there is little uncertainty about what needs to be done about major problems such as overfishing (i.e., reduce exploitation) and wetlands loss (i.e., reduce filling, channeling, and loss of sediment and water flow).

For many ecosystems, it should be possible to begin to synthesize systemwide patterns of change by combining information obtained across various sectors, such as fisheries, energy production, water quality changes, and coastal development. This amounts to a baseline assessment of the status of an ecosystem that can be used to evaluate the likely effects of new activities and the potential impact of changes to management (USCOP 2004; DEFRA 2005). It is also important to not only be forward-thinking, but also develop a historical perspective of ecosystem changes (e.g., Jackson et al. 2001; Rosenberg et al. 2005), rather than assume the current state of the environment is all that can be achieved. This is particularly critical as rebuilding and recovery plans are formulated for various ecosystems.

Many sectors of human activity have extensive infrastructures for science advice to guide management. One of the most highly developed systems provides science advice to support fishery management. The field of fisheries science is rather small, with only perhaps a few hundred scientists providing advice for marine fisheries nationwide within the United States and a few thousand engaged in science and research on fisheries within academia and all levels of government. However, the federal, state, and tribal agencies responsible for fisheries almost invariably include a scientific advisory staff and, often, active researchers. The federal government has regional fishery science centers around the country that include research, monitoring, and management advice as part of their missions. In fact, NOAA's National Marine Fisheries Service (NMFS), the primary federal agency responsible for marine fisheries, marine protected species, and fishery habitat conservation, is composed roughly of two-thirds science personnel and associated budget to support them and only one-third management and policy implementation. NMFS staff routinely collaborate with researchers in other federal agencies (e.g., the US Fish and Wildlife Service and the US Geological Survey) and with those in state and tribal agencies and in academia.

The federal, state, and tribal scientists work together to provide advice to federally managed fisheries and for a number of other fisheries managed by interstate bodies such as the Atlantic States Marine Fisheries Commission. For federally managed fisheries, the management plans are crafted by eight regional fishery management councils made up of state and federal officials and knowledgeable constituents from the region, with scientific support provided by a mature network of federal, state, tribal, and academic fisheries scientists and a growing number of social scientists. Those management plans are submitted by the councils to NMFS for approval and implementation. Approval is contingent upon the plans meeting, in the judgment of the US secretary of commerce through NMFS, ten national standards for management and other provisions laid out in the MSA, which was updated and reauthorized in early 2007. Several of the national standards and provisions of this law relate directly to scientific advice, including the mandate to use the "best science available" in setting management goals, strategies, and tactics. For example, the MSA requires the plans to prevent overfishing and, in cases where overfishing has occurred, to rebuild overfished stocks to a level able to provide maximum sustainable yield on a continuing basis. Fisheries scientists use well-developed methods for estimating maximum sustainable yield or related reference points and for estimating stock status. Improved methods are being developed and implemented on an ongoing basis as they are accepted within the scientific community. In addition, the MSA requires consideration of impacts on habitat from fishing and other activities as part of considering ecosystem-level effects, but it does not require implementation of an EBM approach to fisheries management. However, it does now mandate a study of how EBM might be applied.

The scientific advice for each management plan is required to undergo extensive peer review before it is deemed the "best available." To this end, there are regional peer review workshops, panel reviews, and sometimes special external reviews, including some by the National Academy of Sciences, that evaluate the scientific advice for each fishery in great detail. These peer reviews are far more extensive than the usual academic review of journal articles with panels of four to ten or more scientists considering the methodology, data, and conclusions in detail. Alternative data and analyses are considered, public comment and scrutiny are included as well as comment and submissions from constituent groups, and in general the reviews are held to a very high standard. Each of the regional fishery management councils has a science and statistical committee (SSC) to review the results of the peer reviews and ensure that the council is receiving the best advice possible. Recent changes to the MSA require the use of these SSCs in setting conservation limits for each target stock of fish. However, the advice provided by the SSCs is typically limited to the fishery sector, with little consideration of information from other sectors such as coastal zone management, port management, coastal development, and so forth. While this may be changing incrementally in practice (e.g., see SAFMC 2004), there is as yet no policy requirement or established procedure to incorporate scientific information from these other sectors into fishery management decision making or vice versa.

This is a very highly developed system that is costly, laborious, and reasonably effective, but still fisheries management is very far from perfect. Stocks are still overfished, advice is not always followed (e.g., Sandifer and Rosenberg 2005), and it is sometimes incorrect. The review system notwithstanding, data are often inadequate despite the infrastructure, and everyone working within the system would probably agree that the scientific enterprise to support fishery management is substantially underfunded. Despite these problems, fisheries management has begun to turn around some of the long-standing problems of overfishing and stock rebuilding. Although such change is happening more slowly than many would like, it is based upon clear and supportable scientific advice (Rosenberg et al. 2006).

There is no comparable scientific advisory system for other important marine sectors, such as coastal development, habitat management, coastal tourism, transportation management, or even energy development. That is not to say there is no science advice for these sectors, but the scientific structure underpinning decision making is far less developed and institutionalized, and there is less infrastructure to provide supporting research, monitoring, evaluation, and the preparation of targeted scientific advice in a standard way. Also, there are no clear parallel institutions to NOAA's regional fishery science centers, not even for something as fundamental and important as water quality or energy. Of course, the Environmental Protection Agency has laboratories and monitoring systems and works directly with state monitoring programs, but these are not integrated into the management process regionally or nationally as is the case for fisheries.

The intent of the foregoing brief description of the fishery management system is not to laud or defend fishery science and management, but rather to provide context for consideration of the challenges to developing the scientific infrastructure for broader ecosystem-based management. Creating a system for the provision of advice for ecosystem-based management across a number of sectors will require a number of changes, including those listed in box 2.3. Much of the expertise for such a science advisory process already exists, and many of the activities already under way have laid a foundation. The institutional infrastructure can be virtual, but it must exist in some form and will take resources and an ongoing commitment from government to make it work. There are clearly new areas to develop, including the creation of ecosystem scenario analyses that can be used to guide management. Perhaps most importantly, EBM scientific advice must be interdisciplinary to a greater extent than has been attempted in the past within any one sector (Leslie et al. 2008).

For example, exploring the interactions and trade-offs between management policy options requires a significant interdisciplinary effort among numerous and diverse areas of both natural (e.g., biology, ecology, fisheries, genomics, oceanography) and social (e.g., anthropology, economics, political science, sociology) sciences. This is a critical area of work for implementation of EBM because one of the reasons for moving toward a more comprehensive management approach is in order for trade-offs to be considered explicitly, rather than having the trade-offs continue to occur as the unintended consequences of various management actions. One option is to value the services, using a common methodology or ecological "currency," and then develop decision tools that allow relative values of different policy scenarios to be compared (see Wainger and Boyd, Chap. 6 of this volume). But the valuation task itself is complex, as many ecosystem services do not yet have established market values, and some may never have such values. Development of a process to place widely accepted market values on a variety of ecosystem services is a very active area of research combining economics and other social sciences with the natural sciences of ecosystem dynamics (e.g., Meyerson et al. 2005). Another approach would be to consider one key property, such as biodiversity, as a common focus of management and basis for comparison of management alternatives (Palumbi et al. 2009). Recent work has related conservation of natural biodiversity to maintenance of several important ecosystem services, including fishery productivity and resilience in both marine (Worm et al. 2006; Palumbi et al. 2009) and terrestrial (e.g., Butler et al. 2007) systems.

Recent Progress and Next Steps

There is no current policy mandate to synthesize scientific knowledge across sectors to support EBM, and no comprehensive administrative framework exists to support integrated ocean management in the United States despite calls for such from both ocean commissions (POC 2003; USCOP 2004). The Bush Administration's US Ocean Action Plan (US Executive Branch 2004) response to the USCOP (2004) report included an explicit commitment to ecosystem-based resource management. Additionally, the newly reauthorized and amended MSA requires some evaluation of EBM for application to fisheries, and a comprehensive system could be developed using the fisheries advisory system as a model. Creating ecosystem assessments using the scientific infrastructure for EBM suggested above is an important step in synthesizing our current state of knowledge and will provide a basis for management and for planning a future research agenda (USCOP 2004; DEFRA 2005).

One ray of hope is provided by the recently released national Ocean Research Priorities Plan and Implementation Strategy developed by the US Joint Subcommittee on Ocean Science and Technology (JSOST 2007). This is the first comprehensive ocean research plan that involves every agency of the US government concerned in any way with ocean research. The overarching goal of the plan is "to provide the guidance to build the scientific foundation to improve society's stewardship and use of, and interaction with, the ocean." The plan focuses on three central elements of ocean science and technology: (1) capability to forecast ocean and ocean-influenced processes and phenomena, (2) development of scientific support for EBM, and (3) deployment of an ocean-observing system. These three—ocean forecasting, EBM, and ocean observing—permeate the entire document and its twenty national research priorities organized within six societal themes. The JSOST (2007) further recognized the breadth of scientific support and integration that would be needed to implement EBM, stating that "a multi-dimensional, multi-disciplinary effort to enhance current understanding of ecosystem processes, determine which interactions are most critical, and assess the dynamics of the natural and human factors affecting those interactions" would be necessary (see also box 2.4). Full development and implementation of the Integrated Ocean Observing System (IOOS) and other ocean and coastal observatories will provide a foundation of monitoring data that could enable and enhance management in many ocean sectors. An IOOS that includes not only high-resolution measurements in time and space of the physical and chemical properties within an ecosystem, but also biological attributes, and that incorporates high-resolution data on human activities within that ecosystem, would open up an entirely new world of information for management. Such a system and its related information flows would enable forecasting of ocean processes and phenomena, including severe storms, currents, status of fishery stocks and other biological resources, and human health risks. The requisite tools are rapidly becoming available, with a variety of data collection methods—from measurements of waves, temperature, currents, and productivity in real time with buoys, satellites, and radar to the monitoring of fishing activity with vessel-monitoring systems, and shipping with automated identification systems (USCOP 2004), and even development of a wide range of biological sensors (JSOST

Box 2.3. Some Requisites for Effective Science Support of Marine EBM

1. **Involvement of a wide range of professionals**: assessing the status and trends of major ecosystem services within reasonable ecosystem boundaries (such as the fishery management council regions; see fig. 2.1).
2. **Synthesis of information across sectors**: within each ecoregion and the creation of a dynamic data management system to track the latest observations of the natural system and associated human activities. These data need to be geographically referenced so that they can be overlaid on the ecosystem.
3. **Mechanisms and models for projecting changes**: due to management policies within each sector on all of the services within the region so that policy options can be explored in scenario analyses for the ecosystem.
4. **Peer review mechanisms to validate analyses**: in the synthesis and the scenarios (2 and 3 above; note peer review of 1 should be done within the sectoral science process) and peer review of the best advice and its translation into language readily understandable by managers, resource users, and the public.

Box 2.4. Data Integration for Marine EBM

Since many scientific efforts are focused within a given sector, assembling data for a more comprehensive view of a marine ecosystem and the human activities within that ecosystem is remarkably difficult. Sampling frames are disparate, time series are fragmentary, and data, particularly on business activities, are often considered confidential. Because so much of scientific recognition is based on publication, scientists are also often reluctant to make data they have collected freely available until they have published all the results they intend to publish, which may take a lifetime in some cases. Overcoming the hurdle of assembling comprehensive data is a major issue for implementation of EBM.

Data sets covering as many aspects of the natural system as possible, as well as the human activities within that system, must be geographically specific with well-accepted and documented procedures for interpolation to provide a clear picture of ecosystems in time and space. A dynamic atlas of each ecosystem will ultimately need to be assembled in order to support EBM.

2007; Sandifer et al. 2007). Using these new tools would allow management to operate on a spatial and temporal resolution that has never before been possible. However, developing new management strategies and tactics that can take advantage of such high-resolution information is an important area for growth and research.

A Way Forward

Implementing an EBM approach in practice is a major, revolutionary change in many aspects of ocean policy. Integration, focus on sustainability of services and resilience, and consideration of trade-offs between sectors reframe many of the long-standing debates over management measures for ocean activities. While incremental steps to improve management and account for the broader impacts of human activities within sectors of marine ecosystems are important and can be made, these are not a substitute for implementing an ecosystem-based approach to management. EBM will require major, overarching changes because integration and negotiating trade-offs are fundamentally different from tinkering around the edges of the existing management process. This does not mean that there is no role for sectoral management. Some fundamentals still pertain: Overfishing must still be prevented; contaminants and habitat damage must be minimized; protected species must still be protected. These are societal choices that will not be eliminated by an ecosystem-based approach. But a broader set of societal choices can now be made under EBM because interactions, cumulative effects, and trade-offs are of primary concern.

To implement EBM effectively, managers will need at least five things:

1. A comprehensive and clear legal mandate requiring the application of ecosystem-based management in the sector for which they have management responsibility and authority, and a mechanism to address cross-sectoral impacts, interactions, and trade-offs—this mandate must include the process of goal setting and a requirement to use a precautionary approach with respect to maintaining the delivery of ecosystem services.
2. An outline of what is expected of a manager operating under EBM requirements—this road map needs to be clear and as unambiguous as possible.
3. Summaries of the factors that might affect or be affected by each sector (e.g., fish, energy, water quality, tourism, coastal development, ports, transportation) in instructive and usable formats—it is critical that such information be made available to sectoral managers in action-relevant form through a structured process that includes peer review.

4. Scientific information that is cross-sectoral but also has clear relevance to a given sector (e.g., how water quality affects fisheries; how coastal development affects fish habitat)—such information must be provided, to the greatest extent possible, in geographically specified format so managers and their constituents can readily determine the potentially affected places, regardless of whatever inferences or generalizations may be drawn from it. Uncertainties associated with the data also need to be clearly stated.
5. A forum for comprehensive ocean planning—such planning must bring together the goals, scientific information, management efforts, and implementation in an integrated whole (fig. 2.3).

Are managers ready to use such advice as it develops? Many are already beginning to do so. For fisheries, the mandates to protect habitat (Rosenberg et al. 2000) and bycatch of nontarget species (Harrington et al. 2005) have helped set the stage for consideration of ecosystem-based effects more broadly (e.g., see

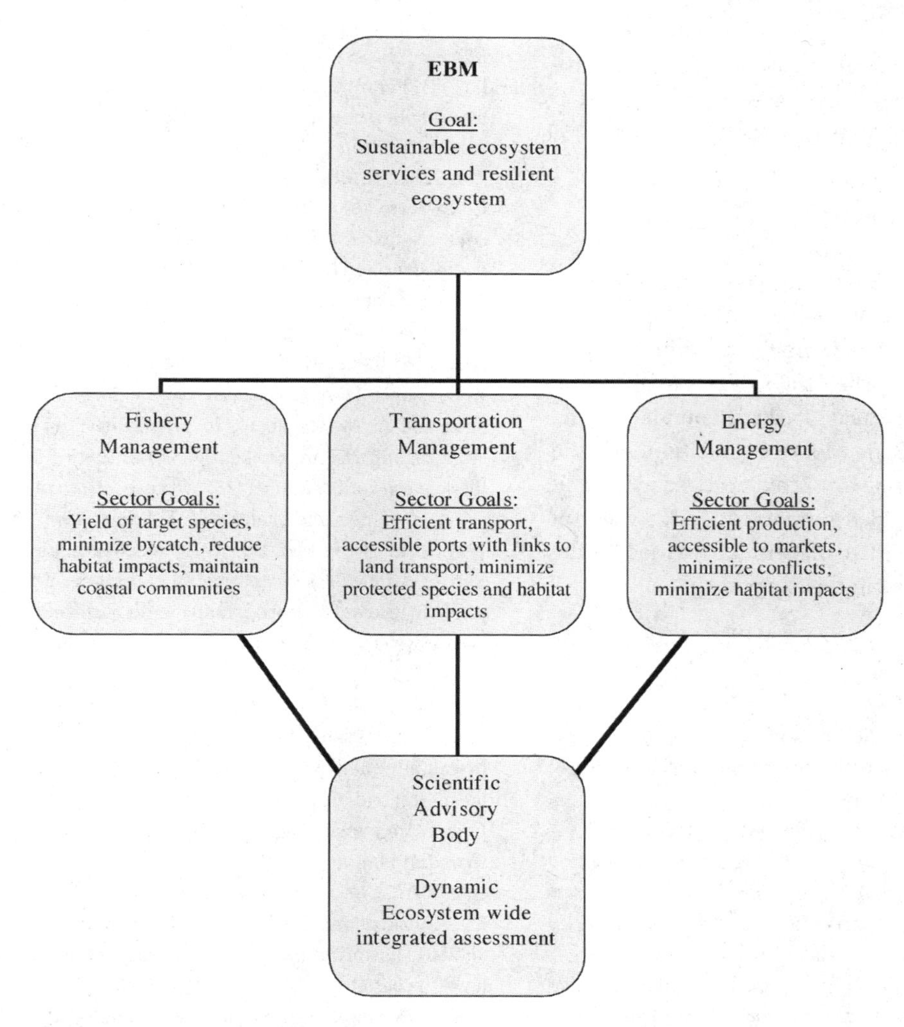

Figure 2.3 Framing ecosystem-based management (EBM) goals across sectors.

NOAA 2006; Merrick et al. 2007). In the energy, coastal development, and pipeline and cabling sectors, consideration of mitigation of impacts on habitat has been part of the management process for several years. These within-sector approaches, while not fully developed EBM, are the first steps toward a more comprehensive system.

Despite the difficulties that come with change, for scientists, managers, and constituents, EBM is clearly an approach whose time has come. We know enough to begin, and we have tried some of the first steps. Perhaps most importantly, there is wide recognition that we need to go beyond the conventional approaches to management of marine resources because sector-by-sector efforts have too often failed to prevent ecosystem degradation (MA 2005).

Acknowledgments

We thank Drs. Steven Murawski, George Sedberry, and Pace Wilber, all of NOAA, as well as anonymous reviewers for providing helpful comments on earlier drafts of this manuscript. The David and Lucile Packard Foundation, the Gordon and Betty Moore Foundation, and the National Center for Ecological Analysis and Synthesis provided support for AAR's work on ecosystem-based management that led to the ideas developed in this chapter.

References

Butler, S. J., J. A. Vickery, and K. Norris. 2007. Farmland biodiversity and the footprint of agriculture. *Science* 315:381–84.

Cairns, J., and T. V. Crawford, ed. 1991. *Integrated environmental management*. Chelsea, MI: Lewis.

Cicin-Sain, B., and R. Knecht. 1998. *Integrated coastal and ocean management: Concepts and practices.* Washington, DC: Island Press.

Clark, T. W. 2002. *The policy process: A practical guide for natural resource professionals.* New Haven, CT: Yale University Press.

Collie, J. S., J. M. Hermsen, P. C. Valentine, and F. P. Almeida. 2005. Effects of fishing on gravel habitats: Assessment and recovery of benthic megafauna on Georges Bank. *American Fisheries Society Symposium* 41:325–43.

Crowder, L. B., G. Osherenko, O. R. Young, S. Airame, E. A. Norse, N. Baron, J. C. Day et al. 2006. Resolving mismatches in US ocean governance. *Science* 313:617–18.

DEFRA (Department for Environment, Food and Rural Affairs). 2005. *Charting progress: An integrated assessment of the state of UK seas.* London: DEFRA. http://www.defra.gov.uk.

Fanning, L., R. Mahon, P. McConney, J. Angulo, F. Burrows, B. Chakalall, D. Gil et al. 2007. A large marine ecosystem governance framework. *Marine Policy* 31(4):434–43.

Frid, C., O. Paramor, and C. Scott. 2005. Ecosystem-based fisheries management: Progress in the NE Atlantic. *Marine Policy* 29:461–69.

Garcia, S. M., A. Zerbi, C. Aliaume, T. Do Chi, and G. Lasserre. 2003. *The ecosystem approach to fisheries: Issues, terminology, principles, institutional foundations, implementation, and outlook.* Fisheries Technical Paper no. 443. Rome: Food and Agriculture Organization of the United Nations (FAO).

Grizzle, R. E., L. G. Ward, L. A. Mayer, M. A. Malik, A. B. Cooper, H. A. Abeels, J. K. Greene, M. A. Brodeur, and A. A. Rosenberg. In press. Effects of a large fishing closure on benthic communities in the western Gulf of Maine, USA: Recovery from the effects of gillnets and otter trawls. *Fishery Bulletin.*

Halpern, B. S., K. L. McLeod, A. A. Rosenberg, and L. B. Crowder. 2008. Managing for cumulative impacts in ecosystem-based management through ocean zoning. *Ocean and Coastal Management* 51:203–11.

Harrington, J. M., R. A. Myers, and A. A. Rosenberg. 2005. Wasted fishery resources: Discarded by-catch in the USA. *Fish and Fisheries* 6:350–61.

Hilborn, R., and C. J. Walters. 1992. *Quantitative fisheries stock assessment methods.* New York: Chapman and Hall.

Jackson, J. B. C., M. X. Kirby, W. H. Berger, K. A. Bjorndal, L. W. Botsford, B. J. Bourque, R. H. Bradbury et al. 2001. Historical overfishing and the recent collapse of coastal ecosystems. *Science* 293:629–38.

JSOST (Joint Subcommittee on Ocean Science and Technology). 2007. *Charting the course for ocean science in the United States for the next decade: An Ocean Research Priorities Plan and Implementation Strategy*. Washington, DC: National Science and Technology Council.

Juda, L. 1999. Considerations in developing a functional approach to the governance of large marine ecosystems. *Ocean Development and International Law* 30:89–125.

Lafolley, D. d'A., E. Maltby, M. A. Vincent, L. Mee, E. Dunn, P. Gilliland, J. P. Hamer, D. Mortimer, and D. Pound. 2004. *The ecosystem approach: Coherent actions for marine and coastal environments. A Report to the UK Government*. Peterborough, UK: English Nature.

Leslie, H. M., and K. L. McLeod. 2007. Confronting the challenges of implementing marine ecosystem-based management. *Frontiers in Ecology and the Environment* 5(10):540–48.

Leslie, H. M., A. A. Rosenberg, and J. Eagle. 2008. Is a new mandate needed for marine ecosystem-based management? *Frontiers in Ecology and the Environment* 6(1):43–58.

MA (Millennium Ecosystem Assessment). 2005. *Ecosystems and human well-being: Synthesis*. Washington, DC: Island Press. http://www.maweb.org.

McLeod, K. L., J. Lubchenco, S. R. Palumbi, and A. A. Rosenberg. 2005. *Scientific consensus statement on marine ecosystem-based management*. The Communication Partnership for Science and the Sea (COMPASS). Signed by 221 academic scientists and policy experts with relevant expertise. http://www.compassonline.org/pdf_files/EBM_Consensus_Statement_v12.pdf.

Merrick, R., R. Kelty, T. Ragen, T. Rowles, P. Sandifer, B. Schroeder, S. Swartz, and N. Valette-Silver. 2007. *Report of the Protected Species SAIP Tier III Workshop, 7–10 March 2006, Silver Spring, MD*. NOAA Tech Memo NMFS-F/SPO-78. Washington, DC: US Dept. of Commerce.

Meyerson, L. A., J. Baron, J. M. Melillo, R. J. Naiman, R. I. O'Malley, G. Orians, M. A. Palmer, S. P. Pfaff, S. W. Running, and O. E. Sala. 2005. Aggregate measures of ecosystem services: Can we take the pulse of nature? *Frontiers in Ecology and the Environment* 3:56–59.

NMFS (National Marine Fisheries Service). 1996. Magnuson–Stevens Fishery Conservation and Management Act of 1996. NOAA Tech Memo NMFS-F/SPO 23. Washington, DC: US Government Printing Office.

NOAA (National Oceanic and Atmospheric Administration). 2006. *Fisheries ecosystem planning for Chesapeake Bay*. Chesapeake Bay Fisheries Ecosystem Advisory Panel. Bethesda, MD: American Fisheries Society.

NOAA. (National Oceanic and Atmospheric Administration). 2004. *Report on the delineation of regional ecosystems*. Regional Ecosystem Delineation Workshop, Charleston, SC, 31 August–1 September 2004. Washington, D.C: NOAA.

NRC (National Research Council). 1998. *Improving fishery stock assessment methods*. Washington, DC: National Academies Press.

Palumbi, S. R., P. A. Sandifer, J. D. Allan, M. W. Beck, D. G. Fautin, M. J. Fogarty, B. S. Halpern et al. 2009. Managing for ocean biodiversity to sustain marine ecosystem services. *Frontiers in Ecology and the Environment* 2009; 7. doi: 10.1890/070135.

POC (Pew Oceans Commission). 2003. *America's living ocean: Charting a course for sea change. A report to the nation*. Washington, DC: Pew Trusts.

Recchia, C., S. Farady, J. Sobel, and J. Cinner. 2001. *Marine and Coastal Protected Areas in the U.S. Gulf of Maine Region*. Washington, DC: The Ocean Conservancy.

Rosenberg, A. A., T. E. Bigford, S. Leathery, R. L. Hill, and K. Bickers. 2000. Ecosystem approaches to fishery management through essential fish habitat. *Bulletin of Marine Science* 66:535–42.

Rosenberg, A. A. 2007. Fishing for certainty. *Nature* 449:989.

Rosenberg, A. A., and K. L. McLeod. 2005. Implementing ecosystem-based approaches to management for the conservation of ecosystem services. In Politics and socio-economics of ecosystem-based management of marine resources, ed. H. I. Browman and K. I. Stergiou. *Marine Ecology Progress Series* 300:270–74.

Rosenberg, A. A., J. H. Swasey, and M. B. Bowman. 2006. Rebuilding US fisheries: Progress and problems. *Frontiers in Ecology and the Environment* 4:303–8.

Rosenberg, A. A. 2002. The precautionary approach from a manager's perspective. *Bulletin of Marine Science* 70:577–88.

SAFMC (South Atlantic Fishery Management Council). 2004. *Ecosystem-based management: Evolution from the habitat plan to a fishery ecosystem plan.* SAFMC action plan. Charleston, SC: SAFMC.

Sandifer, P., C. Sotka, D. Garrison, and V. Fay. 2007. *Interagency oceans and human health research implementation plan: A prescription for the future.* Washington, DC: Interagency Working Group on Harmful Algal Blooms, Hypoxia and Human Health of the Joint Subcommittee on Ocean Science and Technology.

Sandifer, P. A., and A. A. Rosenberg. 2005. Practical recommendations for improving the use of science in marine fisheries management. In *Managing our nation's fisheries II: Focus on the future,* ed. D. Witherell, 197–210. Proceedings of a conference of fisheries management in the United States held in Washington, DC, 24–26 March 2005. http://www.managingfisheries.org.

Shepherd, G. 2004. *The ecosystem approach: Five steps to implementation.* Gland, Switzerland: International Union for Conservation of Nature (IUCN).

Sherman, K., and H. R. Skjoldal, ed. 2002. *Large marine ecosystems of the North Atlantic: Changing states and sustainablility.* Amsterdam: Elsevier.

USCOP (US Commission on Ocean Policy). 2004. *An ocean blueprint for the twenty-first century.* Final report. Washington, DC: USCOP. ISBN 0 9759462 0 X.

US Executive Branch. 2004. *US Ocean Action Plan. The Bush Administration's response to the US Commission on Ocean Policy.* Washington, DC. http://ocean.ceq.gov/actionplan.pdf.

Walker, B., and D. Salt. 2006. *Resilience thinking: Sustaining ecosystems and people in a changing world.* Washington, DC: Island Press.

Worm, B., E. B. Barbier, N. Beaumont, J. E. Duffy, C. Folke, B. S. Halpern, J. B. C. Jackson et al. 2006. Impacts of biodiversity loss on ocean ecosystem services. *Science* 314:787–90.

Young, O. R., G. Osherenko, J. Ekstrom, L. B. Crowder, J. Ogden, J. A. Wilson, J. C. Day et al. 2007. Solving the crisis in ocean governance: Place-based management of marine ecosystems. *Environment* 49(4):20–32.

PART 2
Conceptual Basis for Ecosystem-Based Management

CHAPTER 3
The Oceans as Peopled Seascapes

Janna M. Shackeroff, Elliott L. Hazen, and Larry B. Crowder

To many, the status of the world's oceans is both troubling and inspiring. Mounting evidence has documented declines across all reaches of the oceans, from the poles to the tropics, the shoreline to the deep sea. Cross-disciplinary, international efforts have unequivocally shown the influence of humans on the ocean's capacity to provide essential services (e.g., MA 2005a). Natural and human factors causing changes to and often diminishing the capacity of the production of ecosystem services are increasingly global and synergistic. Yet the growing awareness of the declining health of the oceans has also propelled advances in the theories, practices, and strategies in support of marine ecosystem-based management (EBM). In this chapter, we describe the status of the world's oceans, as well as biophysical, social, and integrated drivers of change in ocean systems in the past, present, and future.

We begin with a review of current approaches to assessing the status of coastal and ocean systems. Then we examine the same issues from the perspective of coupled social–ecological systems (SES), an alternative way of thinking about human interactions with the marine environment. In doing so, we emphasize a broader view of marine SESs: one where humans and their activities are fully integrated into marine EBM, or, where oceans are treated as *peopled seascapes.* We then discuss six implications of approaching the status of marine ecosystems from the perspective of "peopled oceans." Basic assumptions about the relationship between human social and ocean systems need to be reconsidered, as we cannot continue to treat humans as largely exogenous, negative forces on the environment. Interdisciplinary cooperation is essential to the research and management of coupled SESs. In designing marine management strategies, the context, history, and human dimensions at local scales must be kept in mind, particularly when faced with global drivers of change. Finally, power relations among individuals and groups of people can help us better understand the nature and outcomes of human–environment interactions. We advocate recognizing people's varied perspectives and experiences with the marine realm and building these diverse perspectives into ocean management strategies.

The Status of Marine Ecosystems

Identifying key components of marine ecosystems from a biophysical perspective and describing the status of these ecosystems is a challenging but important topic to address before we can fully understand the effects of environmental variability and anthropogenic drivers of change on ocean systems. The status of marine ecosystems can be defined (1) at the ecosystem level (e.g., watersheds, coral reefs, the open ocean, deep sea hydrothermal vents) or (2) based on constituent parts of ecosystems, such as components (e.g., species, populations, communities), patterns (e.g., distribution, genetic variability, species richness, food webs), or processes (e.g., oceanographic linkages, dispersal of organisms, seascape connectivity). Differentiation among the components, patterns, and processes can be made on temporal as well as geographic scales: for example, coral reefs at the scales of leeward Maui, the Hawaiian Archipelago, or the Indo-Pacific across ten thousand, or a hundred thousand, years. Moreover, status can also be discussed with respect to connectivity between systems, such as watersheds upstream of coral reefs, offshore dynamics, or the global climate. Ecosystem services also can be examined to assess the status of ocean ecosystems and implications of changes to that status for human well-being (see also McLeod and Leslie, chap. 1 of this volume; Wainger and Boyd, chap. 6 of this volume; Barbier, chap. 8 of this volume).

Coastal and ocean habitats are in varying degraded states due to the increasing intensity and frequency of anthropogenic activities, coupled with natural environmental variation. For our purposes, the coast begins 100 km inland and extends to the start of the continental shelf, while open ocean systems are defined as waters greater than 50 m in depth (MA 2003). Coastal ecosystems, though more resilient to disturbance than the open ocean due to high natural variability in coastal processes, are more proximate to and are more affected by human activities (Lotze et al. 2006). The coastal zone occupies less than 11% of the world's oceans but accounts for 90% of marine fisheries landings (MA 2005a). Localized depletion of coastal species, particularly large vertebrates, occurred for thousands of years prior to the onset of ecological research (Jackson et al. 2001). The ecological extinctions of numerous species have likely decreased the resilience of these ecosystems to both natural and human-caused disturbances (Jackson et al. 2001). In wetlands and estuaries, polluted runoff from agriculture and urban areas impacts water quality, fisheries, and entire ecosystems and fisheries both acutely and chronically (MA 2005b). Hypoxic bottom waters have increased worldwide between the 1950s and the 1980s, and there has been no global trend toward reversal (Diaz 2001). Some 19% of the world's coral reef areas have been lost due to habitat degradation and loss and pollution, and 15% more are under imminent threat of being lost in the next 10 to 20 years (Wilkinson 2008). Sea grass beds and kelp forests—both highly productive ecosystems—show worldwide declines from pollution (Duarte 2002) and a loss of key consumers due to overfishing (Dayton 2003). Reports of sea grass bed loss have increased tenfold over the past four decades, mirroring similar coastal ecosystems (Orth et al. 2006). Globally, these highly productive coastal ecosystems have all been impacted, with over half of all coral reef and mangrove habitats experiencing medium-high to very high impact scores (Halpern et al. 2008). Mangrove forests have been lost at a rate of 2.1% per year over the past two decades—a total loss of 35% (Valiela et al. 2001). This rate has decreased globally to 0.66% through 2005; however, mangrove habitat is still decreasing in all continents, with an estimated loss of 500,000 ha worldwide during the period of 2000–2005 (FAO 2007). These examples highlight the degree to which a growing number of threats, including overexploitation, pollution, invasive species, climate change, diseases, and habitat loss, have compromised the resilience of coastal and marine systems and their ability to continue to provide services that people need and value (UNEP 2006).

Compared with coastal areas, open ocean ecosystems have experienced significantly less human activity, and we know less about them due to distance from land, rougher seas, and the obvious challenges associated with exploring deeper waters. Open ocean ecosystems are known principally through fishery landings data and patterns of fisheries' activities. Relatively less is known about biodiversity, pollution, and many open ocean ecosystem components and human activities therein (MA 2005a; UNEP 2006). An increase in fishing effort offshore and in deeper waters, combined with a leveling off of global fishery landings since the mid-1980s, indicates that pelagic communities have been impacted spatially through localized depletion (Myers and Worm 2003). A displacement of artisanal and local fisheries, which was masked for years by relatively constant landings globally (Watson and Pauly 2001), highlights the historical and contemporary changes in spatial distribution of fisheries, the loss of local ecological knowledge and practice, and localized changes in biomass, populations, and food webs, among others. The recent trend toward fishing at lower trophic levels suggests that humans may already have depleted many top predators (Pauly 1995; Myers and Worm 2003; Myers et al. 2007), while escalating demand has also increased fishing pressure on lower trophic levels (Essington et al. 2006). This results in severely impacted patterns (food web functioning, genetic variability, biodiversity) and processes (dispersal, behavior) of marine ecosystems (Worm et al. 2006). Populations of long-lived species such as the orange roughy (*Hoplestethus atlanticus*) and pelagic armorhead (*Pseudopentaceros wheeleri*) are showing signs of severe impacts from fishing (Roberts 2002). Moreover, the open ocean and deep sea are seen as the "next frontier" for energy resources (e.g., gas, wind, and tidal), deep sea mineral extraction, and aquaculture. Just as in coastal environments, we have no organized way in which to deal with these diverse and sometimes conflicting human activities.

Drivers of Change in Marine Ecosystems

Biophysical portions of marine ecosystems and their services are continually in flux due to environmental variation and anthropogenic drivers of change. A *driver of change* can be defined as "any natural or human-induced factor that directly or indirectly causes a change in the ecosystem" (MA 2003). Drivers of change in coastal and marine ecosystems can be direct or indirect (box 3.1). Direct drivers, such as the extraction of fish by the groundfish fishery in New England, act unequivocally on a resource. Indirect drivers operate more diffusely by influencing one or more direct drivers; for example, a change in market demand for groundfish may lead to increased or decreased fishing activity. Generally, an endogenous driver is one that can be directly influenced by a resource manager, while an exogenous one cannot. Complex relationships exist among humans, drivers of change, ecosystems, and ecosystem services (MA 2003; Leslie and Kinzig, chap. 4 of this volume). Without an understanding of the range of influential interactions that ultimately affect ecosystem services, it is impossible to effectively manage at the ecosystem scale.

Direct drivers of change can affect an individual species, a food web, or an entire ecosystem, and processes develop and respond at a variety of temporal and spatial scales. Examples include fisheries harvest, habitat change and land-based pollution, climate change, and invasive species (box 3.1). One of the most significant direct drivers of change in biophysical ocean ecosystems is fishing (commercial, recreational, and artisanal). In addition to direct extraction of a target species, nontarget species are often taken as bycatch, which contributes significantly to declines in seabird, marine mammal, sea turtle, and shark populations (Lewison et al. 2004). With declines in populations of target species and shifts to short-lived species, fishing effort has increased in many fisheries. Because bycatch is proportional to fishing effort, not landings, long-lived species taken inadvertently as bycatch are extremely adversely affected by increased fishing effort. Even small-scale fisheries can have strongly negative impacts (Peckham et al. 2007). Bottom trawling can significantly damage benthic habitat (Turner et al. 1999; Thrush and Dayton 2002), and lost gear from fisheries can lead to entanglement. Across the oceans, food webs have been significantly altered by overfishing (Jackson et al. 2001; Christensen et al. 2003), with as many as 90% of all fisheries fished at an unsustainable rate (Myers and Worm 2003). The resulting loss of biodiversity is substantial and may affect system resilience (Sala and Knowlton 2006; Worm et al. 2006). Relatively minor disturbances in one species can catalyze dramatic regime shifts toward less-productive environmental states. By steadily depleting the largest organisms in the population, fisheries may have evolutionary effects as well. Large-scale fishing for Atlantic cod (*Gadus morhua*) in New England has led to the rapid evolution of reduced size at age (Law 2000; Olsen et al. 2004). In heavily overfished ecosystems, population-level effects can last well after fishing pressure is reduced, reduced biodiversity can facilitate the establishment of invasive species, and loss of keystone species can have cascading effects throughout the food web. Decreased commercial catches have led to financial subsidies for unprofitable fisheries, which can exacerbate overfishing as well as create an oversaturation of the market and increase costs for consumers (Munro and Sumaila 2002; Pauly et al. 2002). Additional indirect, exogenous drivers of change associated with fishing include changes in market demand, technological advances in fishing, and globalization (box 3.1).

Population growth, particularly in coastal areas, leads to habitat alteration and increased nutrient loading of marine systems. Nitrogen and phosphorus runoff originates primarily from agricultural fertilizers and waste from farmed animals but is compounded by many other nonpoint sources (Rabalais et al. 2002). Rivers transport the nutrients downstream directly to estuaries, where eutrophication is a growing concern. Moderate levels of coastal eutrophication can act as a subsidy on the system by providing more resources for top-down-

Box 3.1. Direct and Indirect Drivers of Change in Coastal and Marine Ecosystems (adapted from MA 2005a)

<table>
<tr><th></th><th>Type</th><th>Coastal</th><th>Oceanic</th></tr>
<tr><td>Direct</td><td>Endogenous</td><td colspan="2">Fishing
Resource harvesting
Military activities</td></tr>
<tr><td></td><td></td><td>Pollution</td><td></td></tr>
<tr><td></td><td>Exogenous</td><td>Disease
Invasive Species</td><td></td></tr>
<tr><td>Indirect</td><td>Endogenous</td><td>Land use changes</td><td></td></tr>
<tr><td></td><td>Exogenous</td><td colspan="2">Fishing subsidies
Climate change
Technological change
Fishery demand
Resource demand
Globalization
Population growth
Disease</td></tr>
</table>

regulated systems (Cloern 2001), but extreme eutrophication results in widespread bottom-water hypoxia (Rabalais et al. 2001). The Gulf of Mexico along the Louisiana coast receives nutrient inputs from 42% of the continental United States and experiences bottom-water hypoxia covering up to 26,000 km^2 (Rabalais et al. 2002; LUMCON 2008).

An equally important driver—or really suite of drivers—is global climate change, which clearly acts over broad spatial and temporal scales. Impacts of climate change include sea level rise, changes in ocean chemistry and circulation, and shifts in the distribution and ecology of key species (Harley et al. 2006). Anthropogenically driven, directional climate change is confounded by natural climate variation (e.g., the El Niño Southern Oscillation [ENSO], Pacific Decadal Oscillation, and North Atlantic Oscillation), which makes it even more challenging to predict specific causal pathways. Nonetheless, the ecological and social effects of global climate change have already been detected and are predicted to be even more significant in coming decades (IPCC 2007). For example, in the case of coral reef ecosystems, the synergistic effects of rising temperatures and ocean acidification in particular are predicted to result in loss of carbonate reef structure and declines in reef community diversity (Hoegh-Guldberg et al. 2007) as well as significant impacts on dependent human populations (IPCC 2007).

Global temperatures have risen between 0.3°C and 0.6°C in the past century and are predicted to rise between 1.3°C and 3°C over the next century from greenhouse gases in the atmosphere (IPCC 2007). Increases in global temperatures will have a multitude of effects on marine communities and biodiversity. For example, increased ocean temperatures are one of the primary causes of coral bleaching (Knowlton 2001) and of decreased recruitment success of important forage species such as walleye pollock in Alaska (Wespestad et al. 2000) and krill in the Southern Ocean (Loeb et al. 1997). While mobile organisms (e.g., tuna) might alter their migration patterns with changes in ocean temperatures or prey distributions (Polovina 2005), long-lived community building blocks such as kelp forests and coral reefs cannot move rapidly (Steneck et al. 2002; Walther et al. 2002; Bellwood et al. 2004). Changes to ocean circulation patterns could negatively affect the recruitment of species such as corals (Harley et al 2006), resulting in a decrease in biodiversity. Most research has focused on the linear effects of climate change (e.g., sea level rise and temperature increases, Harley et al. 2006), but given the documented occurrence of nonlinear ecological responses to biotic and physical changes in the environment (i.e., regime shifts, Leslie and Kinzig, chap. 4 of this volume), strategies for managing nonlinear responses to climate change and other drivers must be developed.

Invasive marine species are another important driver of change in marine systems. The increase in global ship traffic has increased the transport vectors for invasive species. Marine invaders can impact vulnerable native species leading to decreased biodiversity and often extinction (Mack et al. 2000). Because systems with reduced biodiversity have greater available niche space, strong competitors such as invaders can dominate the landscape (Stachowicz et al. 1999). The Mediterranean Sea now contains over eighty-five species of introduced algal macrophytes, including large monocultures up to 30,000 ha, where substantial diversity once existed (Sala and Knowlton 2006). The number of successful invaders will continue to rise, considering that increases in global shipping and thus transport of invasives are likely, as are conditions like land-based pollution, which stress coastal ecosystems and therefore aid invaders' successful establishment.

As spatial scales increase, the effects of drivers of change take longer to propagate, and more time is needed to study them and to provide useful information for managers. In other words, fine-scale drivers have shorter feedback loops than broader processes. Many drivers that primarily impact coastal habitats (due to the proximity to humans) can eventually propagate to the open ocean. For example, destruction of mangroves and wetlands for aquaculture can reduce critical nursery habitat, thereby effectively reducing recruitment of new individuals to open ocean populations (Valiela et al. 2001; FAO 2007).

The disparity in scale between large-scale drivers and that of management exacerbates the already challenging task of managing human interactions with ecosystems. While endogenous drivers often operate at spatial and temporal scales similar to those of management activities, the feedback effects can be delayed (see also Guichard and Peterson, chap. 5 of this volume). While an increase in fishing rates might cause an initial decline in population abundance, long-term cascading effects for entire ecosystems are more difficult to forecast (Dayton et al. 2002). Exogenous and, often, indirect drivers such as globalization affect the ecosystem in ways that are more difficult to predict (MA 2003). Advances in technology allow more-efficient harvest of marine resources as well as easier transport of them around the globe (Berkes et al. 2006). As the distance increases between the location of harvest and where the resource is used, the effects become more dilute and the awareness of the eventual consumer decreases. In sum, biophysical as well as social and institutional processes can act at multiple scales, many of which extend beyond the geographic or authoritative jurisdiction of resource management. Consequently, effective management of human interactions with coastal and ocean environments demands that management institutions and the social and scientific infrastructure that support them (see Guichard and Peterson, chap. 5 of this volume; Rosenberg and Sandifer, chap. 2 of this volume) be able to adapt to changing conditions, and to learn from the past.

Synergistic Effects of Drivers of Change

Most marine systems are adapted to some level of disturbance, but oceans today are subject to increasing, cumulative, and synergistic effects of both natural and anthropogenic factors (UNEP 2006; Leslie and Kinzig, chap. 4 of this volume). Coastal systems in particular are at high risk from multiple stressors due to their proximity to humans and the prolonged timescale of degradation (Lotze et al. 2006). Synergism between increased temperatures and high levels of eutrophication has led to toxic dinoflagellate blooms and resultant fish kills. Often coastal properties are left with tons of rotting fish, and coastal food webs are impaired (Boesch et al. 2001). Disasters such as the Exxon *Valdez* oil spill in March of 1989 can lead to massive mortalities throughout the food web (Paine et al. 1996). While some species have recovered, chronic, long-lasting effects are still visible in the loss of kelp habitat, as well as reduced growth and physical deformities in fish with high biomarkers (Peterson et al. 2003).

Exemplifying cumulative and synergistic drivers, overharvest, sediment and nutrient pollution, disease, and global climate change have led to mass bleaching and shifts in species composition in coral reef ecosystems (Bellwood et al. 2004). The number of reefs damaged has increased exponentially since 1980 (Bellwood et al. 2004) due to the synergism of these anthropogenic disturbances (e.g., disease and overfishing in Discovery Bay, Jamaica, Knowlton 2001). The additional stressors of ocean warming and acidification could cause major changes in coral species composition, in many cases decreasing the capacity of reefs to provide social and economic services (Hughes et al. 2003). None of the aforementioned drivers act in isolation, but instead they have compounded effects. Elevated global temperatures have increased the intensity and frequency of coral bleaching, extensive nutrient loading has facilitated algal overgrowth of corals, and fishing has released predation pressure on coral grazers and removed significant herbivore biomass. The interactions among these impacts need to be considered to adequately manage marine ecosystems.

Synergistic effects require management strategies that deal with the entire ecosystem, multiple species, and drivers of change and account for natural environmental variability. Part 4 of this book offers examples of such management strategies.

Oceans as Peopled Seascapes

How inappropriate to call this planet Earth, when it is clearly Ocean.

Arthur C. Clark

We suspect that, like us, the readers of this book can quickly conjure a mental image of a coastline, or a little stretch of sea, that holds deep personal meaning. At the level of the individual, the idea of a sense of place (Relph 1976; Tuan 1977) helps us describe what it means for oceans to be "peopled." People develop a complex fabric of meanings of, relationships to, and interactions with the oceans. Consider, for instance, the different roles of the ocean in the lives of a big-wave surfer, a third-generation Sri Lankan fisherman, a sea turtle biologist, or a seafood restaurateur in China. These exemplify, on a micro scale, the textured human experiences with and relationships to the oceans—or what is referred to as the ocean's *human dimensions*. Human dimensions refer to the ways in which people affect, and are affected by, the oceans (NCCOS 2007). Herein, we suggest that the very richness and complexity of the oceans' human dimensions have long been underestimated and unwritten, yet are central to a discussion of drivers of change (fig. 3.1). They relate inextricably to *why people do the things they do* and, by extension, the small- to large-scale dynamics of the social–ecological system.

There is an alternative way of thinking about human impacts on, and interactions with, the marine environment, one in which human systems and the environment are inextricably linked and coevolved (see also McLeod and Leslie, chap. 1; Leslie and Kinzig, chap. 4 of this volume). We refer to these coupled systems, as mentioned above, as social–ecological systems, or SESs. The concept of SESs resonates with indigenous worldviews of human–environment interactions (Berkes 1999); for example, Native Hawaiian people conceive of the *'aina* (land) not simply as the soil. Rather, it is a place of shared lineage, encompassing generations of physical and metaphysical beings, including rocks, streams, trees and plants, air, earth, seas, fishes, humans, ancestors, and gods (Kamakau 1964). From the small to large scale, both social and ecological systems are complex and coevolved. Individual people, social networks, and institutions continually affect and are affected by ecological systems across local, regional, and global scales (Kinzig et al. 2006; Leslie and Kinzig, chap. 4 of this volume). A marine social–ecological system is thus multidimensional and integrative of people, their institutions, and economies as well as the biophysical system. In other words, the *oceans are peopled seascapes.*

Scientific intellectual traditions, until very recently, separated nature and society, creating significant intellectual and practical obstacles to managing the two as integrated domains. For example, biologists (or other physical scientists) typically see humans as exogenous forces of destruction on ecosystems, and applications of social science are often overly simplistic. Similar critiques have been made of investigations led by social scientists, where the environmental elements of the problem may be caricatures of reality. This is of course a challenge in any scientific study: Abstraction is vital for understanding of the most influential patterns and processes. Interdisciplinary approaches that capture both the ecological and sociocultural dynamics of interest

(A)

(B)

Figure 3.1 Mosaic of a Hawaiian coastal system, illustrating direct and indirect, endogenous and exogenous, drivers of change and signifying features of the social–ecological system less easy to illustrate, such as people's spiritual connection to oceans, ecological knowledge, and marine tenure practices. (A) Marine debris on isolated Kamilo Beach, Hawaii Island, originates from nations across the Pacific. (B) A multigenerational small-boat tuna fisherman heading out to ancient fishing grounds, Hilo, Hawaii. (C) Ancient Native Hawaiian *heiau* (place of worship) protected in a national park, a subsistence fisherman "picking seaweed," and a contemporary fishing canoe. (D) Young surfers passing the Waikiki surf racks. All photos courtesy of J. M. Shackeroff.

(C)

(D)

are proliferating, advancing the scholarship of coevolved systems, and beginning to offer guidance for more integrated, ecosystem-based management approaches (e.g., Turner et al. 2003; Hughes et al. 2005). While it is too early to have a complete understanding of social–ecological synergy, status, and drivers in the oceans, we can offer a framework for understanding key linkages and indicate some areas for future research and practice.

Below we identify six issues, which identify knowledge gaps, discuss complexities, and elucidate the links between oceans and humans. Tending to these issues will prepare us to identify drivers of change of coastal and marine social–ecological systems and thus aid in the implementation of true marine ecosystem-based management.

1. Overcoming an Unwritten History

First, in the writing of history and generation of academic knowledge, the oceans have not been treated as peopled. Indeed, as Arthur C. Clark's words at the beginning of this passage indicate, we are a terrestrial people, and perhaps naturally terra-centric. Many maps of the world, for instance, portray the ocean as a featureless blue space. While the ocean has a history in which people have been deeply involved, this history has remained largely uninvestigated and unwritten. Bolster (2006) provides a sense of what a historicizing of the oceans may reveal and what it means for oceans to be peopled:

> We need to better understand many things: how different groups of people made themselves in the context of marine environments, how race, class, fashion, and geo-politics influenced the exploitation and conservation of marine resources, how individual and community identities (and economies) changed as a function of the availability of marine resources, how technological innovation frequently masked declining catches, how fishermen's knowledge of localized depletions accumulated in the past, how public policy debates revealed historically specific values associated with the ocean, how collaboration between (and then antagonism among) fishermen and scientists affected marine environments, how faith in the certainty of marine science waxed and waned, how different cultures perceived the ocean at specific times, and—when possible—how past marine environments looked in terms of abundance and distribution of important species. These are the constituent parts that get to a deeper historical question: the nature of the greatest sea change in human history (Bolster 2006).

The richness of human context in the oceans is a vast area in need of exploration. As illustrated above, the social has historically been separated from the ecological in academia, research, management, and practice. Within Western traditions, many self-reinforcing mechanisms have fostered this divide (distinct epistemologies, departmentalization, differing expectations within literature, publishing, tenure, academic jargon). Yet, human dimensions of marine systems have been identified as key knowledge gaps in multiple fields: political ecology (Bryant 1998), historical ecology (van Sittert 2005), marine environmental history (Bolster 2006), and resilience theory (Walker et al. 2006), to name a few. Until intellectual traditions weave together, in a sophisticated manner, the complex science of social and ecological systems, we risk making management decisions based on incomplete knowledge of the linkages among people and ocean places.

2. Humans as Deterrents versus Humans as Coevolved

Second, part of understanding the human dimensions of the oceans will involve reconceptualizing the relationship between oceans and society. According to conventional wisdom, people threaten the oceans. Human beings pollute and build in the coastal zone, consume seafood and thereby support excessive fisheries extractions, and burn fossil fuels and contribute to global climate change (MA 2005a; UNEP 2006). While we are responsible for these impacts, this perspective casts humans as exogenous to the natural world. It is much harder to get at the question, Under what historic or sociopolitical circumstances are fisheries sustainable? when wisdom suggests humans are external agents rather than part of the marine food web. The conventional wisdom also suggests that human–marine environmental interactions are largely negative—whereas we can have positive impacts on ecosystem structure and functioning through ecosystem engineering, restoration, and other management interventions (e.g., Long et al. 2003; Olsson et al. 2004; Palmer et al. 2004). Such assumptions pervade management and are often paired with a classical economic assumption of the "tragedy of the commons," which views people as rational actors motivated solely by self-interest and invariably overexploiting resources held in common (Hardin 1968).

Yet the prevailing assumptions of ocean–society relationships are being challenged and, in some cases, overturned. Marine management successes and failures are found in all private, state, and communally based property rights regimes (e.g., McCay 1995, 1996; Ostrom et al. 2002). Many cases of self-restraint and sustainable practices by marine stakeholders exist (e.g., see Basurto 2005), and the tragedy of the commons does not always come to pass (Berkes et al. 1989). For example, St. Martin's (2001) work mapping fisheries based on fishers' perceptions revealed landscapes different from what was assumed based on fish stock and fishing-effort numerical data. St. Martin showed that, rather than behaving individually in a homogeneous and unbounded commons, fishers cooperate and form communities and can even act as the basis for more-formal forms of resource management that avoid depletion and sustain their equitable distribution. In summary, social–ecological relationships are proving more complex than previously considered. In addition to *people as culprits of ocean change*, we can describe *people as cooperative, recuperative, restorative agents of ocean change*. People affect, and are affected by, the oceans in positive, negative, and neutral ways. Moreover, humans are not exogenous; rather, they are no less a part of the ocean system than the California grunion, San Francisco Bay, or ENSO.

3. A New Interdisciplinary Marine Research Agenda

Third, managing marine systems in an integrated manner will require a research agenda that is sophisticated in its treatment of *both* human dimensions and the biophysical environment. Several interdisciplinary research fields—political ecology, resilience science, common pool resources, and historical ecology—are particularly well poised to answer key questions about marine social–ecological systems. Other research fields also have made substantial contributions to understanding the connections between coastal and marine ecosystems and associated human communities, including ecological economics, human ecology, political economy, environmental philosophy, and cultural ecology. We highlight the following four because they explore connections among (1) marine ecosystems, (2) local perspectives on human dimensions, and (3) broader political and economic processes, across scales of space and time.

Scholars of political ecology employ a multiscalar approach to examine power relationships among diverse communities and how these relationships relate to broader ecological, political, and economic events. Political ecologists are particularly cognizant of the social and institutional processes by which decisions are made, and how the resulting benefits are distributed among actors (individuals, households, or other social entities) that vary in their degree of vulnerability and marginality. Reviews of the political ecology literature include work by Bryant (1998) and Castree and Braun (2001), and recent ocean-related scholarship has ranged from investigations of fisheries and aquaculture production to endangered species conservation and research (Hanna 1998; Campbell et al. 2002; Young 2003; Mansfield 2004; Campbell 2007).

Resilience science focuses on the dynamics of complex systems, with an emphasis on the coupling and coevolution of social and ecological systems. Leslie and Kinzig (chap. 4 of this volume) expand on this area, so we will not review details here. Ocean-specific applications of resilience science have been focused primarily in coral reef and other tropical environments, as well as with Arctic indigenous societies (e.g., Berkes 1998; Berkes et al. 2003; Hughes et al. 2003, 2005; Wilson 2006).

Historical ecology emerged concurrently from both the social and natural sciences. Both approaches are focused on interdisciplinary investigations of the role of humans in global change. Social scientists focus on the coevolution of human communities and (largely terrestrial) ecosystems (e.g., Crumley 1994; Balée 1998), while biophysical scientists tend to emphasize the reconstruction of past ecosystem conditions based on diverse data sets (e.g., Jackson et al. 2001; Rosenberg et al. 2005).

Finally, scholars of common pool resources examine the links between resource management and social organization. Using approaches from political science, economics, anthropology, and other social science disciplines, they investigate how institutions and property-rights systems negotiate and, in some cases, overcome the tragedy of the commons (McCay 1995, 1996; Ostrom et al. 1999, 2002; St. Martin 2001).

Compared with oceans, human–environment studies are quite advanced in terrestrial environments, as is the use of human dimensions in land resource management (e.g., Zimmerer and Bassett 2003). We believe that the very features that define oceans (three-dimensional, saline, aquatic environment) and that have contributed to its characterization as unpeopled also ensure that humans interact with the ocean in fundamentally different ways than they do with the land. This conclusion corresponds with that of Carr and colleagues (2003), who argued that terrestrial principles of protected area networks should not be applied ad hoc to marine ecosystems and the development of no-take marine reserves. Limitations in data through space and time, differences in the biophysical environment of oceans versus terrestrial ecosystems, and fundamental differences between human–environment relationships on land and in the sea motivate our call for the development of a distinct approach to oceans as peopled seascapes (box 3.2).

4. The Case for Balanced Interdisciplinary Engagement

Fourth, shifting ocean research and management into a framework of "oceans as peopled seascapes" should come, at least in part, from balanced interdisciplinary engagement. By this we mean that biological, social, and interdisciplinary experts all participate from the earliest design and research phases of a management action. Interdisciplinary engagement entails cooperative teams of people trained in various disciplines *and* interdisciplinary research. Without interdisciplinarity and balance across disciplines, marine management actions have been shown to fail. For example, marine protected areas (MPAs), an important EBM tool, are evaluated principally based on biological criteria. While an MPA may be biologically successful, it can fail socially, which may lead

Box 3.2. Terrestrial Approaches Are Necessary but Not Sufficient for Understanding Oceans as Peopled Seascapes

While it is not necessary to reinvent the wheel—many human–environment relationships in terrestrial environments may indeed be relevant to oceans—we suggest terrestrial approaches should be applied with caution. Given the complexity of human systems and marine systems alone, we suggest that we must constantly ask of a marine EBM approach, Is it *marine* enough? Is it *peopled* enough? In particular, the following should be considered:

1. *Limitations of data are more of an issue in marine environments than in terrestrial environments.* In the generation of scientific knowledge, marine ecologists are generally limited to various forms of remote sensing and underwater sampling technologies, due to our land-centric physiologies. Reliance on technology has meant that the span of ecological data tends to be far shorter in duration than for those on land. The origin of the longest coral reef data sets, for instance, almost directly coincides with the advent of scuba technology in the 1960s, long after the onset of ecological change to these systems (Knowlton and Jackson 2001). Today, novel approaches to reconstructing ecological "baselines" are necessary to understand *relatively* pristine, normal ecological conditions (e.g., Jackson 1997, 2001; Rosenberg et al. 2005; HMAP 2008). While these novel approaches can fill some gaps, data paucity remains a problem.
2. *Ocean ecosystems exhibit fundamental ecological and evolutionary differences from terrestrial systems.* The "openness" of transport of nutrients, materials, organisms, and reproductive propagules in the oceans connects coastal and open ocean ecosystems. Moreover, the aquatic, three-dimensional environment is driven by physical processes operating on multiple spatial and temporal scales. These drivers in turn influence the ecology of ocean ecosystems (Carr et al. 2003; see also Guichard and Peterson, chap. 5 of this volume).
3. *Human–environment interactions are fundamentally different on land and in the sea.* Changes to the landscape are more visible to air-breathing humans than changes to the seascape. As people are unable to directly observe most marine processes without assistance (pole and line, scuba, satellite imagery), human–ocean interactions are often mediated by technology. Culturally, the characteristics of fishing communities and fishers around the globe seem to share more similarities with each other than with the cultures of the nations in which they reside (McGoodwin 2001). Historically, the oceans were unhistoried, forgotten in the writings of history compared with the land (Bolster 2006). Moreover, tenure, use, and property rights related to ocean resources differ quite significantly from those in terrestrial environments (Hanna et al. 1996; Osherenko et al. 2006). These observations suggest something different about how people interact with, come to know, perceive, and respond to the oceans.

to dwindling biological success in the long term (e.g., Christie et al. 2003).

In some cases, problems in interdisciplinary engagement relate to how issues are addressed, and in others, problems are about misunderstandings of what social science is and can do. Including social scientists from the outset of research and management initiatives can offer a way to better understand the people affected by management, including their motivations, cultural heritage, social and economic situations, local or indigenous expert knowledge, and cooperative management. Lessons learned from the fisheries council processes indicate that cooperative or joint research efforts with fishers, particularly those impacted by management decisions, can improve trust, communication, and good faith among constituencies (Kaplan and McCay 2004). If social science approaches to working with local communities are implemented from the outset, as a means rather than an end, compliance and communication with stakeholders may improve (see also Kliskey et al., chap. 9 of this volume). Another example comes from Campbell (2005), a social scientist who frequently participates in interdisciplinary marine research. Some of her natural science colleagues have thought that her role should include environment education and remediation of social and economic problems, neither of which is a role of a social scientist (Campbell 2005).

Social science and interdisciplinary efforts can foster successful marine management—but only if approached in a way that respects the social scientific process as well as the communities within which social science is achieved. Recent growth in interdisciplinary engagement is evident in many areas, including interdisciplinary teams of biophysical and social scientists (see Pomeroy et al. 2004), changing research agendas (e.g., Maurstad 2000; St. Martin 2001; Mansfield 2003, 2004), and interdisciplinary training programs for scientists and practitioners. These changes suggest that the concerns enumerated above will fade over time.

5. Keeping Sight of the "Local" when Facing Global-Scale Change

Fifth, the local scale is particularly important to consider in new approaches to peopled oceans. Scale is a theme that arises throughout this volume because cross-scale interactions within and between social–ecological systems are considered a pressing marine management challenge. The rising prevalence of global issues such as climate change is increasingly demanding a large-scale emphasis, as reflected in the tendency to focus on broadscale processes and change in oceans. This is despite burgeoning work in marine political ecology, which tends to look at smaller scales (as well as across scales), and attention to the local context of common pool resources. Few case studies examine integrated marine systems, and even fewer using a multiscalar approach. Yet terrestrial literatures and marine case studies that do take a multiscalar approach find that the complexity of local communities and human experiences are extremely important in explaining broader processes (terrestrial overviews, Leach and Mearns 1996b and Zimmerer and Bassett 2003; for marine case studies, see below).

Keeping the local scale in mind is important in any approach (Campbell 2007) and has often been the focus of terrestrial interdisciplinary studies, to the point that scholars have begun to criticize an *over*-emphasis on the local (Brown and Purcell 2005). But the opposite has tended to be the case with oceans, where the local is often invisible, as people are absent or transitory. Many processes cannot be understood outside of the context (geographic, social, cultural, political, economic, psychological, etc.) in which they exist. Just as fine-scale ecological patterns of recruitment, wave exposure, and habitats are important in designing networks of MPAs, the social, cultural, and economic complexities in human communities are critical to explain patterns and implement effective management measures. The local scale provides the context (political, economic, social, ecological, etc.) to implement scientific knowledge and management practices meaningfully.

Local knowledge is receiving increased consideration for its potential contributions to marine science and management (Davis and Wagner 2003; Kliskey et al., chap. 9 of this volume), including new insights into marine ecosystem dynamics, human dimensions, and local social context (Berkes 1999). Local "eyes" on the system may provide efficient feedback and response to environmental change. Local people, knowledge, and community structure may offer new ideas for management strategies attuned to the local social–ecological context (McGoodwin 2001; Berkes 2003a), as well as help bring attention to what local communities can do to achieve effective management (Olson 2005). When looking closely at fishing communities (McGoodwin 2001) or fishers' knowledge (Berkes 2003a), complex stories of culturally evolved environmental management practices can offer great insights into human dimensions and potential management strategies allowing for small-scale use (St. Martin 2001). These all reflect the benefits of a mostly local-scale approach.

To give an interdisciplinary, cross-scalar research example, Mansfield (2003) traced how different people and groups make distinctions about the biophysical world. Debates surrounding which places were more appropriate than others for certain kinds of production activities—such as whether imported seafood from Vietnam could be labeled "catfish" or whether farmed shellfish could be labeled "organic"—showed that the biophysical world, as manifest in individuals and their cultures, is related to international relations and trade (Mansfield 2003). Examining local complexities has enabled researchers to identify and describe why particular fisheries are sustainable and how sustainability is achieved. Because of the local sociopolitical context and the Maine lobster's unique biology, Acheson (1997) suggests, the Maine lobster fishery instigated shifts toward sustainable practices following periods of decline, three separate times across a century.

Heterogeneity in local contexts must be taken into account, particularly in considering the scale of management actions. Campbell and colleagues (2002) examine the scalar mismatch between an international treaty and community-based conservation. In their example, the use of an international instrument to eliminate the local use of marine turtles, without considering whether turtle use might be sustainable in specific communities, ultimately may undermine the treaty's effectiveness. This international instrument failed to reflect the local emphasis in current conservation thinking, and it seems to demonstrate that both international treaties and local efforts may be effective in some ways—and ineffective in others (Campbell et al. 2002). Indeed, institutional diversity is now considered as essential as biological diversity (Ostrom et al. 1999), and emergent place-based approaches emphasize the importance of local context to overcome mismatched scales of governance and ocean management issues (Young et al. 2007).

Of the many ways that cross-scalar interactions are implicated in marine systems, the local scale is emphasized here because it is so often oversimplified or reduced to technicalities in oceans. The small, local scale of human beings is a fundamental variable in describing drivers—why we do the things we do—and iterative relationships between ecological change and human change.

6. A Multiplicity of Perspectives and Power Relations

The sixth and final implication of oceans as peopled seascapes is one that underlies all of those previously discussed. Many people hold understandings of the oceans different from those described in science; these represent a "multiplicity of perspectives." Suggesting both practical and philosophical reasons for learning about these diverse perspectives, and applying them toward environmental management and conservation, Berkes and colleagues (2003, p. 8) state:

> The need to use a multiplicity of perspectives follows from complex systems thinking. Because of a multiplicity of scales, there is no one "correct" and all-encompassing perspective on a system. . . . Especially with social systems, it is difficult or impossible to understand a system without considering its history, as well as its social and political contexts. . . . A complex social–ecological system cannot be captured using a single perspective.

The single "all-encompassing perspective" to which these authors refer is science. Outside of science there are many environmental perspectives that tell stories of the environment that differ from conventional wisdom (e.g., Leach and Mearns 1996b; Bryant 1998). Science, particularly Western science, has had more direct access to environmental policy and management than other perspectives (Forsyth 2003; Nader 1996). Pragmatically, capturing more perspectives on environmental change will help address data gaps, as well as recognize the "social" as the complex system that it is. In addition, it will better enable managers to *situate*, or contextualize, marine management strategies socially, historically, and geographically. Many other pragmatic justifications for attending to multiple perspectives are addressed elsewhere in this chapter.

In addition to more-practical issues, there is a philosophical argument relating to the need to recognize local and nonscientific knowledge as legitimate. Historically having the most direct access to environmental discourse, management, and policy (Forsyth 2003), science has excluded many other perspectives and knowledge systems (Raffles 2002; Kliskey et al., chap. 9 of this volume). Being excluded historically from the dominant discourse remains a "deeply remembered" aspect of many indigenous peoples' cultural memories (Tuhiwai-Smith 1999). Efforts to reassert their "contested stories," or alternative perspectives on history, have driven political movements and, according to an indigenous scholar/advocate, influence indigenous peoples' perceptions of science and researchers today (Tuhiwai-Smith 1999). Reflecting the science–local knowledge tension, many indigenous communities have written ethical protocols that scientists must adhere to when conducting research in their communities (e.g., Desert Knowledge 2008). Intellectual property rights are also addressed within traditional ecological knowledge research (e.g., Hansen and VanFleet 2003; Gibbs 2001).

Power relations between science and other perspectives not only underlie these issues, but also offer compelling explanations for human–environmental dynamics (Leach and Mearns 1996b). Due to differences in power, equity, access to political power and voice, and level of vulnerability and marginalization, people are affected differently by environmental change, leading to such different perspectives (Leach and Mearns 1996a; Bryant 1998). In sub-Saharan Africa, conventional wisdom long considered pastoralists to be the primary agents of deforestation, despite local peoples' contesting otherwise. This untested assumption pervaded science, policy, and public perception of peoples of sub-Saharan Africa for many years. Examining aerial photos and geology and interviewing local people indeed demonstrated pastoralists were agents of *re*-forestation (Leach and Mearns 1996a). Power and legitimization issues are very much enveloped in considerations of scale—local interests have been shown to struggle as marine management is implemented at broader scales. Across scales, ecological arguments are employed to promote certain types of conservation interests—with consequences to the local rights of access to the resource (Campbell 2007). In certain circumstances worldwide, there is evidence to suggest local knowledge is still considered less legitimate than science, suggesting there is still much progress to be made in attending to local perspectives (e.g., Campbell and Vainio-Mattila 2003).

Scientific knowledge of marine ecosystems will only be as powerful as it is inclusive of other ways of knowing the oceans. For example, in a case study by a Finnish researcher, despite the existence of highly sophisticated silvicultural expertise in Finland, the main reason for the relative health of Finnish forests is that conservation practices are based on traditional knowledge passed from one generation to another (Oksa 1993, as cited in Campbell and Vanio-Mattila 2003). Similarly, Berkes (2003b) suggests a strong need to link community-based conservation to the livelihoods, knowledge, and interests of local people. Small-scale farmers, fishers, and forest users may be the best allies for conservationists (Alcorn 1993, cited in Berkes 2003). In marine management, local, often indigenous, knowledge is increasingly given consideration (Davis and Wagner 2003; Kliskey et al., chap. 9 of this volume). For example, Native Hawaiian traditional ecological knowledge and traditional tenure practices are explicitly recognized in recent reauthorizations of the Hawaii state ocean plan and federal Magnuson–Stevens Fishery Conservation and Management Act (1996), as well as in the development of the Northwestern Hawaiian Islands Coral Reef Ecosystem Reserve (see also Kliskey et al., chap. 9 of this volume). Attending to local peoples' perspectives has the potential to empower people. However, empowerment is not always inherent, as commonly assumed; rather, it must be fostered and achieved (Davis and Wagner 2003). Other chapters in this volume describe *how* to elicit diverse perspectives and *why* they are so important to marine management (e.g., Kliskey et al., chap. 9 of this volume; Moore and Russell, chap. 18 of this volume). We simply state here that there is much to be learned and gained by an inclusive frame.

Closing Thoughts

Throughout this chapter, we have argued that the oceans are peopled seascapes, and thus understanding the drivers of change in coastal and marine systems requires understanding coupled social–ecological systems. Interdisciplinary marine research will continue to unravel the synergistic relationship between oceans and society. Solutions to what ails the oceans may lie in collaborative efforts between local people and managers to learn about, work with, and work within local communities. Management approaches that better reflect local communities' diversity, a multiplicity of perspectives, and power relations are needed in order to achieve true ecosystem-based management of the oceans.

Acknowledgments

We would like to thank Lisa M. Campbell for offering invaluable insights into human aspects of ocean and coastal systems, as well as Zoë Meletis, Noella Gray, Myriah Cornwell, Bethany Haalboom, Noelle Boucquey, and Ed Glazier for their editorial comments. Our appreciation is extended to Fred Guichard, the editors, and two anonymous reviewers for their assistance.

References

Acheson, J. M. 1997. The politics of managing the Maine lobster industry: 1860 to the present. *Human Ecology* 25:3–27.

Alcorn, J. B. 1993. Indigenous peoples and conservation. *Conservation Biology* 7:424–26.

Balée, W. 1998. Historical ecology: Premises and postulates. In *Advances in historical ecology*, ed. W. Balée, 13–29. New York: Columbia University Press.

Basurto, X. 2005. How locally designed access and use controls can prevent the tragedy of the commons in a Mexican small-scale fishing community. *Society & Natural Resources* 18(7):643–59.

Bellwood, D. R., T. P. Hughes, C. Folke, and M. Nystrom. 2004. Confronting the coral reef crisis. *Nature* 429:827–33.

Berkes, F. 2003a. Alternatives to conventional management: Lessons from small-scale fisheries. *Environment* 31:5–20.

Berkes, B., T. P. Hughes, R. S. Steneck, J. A. Wilson, D. R. Bellwood, B. Crona, C. Folke et al. 2006. Globalization, roving bandits and marine resources. *Science* 311:155–58.

Berkes, F. 1998. Indigenous knowledge and resource management systems in the Canadian subarctic. In *Linking social and ecological systems: Management practices and social mechanisms for building resilience*, ed. F. Berkes and C. Folke, 98–128. Cambridge, UK: Cambridge University Press.

Berkes, F., J. F. Colding, and C. Folke, ed. 2003. *Navigating social–ecological systems: Building resilience for complexity and change*. New York: Cambridge University Press.

Berkes, F. 1999. *Sacred ecology: Traditional ecological knowledge and resource management*. Philadelphia: Taylor & Francis.

Berkes, F. 2003b. Re-thinking community-based conservation. *Conservation Biology* 18:621–30.

Berkes, F., D. Feeny, B. J. McCay, and J. M. Acheson. 1989. The benefits of the commons. *Nature* 340:91–93.

Boesch, D. F., R. H. Burroughs, J. E. Baker, R. P. Mason, C. L. Rowe, and R. L. Siefert. 2001. *Marine pollution in the United States*. Arlington, VA: Pew Oceans Commission.

Bolster, W. J. 2006. Opportunities in marine environmental history. *Environmental History* 11:567–97.

Brown, J. C., and M. Purcell. 2005. There's nothing inherent about scale: Political ecology, the local trap, and the politics of development in the Brazilian Amazon. *Geoforum* 36:607–24.

Bryant, R. L. 1998. Power, knowledge and political ecology in the third world: A review. *Progress in Physical Geography* 22:79–94.

Campbell, L. M., M. H. Godfrey, and O. Drif. 2002. Community-based conservation via global legislation? Limitations of the Inter-American Convention for the Protection and Conservation of Sea Turtles. *Journal of International Wildlife Law and Policy* 5:121–43.

Campbell, L. M. 2007. Local conservation practice and global discourse: A political ecology of sea turtle conservation. *Annals of the Association of American Geographers* 97:331–34.

Campbell, L. M. 2005. Obstacles to interdisciplinary research. *Conservation Biology* 19:574–77.

Campbell, L. M., and A. Vainio-Mattila. 2003. Participatory development and community-based conservation: Opportunities missed for lessons learned? *Human Ecology* 31:417–37.

Carr, M. H., J. E. Neigel, J. A. Estes, S. Andelman, R. R. Warner, and J. L. Largier. 2003. Comparing marine and terrestrial ecosystems: Implications for the design of coastal marine reserves. *Ecological Applications* 13:S90–S107.

Castree, N., and B. Braun, ed. 2001. *Social nature: Theory, practice and politics*. Oxford: Blackwell.

Christensen, V., S. Guenette, J. J. Heymans, C. J. Walters, R. Watson, D. Zeller, and D. Pauly. 2003. Hundred-year decline of North Atlantic predatory fishes. *Fish and Fisheries* 4:1–24.

Christie, P., B. J. McCay, M. L. Miller, C. Lowe, A. T. White, R. Stoffle, D. L. Fluharty et al. 2003. Toward developing a complete understanding: A social science research agenda for marine protected areas. *Fisheries* 28:22–26.

Cloern, J. E. 2001. Our evolving conceptual model of the coastal eutrophication problem. *Marine Ecology Progress Series* 210:223–53.

Crumley, C. L. 1994. *Historical ecology: Cultural knowledge and changing landscapes*. Santa Fe, NM: School of American Research Press.

Davis, A., and J. R. Wagner. 2003. Who knows? On the importance of identifying "experts" when researching local ecological knowledge. *Human Ecology* 31:463–89.

Dayton, P. 2003. The importance of the natural sciences to conservation. *American Naturalist* 162:1–13.

Dayton, P. K., S. Thrush, and F. C. Coleman. 2002. *Ecological effects of fishing in marine ecosystems of the United States*. Arlington, VA: Pew Oceans Commission.

Desert Knowledge. 2008. http://www.desertknowledgecrc.com.au/socialscience/socialscience.html.

Diaz, R. J. 2001. Overview of hypoxia around the world. *Journal of Environmental Quality* 30:275–81.

Duarte, C. M. 2002. The future of seagrass meadows. *Environmental Conservation* 29:192–206.

Essington, T. E., A. H. Beaudreau, and J. Wiedenmann. 2006. Fishing through marine food webs. *Proceedings of the National Academy of Sciences* 103:3171–75.

FAO. 2007. The world's mangroves 1980–2005. Rome: FAO. http://www.fao.org/forestry/site/1720/en.

Forsyth, T. 2003. *Critical political ecology: The politics of environmental science.* New York: Routledge.

Gibbs, M. 2001. Toward a strategy for undertaking cross-cultural collaborative research. *Society & Natural Resources* 14:673–87.

Halpern, B. S., S. Walbridge, K. A. Selkoe, C. V. Kappel, F. Micheli, C. D'Agrosa, J. F. Bruno et al. 2008. A global map of human impacts on marine ecosystems. *Science* 319:948–52.

Hanna, S. S. 1998. Managing for human and ecological context in the Maine soft shell clam fishery. In *Linking social and ecological systems: Management practices and social mechanisms for building resilience,* ed. F. Berkes and C. Folke,190–211. Cambridge, UK: Cambridge University Press.

Hanna, S., C. Folke, and K. G. Mäler. 1996. *Property rights and the natural environment.* Washington, DC: Island Press.

Hansen, S. A., and J. W. VanFleet. 2003. *Traditional knowledge and intellectual property: A handbook on issues and options for traditional knowledge holders in protecting their intellectual property and maintaining biological diversity.* Washington, DC: American Association for the Advancement of Science (AAAS) Science and Human Rights Program. http://shr.aaas.org/tek/handbook/handbook.pdf.

Hardin, G. 1968. The tragedy of the commons. *Science* 162:1243–48.

Harley, C. D. G., A. R. Hughes, K. M. Hultgren, B. J. Miner, C. J. B. Sorte, C. S. Thornber, L. F. Rodriguez, L. Tomanek, and S. L. Williams. 2006. The impacts of climate change in coastal marine systems. *Ecology Letters* 9:228–41.

HMAP (History of Marine Animal Populations). 2008. http://www.hmapcoml.org/.

Hoegh-Guldberg, O., P. J. Mumby, A. J. Hooten, R. S. Steneck, P. Greenfield, E. Gomez, C. D. Harvell et al. 2007. Coral reefs under rapid climate change and ocean acidification. *Science* 318:1737–42.

Hughes, T. P., A. H. Baird, D. R. Bellwood, M. Card, S. R. Connolly, C. Folke, R. Grosberg et al. 2003. Climate change, human impacts, and the resilience of coral reefs. *Science* 301:929–33.

Hughes, T. P., D. R. Bellwood, C. Folke, R. S. Steneck, and J. Wilson. 2005. New paradigms for supporting the resilience of marine ecosystems. *Trends in Ecology and Evolution* 20:380–86.

IPCC. 2007. *Climate change 2007: Synthesis report.* Contribution of Working Groups I, II, and III to the Fourth Assessment Report of the Intergovernmental Panel on Climate Change (IPCC). R. K. Pachauri and A. Reisinger, ed. Geneva: IPCC.

Jackson, J. B. C., M. X. Kirby, W. H. Berger, K. A. Bjorndal, L. W. Botsford, B. J. Bourque, R. H. Bradbury et al. 2001. Historical overfishing and the recent collapse of coastal ecosystems. *Science* 293:629–38.

Jackson, J. B. C. 1997. Reefs since Columbus. *Coral Reefs* 16:S23–S32.

Kamakau, S. 1964. Ka Po'e Kahiko, the people of old, trans. M. K. Pukui, from the newspaper *Keau'oko'a.* Honolulu: Bishop Museum Press.

Kaplan, I. M., and B. J. McCay. 2004. Cooperative research, co-management and the social dimension of fisheries science and management. *Marine Policy* 28:257–58.

Kinzig, A. P., P. Ryan, M. Etienne, H. Allison, T. Elmqvist, and B. H. Walker. 2006. Resilience and regime shifts: Assessing cascading effects. *Ecology and Society* 11(1):20.

Knowlton, N. 2001. The future of coral reefs. *Proceedings of the National Academy of Sciences* 98:5419–25.

Knowlton, N., and J. B. C. Jackson. 2001. The ecology of coral reefs. In *Marine community ecology,* ed. M. D. Bertness, S. D. Gaines, and M. E. Hay, 395–422. Sunderland, MA: Sinauer Associates.

Law, R. 2000. Fishing, selection, and phenotypic evolution. *ICES Journal of Marine Science* 57:659–68.

Leach, M., and R. Mearns. 1996a. Environmental change and policy: Challenging received wisdom in Africa. In *The lie of the land: Challenging received wisdom on the African environment,* ed. M. Leach and R. Mearns, 1–33. Oxford: James Currey.

Leach, M., and R. Mearns, ed. 1996b. *The lie of the land: Challenging received wisdom on the African environment*. Oxford: James Currey.

Lewison, R. L., L. B. Crowder, A. J. Read, and S. A. Freeman. 2004. Understanding impacts of fisheries bycatch on marine megafauna. *Trends in Ecology and Evolution* 19:598–604.

Loeb, V., V. Siegel, O. Holm-Hansen, R. Hewitt, W. Fraser, W. Trivelpiece, and S. Trivelpiece. 1997. Effects of sea-ice extent and krill or salp dominance on the Antarctic food web. *Nature* 387:897–900.

Long, J., A. Tecle, and B. Burnette. 2003. Cultural foundations for ecological restoration on the White Mountain Apache Reservation. *Conservation Ecology* 8(1):4.

Lotze, H. K., H. S. Lenihan, B. J. Bourque, R. H. Bradbury, R. G. Cooke, M. C. Kay, S. M. Kidwell, M. X. Kirby, C. H. Peterson, and J. B. C. Jackson. 2006. Depletion, degradation, and recovery potential of estuaries and coastal seas. *Science* 312:1806–9.

LUMCON (Louisiana University Marine Consortium). 2008. http://www.gulfhypoxia.net/.

MA (Millennium Ecosystem Assessment). 2003. *Ecosystems and human well-being: A framework for assessment*. Washington, DC: Island Press.

MA (Millennium Ecosystem Assessment). 2005a. *Ecosystems and human well-being: Current state and trends*. Washington, DC: Island Press.

MA (Millennium Ecosystem Assessment). 2005b. *Ecosystems and human well-being: Wetlands and water synthesis*. Washington, DC: World Resources Institute.

Mack, R. N., D. Simberloff, W. M. Lonsdale, H. Evans, M. Clout, and F. A. Bazzaz. 2000. Biotic invasions: Causes, epidemiology, global consequences, and control. *Ecological Applications* 10:689–710.

Mansfield, B. 2003. From catfish to organic fish: Making distinctions about nature as cultural economic practice. *Geoforum* 34:329–42.

Mansfield, B. 2004. Rules of privatization: Contradictions in neoliberal regulation of North Pacific fisheries. *Annals of the Association of American Geographers* 94:565–84.

Maurstad, A. 2000. To fish or not to fish: Small-scale fishing and changing regulations of the cod fishery in northern Norway. *Human Organization* 59:37–47.

McCay, B. J. 1995. Common and private concerns. *Advances in Human Ecology* 4:89–116.

McCay, B. J. 1996. Common and private concerns. In *Rights to nature: Ecological, economic, cultural, and political principles of institutions for the environment*, ed. S. S. Hanna, C. Folke, and K. G. Maler, 111–20. Washington, DC: Island Press.

McGoodwin, J. R. 2001. *Understanding the cultures of fishing communities: A key to fisheries management and food security*. Rome: Food and Agriculture Organization of the United Nations (FAO).

Munro, G. R., and U. R. Sumaila. 2002. The impact of subsidies upon fisheries management and sustainability: The case of the North Atlantic. *Fish and Fisheries* 3:1–18.

Myers, R. A., J. K. Baum, T. D. Shepherd, S. P. Powers, and C. H. Peterson. 2007. Cascading effects of the loss of apex predatory sharks from a coastal ocean. *Science* 315(5820):1846–50.

Myers, R. A., and B. Worm. 2003. Rapid worldwide depletion of predatory fish communities. *Nature* 423:280–83.

Nader, L., ed. 1996. *Naked science: Anthropological inquiry into boundaries, power, and knowledge*. New York: Routledge.

NCCOS (National Centers for Coastal Ocean Science). 2007. *NCCOS human dimensions strategic plan. FY2009–FY2014*. Silver Spring, MD: National Oceanic and Atmospheric Administration, National Ocean Service, NCCOS.

Oksa, J. 1993. The benign encounter: The Great Move and the role of the state in Finnish forests. In *Who will save the forests: Knowledge, power and environmental destruction*, ed. T. Banuri and F. Apffel-Marglin, 114–41. London: Zed Books.

Olsen, E. M., M. Heino, G. R. Lilly, M. J. Morgan, J. Brattey, B. Ernande, and U. Dieckmann. 2004. Maturation trends indicative of rapid evolution preceded the collapse of northern cod. *Nature* 428:932–35.

Olson, J. 2005. Re-placing the space of community: A story of cultural politics, policies, and fisheries management. *Anthropological Quarterly* 78:247–68.

Olsson, P., C. Folke, and F. Berkes. 2004. Adaptive co-management for building resilience in social–ecological systems. *Environmental Management* 34:75–90.

Orth, R. J., T. J. B. Carruthers, W. C. Dennision, C. M. Duarte, J. W. Fourqurean, K. L. Heck Jr, A. R. Hughes et al. 2006. A global crisis for seagrass ecosystems. *BioScience* 56:987–96.

Osherenko, G., O. R. Young, L. B. Crowder, J. A. Wilson, and E. A. Norse. 2006. Property rights and ocean governance. *Science* 314:593.

Ostrom, E., J. Burger, C. B. Field, R. B. Norgaard, and D. Policansky. 1999. Revisiting the commons: Local lessons, global challenges. *Science* 284:278–82.

Ostrom, E., T. Dietz, N. Dolsak, P. C. Stern, S. Stonich, and E. U. Weber, ed. 2002. *The drama of the commons*. Washington, DC: National Research Council, National Academies Press.

Paine, R. T., J. L. Ruesink, A. Sun, E. L. Soulanille, M. J. Wonham, C. D. G. Harley, D. R. Brumbaugh, and D. L. Secord. 1996. Trouble on oiled waters: Lessons from the Exxon Valdez oil spill. *Annual Review of Ecology and Systematics* 27:197–235.

Palmer, M. A., E. Bernhardt, E. Chornesky, S. Collins, A. Dobson, C. Duke, B. Gold et al. 2004. Ecology for a crowded planet. *Science* 304:1251–52.

Pauly, D. 1995. Anecdotes and the shifting baseline syndrome of fisheries. *Trends in Ecology and Evolution* 10:430.

Pauly, D., V. Christensen, S. Guenette, T. J. Pitcher, U. R. Sumaila, C. J. Walters, R. Watson, and D. Zeller. 2002. Toward sustainability in world fisheries. *Nature* 418:689–95.

Peckham, S. H., D. M. Diaz, A. Walli, G. Ruiz, L. B. Crowder, and W. J. Nichols. 2007. Small-scale fisheries bycatch jeopardizes endangered Pacific loggerhead turtles. *PLoS One* 2(10):e1041.

Peterson, C. H., S. D. Rice, J. W. Short, D. Esler, J. L. Bodkin, B. E. Ballachey, and D. B. Irons. 2003. Long-term ecosystem response to the Exxon Valdez oil spill. *Science* 302:2082–86.

Polovina, J. J. 2005. Climate variation, regime shifts, and implications for sustainable fisheries. *Bulletin of Marine Science* 76:233–44.

Pomeroy, R. S., P. McConney, and R. Mahon. 2004. Comparative analysis of coastal resource co-management in the Caribbean. *Ocean and Coastal Management* 47:429–47.

Rabalais, N. N., R. E. Turner, and W. J. Wiseman. 2002. Gulf of Mexico hypoxia, aka "The dead zone." *Annual Review of Ecology and Systematics* 33:235–63.

Rabalais, N. N., R. E. Turner, and W. J. Wiseman. 2001. Hypoxia in the Gulf of Mexico. *Journal of Environmental Quality* 30:320–29.

Raffles, H. 2002. Intimate knowledge. *International Social Science Journal* 54:325–35.

Relph, E. 1976. *Place and placelessness*. London: Pion.

Roberts, C. M. 2002. Deep impact: The rising toll of fishing in the deep sea. *Trends in Ecology and Evolution* 17:242–45.

Roberts, C. M., C. J. McClean, J. E. N. Veron, J. P. Hawkins, G. R. Allen, D. E. McAllister, C. G. Mittermeier et al. 2002. Marine biodiversity hotspots and conservation priorities for tropical reefs. *Science* 295:1280–84.

Rosenberg, A. A., W. J. Bolster, K. E. Alexander, W. B. Leavenworth, A. B. Cooper, and M. G. McKenzie. 2005. The history of ocean resources: Modeling cod biomass using historical records. *Frontiers in Ecology and Environment* 3:84–90.

Sala, E., and N. Knowlton. 2006. Global marine biodiversity trends. *Annual Review of Environment and Resources* 31:93–122.

Stachowicz, J. J., R. B. Whitlatch, and R. W. Osman. 1999. Species diversity and invasion resistance in a marine ecosystem. *Science* 286:1577–79.

Steneck, R. S., M. H. Graham, B. J. Bourque, D. Corbett, J. M. Erlandson, J. A. Estes, and M. J. Tegner. 2002. Kelp forest ecosystems: Biodiversity, stability, resilience and future. *Environmental Conservation* 29:436–59.

St. Martin, K. 2001. Making space for community resource management in fisheries. *Annals of the Association of American Geographers* 91:122–42.

Thrush, S. F., and P. K. Dayton. 2002. Disturbance to marine benthic habitats by trawling and dredging: Implications for marine biodiversity. *Annual Review of Ecology and Systematics* 33:449–73.

Tuan, Y. F. 1977. *Space and place: The perspective of experience*. Minneapolis: University of Minnesota.

Tuhiwai-Smith, L. 1999. *Decolonizing methodologies: Research and indigenous peoples*. New York: Zed Books.

Turner, B. L., II, R. E. Kasperson, P. A. Matson, J. J. McCarthy, R. W. Corell, L. Christensen, N. Eckley et al. 2003. A framework for vulnerability analysis in sustainability science. *Proceedings of the National Academy of Sciences* 100:8074–79.

Turner, S. J., S. F. Thrush, J. E. Hewitt, V. J. Cummings, and G. Funnell. 1999. Fishing impacts and the degradation or loss of habitat structure. *Fisheries Management and Ecology* 6:401–20.

UNEP (United Nations Environment Programme). 2006. *Marine and coastal ecosystems and human wellbeing: A synthesis report based on the findings of the Millennium Ecosystem Assessment*. Nairobi: UNEP.

Valiela, I., J. L. Bowen, and J. K. York. 2001. Mangrove forests: One of the world's threatened major tropical environments. *BioScience* 51:807–15.

van Sittert, L. 2005. The other seven-tenths. *Environmental History* 10:206–9.

Walker, B. H., L. H. Gunderson, A. P. Kinzig, C. Folke, S. R. Carpenter, and L. Schultz. 2006. A handful of heuristics and some propositions for understanding resilience in social–ecological systems. *Ecology and Society* 11(1):13.

Walther, G. R., E. Post, P. Convey, A. Menzel, C. Parmesan, T. J. C. Beebee, J. M. Fromentin, O. Hoegh-Guldberg, and F. Bairlein. 2002. Ecological responses to recent climate change. *Nature* 416:389–95.

Watson, R., and D. Pauly. 2001. Systematic distortion in world fisheries catch trends. *Nature* 414: 534–36.

Wespestad, V. G., L. W. Fritz, W. J. Ingraham, and B. A. Megrey. 2000. On relationships between cannibalism, climate variability, physical transport, and recruitment success of Bering sea walleye pollock (*Theragra chalcogramma*). *ICES Journal of Marine Science* 57:272–78.

Wilkinson, C. R., ed. 2008. *Status of the coral reefs of the world: 2008*. Townsville, Australia: Global Coral Reef Monitoring Network and Reef and Rainforest Research Centre. http://www.reefbase.org/resource_center/publication/statusreport.aspx.

Wilson, J. A. 2006. Matching social and ecological systems in complex ocean fisheries. *Ecology and Society* 11(1):9.

Worm, B., E. B. Barbier, N. Beaumont, J. E. Duffy, C. Folke, B. S. Halpern, J. B. C. Jackson et al. 2006. Impacts of biodiversity loss on ocean ecosystem services. *Science* 314:787–90.

Young, E. H. 2003. Balancing conservation with development in marine-dependent communities: Is ecotourism an empty promise? In *Political ecology: An integrative approach to geography and environment-development studies*, ed. K. S. Zimmerer and T. J. Bassett, 29–49. New York: Guilford Press.

Young, O. R., G. Osherenko, J. Ekstrom, L. B. Crowder, J. Ogden, J. A. Wilson, J. C. Day et al. 2007. Solving the crisis in ocean governance: Place-based management of marine ecosystems. *Environment* 49(4):20–32.

Zimmerer, K. S., and T. J. Bassett. 2003. *Political ecology: An integrative approach to geography and environment-development studies*. New York: Guilford Press.

CHAPTER 4
Resilience Science

Heather M. Leslie and Ann P. Kinzig

A key feature of ecosystem-based management (EBM) is the ability to manage or adapt to change. Embracing change requires us to better understand what influences the responses of systems, both human and natural, to a range of natural and human-caused disturbances. Will a system resist disturbance, rebound quickly, slowly degrade, or shift to a completely new state? Once a threshold is crossed, is it possible for a system to return to a preexisting state; in other words, is the change reversible, and with what effort? Thus, understanding *resilience*—the extent to which a system can maintain its structure, function, and identity in the face of disturbance—can enable us to better predict how systems will respond not only to a growing array of perturbations, but also to a range of management strategies.

Resilience science is a conceptual framework that can contribute to more sustainable interactions between people and terrestrial, freshwater, and marine ecosystems throughout the world (Walker and Salt 2006). With roots in ecology and complex systems science, the framework arose from the recognition that a given ecological system could exhibit two or more fundamentally different states (Holling 1973). For example, in the coastal marine environment, reefs in the Aleutian Islands have been shown to alternate between kelp-dominated and urchin-dominated systems (fig. 4.1; Estes and Duggins 1995). For a system to absorb perturbations and remain in a functionally similar state requires that there be a threshold between the alternatives—a set of conditions that, once exceeded, destines the system to move from its earlier state toward a fundamentally new configuration. Such thresholds imply nonlinear behavior of the system, that is, dynamics that are qualitatively different from those experienced before. Such "lurches" or "flips" in system behavior may occur rapidly, although they needn't. Resilience in this context, then, means the capacity to avoid breaching thresholds between alternative states.

While the resilience framework has its origins in ecology, the focus is now on coupled social–ecological systems (see also McLeod and Leslie, chap. 1 of this volume; Shackeroff et al., chap. 3 of this volume). Ecological dynamics cannot be understood apart from the human activities and decisions that influence ecosystems and their functioning. Ultimately, we are interested in how resilience science can be applied to coupled systems, rather than to solely the ecosystem or the social system. However, that does not obviate the need to understand the resilience of one or the other domain, as we will sometimes do in this chapter.

To explore the benefits of resilience science for marine ecosystem-based management, we first discuss key elements of resilience science, including the coupled and dynamic nature of social and ecological systems and characteristics of these systems that contribute to their resilience, specifically diversity, disturbance regimes, and interactions (box 4.1 and below). We then explore how resilience science can help inform the development and implementation of marine ecosystem-based management efforts. These elements are further illustrated by the case studies of ongoing ecosystem-based management efforts in part 4 of the volume.

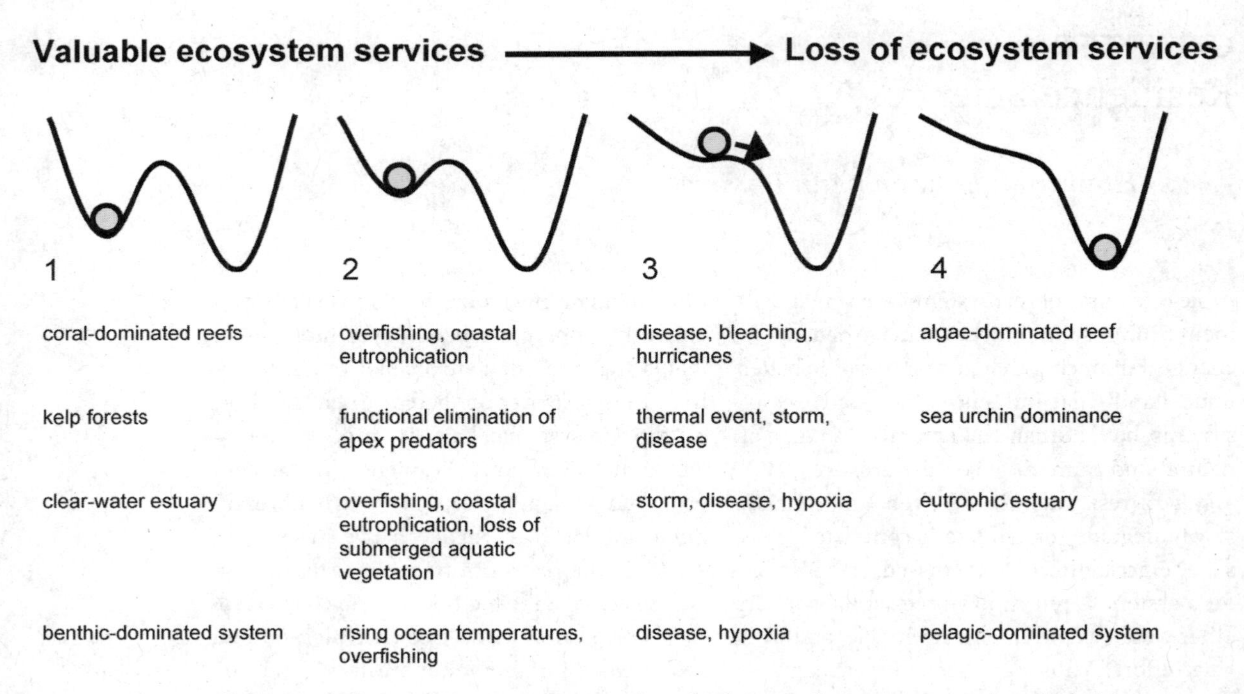

Figure 4.1 Shifts in marine ecosystems with multiple states from a more- to a less-desirable state (1, 4), resulting in a loss of valuable ecosystem services. (2) Drivers of such state shifts include both natural and anthropogenic causes that erode resilience. (3) Perturbations that subsequently "trigger" state shifts (as indicated by the arrow) could have been absorbed by a more resilient system. Modified with permission from Elmqvist et al. 2003 and Folke et al. 2004.

The Coupled and Dynamic Nature of Social and Ecological Systems

In many modern natural resource systems, management strategies have focused narrowly on the production of a particular commodity, such as timber or fish, by attempting to reduce the variability of the resource so that maximum yields can be realized consistently through time (Holling and Meffe 1996). In systems with a single state or functional regime, such strategies may make some sense, since the system would be expected to recover eventually from any management mistakes (e.g., overexploitation), with original resource levels restored. But many social and ecological systems exhibit more than one state. For those systems, such management approaches tend to be problematic over the long term. Narrowly controlling for maximum yield and stability of a particular component of the system usually *reduces* the resilience of the system, altering the position of thresholds in the system and making a "flip" to a new state more likely (Holling and Meffe 1996; Folke et al. 2004; MA 2005). This new state is often less desirable, that is, characterized by depleted stocks of economically or ecologically valuable species or reduced ecosystem services. For example, Boesch and Goldman (chap. 15 of this volume) describe how the synergistic impacts of nutrient pollution, overexploitation, and other human impacts have resulted in depleted fisheries and diminished recreational opportunities in Chesapeake Bay. In the Gulf of California (chap. 13 of this volume), Ezcurra and colleagues observe how degradation of coastal mangroves associated with changing land uses (e.g., aquaculture, agriculture, urbanization) has led to a shift in ecosystem states, and subsequent changes in the services provided by these coastal areas. Thus, understanding the drivers of multistate dynamics in both marine ecosystems and human communities is vital to ensure the long-term provision of the full range of benefits that coastal and marine ecosystems provide.

The evidence for multiple states in coastal and marine systems has been growing for some time. We now know that both natural and anthropogenic drivers of change can contribute to such shifts (Scheffer and Carpenter 2003; Folke et al. 2004; Knowlton 2004; Hughes et al. 2005). To return to the example of kelp- and urchin-dominated reefs noted above, changes in multiple biotic and abiotic parameters can contribute to the shift between the two states. Predation by the sea otter can drastically reduce the number of urchins on a reef, enabling kelp to thrive in the absence of one of its primary consumers. But other stressors also can mediate the shift in the nearshore system, including the variation in recruitment of new urchins to the reef or the addition of novel predators (Estes and Duggins 1995; Estes et al. 1998). When killer whales began preying upon sea otters in the early 1990s, nearshore community dynamics changed markedly within a decade (Estes et al. 1998). Kelp—a structurally important, foundational species that provides habitat for many other marine species—declined dramatically throughout western Alaska as the number of sea urchins (and the intensity of their grazing) in nearshore communities increased.

Another local-scale example of alternate states comes from tropical coral reefs. There, ecologists have documented shifts caused by synergistic interactions between overexploitation, land-based pollution, and disease. These forces, acting in combination, can drive a reef from a state dominated by extensive healthy coral cover and an intact community of fish and invertebrate consumers to one dominated by macroalgae and a depauperate group of consumers (fig. 4.1; Knowlton 1992; Hughes 1994; McClanahan et al. 2002). Finally, eutrophication of coastal waters due to changes in land use and increased nutrient inputs can lead to a radically different biophysical state than that characterizing a healthy coastal ecosystem, particularly in bays and semienclosed seas (Gunderson and Pritchard 2002; Lotze et al. 2006; Boesch and Goldman, chap. 15 of this volume).

At larger spatial scales, these types of ecological shifts are potentially no less prevalent, but they are harder to detect and document. Consequently, investigators rely more on long-term observations and models to capture shifts in the composition and functioning of the system, as opposed to local-scale field

Box 4.1. Key Elements of Resilience Science

1. Human and ecological systems are closely linked, and understanding these connections can help inform environmental decision making.
2. Many systems exhibit alternate states, and changes between them can be abrupt.
3. Shifts between alternate states may be irreversible, particularly on policy-relevant time scales.
4. Given the ever-changing nature of both social and ecological systems (and our uncertain knowledge of them), management goals should focus on maintaining options into the future, rather than maximizing production of a particular commodity or ecosystem service.
5. Management strategies should focus on key system characteristics that are sources of resilience, for example, diversity, disturbance regimes, and interactions.
6. Resilience cannot be maintained at all scales at all times—vulnerabilities will remain and should be incorporated into management efforts.
7. Assessing the success of management efforts and learning from those evaluations is critical.

experiments. In the Pacific Ocean, shifts in zooplankton and fish guild structure, as well as changes in physical variables, serve as indicators of the Pacific Decadal Oscillation, a multidecadal-scale regime change in the Pacific Ocean (Hare and Mantua 2000). In the Northwest Atlantic, Collie and colleagues (2004) combined data on fisheries landings and ocean conditions with ecological models to detect a discontinuous oceanic regime shift, one of the few rigorously documented in an open ocean ecosystem. Additional examples of large-scale ecological shifts include the anchovy and sardine fisheries throughout the Pacific (Chavez et al. 2003); the decline of the Northwest Atlantic groundfish assemblage, including Atlantic cod (Choi et al. 2004; Steneck et al. 2004); and shifts from vertebrates to invertebrates and benthic to pelagic species in Narragansett Bay and Rhode Island Sound (Wood et al. 2008). In summary, scientists have observed alternative ecological states involving a diversity of marine organisms (algae, invertebrates, fish, marine mammals, and humans) in both nearshore and open ocean ecosystems, and these alternate states often generate different sets of ecosystem services. For recent reviews of alternate states in marine systems, see Knowlton (2004) and deYoung and colleagues (2008).

In the social context, the shift in local economies and social networks that occurs when a fishery- or forestry-dependent community loses access to primary resources may be analogous to the ecological shifts detailed above (Kinzig et al. 2006; Nelson et al. 2007). For example, following the collapse of the northern cod fishery off the Grand Banks of Canada, changes in the social and economic structure of coastal human communities in Newfoundland were observed (Hamilton and Butler 2001; Hamilton et al. 2004). Specifically, with the 1994 moratorium on cod fishing and subsequent closure of the fishery in 2003, hundreds of fish plant workers lost their jobs, as did fishermen themselves. Subsequently, many young people left the area to pursue education or employment elsewhere (Hamilton et al. 2004). This story has been repeated in resource-dependent communities throughout the world (Nelson et al. 2007).

Importantly, from a management perspective, the new ecological or social state is often just as resilient as the previous state, creating challenges for managers interested in reversing the situation. To return to the Newfoundland coastal community, the conditions that enabled the development of a healthy coastal economy—for example, abundant fishery resources, access to markets, dockside infrastructure, and functioning social networks—may no longer be in place, creating an essentially irreversible situation on timescales of interest to most policymakers and managers. Other livelihoods and local economies would have to be developed to replace the former opportunities. In other words, resilience is not always a positive phenomenon. Many undesirable and degraded states are highly resilient (e.g., eutrophic estuaries, expansive bureaucracies), and the management challenge in such cases is to erode resilience by breaching a threshold and accessing a more desirable state. In some cases, through management of human activities, we seek to maintain or restore the resilience of ecological or social states, while in others, we implement strategies so as to reduce their resilience. Which path is pursued will depend on which states are valued by society and why (see also Kinzig et al. 2006; Moore and Russell, chap. 18 of this volume).

Documenting shifts in coupled social–ecological systems presents considerable challenges scientifically. Experiments are often out of the question, both logistically and ethically, which makes determination of potential alternate states and underlying mechanisms difficult (Scheffer and Carpenter 2003; Walker and Meyers 2004). Moreover, in a review of state shifts in social and ecological systems, Walker and Meyers (2004) found no examples where a novel threshold was predicted before it had been experienced, suggesting that further research is needed to develop tools to detect and avoid system shifts (see also deYoung et al. 2008). The development of tools to compare the impacts of system shifts, for example, to assess trade-offs inherent in different management approaches, is further along and equally important (see Wainger and Boyd, chap. 6 of this volume; McLeod and Leslie, chap. 19 of this volume). In summary, the challenges associated with anticipating and documenting system shifts highlight the importance of managing in light of resilience, that is, managing human interactions with coastal and marine ecosystems in ways that avoid shifts to undesirable ecosystem or social states, when possible. We offer suggestions for how to do this later in the chapter (see box 4.2).

This discussion about multiple states raises the question of what relevance the above discussion has for those systems that *do not* exhibit such states. First, in many cases, there may be little operational difference between a system that exhibits multiple alternative states and one that experiences a slow, albeit deterministic, recovery to the "original" state (Knowlton 2004). Moreover, the precautionary principle makes it clear that we need to recognize the potential for the existence of multiple states and act accordingly. That is, where an action threatens to do serious or irreversible damage, lack of scientific certainty should not be used as a reason to postpone actions that would prevent environmental degradation (UNCED 1992 principle 15).

As an example, consider the water filtration service provided by estuarine shellfish. This service is provided by an ecosystem state characterized by robust shellfish populations of sufficient abundance and individual filtering capacity. When shellfish populations decrease through time due to the cumulative impacts of multiple natural and anthropogenic factors, as in the Chesapeake Bay, the filtration service is impacted as well (Jackson et al. 2001; Boesch and Goldman, chap. 15 of this volume). That is, the filtration rate will be low when shellfish biomass is low, regardless of whether the system exhibits a single state or multiple states. Which situation applies, however, may play a significant role in the success of particular restoration strategies—and thus articulating these alternatives is an important consideration for management. Resilience science offers a conceptual framework for articulating these alternatives and also suggests strategies for moving forward with marine EBM in the face of such uncertainty. We will return to this issue later in the chapter.

Sources of Resilience in Ecological and Social Systems

Case studies and modeling of coupled social–ecological systems throughout the world have yielded insights on the characteristics of these systems that contribute to resilience (table 4.1). Here we highlight several that are particularly relevant to the development of ecosystem-based management in coastal and ocean areas: diversity, disturbance regimes, and interactions within and across scales. Knowledge of how and why these factors operate can help inform the development of effective management strategies and thus contribute to one of the overarching goals of marine EBM: to maintain an ecosystem in a healthy, productive, and resilient condition so that it can provide the services humans want and need (McLeod et al. 2005).

Diversity

In general, across terrestrial, freshwater, and marine systems, we see that increased biological diversity leads to increased production of ecosystem services or functioning (Hooper et al. 2005). In ecosystems, diversity is often

Table 4.1. Characteristics that contribute to ecological and social resilience

Ecological systems

Biological diversity[1]

- Genetic diversity[2]
- Species-rich biological communities[3]
- Functional diversity[1,4]
- Response diversity[1,5]

Biological legacies[6]

Intact disturbance regimes[7]

Intact interactions and feedback loops[8]

- Among species[9]
- Among biological communities and ecosystems[6,10]
- Across spatial, temporal, and organizational scales[11,12]

Modularity among populations and ecosystems[8,13]

Social systems

Adaptability of individuals and institutions[14,15]

Bridging organizations[16]

Capacity for collective action[17,18]

Clearly defined ecological boundaries[17]

Clearly defined property rights[17,19]

Diversity[14]

- Institutional[17,20]
- Livelihood options[21,22]
- Sources of knowledge[23]

Equitable distribution of benefits and costs[17,24]

Leaders and other key actors[14,16,25]

Multiscale governance[14,17]

Social memory[23]

Social learning[16]

Social networks[16]

Social capital (e.g., trust, reciprocity)[16,17]

Sources: The above information was synthesized from diverse sources, including Adger 2000; Folke et al. 2004, 2005; Nelson et al. 2007; Olsson et al. 2004; and Ostrom 1990. References were selected to be illustrative rather than comprehensive.

[1]Folke et al. 2004; [2]Hilborn et al. 2003; [3]Worm et al. 2006; [4]Hughes et al. 2003; [5]Bellwood et al. 2006; [6]Nyström and Folke 2001; [7]Holling and Meffe 1996; [8]Hughes et al. 2005; [9]Sandin et al. 2008; [10]Mumby et al. 2004; [11]Roberts 1997; [12]Elmqvist et al. 2003; [13]Levin 1999; [14]Folke et al. 2005; [15]Nelson et al. 2007; [16]Olsson et al. 2004; [17]Ostrom 1990; [18]Acheson 2006; [19]Berkes and Folke 1998; [20]Low et al. 2003; [21]Pomeroy et al. 2006; [22]Hilborn 2004; [23]Olsson and Folke 2001; [24]Adger et al. 2001; [25]Gladwell 2000.

equated with the number of species present in a given area, that is, species richness. While the majority of investigations of diversity and ecosystem functioning have focused on terrestrial ecosystems, we have an increasing number of marine cases from which to draw. These include work on how diversity influences primary production (Callaway et al. 2003), secondary production (Duffy et al. 2003), nutrient cycling (Emmerson et al. 2001; Bracken et al. 2008), and invasion resistance (Stachowicz et al. 2002). There is also evidence that diversity increases resilience to overexploitation: Worm and colleagues (2006) analyzed fisheries catch data from sixty-four large marine ecosystems worldwide and found that recovery from overexploitation was positively correlated with fish species diversity and that collapses were more likely to occur in relatively species-poor ecosystems.

In addition to species richness, two other forms of diversity are particularly important when considering ecological or social resilience: functional diversity and response diversity (Folke et al. 2004). Functional diversity refers to the variety of functions or activities occurring within a given system. In coral reef ecosystems, for example, herbivorous fish consume macroalgae and thus enable continued coral recruitment, while symbiotic zooxanthellae provide their coral hosts with energy. In estuarine environments, oysters and other reef-building shellfish that filter particulates and create habitat for other species play important and functionally distinct roles (Coen et al. 2007). Thus, resilient coral reef and estuarine ecosystems—that is, those that are able to maintain themselves through time—include these arrays of species (and functions). The functional groups described above enable production of valued ecosystem services (including tourism opportunities and seafood production, respectively).

Response diversity, or redundancy within functional groups or species, also has been demonstrated to contribute to the functioning of ecosystems and continued production of services in the face of disturbance, that is, to resilience. This redundancy within groups increases the possibility that at least one species within a functional group will be well adapted to current conditions (Walker 1995; Elmqvist et al. 2003). Work by Ray Hilborn and colleagues in Bristol Bay, Alaska, has shown that the robust yields of sockeye salmon in the bay over a number of decades can be linked with the varied performance of subpopulations of salmon within the bay (Hilborn et al. 2003). Different populations respond in a variable fashion to environmental change, particularly the Pacific Decadal Oscillation.

In the social context, functional and response diversity are also both important and have been shown to enable human communities to buffer themselves from disturbance or adapt as necessary. For example, in their study of the post-2004 tsunami response in rural coastal communities in Southeast Asia, Pomeroy and colleagues (2006) contend that those households with diversified livelihood strategies, where people had varied sources of income (including but not limited to fishing), were better able to adapt to the changing ecological and economic conditions after the disaster. Also, in response to environmental variability and fluctuations in target species, Pacific coast fishermen historically shifted from one species to another, including salmon, halibut, herring, shrimp, groundfish, and crab, during the course of a year, thereby diversifying their economic opportunities (Hilborn 2004).

The rule of thumb may be that more diversity is better, but the details of this need to be examined within a particular system. Most ecologists would not agree that importing nonnative species to increase diversity would necessarily increase functioning or resilience; the relationships among organisms can be critical, and at least some low-diversity systems are highly resilient (e.g., desertified savannas or rice paddies that have been harvested for millennia). Moreover, there are likely to be tradeoffs among the different types of diversity; for example, enhancing functional diversity may mean decreasing response diversity or altering interactions across scales or other attributes of the system that are desirable (Levin 1999).

In sum, the importance of diversity is multifaceted: While intact biological communities (and thriving societies) both benefit from a sheer variety of components (or opportunities), the functional roles played by particular entities within the system and their range of responses to change are also critical.

Disturbance Regimes

By *disturbance regimes* we are referring to frequency and impacts of biogeophysical processes (e.g., hurricanes, coastal flooding, bioturbation) as well as those generated by human activities (e.g., offshore oil development, fishing, removal of drift materials from barrier beaches). Human activities often alter the disturbance regimes under which ecological systems have evolved (MA 2005). Based on investigations of fossilized coral reefs, John Pandolfi and colleagues (2006) found that the corals in Papua New Guinea experienced mass mortality less than once every fifteen hunded years—a disturbance frequency that contrasts quite markedly with the episodic bleaching events of recent decades caused by global warming. Worldwide, an estimated 40% of marine ecosystems are strongly impacted by human activities ranging from fishing to coastal pollution to climate change; no place in the world's oceans is untouched by human activities (fig. 4.2; Halpern et al. 2008a). These increases in the frequency, magnitude, and extent of disturbances to ocean ecosystems raise questions: How are marine systems responding to these changed disturbance regimes? Are they recovering? If not, what management strategies might enable restoration of functioning?

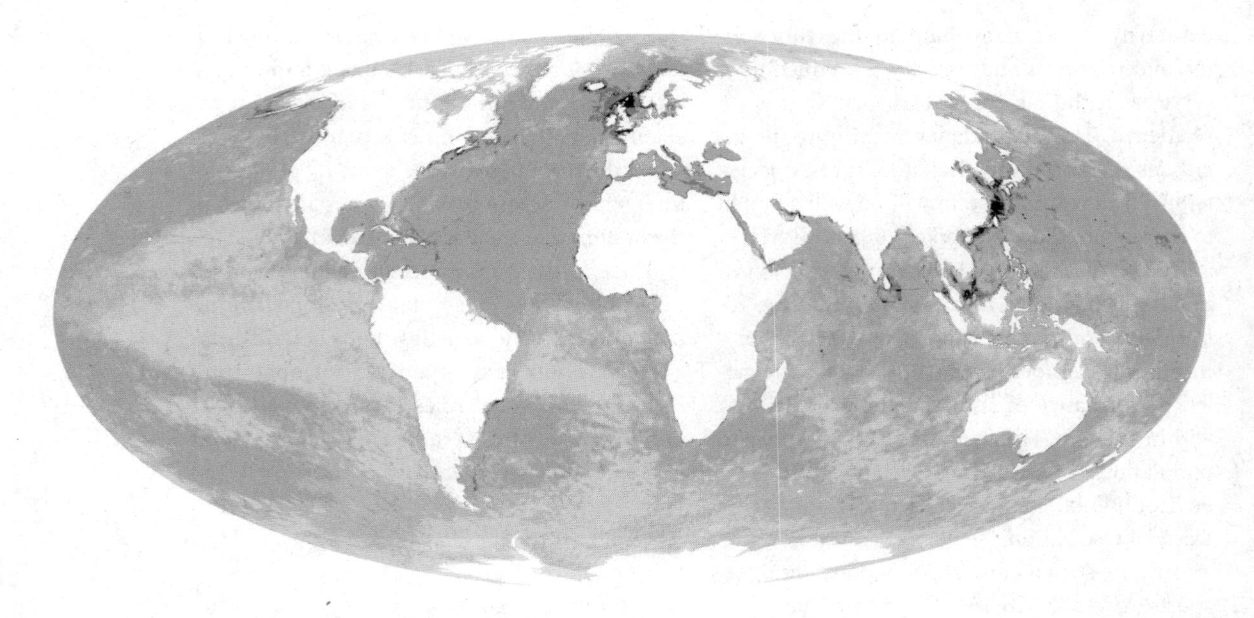

Figure 4.2 Cumulative impacts of human activities on the Earth's coastal and marine ecosystems, where the darker shading indicates the heaviest impacts overall. Adapted from Halpern et al. 2008a and reprinted with the permission of the American Association for the Advancement of Science (AAAS).

Systems may be characterized as having specific or general resilience to a particular set of stressors or disturbances (Levin 1999; Anderies et al. 2006; Walker et al. 2006). Specific resilience refers to the capacity of a system to absorb a particular perturbation, or set of perturbations, and continue to remain in a functionally similar state. The system may have high resilience to that particular perturbation, but not others. General resilience refers to the capacity of a system to absorb a broader range of perturbations, including novel or surprising perturbations. General resilience is achieved by investing in adaptive capacity or flexibility, while specific resilience is achieved by investing in the properties of a system that confer robustness against a particular stressor. Investing in general resilience will increase capacity to deal with a wide range of perturbations, but the capacity to deal with any particular perturbation will be lower than if one had invested in specific resilience toward that perturbation (Levin 1999; Nelson et al. 2007; Guichard and Peterson, chap. 5 of this volume). Conversely, investing in specific resilience may confer high resilience with respect to one particular perturbation but lower the system's capacity to deal with other, qualitatively different disturbances (Walker et al. 2006). This can be true across spatial or organizational scales as well; investing in increasing resilience at one scale may mean reducing it at others (Levin 1999).

Thus, not all functions or features of a system can be maintained in the face of disturbance, and not all functions or features will be equally valued or valuable. Some features may need to give way in order for others to be maintained. Thus, one must always ask, Resilience of what, to what? (Carpenter et al. 2001). If disturbance regimes are well known and fairly consistent over time, then one should manage human interactions with marine ecosystems to build specific resilience vis-à-vis those disturbances that most threaten valued components.

If disturbance regimes are changing and uncertain, one might more fruitfully manage for general resilience over specific resilience, or invest in some combination of general and specific resilience (they aren't discrete categories, after all, but rather a continuum). Moreover, if a system were impacted by a large array of disturbances, particularly if they were potentially interacting in a synergistic or otherwise unpredictable fashion, one would want to proceed in a precautionary fashion and manage for general resilience (see Halpern et al. 2008b for further discussion of the interacting cumulative impacts of multiple stressors). The point is that not all hazards can be avoided; there will always be trade-offs in management and disturbances to which the system will be vulnerable (Anderies et al. 2006). The only attainable goal is to reduce those vulnerabilities, rather than to eliminate them. One example of how to do this comes from McClanahan and colleagues (2008), who offer an innovative framework for integrating knowledge of ecological and social vulnerabilities into coral reef conservation by combining multiscale information on bleaching and adaptive capacity.

In terms of operationalizing these concepts, the approach of The Nature Conservancy and partners in coral reef systems is instructive. They are focused on the threat posed by coral bleaching and have published a handbook to guide managers and other practitioners in managing coral reefs so that they are "specifically" resilient to this type of disturbance (Grimsditch and Salm 2005). In this publication, Grimsditch and Salm highlight several mechanisms by which coral reef ecosystems may avoid or resist bleaching events, as well as factors that contribute to ecosystem recovery following bleaching events. They suggest that conservation practitioners focus protection on reefs that exhibit these characteristics and thus are less prone to bleaching or more likely to successfully recover. Where multiple types of perturbations (e.g., habitat degradation, pollution, and overexploitation) are of equal concern, developing strategies focused on general resilience (e.g., no-take marine reserves and other types of protected areas [PISCO 2007]) may be more appropriate.

Uncertainty about the future behavior of many coupled social–ecological systems—both in terms of large-scale drivers such as climate change and globalization and as responses of humans and other strongly interacting species—shows that a focus on general resilience is warranted. In some cases, however, as illustrated by the coral reef example above, a consensus emerges on what the primary threats to a system are. *If* these assessments are correct (see Boesch and Goldman, chap. 15 of this volume, for a more complicated story from the Chesapeake Bay), a focus on building specific resilience of desirable ecological and social states to those threats may be in order. Realistically, however, management strategies will be developed in the face of multiple uncertainties related to ecosystem responses, human behavior, and other factors. Consequently, it is best to proceed within an adaptive management framework, that is, where management interventions are implemented and evaluated as large-scale "experiments" (see also Guichard and Peterson, chap. 5 of this volume; Kaufman et al., chap. 7 of this volume). Articulating the trade-offs between managing for specific and general resilience—and doing so adaptively—is an area where little systematic work has been done, and where collaboration among scientists, managers, and resources users would be particularly fruitful.

Interactions

The resilience of a social or ecological system is very much a product of interactions across scales of time, space, and organization—from individual ecosystems to large marine ecosystems and from individuals and households to entire communities (see fig. 1.1; McLeod and

Leslie, chap. 1 of this volume). Here we highlight several types of interactions of particular importance to system resilience: interactions between rapidly and slowly changing characteristics of a system, interactions between social and ecological systems and across multiple geographic scales, and feedback loops.

The interaction between rapidly and slowly changing characteristics of a given system can help inform understanding of a system's resilience at a given point in time, and how it is likely to change (Levin 1999; Carpenter et al. 2001; Carpenter 2003; Folke et al. 2004). For example, investigations of the sources of ecological resilience in lakes indicate that changes in the water clarity and biological community of the lake are impacted not only by rate of nutrient loading into the lake from agriculture and other land use activities, but also by longer-term changes in the nutrient storage capacity of lake sediments (Carpenter 2003). Thus, to maintain the specific resilience of the lake system to nutrient pollution, managers and stakeholders need to consider not only the current rates of nutrient loading and the factors influencing those rates, but also how the concentration of phosphorus in the sediments is changing and ways to alter that long-term storage capacity. While understanding of the importance of slowly changing variables is growing in marine systems as well (see, for example, the case studies in part 4), this knowledge has not been as widely applied as in freshwater and terrestrial systems.

Interactions across multiple geographic scales also influence resilience. It is for this reason, in part, that scientists recommend that no-take marine reserves and other marine protected areas be implemented in a network (Lubchenco et al. 2003 and papers therein). Not only is it vital to maintain the flows of materials and propagules among local ecosystems, but, in addition, many marine populations are structured as metapopulations. In other words, populations are maintained on a regional scale through the dispersal and movement of individuals among local habitats (Kritzer and Sale 2006). Even though individual populations go extinct, the regional population is maintained through time by the movement of individuals among local habitats. Note that modularity, or spatially distinct populations and ecosystems, also has the potential to confer resilience (Levin 1999; Hughes et al. 2005). Allison and colleagues provide guidance on how to operationalize this knowledge for reserve network design (Allison et al. 2003).

Interactions between ecological and social systems can affect the resilience of coupled systems, as well. The dynamics, disturbances, and performance of ecological systems cannot be readily parsed from the dynamics, disturbances, and performance of social and economic systems. Disturbances that originate in the social or economic realm—even at small scales—can have profound consequences for the ecological domain (Kinzig et al. 2006). For example, the dry forests of the Androy region in southern Madagascar are becoming increasingly fragmented, in part due to changes in land use, demographics, and ultimately, decline in the traditional belief system and informal institutions linked with forest stewardship. With loss of forest cover, highly valued ecosystem services (e.g., agriculture crops and habitat for biodiversity) are in turn declining at the local scale. Kinzig and colleagues (2006) suggest that such local-level changes in the ecological and social domains could have larger-scale consequences for the functioning of the region's ecological and human systems.

Mismatches of ecological and institutional scales can lead to loss of resilience and subsequent decline in ecosystem services (Cumming et al. 2006; Wilson 2006). Wilson (2006), for example, describes how marine fisheries in the Northwest Atlantic tend to be managed species by species and at a broad, single spatial scale. Yet many exploited species (e.g., Atlantic cod, longfin squid) are composed of spatially structured populations (Ames 2004; Buresch et al. 2006), making the assumption of a single, large stock implausible. The spatial heterogeneity of fisheries and marine ecosystems has implications for the magnitude and spatial distribution of fishing effort (e.g., see Wilson et al. 2007) and thus for food production and the provision of other marine ecosystem services. Wilson (2006) proposes that management institutions be reconfigured so as to incorporate such cross-scale ecological dynamics.

Feedback loops are interactions of particular importance to system resilience (Levin 1999). Feedback loops can exist within or between trophic levels, between biotic and abiotic components of a system, across social and ecological domains, and across scales of space and time. They can be either positive or negative—with positive feedbacks serving to reinforce an initial disturbance, and negative feedbacks serving to dampen them. Feedbacks can also be tight or loose; tight feedbacks have consequences (positive or negative) that reverberate quickly through the system. Loose feedbacks may only create noticeable changes in more distant time. In general, feedbacks between ecological and social systems have become looser in recent times—for example, due to globalization (Levin 1999).

In the ecological context, for example, the loss of herbivorous fish can facilitate settlement and growth of macroalgae, which in turn overgrow living coral. In addition, the spread of macroalgae on the reef can inhibit recruitment of new corals to the reef (Hughes 1994; McClanahan et al. 2002). Thus the initial "success" of the macroalgae is reinforced through a positive feedback loop, enhancing the abundance of the algae and leading to further decline in coral cover.

Socioeconomic feedback loops are often more complex, as they can change with institutional and economic contexts. For example, in the face of declining fish catches, fishermen may reduce fishing effort or shift to another target stock. The reduced pressure on the stock often leads to a rebound in fish abundance, resulting in a negative feedback loop. The more common case in response to declining catches, however, is an increase in fishing effort, which is often driven by subsidies, market prices, or other incentives (e.g., see Berkes et al. 2006). This latter response generates a positive feedback whereby the stock declines further. The phenomenon of "shifting baselines" is another example of a loose positive feedback loop: Management decisions are made based on the current state of the ecosystem, as people do not recognize that the current state is a degraded shadow of the original system (see Pauly 1995; Jackson et al. 2001; but also Campbell et al. 2009). As time passes, the "bar" for ecosystem restoration may fall lower and lower (Pauly and Maclean 2003).

Implications of Resilience Science for Marine Ecosystem-Based Management

Some ecological and social states are more desirable than others, as they enable the delivery of ecosystem services that people value. Given the dynamic nature of ecological and social systems, it is advisable to implement management strategies that preserve the integrity and resilience of desirable states. So then the central question emerges: How can decision makers develop marine policies and management programs that preserve the integrity and resilience of valued states? We see opportunities for integrating resilience science into all stages of marine ecosystem-based management, including planning, implementation, and evaluation (see box 4.2).

Planning

Marine EBM recognizes that people are part of ecosystems and that, ultimately, we are managing human interactions with coastal and

Box 4.2. Applying Resilience Science for Marine Ecosystem-Based Management

Planning

1. Integrate knowledge of the ecological and social domains.
2. Identify sources of resilience in the social and ecological systems of interest.
3. Use multiple approaches to anticipate shifts in system states.
4. Consider a range of management strategies to build or erode resilience, as appropriate.
5. Develop both mitigation and adaptation strategies.

Implementation

6. Proceed in an adaptive fashion.
7. Prepare for and capitalize on opportunities for change.

Evaluation

8. Assess progress in an integrated manner, using biogeophysical, social, economic, and institutional information.

marine environments, rather than ecosystems themselves (McLeod et al. 2005). Consequently, integrating knowledge of the ecological and human domains in a particular region is vital in order to develop effective management strategies and to assess them accordingly (see fig. 1.1 in McLeod and Leslie, chap. 1 of this volume). Knowledge integration is best done by an interdisciplinary team of scientists, practitioners, and stakeholders, and it is particularly important that the relevant social sciences (e.g., economics, anthropology, sociology, geography), humanities (e.g., history, philosophy), and sources of local and indigenous knowledge (Kliskey et al., chap. 9 of this volume) are engaged from the outset of the process. This way, the EBM planning process will take full advantage of the relevant sources of knowledge for the design and implementation of more ecosystem-based approaches to ocean management.

Stakeholders engaged in marine EBM, as in all management processes, need to articulate a vision of what they are working to achieve. Scientists can help stakeholders understand the plausibility and likely scenarios to achieve such visions. Resilient ecosystems and human communities are often part of such visions (POC 2003; USCOP 2004; case studies in part 4 of this volume). In order to manage for resilience of coupled social–ecological systems, we need to understand the system characteristics that contribute to or inhibit system resilience. We focused on several key ones here, including diversity, disturbance regimes, and interactions (see also table 4.1). A formal resilience assessment can provide an opportunity to explore these elements in greater depth (RA 2007).

To anticipate and potentially avoid shifts to alternate (particularly undesirable) ecological or social states, it can be helpful to try to identify thresholds between these states and the drivers that contribute to such shifts. A number of statistical techniques can be used to identify changes in system state from field observations, but importantly, these cannot be used to establish causality among different variables (Scheffer and Carpenter 2003; Kemp and Goldman 2008). Consequently, field experiments and conceptual and quantitative models are critical complements to observational data. To date, these techniques have primarily been used to document shifts that have already occurred, rather than to forecast future changes. But Carpenter and Brock (2006) illustrate how increasing variance of a slowly changing variable (in this case, lake-water phosphorus) can be used to predict ecosystem shifts in advance. This area is strongly in need of further research and application.

In the selection of management strategies, we urge that both mitigation and adaptation strategies be considered. Mitigation refers to the capacity of managers to avoid the probability of a particular outcome. It requires that those forces influencing the system be under at least partial managerial control. Managers can, for instance, reduce the probability of collapse in fishing stocks by controlling fishing pressure; this would be mitigation. Adaptation refers to actions taken to alter the impact, or value, of an outcome, rather than its probability. For instance, local or regional managers can do little to influence the global course of climate change, that is, to alter the probability that it will occur. But they may be able to alter the impact climate change will have on their system, for example, by establishing a marine reserve, building general resilience, or reducing other, more locally driven pressures. This would be adaptation. In other words, mitigation is possible in the domains and on the scales that managers can directly influence. In other cases, when global forces influence local dynamics, local managers must plan for adaptation and work with higher levels of government to implement mitigation measures. We see this dual-strategy approach in the climate change community, where policymakers recognize the need to develop mitigation measures that reduce carbon emissions, as well as adaptive strategies like improved coastal responses to flooding.

Implementation

Resilience science can provide guidance during the implementation stage of ecosystem-based management, as well. Given the challenges of predicting ecosystem responses to management, let alone human behavior in the face of local-scale management interventions as well as climate change, globalization, and other large-scale drivers of change (see also Shackeroff et al., chap. 3 of this volume), we suggest that maintaining system resilience should be an explicit objective of EBM. The strategies used to build system resilience (e.g., habitat restoration, no-take marine reserves and other protected areas, ocean zoning, and outreach and education), are no different from those used in more traditional ocean management. The difference lies in the formulation of the objectives, the ways in which progress toward those objectives is assessed, and ultimately, improvements in the condition of the social and ecological systems of concern.

Management interventions should be developed and implemented in ways that recognize uncertainties and enable learning and change as needed (e.g., see Gunderson 1999; Sainsbury et al. 2000). In other words, management should proceed in an adaptive fashion (Lee 1999; Guichard and Peterson, chap. 5 of this volume). Adaptive management is a deliberate, structured process of formulating and testing hypotheses about how the world works—in this case, how ecological and social systems are connected and influence one another. Management interventions are viewed as "experiments" from which we can learn in order to improve management in the future. However, before engaging in adaptive management, it is vital that participants agree on the outstanding questions that need to be answered (Lee 1999). In addition, it is important to realize that EBM is an iterative process that requires flexibility and adaptation, as many things change over time, including knowledge, the state of the system, and the values people place on different ecosystem services.

Moreover, the resilience framework illustrates the value of identifying windows of opportunity in order to bring about change. Gunderson and Light (2006) identify three types of "policy windows" that can precipitate changes in social norms or public policy: ecological, political, and epistemic windows. The first results from an ecological state shift or following a large-scale perturbation (e.g., drought or hurricane). Political windows are created by stakeholders, managers, and other actors seeking to alter policy. And epistemic windows are created by people seeking new understanding or ways of interacting with the coupled systems of interest. As Gunderson and Light (2006, p. 332) explain, "These groups are able to suspend extant beliefs, question mental models, contrast possible futures and other such rules that allow for exploration of new and novel system configurations." An example to illustrate such windows of opportunity is provided by Boesch and Goldman (chap. 15 of this volume). The record floods of 1972 in the Chesapeake Bay region—and the slow ecological recovery that followed—raised public awareness of the failing health of the bay. Ten years later, in 1983, the Chesapeake Bay Program was established to coordinate the improvement and protection of water quality and living resources of the Chesapeake Bay estuarine system.

While the occurrence of such windows is often difficult to predict precisely, it is possible to be prepared for them. For example, Ezcurra and colleagues (chap. 13 of this volume) highlight the growing role in the Gulf of California of *Noroeste Sustentable*, a group of prominent businesspeople in the region who are facilitating a regional dialogue and building capacity to help balance environmental and development concerns in northwest Mexico. Similarly, in the Puget Sound region, Governor Gregoire and the state legislature have initiated the Puget Sound Partnership, to bring together people from public and private institutions, tribes, and citizen groups to restore and protect Puget Sound (Ruckelshaus et al., chap. 12 of this volume).

Evaluation

Given the connections between marine ecosystems and the human communities that depend on them, expanded monitoring and evaluation of social, economic, and institutional variables—as well as biogeophysical ones—is needed. Such an integrated program will increase the scientific and practitioner communities' understanding of the dynamics and drivers of the systems of interest and provide more vital information that can be used to develop scenarios for the future behavior of the coupled systems (Boesch and Greer 2003; Peterson et al. 2003).

The development of more resilience-based EBM also requires broadening monitoring and evaluation, in order to document the crossing of thresholds and, ideally, to anticipate and respond to such shifts. Thus, ecological monitoring programs should include traditional indicators of physical and chemical processes, community structure, and biomass and relative abundance of ecologically and commercially important species, but they also need to incorporate information on the ecological processes that sustain biodiversity patterns (e.g., recruitment, dispersal, and cross-scale interactions) as well as information on the relative abundances and composition of functional groups that have strong effects on ecosystem functioning. Like all indicators, resilience indicators should be clearly defined and relevant to key stressors or drivers in the system of interest (Carpenter et al. 2001, 2005). Carpenter and colleagues (2001) note that indicators of resilience often are slowly changing variables, for example, the long-term nutrient storage capacity of a lake or the extent of social networks that can be used to address natural resource dilemmas. Ideally some indicators will provide "early warning" of potential changes in the system and enable managers to anticipate that the ecosystem or social system of interest is approaching a state where abrupt, nonlinear responses to policy and management actions are more likely. Needless to say, such expanded monitoring and evaluation programs must be implemented with awareness of the relevant institutional and scientific constraints (see the case studies in part 4 and Kaufman et al., chap. 7 of this volume for further discussion of monitoring programs for marine EBM).

Concluding Thoughts

In looking to the future and how to integrate resilience and related ideas into the science and practice of EBM, three areas stand out. First, there is a need for a broader dialogue within and among the scientific, policy, and stakeholder communities about what EBM outcomes should be (see also Leslie and McLeod 2007; Moore and Russell, chap. 18 of this volume). Without clear goals, it is very difficult to track progress and to adapt policy and management efforts as needed. Second, for resilience science to be applied, scientists need to engage with other stakeholders in applying these concepts, for example, through participatory scientific investigations and synthesis and communication of existing knowledge. A participatory mode of doing science is not the standard one and will be aided by changes in expectations and norms within the scientific community itself (Leslie and McLeod 2007) as well as by the development of entities like boundary institutions, which can help bridge the gaps in knowledge, culture, and values between diverse communities of scientists, policymakers, and other stakeholders (Cash et al. 2003). Moreover, by engaging stakeholders in the practice and translation of science, scientists will become more familiar with marine policy and management processes and with the constraints and opportunities facing resource users, managers, and policymakers. Finally, in the course of moving from broad theory to place-based science and practice, it is vital to continue to develop the theoretical framework and tools that will enable the empirical data from diverse situations to be critically evaluated and synthesized. While examples of key characteristics of systems that contribute to resilience are available for some ecosystems (e.g., freshwater lakes, coral reefs, rangelands), their roles have not been evaluated comprehensively. Moreover, understanding of social resilience has lagged behind that of ecological resilience. Through interplay between empiricism and theory, scholarship and practice, we will be in a better position to contribute to sustaining the coasts, oceans, and other vital ecosystems on which human life depends and to contribute to effective marine ecosystem-based management efforts.

Acknowledgments

Thanks to D. Wendt, C. Krenz, K. Lee, S. Levin, K. McLeod, K. Dean Moore, J. Rich, and three anonymous reviewers for comments on earlier drafts of this chapter. HL gratefully acknowledges support from the David and Lucile Packard Foundation, the James S. McDonnell Foundation, the Santa Fe Institute, and the Princeton Environmental Institute. AK thanks the Packard Foundation as well as colleagues in the Resilience Alliance who continue to stimulate her thinking on these issues.

References

Acheson, J. M. 2006. Lobster and groundfish management in the Gulf of Maine: A rational choice perspective. *Human Organization* 65(3):240–52.

Adger, W. N., P. M. Kelly, N. H. Ninh, and N. C. Thanh. 2001. Property rights, institutions and resource management: Coastal resources under doi moi. In *Living with environmental change: Social vulnerability, adaptation and resilience in Vietnam*, ed. W. N. Adger, P. M. Kelly, and N. H. Ninh, 79–92. London: Routledge.

Adger, W. N. 2000. Social and ecological resilience: Are they related? *Progress in Human Geography* 24:347–64.

Allison, G. W., S. D. Gaines, J. Lubchenco, and H. P. Possingham. 2003. Ensuring persistence of marine reserves: Catastrophes require adopting an insurance factor. *Ecological Applications* 13:S8–S24.

Ames, E. P. 2004. Atlantic cod stock structure in the Gulf of Maine. *Fisheries* 29:10–28.

Anderies, J. M., B. H. Walker, and A. P. Kinzig. 2006. Fifteen weddings and a funeral: Case studies and resilience-based management. *Ecology and Society* 11(1):21.

Bellwood, D. R., T. P. Hughes, and A. S. Hoey. 2006. Sleeping functional group drives coral-reef recovery. *Current Biology* 16:2434–39.

Berkes, B., T. P. Hughes, R. S. Steneck, J. A. Wilson, D. R. Bellwood, B. Crona, C. Folke et al. 2006. Globalization, roving bandits and marine resources. *Science* 311:1557–58.

Berkes, F., and C. Folke. 1998. *Linking social and ecological systems: Management practices and social mechanisms for building resilience.* Cambridge, UK: Cambridge University Press.

Boesch, D. F., and J. Greer, ed. 2003. *Chesapeake futures: Choices for the twenty-first century.* Edgewater, MD: Chesapeake Bay Program.

Bracken, M. E. S., S. E. Friberg, C. A. Gonzalez-Dorantes, and S. L. Williams. 2008. Functional consequences of realistic biodiversity changes in a marine ecosystem. *Proceedings of the National Academy of Sciences* 105:924–28.

Buresch, K. B., G. Gerlach, and R. T. Hanlon. 2006. Multiple genetic stocks of the longfin inshore squid *Loligo pealeii* in the NW Atlantic: Stocks segregate inshore in summer, but aggregate offshore in winter. *Marine Ecology Progress Series* 310:263–70.

Callaway, J. C., G. Sullivan, and J. B. Zedler. 2003. Species-rich plantings increase biomass and nitrogen accumulation in a wetland restoration experiment. *Ecological Applications* 13:1626–39.

Campbell, L. M., N. J. Gray, E. Hazen, and J. Shackeroff. 2009. Beyond baselines: Rethinking priorities for ocean conservation. *Ecology and Society* 14(1).

Carpenter, S., B. Walker, J. M. Anderies, and N. Abel. 2001. From metaphor to measurement: Resilience of what to what? *Ecosystems* 4:765–81.

Carpenter, S. R. 2003. *Regime shifts in lake ecosystems: Pattern and variation.* Excellence in Ecology Series no. 15. Oldendorf/Luhe, Germany: Ecology Institute.

Carpenter, S. R., and W. A. Brock. 2006. Rising variance: A leading indicator of ecological transition. *Ecology Letters* 9:311–18.

Carpenter, S. R., F. Westley, and M. G. Turner. 2005. Surrogates for resilience of social–ecological systems. *Ecosystems* 8:941–44.

Cash, D. W., W. C. Clark, F. Alcock, N. M. Dickson, N. Eckley, D. H. Guston, J. Jager, and R. B. Mitchell. 2003. Knowledge systems for sustainable development. *Proceedings of the National Academy of Sciences* 100:8086–91.

Chavez, F. P., J. Ryan, S. E. Lluch-Cota, and M. Ñiquen. 2003. From anchovies to sardines and back: Multidecadal change in the Pacific Ocean. *Science* 299:217–21.

Choi, J. S., K. T. Frank, W. C. Leggett, and K. Drinkwater. 2004. Transition to an alternate state in a continental shelf ecosystem. *Canadian Journal of Fisheries and Aquatic Sciences* 61:505–10.

Coen, L. D., R. D. Brumbaugh, D. Bushek, R. Grizzle, M. W. Luckenbach, M. H. Posey, S. P. Powers, S. G. Tolley. 2007. Ecosystem services related to oyster restoration. *Marine Ecology Progress Series* 341:303–7.

Collie, J. S., K. Richardson, and J. H. Steele. 2004. Regime shifts: Can ecological theory illuminate the mechanisms? *Progress in Oceanography* 60:281–302.

Cumming, G. S., D. H. M. Cumming, and C. L. Redman. 2006. Scale mismatches in social–ecological systems: Causes, consequences, and solutions. *Ecology and Society* 11:14.

de Young, B., M. Barange, G. Beaugrand, R. Harris, R. I. Perry, M. Scheffer, and F. Werner. 2008. Regime shifts in marine ecosystems: Detection, prediction and management. *Trends in Ecology and Evolution* 23:402–9.

Duffy, J. E., J. P. Richardson, and E. A. Canuel. 2003. Grazer diversity effects on ecosystem functioning in seagrass beds. *Ecology Letters* 6:637–45.

Elmqvist, T., C. Folke, M. Nystrom, G. Peterson, J. Bengtsson, B. Walker, and J. Norberg. 2003. Response diversity, ecosystem change, and resilience. *Frontiers in Ecology and the Environment* 1:488–94.

Emmerson, M. C., M. Solan, C. Emes, D. M. Paterson, and D. Raffaelli. 2001. Consistent patterns and the idiosyncratic effects of biodiversity in marine ecosystems. *Nature* 411:73–77.

Estes, J. A., M. T. Tinker, T. M. Williams, and D. F. Doak. 1998. Killer whale predation on sea otters linking oceanic and nearshore ecosystems. *Science* 282:473–76.

Estes, J. A., and D. O. Duggins. 1995. Sea otters and kelp forests in Alaska: Generality and variation in a community ecological paradigm. *Ecological Monographs* 65:75–100.

Folke, C., T. Hahn, P. Olsson, and J. Norberg. 2005. Adaptive governance of social–ecological systems. *Annual Review in Environment and Resources* 30:441–73.

Folke, C., S. Carpenter, B. Walker, M. Scheffer, T. Elmqvist, L. Gunderson, and C. S. Holling. 2004. Regime shifts, resilience, and biodiversity in ecosystem management. *Annual Review of Ecology and Systematics* 35:557–81.

Gladwell, M. 2000. *The tipping point: How little things can make a big difference.* Boston: Little, Brown.

Grimsditch, G. D., and R. V. Salm. 2005. *Coral reef resilience and resistance to bleaching.* International Union for Conservation of Nature (IUCN) Resilience Science Group Working Paper Series, no. 1. Gland, Switzerland: The Nature Conservancy and IUCN. http://data.iucn.org/dbtw-wpd/edocs/2006-042.pdf.

Gunderson, L. H., and S. S. Light. 2006. Adaptive management and adaptive governance in the Everglades ecosystem. *Policy Science* 39:323–34.

Gunderson, L. H., and L. Pritchard, ed. 2002. *Resilience and the behavior of large-scale systems.* Washington, DC: Island Press.

Gunderson, L. 1999. Resilience, flexibility and adaptive management: Antidotes for spurious certitude? *Conservation Ecology* 3:7.

Halpern, B. S., S. Walbridge, K. A. Selkoe, C. V. Kappel, F. Micheli, C. D'Agrosa, J. F. Bruno et al. 2008a. A global map of human impacts on marine ecosystems. *Science* 319:948–52.

Halpern, B. S., K. L. McLeod, A. A. Rosenberg, and L. B. Crowder. 2008b. Managing for cumulative impacts in ecosystem-based management through ocean zoning. *Ocean and Coastal Management* 51:203–11.

Hamilton, L. C., R. L. Haedrich, and C. M. Duncan. 2004. Above and below the water: Social/ecological transformation in northwest Newfoundland. *Population and Environment* 25(3):195–215.

Hamilton, L. C., and M. J. Butler. 2001. Outport adaptations: Social indicators through Newfoundland's cod crisis. *Human Ecology Review* 8:1–11.

Hare, S. R., and N. J. Mantua. 2000. Empirical evidence for North Pacific regime shifts in 1977 and 1989. *Progress in Oceanography* 47:103–45.

Hilborn, R. 2004. Are sustainable fisheries achievable? In *Marine Conservation Biology*, ed. E. A. Norse and L. B. Crowder, 247–59. Washington, DC: Island Press.

Hilborn, R., T. P. Quinn, D. E. Schindler, and D. E. Rogers. 2003. Biocomplexity and fisheries sustainability. *Proceedings of the National Academy of Sciences* 100:6564–68.

Holling, C. S., and G. K. Meffe. 1996. Command and control and the pathology of natural resource management. *Conservation Biology* 10:328–37.

Holling, C. S. 1973. Resilience and stability of ecological systems. *Annual Review of Ecology and Systematics* 4:1–23.

Hooper, D. U., F. S. Chapin III, J. J. Ewel, A. Hector, P. Inchausti, S. Lavorel, J. H. Lawton et al. 2005. Effects of biodiversity on ecosystem functioning: A consensus of current knowledge. *Ecological Monographs* 75(1):3–35.

Hughes, T. P. 1994. Catastrophes, phase shifts, and large-scale degradation of a Caribbean coral reef. *Science* 265:1547–51.

Hughes, T. P., A. H. Baird, D. R. Bellwood, M. Card, S. R. Connolly, C. Folke, R. Grosberg et al. 2003. Climate change, human impacts, and the resilience of coral reefs. *Science* 301:929–33.

Hughes, T. P., D. R. Bellwood, C. Folke, R. S. Steneck, and J. Wilson. 2005. New paradigms for supporting the resilience of marine ecosystems. *Trends in Ecology and Evolution* 20:380–86.

Jackson, J. B. C., M. X. Kirby, W. H. Berger, K. A. Bjorndal, L. W. Botsford, B. J. Bourque, R. H. Bradbury et al. 2001. Historical overfishing and the recent collapse of coastal ecosystems. *Science* 293:629–38.

Kemp, W. M., and E. B. Goldman. 2008. *Thresholds in the recovery of eutrophic coastal ecosystems: A synthesis of research and implications for management.* Workshop report from February 2007. UM-SG-TS-2008-01. Maryland Sea Grant and the Chesapeake Bay Program. Annapolis, MD: Chesapeake Bay Program.

Kinzig, A. P., P. Ryan, M. Etienne, H. Allison, T. Elmqvist, and B. H. Walker. 2006. Resilience and regime shifts: Assessing cascading effects. *Ecology and Society* 11(1):20.

Knowlton, N. 2004. Multiple "stable" states and the conservation of marine ecosystems. *Progress in Oceanography* 60:387–96.

Knowlton, N. 1992. Thresholds and multiple stable states in coral reef community dynamics. *American Zoologist* 32:674–82.

Kritzer, J. P., and P. F. Sale. 2006. *Marine metapopulations*. Burlington, MA: Elsevier.

Lee, K. N. 1999. Appraising adaptive management. *Conservation Ecology* 3(2):3.

Leslie, H. M., and K. L. McLeod. 2007. Confronting the challenges of implementing marine ecosystem-based management. *Frontiers in Ecology and the Environment* 5(10):540–48.

Levin, S. A. 1999. *Fragile dominion: Complexity and the commons*. Reading, MA: Perseus Books.

Lotze, H. K., H. S. Lenihan, B. J. Bourque, R. H. Bradbury, R. G. Cooke, M. C. Kay, S. M. Kidwell, M. X. Kirby, C. H. Peterson, and J. B. C. Jackson. 2006. Depletion, degradation, and recovery potential of estuaries and coastal seas. *Science* 312:1806–9.

Low, B., E. Ostrom, C. Simon, and J. Wilson. 2003. Redundancy and diversity: Do they influence optimal management? In *Navigating social–ecological systems: Building resilience for complexity and change*, ed. F. Berkes, J. Colding, and C. Folke, 83–114. Cambridge, UK: Cambridge University Press.

Lubchenco, J., S. R. Palumbi, S. D. Gaines, and S. Andelman. 2003. Plugging a hole in the ocean: The emerging science of marine reserves. *Ecological Applications* 13:S3–S7.

MA (Millennium Ecosystem Assesssment). 2005. *Ecosystems and human well-being: Synthesis*. Washington, DC: Island Press.

McClanahan, T. R., J. E. Cinner, J. Maina, N. A. J. Graham, T. M. Daw, S. M. Stead, A. Wamukota et al. 2008. Conservation action in a changing climate. *Conservation Letters* 1:53–59.

McClanahan, T., N. Polunin, and T. Done. 2002. Ecological states and the resilience of coral reefs. *Conservation Ecology* 6:18.

McLeod, K. L., J. Lubchenco, S. R. Palumbi, and A. A. Rosenberg. 2005. *Scientific consensus statement on marine ecosystem-based management*. The Communication Partnership for Science and the Sea (COMPASS). Signed by 221 academic scientists and policy experts with relevant expertise. http://www.compassonline.org/pdf_files/EBM_Consensus_Statement_v12.pdf.

Mumby, P. J., A. J. Edwards, J. E. Arias-Gonzalez, K. C. Lindeman, P. G. Blackwell, A. Gall, M. I. Gorczynska et al. 2004. Mangroves enhance the biomass of coral reef fish communities in the Caribbean. *Nature* 427:533–36.

Nelson, D. R., W. N. Adger, and K. Brown. 2007. Adaptation to environmental change: Contributions of a resilience framework. *Annual Review of Environment and Resources* 32:395–419.

Nyström M., and C. Folke. 2001. Spatial resilience of coral reefs. *Ecosystems* 4:406–17.

Olsson, P., and C. Folke. 2001. Local ecological knowledge and institutional dynamics for ecosystem management: A study of Lake Racken watershed, Sweden. *Ecosystems* 4:85–104.

Olsson, P., C. Folke, and T. Hahn. 2004. Social–ecological transformation for ecosystem management: The development of adaptive co-management of a wetland landscape in southern Sweden. *Ecology and Society* 9(4):2.

Ostrom, E. 1990. *Governing the commons: The evolution of institutions for collective action*. New York: Cambridge University Press.

Pandolfi, J. M., A. W. Tudhope, G. Burr, J. Chappell, E. Edinger, M. Frey, R. Steneck et al. 2006. Mass mortality following disturbance in Holocene coral reefs from Papua New Guinea. *Geology* 34:949–52.

Pauly, D. 1995. Anecdotes and the shifting baseline syndrome of fisheries. *Trends in Ecology and Evolution* 10(10):430.

Pauly, D., and J. Maclean. 2003. *In a perfect ocean: The state of fisheries and ecosystems in the North Atlantic Ocean*. Washington, DC: Island Press.

Peterson, G. D., G. S. Cumming, and S. R. Carpenter. 2003. Scenario planning: A tool for conservation in an uncertain world. *Conservation Biology* 17:358–66.

PISCO (Partnership for Interdisciplinary Studies of Coastal Oceans). 2007. *The science of marine reserves*, 2nd ed. International version. PISCO. http://www.piscoweb.org.

POC (Pew Oceans Commission). 2003. *America's living ocean: Charting a course for sea change. A report to the nation*. Washington, DC: Pew Trusts.

Pomeroy, R. S., B. D. Ratner, S. J. Hall, J. Pimoljinda, and V. Vivekanandan. 2006. Coping with disaster: Rehabilitating coastal livelihoods and communities. *Marine Policy* 30:786–93.

RA (Resilience Alliance). 2007. *Assessing and managing resilience in social–ecological systems: A practitioners workbook*, vol. 1, version 1.0. The Resilience Alliance. http://www.resalliance.org/3871.php.

Roberts, C. M. 1997. Connectivity and management of Caribbean coral reefs. *Science* 278:1454–57.

Sainsbury, K. J., A. E. Punt, and A. D. M. Smith. 2000. Design of operational management strategies for achieving fishery ecosystem objectives. *ICES Journal of Marine Science* 57:731–41.

Sandin, S. A., J. E. Smith, E. E. DeMartini, E. A. Dinsdale, S. D. Donner, A. M. Friedlander, T. Konotchick et al. 2008. Baselines and degradation of coral reefs in the Northern Line Islands. *PLoS ONE* 3(2):e1548. doi: 10.1371/journal.pone.0001548.

Scheffer, M., and S. R. Carpenter. 2003. Catastrophic regime shifts in ecosystems: Linking theory to observation. *Trends in Ecology and Evolution* 18: 648–56.

Stachowicz, J. J., H. Fried, R. W. Osman, and R. B. Whitlatch. 2002. Biodiversity, invasion resistance, and marine ecosystem function: Reconciling pattern and process. *Ecology* 83:2575–90.

Steneck, R. S., J. Varinec, and A. V. Leland. 2004. Accelerating trophic-level dysfunction in kelp forest ecosystems of the western North Atlantic. *Ecosystems* 7:323–32.

UNCED. 1992. *Report of the United Nations Conference on Environment and Development*. Rio de Janeiro: United Nations.

USCOP (US Commission on Ocean Policy). 2004. *An ocean blueprint for the twenty-first century*. Final report. Washington, DC: USCOP. ISBN 0 9759462 0 X.

Walker, B. H., L. H. Gunderson, A. P. Kinzig, C. Folke, S. R. Carpenter, and L. Schultz. 2006. A handful of heuristics and some propositions for understanding resilience in social–ecological systems. *Ecology and Society* 11(1):13.

Walker, B. H. 1995. Conserving biological diversity through ecosystem resilience. *Conservation Biology* 9:747–52.

Walker, B., and D. Salt. 2006. *Resilience thinking: Sustaining ecosystems and people in a changing world*. Washington, DC: Island Press.

Walker, B., and J. A. Meyers. 2004. Thresholds in ecological and social–ecological systems: A developing database. *Ecology and Society* 9:3.

Wilson, J. A. 2006. Matching social and ecological systems in complex ocean fisheries. *Ecology and Society* 11(1):9.

Wilson, J. A., L. Yan, and C. Wilson. 2007. The precursors of governance in the Maine lobster fishery. *Proceedings of the National Academy of Sciences* 104:15212–17.

Wood, A. D., H. P. Jeffries, J. S. Collie. 2008. Long-term shifts in the species composition of a coastal fish community. *Canadian Journal of Fisheries and Aquatic Sciences* 65: 1325–65.

Worm, B., E. B. Barbier, N. Beaumont, J. E. Duffy, C. Folke, B. S. Halpern, J. B. C. Jackson et al. 2006. Impacts of biodiversity loss on ocean ecosystem services. *Science* 314:787–90.

CHAPTER 5
Ecological Cross-Scale Interactions

Frédéric Guichard and Garry Peterson

The call for ecosystem-based management (EBM) in coastal ecosystems stems from the recognition that all components of coastal ecosystems are connected through their dynamics. The intrinsic complexity of coastal ecosystems and their association with human activities can explain the general failure of management strategies based on managing single aspects of coastal ecosystems independently of others. EBM, for example, predicts that consequences of human management actions are not limited to target species and areas: Interconnected ecosystems can propagate, amplify, or attenuate site-level actions. Similarly, management and conservation strategies are implemented through multiple and interconnected governance levels, from international fisheries and conservation treaties to regional and local marine protected areas. These challenges are problems of scale that can be addressed within a general framework for EBM that recognizes the scale dependence of ecological and governance processes.

Cross-scale dynamics phenomena are fundamental characteristics of nonequilibrium systems where fluctuations can be strong and sudden. They can take the form of regular large-scale patterns in the abundance of subtidal mussels resulting from local competition for food (van de Koppel et al. 2005), or of traveling waves in the dynamics of pest insects resulting from their local dispersal abilities (Bjornstad et al. 2002). In systems where such fluctuations are observed, cross-scale dynamics can reveal the feedbacks that exist between local ecological and governance processes and large-scale ecosystem fluctuations (fig. 5.1) that constitute baseline information for setting up regional policies and global treaties. These feedbacks make it difficult to achieve management success by focusing management strictly on managing patterns, such as protected areas, or by just managing processes, such as fishing effort. Rather, cross-scale interactions between pattern and process require an ecosystem approach to management that integrates multiple scales and recognizes that small perturbations can have long-term and disproportionate impacts on social–ecological systems. These different approaches have characteristic problems related to controllability and uncertainty: We have to adjust to our limited ability to control important managed properties of coupled social–ecological systems, and account for the uncertainty inherent in predicting ecosystem responses to management.

In this chapter, we suggest a theoretical framework of cross-scale phenomena in human-dominated ecosystems. We suggest that the complexity of cross-scale processes can be addressed using simple ecological models that can make testable qualitative predictions and that can be coupled to adaptive management scenarios (box 5.1). We first review scale dependence in the regulation of coastal ecosystems (fig. 5.1B) and show how an explicit integration of cross-scale processes into dynamic models can explain the maintenance of spatiotemporal variability as well as drastic changes in response to perturbations (fig. 5.1C). We then discuss how this variability can interact with management strategies and provide guidelines for the development of current and future EBM theoretical frameworks.

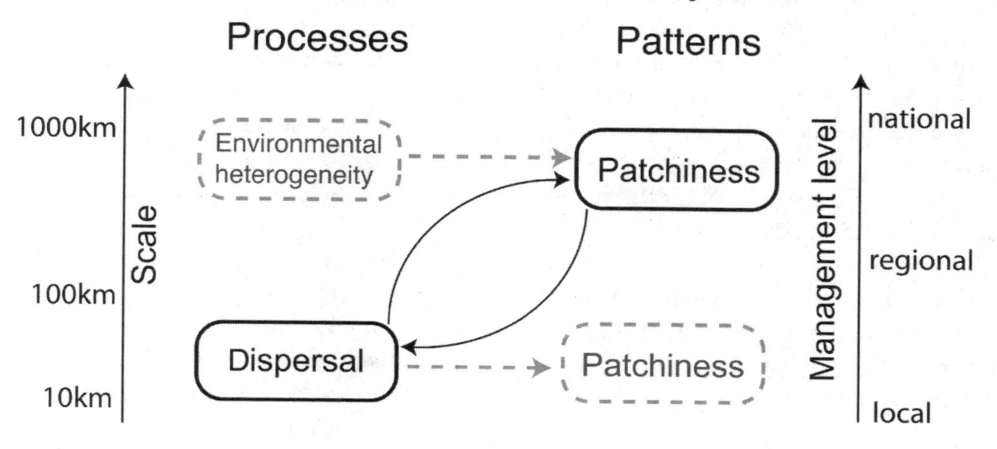

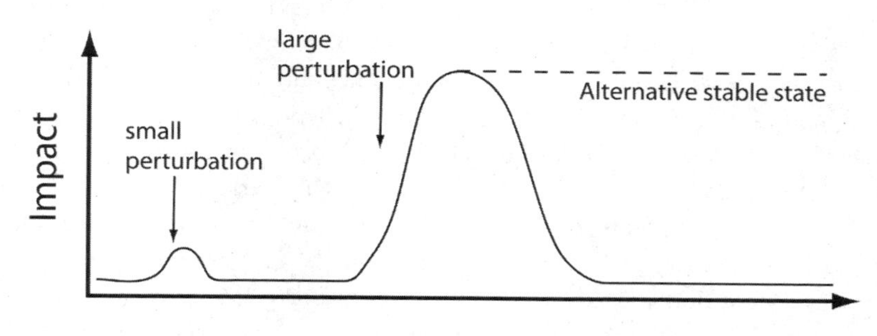

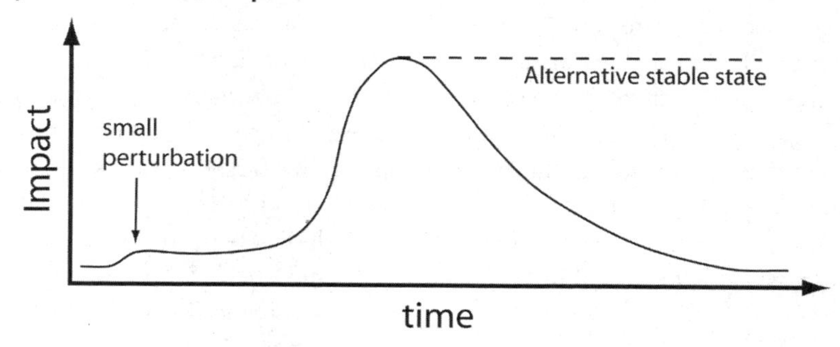

Figure 5.1 (A) Schematic illustration of scale-dependent interactions (dotted gray lines) and reciprocal causality (solid black lines) leading to cross-scale phenomena between ecological processes and patterns. (B, C) Examples of scale-dependent (B) and cross-scale (C) relationships between spatial and temporal scales of perturbation and scales of impacts in ecological systems. In each case, an alternate stable state (dashed line) can result from large impacts.

We argue that ecological models are necessary to understand interactions between patterns and processes that are associated with cross-scale phenomena. These models suggest how managing for resilience can be improved by considering feedbacks linking local ecological processes or management actions to the regional state of managed ecological systems (see also Leslie and Kinzig, chap.4 of this volume). We build this discussion upon concepts of unpredictability and controllability that require management approaches that hedge against uncertainty, build resilience against surprise, and enable learning.

Scale-Dependent Regulation of Coastal Ecosystems

Because they are more tractable than natural systems but can still provide testable predictions, ecological models of coastal ecosystems have recently been adapted to develop management rules in complex coastal systems that still acknowledge fundamental EBM requirements; no-take marine reserve networks are one notable example (Hastings and Botsford 2003). We now review current ecological theories of coastal ecosystems, and more specifically of intertidal communities, that are most directly related to EBM, including problems of scale, larval dispersal, and marine reserve networks. We show how these theories could be further improved by explicitly integrating (1) nonequilibrium spatial and species dynamics and (2) biotic and abiotic processes associated with cross-scale phenomena as a tool for coastal EBM.

Biotic versus Abiotic Control: A Problem of Scale

Recent progress in marine ecology has revealed how biotic (competition, predation, dispersal) and abiotic (climate, circulation) processes can be linked across spatial and temporal scales. However, most simple theoretical frameworks for the study of intertidal communities still assume the scale-dependent control of biotic processes by the (abiotic) environment. Biotic and abiotic processes controlling the presence and abundance of marine species have historically been opposed through the separation of scales (fig. 5.1A) at which these processes are assumed to operate (Menge and Olson 1990). While biotic regulation, such as competition and predation, has been studied at small scales, large-scale regional variability and important discontinuities in the abundance of species are usually attributed to external environmental control such as oceanographic currents or temperature (Gaylord and Gaines 2000; Broitman et al. 2001). In the ocean, climate and oceanographic processes operate at many temporal and spatial scales (Stommel 1963; Levin 1992).

The development of causal relationships between patchiness in the distribution of organisms and the scale of underlying environmental processes has received much attention. The physical environment of coastal areas is highly variable, from daily and localized upwelling fluctuations (Smith 1974), to El Niño and decadal oceanographic oscillations (Chavez et al. 2003; Collins et al. 2003), to long-term and historical climate changes (Smith et al. 2001). In ecological systems, additional complexity arises from the various proximal and scale-dependent effects linking the environment to ecological processes. For example, the effect of large-scale fluctuations in nutrient supply from nearshore waters can be amplified through strong bottom-up effects on zooplankton growth (Feinberg and Peterson 2003). However, the presence of top predators can buffer fast oceanographic fluctuations through their strong control of community responses over longer temporal scales (Paine 1980). The environment can also affect marine ecosystems across levels of organization. For example, the proximal effect of environmental parameters is not limited to a population-level response (e.g., growth rate, fecundity), but instead can directly affect per capita species interactions (Sanford 1999; Sanford et al. 2003). Current theories of the environment therefore recognize the multiplicity of scales of the environment, as well as the similar multiplicity of scales and organizational levels of ecosystem responses.

> **Box 5.1. Messages of This Chapter for Marine EBM**
>
> 1. Strong fluctuations in ecological systems can be explained by cross-scale interactions, which limit controllability and increase uncertainty for managers.
> 2. Simple ecological models illustrate how large-scale resilience can be increased by trading off local resilience and by managing variability within ecological systems across scales.
> 3. Adaptive management provides a tool to manage for resilient cross-scale interactions, rather than managing processes and patterns in isolation.

Equilibrium Theories of Coastal Ecosystems

Cross-scale phenomena can be observed when small-scale nonlinear interactions between organisms and their environment lead to spatial and temporal variability at much larger scales. In order to predict and test the occurrence of such phenomena, marine ecologists have to move away from theories that assume the stable equilibrium of populations and communities with their environment. The basis for such equilibrium theories can be found in the description of demographic processes (reproduction, survival) and of larval dispersal. We review these theories and discuss how the supposition that coastal ecosystems are controlled by their abiotic environment is associated with equilibrium assumptions that dictate how species abundance tracks changes in the environment (Connolly and Roughgarden 1999).

The existence of equilibrium states constitutes the basis and "currency" of predictability in many management frameworks adopting open system models (Leslie et al. 2005). With the assumption of no demographic coupling between larval production and recruitment, marine systems can be approximated as demographically open systems (Johnson 2005). Models of such systems represent larval recruitment as an externality (Roughgarden et al. 1984) or, in well-mixed systems, couple recruitment and production through a global pool of larvae (Iwasa and Roughgarden 1986; Alexander and Roughgarden 1996). Because of the intrinsic spatiotemporal homogeneity often predicted by such models, external fluctuations in the supply of larvae (Alexander and Roughgarden 1996; Gaylord and Gaines 2000) or the rate and nature of local community processes (Johnson 2000) have been progressively integrated to account for the observed heterogeneity in the distribution and abundance of species.

Open system theories applied to coastal population models assume no large-scale demographic coupling between larval production and recruitment within or among populations (Gaines et al. 1985; Gaines and Roughgarden 1985; Roughgarden et al. 1988; Underwood and Fairweather 1989). Equilibrium behavior of such theories has important consequences for management, and more specifically for any ecosystem-based management effort. First, they predict that, irrespective of the scales of oceanographic and climate variability, local understanding of environmental conditions and ecosystem functioning provides the required predictability for local ecosystem-based management: Local predictions aggregate in a linear or nonlinear fashion to regional scales. For example, the impact of starfish predation on mussel prey abundance is predicted to locally depend on larval supply by currents, which decreases equilibrium abundance of adult mussels. From this local equilibrium response to the environment, a latitudinal gradient in larval supply is predicted to result in a corresponding gradient of mussel abundance

(Connolly and Roughgarden 1999). In other words, equilibrium theories predict that controllability gained from site-based efforts can be directly scaled up to regional predictability across many such localities.

Dispersal and Reserve Networks

Recent empirical and modeling efforts have revealed limited dispersal (Jones et al. 1999; Swearer et al. 1999; Kinlan and Gaines 2003; Almany et al. 2007) and the coupling between regional abundance and recruitment (Witman et al. 2003). The many spatial constraints to larval dispersal have forced ecologists to recognize the importance of such limited connectivity in regulating coastal populations. For example, reserve network theories have integrated larval dispersal and large-scale connectivity into equilibrium theories (Botsford et al. 2003), thereby introducing the idea that scales at which biotic processes (including dispersal) drive population dynamics can be large and therefore similar to scales of environmental control.

Within reserve network theory, equilibrium state still remains the currency of most models that predict how to scale up management from local populations to integrate the scale of larval dispersal. Equilibrium theories have, for example, been developed to predict the optimized size and spacing of no-take marine reserves in relation to dispersal distance and to conservation goals (Botsford et al. 2001; Hastings and Botsford 2003). However, some cross-scale integration has been achieved through complex network designs that accommodate dispersal distance of multiple species (Sala et al. 2002; Kinlan et al. 2005).

Despite these efforts, strong fluctuations that can result from cross-scale phenomena have not been integrated into our understanding of ecological systems. Here we show how knowledge of nonequilibrium dynamics and connectivity, which generate cross-scale phenomena and strong fluctuations, contributes important insights to the problem of scale for ecosystem-based management. We illustrate these ideas with a simple model of intertidal ecosystems and discuss the implications of the model for the resilience, predictability, and controllability of managed coastal systems.

Cross-Scale Phenomena: Coastal Ecosystems as Dynamic Systems

Our goal here is to expand current ecological theories to provide the minimum ecological framework for the study of cross-scale phenomena and suggest its integration to current adaptive management approaches. Ecosystem-based management requires decisions whose consequences are difficult to predict, due to the complexity and uncertainty of social–ecological dynamics. As we integrate the dynamic and nonequilibrium nature of ecological systems in models, scale-dependence and cross-scale interactions become model predictions rather than being built in as extrinsic environmental or management properties (fig. 5.1A). One important implication is that cross-scale phenomena need not increase the complexity of models themselves. The maintenance of variability within complex and interconnected coastal systems can be interpreted through cross-scale relationships between processes and patterns of abundance (fig. 5.1A).

We use rocky intertidal communities as an example to introduce a modeling framework of ecological cross-scale phenomena and to illustrate how nonequilibrium and interconnected systems are built on the causal reciprocity between patterns and processes across scales. This causal reciprocity comes from the predicted importance of local processes for large-scale variability and from the reciprocal effects of variability on rates of local processes. For example, while local predation rates can maintain regional variability in prey abundance, this same variability is predicted to feed back to affect local predation rates. This approach contrasts with common management strategies that directly link patterns to causal processes as a way to increase controllability (fig. 5.2). Parallels can be drawn with other coastal systems, such as mangrove-dominated ecosystems (Ezcurra et al., chap. 13 of this volume) and temperate rocky reef ecosystems (Ruckelshaus et al., chap. 12 of this volume), that are characterized by similar spatial connectivity and nonlinear ecological processes such as predation.

A Simple Ecological Model

Cross-scale phenomena can be revealed by simple models that integrate (1) key nonlinear interactions and (2) their basic spatial structure. Such models are not intended to make quantitative predictions and management decisions; they rather serve the purpose of identifying important processes and their potential to generate cross-scale phenomena. Rocky intertidal communities constitute an ideal model system to study spatial connectivity and species interactions. The life cycle of many benthic invertebrates involves a pelagic larval stage. Consequently, connectivity (i.e., exchange of individuals) between local populations can be assumed to result from larval transport at a characteristic spatial scale. At the small scale, sessile adults interact through competition for space and resources and through predations. Such interactions might involve many species and ecological processes, but consider the simple case where a dominant benthic invertebrate species (e.g., a mussel) is locally exposed to wave disturbances from which it recovers through the supply and recruitment of larvae from neighboring populations (Paine and Levin 1981; Paine 1984; Guichard et al. 2004; Guichard 2005). Additional species interactions can capture the facilitation of mussel recovery by opportunistic species such as barnacles that can quickly colonize a bare substratum and

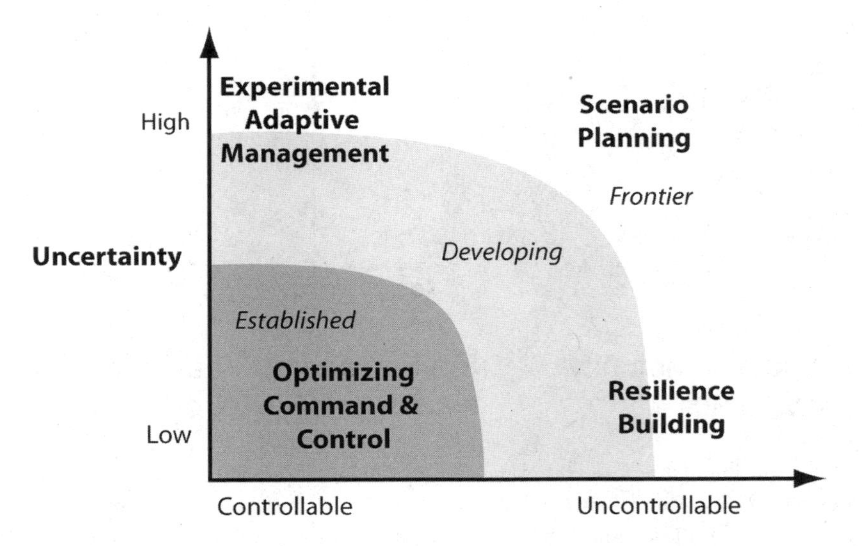

Figure 5.2 Management within a social–ecological system can be represented in a two-dimensional space defined by the uncertainty that surrounds an issue and the degree to which management actions can control the system's behavior. Most resource management theory, such as the concept of maximum sustained yield, has been developed for situations that are relatively certain and controllable; unfortunately, most environmental problems are poorly understood and only weakly controllable. These types of situations represent the frontier of research and practice in EBM.

increase its roughness and thus quality for young mussel recruits (fig. 5.3).

This model system now contains required conditions to study cross-scale phenomena: (1) local (0.1–1 m) nonlinear disturbance and recovery can generate nonequilibrium dynamics that can, in turn, (2) feed back into distant (10–100 km) population dynamics through larval dispersal. In such models the local scale can be arbitrarily small and simply needs to encompass processes of disturbance and recolonization by the dominant species (fig. 5.3). The regional scale is in turn defined by the average larval dispersal distance, which can span tens to hundreds of kilometers for benthic species with pelagic larvae (Becker et al. 2007; fig. 5.3). To assess which scales are important for management, we have to determine the scales at which resilience and variation occur. If mussel larvae disperse over tens of kilometers, should we expect most of the variability in their abundance to be observed at that scale? More importantly, should we hope to improve resilience of intertidal systems by managing human–marine environment interactions at the scale of dispersal? Possible management strategies include marine protected areas, and particularly no-take marine reserves, as well as other means of controlling people's interactions with marine resources and the ecosystems of which they are a part, such as land use and harvesting regulations.

The important result emerging from the study of such models is the possibility for

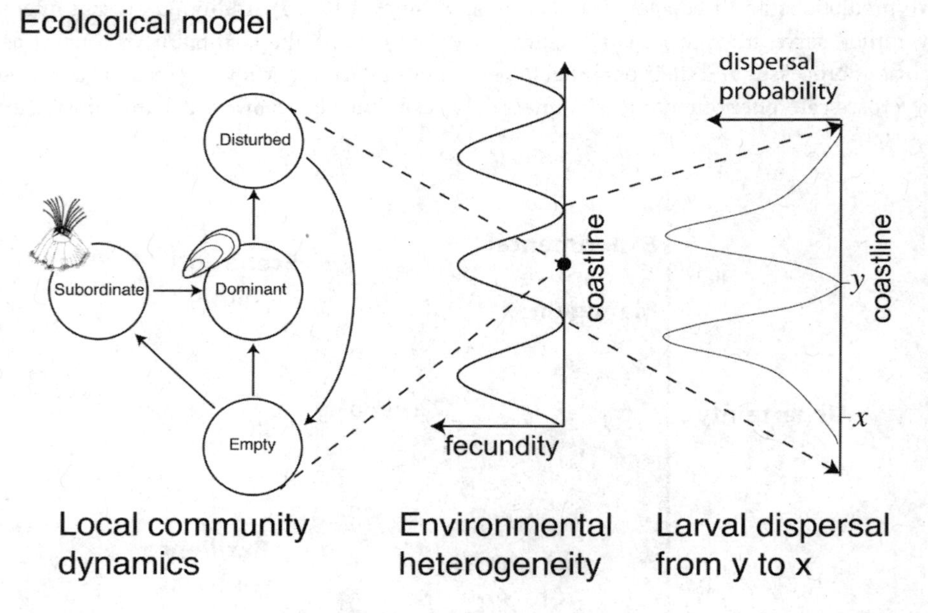

Figure 5.3 Ecological model of rocky intertidal communities. Within-site (< 1 km) community dynamics showing the transitions between subordinate and dominant species that are disturbed by waves. Each "empty" location within a site can be colonized by subordinate (e.g., barnacles) or dominant (e.g., mussels) species, but community dynamics are controlled by the rate at which subordinate species facilitate the colonization by the dominant species. Regional dynamics are controlled by larval dispersal between sites. Larvae are produced as a proportion of within-site abundance and of fecundity. Fecundity itself can be heterogeneous along the coastline and reflect the level of protection of high-fecundity areas in marine reserves. Adapted from Guichard and Steenweg 2008 with permission.

cross-scale phenomena to emerge. That is, spatial and temporal heterogeneity develop and are maintained in (idealized) homogeneous environments at scales much larger than those of the ecological processes (i.e., local disturbance or regional dispersal dynamics; fig. 5.4A). In our model, these cross-scale phenomena take the form of large areas of weak and strong synchronized fluctuations that are much larger than larval dispersal distance, the underlying process of connectivity (fig. 5.4A).

Acknowledging connectivity between ecosystems and between species at the local level has multiple implications for ecosystem-based management. In our model, parameter values of species interactions and dispersal leading to large-scale fluctuations (as in fig. 5.4A) are also associated with (1) maximum regional population density, (2) minimum regional extinction probability, and (3) maximum local extinction probability (Guichard 2005). Contrasting results 1 and 2 with result 3 provides an example of systems where managing for regional persistence and abundance means accepting increased variation and extinction risk at the local level (Holling 1986). This idea of trading off local resilience for regional resilience can also be understood within the more general concept of specific versus general resilience (Leslie and Kinzig, chap. 4 of this volume).

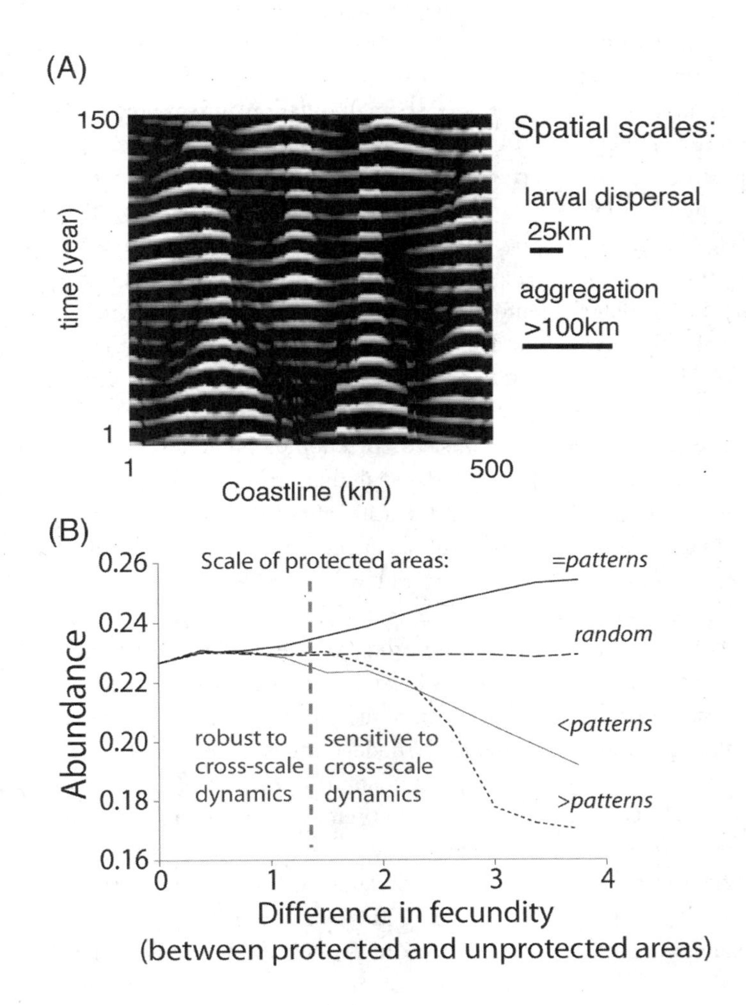

Figure 5.4 (A) Spatiotemporal variability in mussel cover (represented as shades of gray, from dark for low cover to light for high cover) along a coastline predicted under intermediate strength of facilitation between mussels and barnacles. The scale of mussel larval dispersal and of average aggregation of mussel cover are shown. (B) Population response (measured as regional abundance) to increasing difference in fecundity (per capita production of larvae) between protected and unprotected areas and for various scales of protected areas (i.e., number of high-fecundity areas along the coastline; see fig. 5.3 for details). While random distribution of fecundity is averaged out by larval dispersal and has no effect on abundance, the only positive response to increasing habitat differences between protected and unprotected areas occurs when the scale of protected areas matches that of intrinsic patchiness caused by species interactions and dispersal. See text for details. A and B adapted from Guichard and Steenweg 2008 with permission.

Ecological Cross-Scale Interactions and Marine Reserve Design

No-take marine reserves and other types of marine protected areas (MPAs) are a key tool in the marine ecosystem-based management portfolio. Our simple ecological model reveals how cross-scale phenomena—in addition to the scale of ecological processes such as dispersal—can drive the response of marine ecosystems to habitat protection, for example, through MPA implementation. Applying ecological predictions to the design of marine protected areas first requires that we move away from idealized homogeneous environments to consider environmental heterogeneity that can result from habitat fragmentation into protected and unprotected habitats. For example, we must consider how best to fragment coastal ecosystems and select protected areas in the presence of cross-scale phenomena, that is, when the scale of underlying processes such as dispersal is separate from the scale of variability in the abundance of species. We can start answering this question using the model described above and by imposing spatial heterogeneity in the parameters most closely related to the effect of reserve networks. We illustrate this approach by studying our model with heterogeneity in the fecundity (per capita reproductive output) of invertebrates as a proxy for varying habitat quality along the coast (fig. 5.3). Spatial variability in fecundity of intertidal invertebrates has been linked to nearshore and/or onshore habitat quality indicators such as chlorophyll *a* concentration, sea surface temperature, and wave exposure (Bertness et al. 1991; Leslie et al. 2005). We can systematically vary the scale (period) and the strength (amplitude) of habitat fragmentation into reserves as a periodic pattern of fecundity (fig. 5.3). Can we then predict the scale of protected areas having the strongest (negative or positive) impact on population persistence and abundance as we increase the difference in fecundity between protected and unprotected areas?

We can show that implementing protected areas at the scale of intrinsic patterns of abundance described above in homogeneous environments (see fig. 5.4A), rather than at the scale of the underlying process (i.e., dispersal), is again associated with maximum population abundance and persistence (Guichard and Steenweg 2008; fig. 5.4B). More precisely, most spatial scales of protected areas (i.e., areas of high fecundity) across the coastline have a negative effect on abundance (fig. 5.4B) and on persistence (Guichard and Steenweg 2008) as we increase the difference in fecundity between protected and unprotected areas. However, strong habitat differences at the scale of cross-scale phenomena (see above) have a positive effect on abundance (fig 5.4B). These results reveal how ecosystems become sensitive to cross-scale phenomena as we increase the strength of fragmentation into protected and unprotected areas. Equilibrium theories of marine reserve networks predict how matching size and spacing of individual reserves with dispersal distance can be used to maximize overall yield (Hastings and Botsford 2003). Simple models integrating nonequilibrium dynamics, and cross-scale phenomena suggest that size and spacing of reserves (favorable habitats) at the scale of patterns of abundance (a hundred to thousands of kilometers) could actually be more beneficial than reserve networks, considering the scale of ecological processes (e.g., dispersal distance, 10–100 km) in isolation (Guichard and Steenweg 2008).

Applications of these results are not limited to marine protected areas. Most management strategies have scale-dependent and cross-scale impacts on natural and social systems. Simple ecological models predict that management strategies for spatially subdivided fish stocks (e.g., cod fishery in the Northwest Atlantic, see O'Boyle and Worcester, chap. 14 of this volume) and for large-scale ecosystems, such as coral reefs and mangrove habitats (Ezcurra et al., chap. 13 of this volume), will be more effective if management institutions match the scales of cross-scale phenomena such as those described here (Bellwood et al. 2004; Lebel et al. 2006). However, appropriate matching of ecological and institutional scales requires having strong evidence that cross-scale phenomena are real, and not simply model predictions. This situation suggests that regional monitoring programs that adapt to changes in biotic and abiotic variability across scales, rather than permanent monitoring stations, are advisable. Such an adaptive monitoring program could address the challenge of detecting cross-scale phenomena in marine systems. For example, strong shifts in ecosystem processes have been documented in the Chesapeake Bay (Boesch and Goldman, chap. 15 of this volume), but their causal interactions with larger-scale dynamics (e.g., across the whole watershed) have not been resolved.

Our models suggest how building resilience of coastal ecosystems to diverse stressors could also mean moving away from a predictability framework that relies on scale-dependent and directional causality between patterns and processes (see also Leslie and Kinzig's discussion, in chap. 4 of this volume, of generalized resilience). Ecological processes and their associated scales will continue to be important management priorities (Lubchenco et al. 2003). However, when we integrate the problem of scale into management, our prediction is that cross-scale phenomena involving feedbacks between processes and patterns can emerge: Large-scale patterns influence the response of more-local ecological processes in managed habitats (fig. 5.4B). In such cases, no strict priority should be given to patterns or to processes alone. Instead, the full cycle of causal links and feedbacks (fig. 5.1A) across scales needs to be considered explicitly in relation to management. As discussed earlier, the model presented here is meant to provide qualitative predictions, and more-complex models should be explored to directly inform management decisions. Such simple models can provide formal ecological frameworks and be explicitly integrated into adaptive management scenarios to improve our understanding of cross-scale phenomena in coupled social–ecological systems. However, such integration has yet to be achieved.

We now review current frameworks for resilience building in social–ecological systems and discuss their compatibility with ecological cross-scale phenomena discussed above. We specifically suggest that uncertainty and controllability of social–ecological systems characterized by feedbacks between patterns and processes across scales can be resolved by integrating local controls and large-scale resilience building (table 5.1).

Ecosystem-Based Management across Scales and Levels

In addition to the complexity of natural marine systems, ecosystem-based management also confronts social complexity (Shackeroff et al., chap. 3 of this volume). The lack of integration of governance across geographic and institutional scales has repeatedly been suggested as a key factor limiting the success of EBM-related initiatives (Berkes 2002; Hilborn 2004; Agardy 2005). How can management strategies adapt to natural and human systems that are undergoing strong fluctuations? Using the nonequilibrium framework presented above as a starting point, reassessment of uncertainty and controllability is needed, to explicitly incorporate their nonequilibrium and scale-dependent nature. From a governance point of view, such cross-scale phenomena emphasize the importance of reciprocal communication based on dialogue, rather than hierarchical communication based on rigid structures across levels of governance (Peterson 2000).

Uncertainty

Formal approaches to management use mathematical models to cope with uncertainty.

Table 5.1. Different management approaches defined based on their focus on ecological patterns and/or processes

For each approach, trade-offs in terms of system controllability and uncertainty are suggested.

Management approach	*Example*	*Control*	*Uncertainty*
Managing pattern	Defining reserves, or removing invasive species	Difficult to regulate distribution of organisms, but relatively easy to restrict use to different areas	Pattern easier to monitor: importance of pattern changing as pattern changing process
Managing process	Regulating extraction rate, or restricting entry into fishery	Difficult to regulate ecological processes, but easier to regulate rates of extraction	Processes difficult to monitor: pattern altered by changes in process, thereby impacting resilience of ecosystems
Managing cross-scale interactions (adaptive management)	Changing fishing practices over salmon returns, varying fishing practices in space—e.g., historical First Nations salmon fisheries in Pacific Northwest	Cross-scale management difficult to coordinate due to the range of actors	Appropriate approach to build resilience when interactions are very difficult to monitor; theoretical understanding required to identify key pattern–process interactions

Mathematical models are fit to data and compared to competing models to discover models that forecast future behavior better than others. This approach can be effective if the system being modeled continues to operate in a similar fashion. However, if the system has the potential to reorganize—for example, following the introduction of a new species—models based upon past behavior will often fail to predict reorganization (Peterson et al. 2003). This situation is characteristic of nonlinear and nonequilibrium systems where uncertainty is not managed through predictions of equilibrium states, but is, rather, built into the intrinsic variability observed across scales. More specifically, the non-stationarity of such systems shifts predictability from the mean to scale-dependent variance of predicted responses to management. Because not all possible responses can be considered, effective management approaches should integrate intrinsic fluctuations and also expect predictions to occasionally fail and surprises to occur (Holling 1986).

Social dynamics are even more difficult to predict than ecological dynamics, due to the diversity of human values, the currently rapid pace of social change, and the reflexive nature of people (Westley et al. 2002). Policy is often based on existing technologies and values, or the continuation of existing trends. However, frequently new inventions change the way people impact ecosystems. Furthermore, humans are reflexive and therefore consider the consequences of their future behavior, as well as that of others, before making a decision. This reflexivity can make predictions self-fulfilling and self-negating, which makes predicting future behavior more complex. For example, people's behavior changes based upon their beliefs about other people's behavior. The social nature of belief means that while norms of behavior are frequently reinforced, they can also rapidly change (Kuran 1989).

Controllability

Controllability of ecological processes by management action depends on both the nature of the ecological processes involved and the organization of the society being managed. Available knowledge and technology strongly influence management control of both these aspects of controllability. Ecological processes vary in their ability to be controlled depending upon three factors: novelty, visibility, and connections across spatial and temporal scales.

First, the uniqueness of a given ecological situation makes control more difficult, because control techniques are more easily transferred between analogous situations. For example, there are fewer analogues to Florida Bay than to a small boreal forest lake. Second, the ease with which a system's dynamics can be understood depends upon the ease with which ecosystem functional relationships can be detected and separated from environmental variation. This process is difficult when ecological change occurs slowly, when ecological processes are technically challenging to measure or observe, such as in oceanic fish populations (Walters 1986), or when cross-scale phenomena generate feedbacks between internal functional relationships and the environment. For example, such cross-scale phenomena discussed above predict how intrinsic variability (i.e., independent from the environment) can dictate at which scale(s) the environment is most likely to have a measurable impact on ecosystem properties (fig 5.4B). Controllability is also scale-dependent: The controllability of a system being managed is decreased if processes external to it strongly influence the system's dynamics. For example, attempts to restore salmon to a section of river will be influenced by dam management, runoff from surrounding land, changes in fishing techniques, and the ocean temperature patterns in the North Pacific. The presence of cross-scale phenomena further suggests that control at the scales of processes might not be sufficient. Feedbacks between managed processes and patterns therefore constitute a cross-scale cycle that needs to be considered in an attempt to improve their overall controllability. The limitations of controllability across scales are of particular relevance in coastal ecosystems where interactions between terrestrial and oceanic physical systems create extreme ranges of spatial scales and management levels that need to be considered. For example, local competition for space among sessile benthic invertebrates (mussels, barnacles) is mediated by larval transport driven by large-scale currents, and the outcomes of local management of coastal systems are influenced by social and ecological dynamics beyond the boundaries of a specific area.

Along with these ecological features of controllability, society also strongly influences what type of control is possible in a given location. Attributes of management institutions can help or hinder the ability to respond to change. Four attributes of institutions that have been suggested to impact the effectiveness of resource management are shared ecological understanding, effectiveness of collective action, effectiveness of past conflict resolution, and match of scales of organization and ecological processes. Shared understanding among actors of how the natural world works is important because it determines whether people can agree upon an intervention in the system (Holling 1978). Effectiveness of collective action is the degree to which organized groups of people can implement new policies (Bromley 1989). The legitimacy of past conflict resolution shapes the ability of an institution to address new issues (Hoff 1998). Governments have frequently expropriated resources from local people and then had attempts at management fail due to passive or active resistance to management policies from these people (Scott

1998). The fit between institutional and ecological scales represents the degree to which management institutions function at the scales at which important ecological change occurs (Folke et al. 1998; Young 2002; Crowder et al. 2006). These scales have been associated with those of either ecological processes or patterns of distribution. However, as suggested by theories such as those discussed above, cross-scale phenomena require that management institutions match the multiple scales of processes and patterns, but also that their interactions across levels of governance match causal associations between local processes and regional patterns.

The controllability of a social–ecological system is determined, to a large extent, by the group of people involved with using and managing it (Ostrom 1990). By working to increase controllability and reduce uncertainty, people make social–ecological situations more manageable. Increasing social agreement can increase the ability of people to control a system, as can ecological engineering interventions. Similarly, social learning processes, such as adaptive management and participatory modeling efforts that make the links between people and the environment more explicit, can increase understanding.

Adaptive Management

Adaptive management is an approach to management in which policies are treated as hypotheses and management actions as experiments (Holling 1978; Walters 1986). Adaptive environmental management is a structured process of "learning by doing" that aims to reduce the social and ecological costs of management experiments, while increasing opportunities for learning. It aims to facilitate social learning by using a combination of assessment, computer-modeling, and management experimentation to identify critical uncertainties facing managers. During this process, people develop alternate hypotheses that address these uncertainties. Management plans are developed to evaluate these hypotheses by using the human manipulation of ecological processes to strategically probe the functioning of ecosystems. Adaptive management therefore becomes increasingly appropriate as uncertainty increases and as controllability of social and ecological processes decreases (fig. 5.2). From the controllability-uncertainty perspective, adaptive management is more precisely aimed at harnessing system uncertainty through an iterative process that acknowledges low controllability and possibility of management failure.

An adaptive management process typically has three phases: assessment, modeling, and management experimentation. Assessment integrates existing experience, data, and theory. This integrated understanding is then embodied in dynamic computer models that attempt to make predictions about the impacts of alternative policies. The modeling step serves three functions. First, it clarifies the problem and provides a common forum for people to discuss management issues. Second, the model can be used to eliminate policies that are unlikely to be effective. Third, the modeling process identifies which gaps in existing knowledge have the largest consequences for policy. Often these gaps involve poorly understood large-scale processes that cannot be investigated by small-scale experiments. If this is the case, this discovery leads to the design of large-scale management experiments to resolve these key uncertainties at the temporal and spatial scales that are relevant for management.

Despite being initially proposed over twenty-five years ago and widely advocated in the last decade, adaptive management has not been widely practiced (Lee 1999). Barriers to the implementation of adaptive management include the unwillingness of managers or decision makers to confront uncertainty, the desire of vested interests to avoid the change that experimentation may produce, and the cost of monitoring and experimentation (Walters 1997). They are produced by the social and political context in which adaptive management takes place. These barriers indicate that adaptive management is unlikely to be successful if it is imposed on a social–ecological system rather than developed in cooperation with participants in that system. Consequently, many researchers have begun to examine what are the conditions for successful adaptive comanagement (Folke et al. 2005). In the context of cross-scale phenomena, adaptive management has the potential to reconcile management practices with the dynamic and nonequilibrium nature of many ecological systems. Unpredicted variability resulting from cross-scale phenomena can be assessed dynamically and accounted for through adaptive management, which also bridges cross-scale variability in ecological systems with the need for communication across levels of institutions. Both objectives are associated with increased uncertainty and decreased controllability (fig. 5.2) and are best formulated through resilience building.

Resilience Building

Resilience building, enhancing the ability of a system to cope with stress or surprise, is an approach that has been advocated in situations where control is difficult but where there is some understanding about how the system works (Adger et al. 2005a). A simple example is the preservation of mangroves in tropical areas to reduce the vulnerability of coastal areas to flooding (Barbier et al. 2008; but see also Kerr and Baird 2007). Building resilience is equivalent to hedging one's bets or purchasing insurance for unlikely but possible outcomes (Costanza et al. 2000). Resilience depends upon both the properties of a system and the connections between that system and its surrounding area. For example, both diversity of species and some level of functional redundancy or modularity among these species can contribute to increased resilience (see Leslie and Kinzig, chap. 4 of this volume).

One of the key local factors that resilience depends upon is response diversity—the variation of responses to environmental change among species that contribute to the same ecosystem function (Elmqvist et al. 2003). For example, both temperate and tropical coastal systems have experienced large increases in nutrient loading due to land-based activities, including coastal development and agriculture. These inputs can in turn cause eutrophication of coastal waters. The vulnerability of coastal zones to the undesirable consequences of eutrophication appears to have been increased by declines in organisms at higher trophic levels (e.g., Jackson et al. 2001). Equally important, diversity within functional groups—whether they be predators (Sandin et al. 2008) or herbivores (Bellwood et al. 2004)—has been demonstrated to increase the resilience of marine ecosystems to multiple stressors, particularly those associated with global climate change (see also Leslie and Kinzig, chap. 4 of this volume).

The cross-scale connections of an area are also vitally important in determining its resilience. The connections between an area and its support areas are key determinants in the ability of an area to recover following disturbance (Lundberg and Moberg 2003). For example, the presence of intact reefs near damaged reefs can provide a source of coral propagules (Nystrom et al. 2000). Such a source–sink relationship between intact and damaged areas can be highly dynamic and involve highly heterogeneous distributions that are compatible with the regional persistence of coral reefs. This regional persistence in the face of more localized and unpredictable catastrophes illustrates the prediction that managing for resilience of cross-scale phenomena means accepting local changes to foster regional persistence. Similarly, EBM that shares decision-making ability, knowledge, and management actions among local groups, their neighbors, and larger entities can increase the resilience of a coupled social–ecological system. Comanagement processes that further involve a diverse constellation of partners, such as scientific organizations, NGOs, and corporations, have the potential to build the capacity of people to respond to change (Berkes 2002; Adger 2005b).

There are not general well-developed approaches to building resilience in ecosystems, but it is an area of active research and practice (Berkes et al. 2003; Leslie and Kinzig, chap. 4 of this volume). Adaptive management and resilience building are complementary elements of ecosystem-based management. Adaptive management focuses on learning and hedging against uncertainty through experimentation, while resilience building provides a "safe-fail" environment (Redford and Taber 2000) in which there can be experiments without there being devastating consequences.

Conclusions

We suggest that coastal ecosystems can operate as nonequilibrium systems where nonlinear interactions among spatially and functionally structured components lead to internal fluctuations in time and space. We adopted a simple model of intertidal communities to illustrate how such nonequilibrium dynamics can be closely linked to cross-scale phenomena through feedbacks between local processes and regional heterogeneity in the distribution of abundance. We also showed how the presence of such cross-scale cycles can promote ecosystem persistence and dictate scale-dependent ecosystem responses to natural or anthropogenic environmental heterogeneity.

Cross-scale and nonequilibrium phenomena highlight the uncertainty and uncontrollability of coastal ecosystems but also suggest that key scales and key points in time and space can be identified for management and monitoring. We argue that coastal systems and the ecological model presented here provide a quantitative framework for addressing uncertainty as a problem of cross-scale cycles rather than as one of scale-dependent predictions. We hope to have shown how local ecological processes can generate large-scale heterogeneity in coastal systems. This variation, in turn, creates uncertainty that can affect controllability of underlying local processes. From the cross-scale formulation of such uncertainty, we conclude that controllability could be better assessed through adaptive management, which integrates reciprocal communication and power sharing across institutional levels and creates economic and cultural incentives that favor large-scale resilience building.

Acknowledgments

We wish to thank H. Leslie and K. McLeod and two anonymous reviewers for very useful comments on earlier versions of our manuscript. FG also wishes to acknowledge support from the James S. McDonnell Foundation through a 21st Century Initiative award. GP was supported by a Canada Research Chair.

References

Adger, W. N., T. P. Hughes, C. Folke, S. R. Carpenter, and J. Rockström. 2005a. Social–ecological resilience to coastal disasters. *Science* 309:1036–39.

Adger, W. N., K. Brown, and E. L. Tompkins. 2005b. The political economy of cross-scale networks in resource co-management. *Ecology and Society* 10:9.

Agardy, T. 2005. Global marine conservation policy versus site-level implementation: The mismatch of scale and its implications. *Marine Ecology Progress Series* 300:242–48.

Alexander, S. E., and J. Roughgarden. 1996. Larval transport and population-dynamics of intertidal barnacles: A coupled benthic/oceanic model. *Ecological Monographs* 66:259–75.

Almany, G. R., M. L. Berumen, S. R. Thorrold, S. Planes, and G. P. Jones. 2007. Local replenishment of coral reef fish populations in a marine reserve. *Science* 316:742–44.

Barbier, E. B., E. W. Koch, B. R. Silliman, S. D. Hackery, E. Wolanski, J. Primavera, E. F. Granek et al. 2008. Coastal ecosystem-based management with nonlinear ecological functions and values. *Science* 319:321–23.

Becker, B. J., L. A. Levin, F. J. Fodrie, and P. A. McMillan. 2007. Complex larval connectivity patterns among marine invertebrate populations. *Proceedings of the National Academy of Sciences* 104:3267–72.

Bellwood, D. R., T. P. Hughes, C. Folke, and M. Nystrom. 2004. Confronting the coral reef crisis. *Nature* 429:827–33.

Berkes, F. 2002. Cross-scale institutional linkages: Perspectives from the bottom. In *Drama of the commons*, ed. T. D. E. Ostrom, N. Dolsak, P. C. Stern, S. Stonich, and E. U. Weber, 293–321. Washington, DC: National Academies Press.

Berkes, F., J. Colding, and C. Folke, ed. 2003. *Navigating social–ecological systems: Building resilience for complexity and change*. Cambridge, UK: Cambridge University Press.

Bertness, M. D., S. D. Gaines, D. Bermudez, and E. Sanford. 1991. Extreme spatial variation in the growth and reproductive output of the acorn barnacle Semibalanus balanoides. *Marine Ecology Progress Series* 75:91–100.

Bjornstad, O. N., M. Peltonen, A. M. Liebhold, and W. Baltensweiler. 2002. Waves of larch budmoth outbreaks in the European Alps. *Science* 298:1020–23.

Botsford, L. W., A. Hastings, and S. D. Gaines. 2001. Dependence of sustainability on the configuration of marine reserves and larval dispersal distance. *Ecology Letters* 4:144–50.

Botsford, L. W., F. Micheli, and A. Hastings. 2003. Principles for the design of marine reserves. *Ecological Applications* 13:S25–S31.

Broitman, B. R., S. A. Navarrete, F. Smith, and S. D. Gaines. 2001. Geographic variation of southeastern Pacific intertidal communities. *Marine Ecology Progress Series* 224:21–34.

Bromley, D. 1989. *Economic interests and institutions: The conceptual foundations of public policy*. New York: Basil Blackwell.

Chavez, F. P., J. Ryan, S. E. Lluch-Cota, and M. Ñiquen. 2003. From anchovies to sardines and back: Multidecadal change in the Pacific Ocean. *Science* 299:217–221.

Collins, C. A., J. T. Pennington, C. G. Castro, T. A. Rago, and F. P. Chavez. 2003. The California Current system off Monterey, California: Physical and biological coupling. *Deep Sea Research II* 50:2389–404.

Connolly, S. R., and J. Roughgarden. 1999. Theory of marine communities: Competition, predation, and recruitment-dependent interaction strength. *Ecological Monographs* 69:277–96.

Costanza, R., H. Daly, C. Folke, P. Hawken, C. S. Holling, A. J. McMichael, D. Pimentel, and D. Rapport. 2000. Managing our environmental portfolio. *BioScience* 50:149–55.

Crowder, L. B., G. Osherenko, O. R. Young, S. Airamé, E. A. Norse, N. Baron, J. C. Day et al. 2006. Resolving mismatches in US ocean governance. *Science* 313(5787):617–18.

Elmqvist, T., C. Folke, M. Nystrom, G. Peterson, J. Bengtsson, B. Walker, and J. Norberg. 2003. Response diversity, ecosystem change, and resilience. *Frontiers in Ecology and the Environment* 1:488–94.

Feinberg, L. R., and W. T. Peterson. 2003. Variability in duration and intensity of euphausiid spawning off central Oregon, 1996–2001. *Progress in Oceanography* 57:363–79.

Folke, C., T. Hahn, P. Olsson, and J. Norberg. 2005. Adaptive governance of social–ecological systems. *Annual Review in Environment and Resources* 30:441–73.

Folke, C., L. Pritchard Jr., F. Berkes, J. Colding, and U. Svedin. 1998. *The problem of fit between ecosystems and institutions*. International Human Dimensions Programme on Global Environmental Change (IHDP) Working Paper. Bonn: IHDP.

Gaines, S., and J. Roughgarden. 1985. Larval settlement rate: A leading determinant of structure in an ecological community of the marine intertidal zone. *Proceedings of the National Academy of Sciences* 82:3707–11.

Gaines, S., S. Brown, and J. Roughgarden. 1985. Spatial variation in larval concentrations as a cause of spatial variation in settlement for the barnacle, Balanus glandula. *Oecologia* 67:267–72.

Gaylord, B., and S. D. Gaines. 2000. Temperature or transport? Range limits in marine species mediated solely by flow. *American Naturalist* 155: 769–89.

Guichard, F. 2005. Interaction strength and extinction risk in a metacommunity. *Proceedings of the Royal Society B* 272:1571–76.

Guichard, F., and R. Steenweg. 2008. Intrinsic and extrinsic causes of spatial variability across scales in a metacommunity. *Journal of Theoretical Biology* 250(1):113–24.

Guichard, F., S. A. Levin, A. Hastings, and D. A. Siegel. 2004. Toward a dynamic metacommunity approach to marine reserve theory. *BioScience* 54:1003–11.

Hastings, A., and L. W. Botsford. 2003. Comparing designs of marine reserves for fisheries and for biodiversity. *Ecological Applications* 13:S65–S70.

Hilborn, R. 2004. Ecosystem-based fisheries management: The carrot or the stick? *Marine Ecology Progress Series* 274:275–78.

Hoff, M. D., ed. 1998. *Sustainable community development: Studies in environmental, economic and cultural revitalization*. St Lucie, FL: CRC Press.

Holling, C. S., ed. 1978. *Adaptive environmental assessment and management*. London: John Wiley & Sons.

Holling, C. S. 1986. The resilience of terrestrial ecosystems: Local surprise and global change. In *Sustainable develoment of the biosphere*, ed. W. C. Clark and R. E. Munn, 292–317. Cambridge, UK: Cambridge University Press.

Iwasa, Y., and J. Roughgarden. 1986. Interspecific competition among metapopulations with space-limited subpopulations. *Theoretical Population Biology* 29:235–61.

Jackson, J. B. C., M. X. Kirby, W. H. Berger, K. A. Bjorndal, L. W. Botsford, B. J. Bourque, R. H. Bradbury et al. 2001. Historical overfishing and the recent collapse of coastal ecosystems. *Science* 293:629–38.

Johnson, M. 2000. A reevaluation of density-dependent population cycles in open systems. *American Naturalist* 155:36–45.

Johnson, M. P. 2005. Is there confusion over what is meant by "open population"? *Hydrobiologia* 544:333–38.

Jones, G. P., M. J. Milicich, M. J. Emslie, and C. Lunow. 1999. Self-recruitment in a reef fish population. *Nature* 402:802–4.

Kerr, A. M., and A. H. Baird. 2007. Natural barriers to natural disasters. *BioScience* 57(2):102–3.

Kinlan, B. P., S. D. Gaines, and S. E. Lester. 2005. Propagule dispersal and the scales of marine community process. *Diversity and Distributions* 11:139–48.

Kinlan, B. P., and S. D. Gaines. 2003. Propagule dispersal in marine and terrestrial environments: A community perspective. *Ecology* 84:2007–20.

Kuran, T. 1989. Sparks and prairie fires: A theory of unanticipated political revolution. *Public Choice* 61:41–74.

Lebel, L., J. M. Anderies, B. Campbell, C. Folke, S. Hatfield-Dodds, T. P. Hughes, and J. Wilson. 2006. Governance and the capacity to manage resilience in regional social–ecological systems. *Ecology and Society* 11(1):19.

Lee, K. N. 1999. Appraising adaptive management. *Conservation Ecology* 3(2):3.

Leslie, H. M., E. N. Breck, F. Chan, J. Lubchenco, and B. A. Menge. 2005. Barnacle reproductive hotspots linked to nearshore ocean conditions. *Proceedings of the National Academy of Sciences* 102:10534–39.

Levin, S. A. 1992. The problem of pattern and scale in ecology. *Ecology* 73:1943–67.

Lubchenco, J., S. R. Palumbi, S. D. Gaines, and S. Andelman. 2003. Plugging a hole in the ocean: The emerging science of marine reserves. *Ecological Applications* 13:S3–S7.

Lundberg, J., and F. Moberg. 2003. Mobile link organisms and ecosystem functioning: Implications for ecosystem resilience and management. *Ecosystems* 6:87–98.

Menge, B. A., and A. M. Olson. 1990. Role of scale and environmental factors in regulation of community structure. *Trends in Ecology and Evolution* 5:52–56.

Nystrom, M., C. Folke, and F. Moberg. 2000. Coral reef disturbance and resilience in a human-dominated environment. *Trends in Ecology and Evolution* 15:413–17.

Ostrom, E. 1990. *Governing the commons: The evolution of institutions for collective action*. New York: Cambridge University Press.

Paine, R. T. 1984. Ecological determinism in the competition for space. *Ecology* 65:1339–48.

Paine, R. T. 1980. Food webs: Linkage, interaction strength and community infrastructure. *Journal of Animal Ecology* 49:667–85.

Paine, R. T., and S. A. Levin. 1981. Intertidal landscapes: Disturbance and the dynamics of pattern. *Ecological Monographs* 51:145–78.

Peterson, G. D. 2000. Political ecology and ecological resilience: An integration of human and ecological dynamics. *Ecological Economics* 35:323–36.

Peterson, G. D., S. R. Carpenter, and W. A. Brock. 2003. Uncertainty and the management of multistate ecosystems: An apparently rational route to collapse. *Ecology* 84:1403–11.

Redford, K. H., and S. Taber. 2000. Writing the wrongs: Developing a safe-fail culture in conservation. *Conservation Biology* 14:1567–68.

Roughgarden, J., S. Gaines, and Y. Iwasa. 1984. Dynamics and evolution of marine populations with pelagic larval dispersal. In *Exploitation of marine communities*, ed. R. M. May, 111–28. Berlin: Springer.

Roughgarden, J., S. Gaines, and H. Possingham. 1988. Recruitment dynamics in complex life cycles. *Science* 241:1460–66.

Sala, E., O. Aburto-Oropeza, G. Paredes, I. Parra, J. C. Barrera, and P. K. Dayton. 2002. A general model for designing networks of marine reserves. *Science* 298:1991–93.

Sandin, S. A., J. E. Smith, E. E. DeMartini, E. A. Dinsdale, S. D. Donner, A. M. Friedlander, T. Konotchick et al. 2008. Baselines and degradation of coral reefs in the Northern Line Islands. *PLoS ONE* 3(2):e1548. doi: 10.1371/journal.pone.0001548.

Sanford, E., M. Roth, G. C. Johns, J. P. Wares, and G. N. Somero. 2003. Local selection and latitudinal variation in a marine predator–prey interaction. *Science* 300:1135–37.

Sanford, E. 1999. Regulation of keystone predation by small changes in ocean temperature. *Science* 283:2095–96.

Scott, J. C. 1998. *Seeing like a state: How certain schemes to improve the human condition have failed.* New Haven, CT: Yale University Press.

Smith, R. L. 1974. Description of current, wind, and sea-level variations during coastal upwelling off Oregon coast, July–August 1972. *Journal of Geophysical Research* 79:435–43.

Smith, R. L., A. Huyer, and J. Fleischbein. 2001. The coastal ocean off Oregon, from 1961 to 2000: Is there evidence of climate change or only of Los Niños? *Progress in Oceanography* 49:63–93.

Stommel, H. 1963. Varieties of oceanographic experience. *Science* 139(3555):572–76.

Swearer, S. E., J. E. Caselle, and D. W. Lea. 1999. Larval retention and recruitment in an island population of a coral-reef fish. *Nature* 402:799–802.

Underwood, A. J., and P. G. Fairweather. 1989. Supply-side ecology and benthic marine assemblages. *Trends in Ecology and Evolution* 4:16–20.

van de Koppel, J., M. Rietkerk, N. Dankers, and P. M. J. Herman. 2005. Scale-dependent feedback and regular spatial patterns in young mussel beds. *American Naturalist* 165:E66–E77.

Walters, C. J. 1986. *Adaptive management of renewable resources.* New York: McGraw-Hill.

Walters, C. J. 1997. Challenges in adaptive management of riparian and coastal ecosystems. *Conservation Ecology* 1(2):1.

Westley, F., S. R. Carpenter, W. A. Brock, C. S. Holling, and L. H. Gunderson. 2002. Why systems of people and nature are not just social and ecological systems. In *Panarchy: Understanding transformations in human and natural systems*, ed. L. H. Gunderson and C. S. Holling, 103–19. Washington, DC: Island Press.

Witman, J. D., S. J. Genovese, J. F. Bruno, J. W. McLaughlin, and B. I. Pavlin. 2003. Massive prey recruitment and the control of rocky subtidal communities on large spatial scales. *Ecological Monographs* 73:441–62.

Young, O. R. 2002. *The institutional dimensions of environmental change: Fit, interplay, and scale.* Cambridge, MA: Massachusetts Institute of Technology Press.

CHAPTER 6
Valuing Ecosystem Services

Lisa A. Wainger and James W. Boyd

Ecology and economics appear to be heading toward the cross-disciplinary alliance so necessary to environmental analysis. A key to this has been the articulation of ecosystem services as a concept that bridges the two disciplines. Interpretations differ in the details, but the central idea—that nature contributes materially to both our personal well-being and the health of the market economy—is creating new opportunities for combined ecological and economic analysis (e.g., Heal et al. 2005).

The concept of ecosystem services holds the promise of presenting all effects on human welfare that result from resource use decisions within the same framework. Currently, resource managers struggle with combining the results of separate ecological and economic impact assessments and therefore are anxious to have such a framework realized. Much of the previous work in resource management has generated economic impacts assessed in dollars while environmental impacts were assessed using ecological indicators.

While the promise of a cohesive framework for assessing all types of damages in a single unit is not yet operational, many are working toward this goal through more rigorous conceptualization and communication of the links between changes in natural systems and effects on human welfare (Daily 1997; Wainger et al. 2001; Johnston et al. 2002; Ricketts et al. 2004; Polasky et al. 2005; Boyd 2006; Carpenter et al. 2006; Chan et al. 2006; Boyd and Banzhaf 2007; Barbier et al. 2008). In addition, much of economic analysis is concerned with assessing the potential consequences of specific management policies to human well-being. Such analyses may be less concerned with assessing values and more with understanding how to avoid or minimize damages in the first place by understanding and motivating human behaviors (Bromley 1995).

When it comes to assessing values for the goods and services we derive from ecosystems, much of the historical focus of resource economics has been a piecemeal approach to valuing the more tangible outcomes of ecosystems, usually commercial commodities, health outcomes, and recreational uses. The protection of scarce species and ecosystems has received some attention, but what economics has barely begun to evaluate are the values people attach to ensuring long-term functioning of ecosystems and the value of increased resilience. Ecosystem-based management (EBM) emphasizes the need to evaluate ecosystems' ability to rebound from disturbance (McLeod et al. 2005). In principle, economic analysis shares this perspective. At present, however, it has failed to produce practical tools to help managers who wish to justify such a focus.

This chapter provides perspective on what economics can contribute now and in the future to ecosystem-based management. This contribution depends on the adoption of new perspectives and tools, which are currently being debated, within both ecology and economics. We begin with a description of the goals of natural resource management. We then describe a range of existing techniques used to examine the value of natural resources. Finally, we discuss newer approaches that integrate ecological information and analysis with their economic counterparts.

The Economic View of Natural Resource Management

At its core, economics is the study of human well-being. Nature is central to that well-being. As individuals, we get pleasure from nature and satisfaction when we act as its steward. And clearly our economies depend on nature to satisfy most of our material needs. The challenge is how to incorporate ecological science into economic analyses of nature.

What does economics mean by the "value of nature"? The value of something is simply a way to depict its importance or desirability. If you like oranges more than apples, oranges are more valuable to you than apples. If you think preserving endangered species habitat is more important than building another shopping mall, then you *value* that habitat more than the shopping mall. That's it. Economists believe that when we add up all these individual measures of importance and desirability, society should act on the basis of that knowledge: providing the greatest good overall. Thus, the value of something natural is the relative importance we place—collectively—on that thing. Value helps us rank things, weight them, set priorities, and make choices. Whenever we choose one thing over another, as individuals do every day, we are engaged in valuation. It is an inescapable part of our humanity, not a construct imposed on us by economists.

To be sure, the methods and language of economics tend to obscure the concept of value and make it seem more alien than it is. Environmental economists talk about the "willingness to pay" for nature, and their methods place a monetary, dollar value on aspects of nature. This monetization of nature seems more philosophically loaded than the act of ranking, weighting, and setting priorities, as resource managers do. What economists don't communicate clearly enough is that monetizing nature is a practical strategy, not a philosophical imperative. Expressing value in dollars is a method, not a philosophy. It is simply a way to rank the social importance of alternatives.

A few notes about "dollar values": First, economists use dollar values (prices) or other currencies because they are data generated by markets. But much of what we value in nature is not bought and sold in markets. So no price data are available. The absence of these data does not imply that nature is not valuable, of course. It just means that the measurement of value cannot rely directly on price data from market transactions (see box 6.1).

While dollars are simply one kind of a numerical scale for assessing value, they are an attractive one, at least where market commodities are concerned. Market commodities have dollar values already attached to them (prices). Thus, the market's invisible hand not only gives us a ready scale, but also attaches numerical scores to commodities in the market—their prices. This elegantly *practical* scale is why economists gravitate to monetary descriptions of ranks, weights, priorities, and preferences. But what is ultimately important is that we have a scale of some kind, any kind.

Second, prices are a measure of value, but not a perfect measure. Our actual willingness to pay for something may be strictly higher than the market price. In fact, we only buy things when this is true. The difference between what I paid and what I would have been willing to pay is the "benefit" I derive from the purchase. Therefore, the true measure of value is what we would be willing to pay, not what we actually pay. When economic methods are described as *willingness to pay* methods, it signifies the desire to capture this full value.

Third, economics takes care to distinguish between the total value of something and its so-called marginal value. The total value of US wetlands, for example, is what we get when we add up the willingness to pay for all wetlands. The marginal value is our willingness to pay for *one more* wetland. Marginal analysis is important because our willingness to pay for *one more* changes as there is more and more of that thing. The value of a particular wetland depends on how many other wetlands there are. Scarcity typically implies a greater marginal value than abundance.

Nature as Commodity

Nature often benefits us in a shared, collective way. In economic parlance, natural capital and ecosystem goods and services tend to be "public goods." These forms of natural capital are shared, not private, and thus resist market trading. How do we make the right choices if nature is not a conventional commodity? For example, how do we trade off losses in nature against gains in material commodities? Or how do we make a choice between one environmental benefit and another? The economic strategy is to turn nature into a set of "simulated commodities" to which value can be attached in some way. This commoditization requires a philosophical leap: to think about nature as a disaggregated collection of features, qualities, and functions to which discrete values can be attached.

Nature, taken as a whole, is priceless and resists economic analysis. Since all life depends on natural systems, those systems, taken as a whole, have an effectively infinite value. What would we give up in order to avoid the collapse of our biosphere? Presumably, we would give up everything but our lives. So, in this sense, nature is economically unique and can be said to be "priceless."

But while nature as a whole is priceless, its individual, constituent elements are not. The fact is we can, do, and always will choose to live with fewer trees, dirtier air and water, and compromised habitats because nature's components are not *all* society cares about. When economics commoditizes nature (divides it into its constituent parts), it does so in order to systematically depict these social choices.

This analytical process (commoditization) is philosophically offensive to some. It seems to deny nature's wholeness and its deep-seated importance to our lives. If true, the discipline seems particularly unsuited to the analysis of nature, whose role in our lives is often distinctly spiritual, ineffable, collective, and beyond the rough calculus of dollars and cents. Here, two clarifications are important. First, economics in no way denies the comprehensiveness, interconnectedness, and complexity of nature. In fact, economists—schooled as they are in the complex, mysterious, and invisible hand of the market economy—intuitively grasp the complex, mysterious, and invisible hand of nature. Second, as a matter of philosophy, economics cares about *anything* that contributes to human well-being, even things of a spiritual, ineffable, and collective character. Economics places no limits on what people think is important or valuable. The desire for beauty, awe, naturalness, and the protection of species other than our own may be hard to measure, but they are all benefits known to economics and accepted as important social realities that "count." If nature presents us with less tangible, distinct, and clearly identifiable benefits than conventional economic goods and services, that is a *practical* problem for economics, not a philosophical one.

Utilitarianism: The Human Perspective Is What Matters

There is one important limitation to what counts in economics, however. Economics *only* focuses on human well-being. Since many things matter to people (beauty, species existence, a sense of wilderness), this is not a huge limitation. But to some, economics' human-centeredness (utilitarianism) is a philosophical problem (e.g., McCauley 2006; see also Moore and Russell, chap. 18 of this volume). Utilitarianism creates a precise description of what society should do when assessing social benefits, namely, maximize human well-being, whatever that is. But is nature, like life itself, really something to be managed and maximized? Shouldn't we protect nature simply because it is right to do so? For those who see the problem in these terms, it is jarring, if not offensive, to view nature as a set of inputs to a human-welfare machine.

The economist's only response to this critique is pragmatic. If nature has inherent rights of its own, does that mean we can never use it for our own advantage? Where do we draw the line? What if nature's right to exist conflicts with other plausible, inherent rights, such as the human rights to life and liberty? How is that conflict to be resolved? In the absence of philosophical or political clarity on these matters, economists pursue a more practical objective: Achieve the greatest good overall. Human preferences—for nature and all other things—are the economist's guide to what should be done and how it can be done. Preferences not only reveal what people value, but also help us predict how people are likely to respond to policy options—allowing economists to identify effective tools for implementing change.

Valuation Techniques

As economists have tried to assess the economic value of nature's commodities, they have created a variety of techniques to assess goods that are not traded in markets. Two basic approaches are used: stated preference and revealed preference. Stated preference relies on surveys to estimate willingness to pay, while revealed-preference approaches use observable behavior to indirectly derive willingness to pay. The application of both techniques to ecosystem services is fraught with methodological challenges and, more importantly, is time-consuming, is expensive to do well, and has conceptual limitations (Pearce 1998; Shabman and Stephenson 2000). Therefore, most policy analyses rely on benefit transfer techniques to estimate values for goods and services based on previously conducted studies (Desvousges et al. 1992; Rosenberger and Loomis 2001; Ready and Navrud 2005; Spash and Vatn 2006; Iovanna and Newbold 2007). This approach also contains many pitfalls, as described in detail by Plummer (2009). The bottom line is that assigning dollar values to natural resources may not always be the most useful tool for choosing among policy options.

In chapter 8 of this volume, Barbier offers a summary of the methods used to associate dollar values with ecosystem goods and services (including the limitations of these methods) and illustrates their use through an example from mangroves in Thailand. Our role in this chapter is to provide the context for an alternative view.

Alternative Measures of the Value of Ecosystem Services

Nonmonetary metrics are widely used to weight and rank decisions and illustrate the benefits of spending (NRC 2000). In the absence of reliable tools to value ecosystem services using dollars, nonmonetary metrics are crucial to the communication of ecosystem benefits. The keys to their successful use are that they (1) communicate something about value, (2) are quantifiable and allow trade-offs to be compared, and (3) represent a well-reasoned evaluation of conditions that minimize the subjectivity of measurement.

In this section, we discuss techniques for measuring ecological endpoints that can inform management, and we describe a framework for using such endpoints in an assessment of the relative value of the goods and services produced by ecosystems. The goal of the framework is to provide measures that allow actions to be ranked and prioritized in order to sustain desirable ecosystem services while balancing a range of competing needs.

Box 6.1. Economic Methods in Practice—Aspirations Unfulfilled

Dozens, if not hundreds, of methodologically sound studies have been conducted to overcome the "missing prices" problem that bedevils economic assessment of the environment. Collectively, these studies show that aspects of nature are demonstrably valuable. Unfortunately, the relevance of this intellectual activity for many types of real-world environmental management has been limited (Pagiola et al. 2004). Environmental economists and policymakers alike have hoped that, with enough valuation studies, a set of generalizations would emerge of practical relevance to environmental management and public policy. This has not happened, for several reasons.

1. The difficulty of valuation methods is a huge constraint. A single study can easily take years and cost hundreds of thousands of dollars. In the end, the results can be highly controversial.
2. Economists are understandably drawn to studies they can actually accomplish and get published. This leads to the "looking under the lamppost" phenomenon, in which studies focus on exceptionally compelling examples. The more intangible the benefit in terms of people being able to perceive and understand the links, the less likely it is to be the focus of study. Thus, economic analysis has a spotty, rather than comprehensive, quality.
3. Value of ecosystem services is usually location-dependent. It is generally inappropriate to extrapolate from one area to another. If an economist shows that wetlands in Poughkeepsie are worth one million dollars, that tells us almost nothing about what the value of wetlands is in Portland. Why? The scarcity of, substitutes for, and complements to wetlands may be very different in the two places, and thus their social importance will be very different. Also, people in Portland may simply like wetlands more than people in Poughkeepsie. Thus, we cannot easily economize on assessment by "transferring" a small set of benefit estimates across the national landscape.
4. Economic analysis is inherently limited by the lack of ecological production functions and understanding of causality in nature. What actions can society take to protect nature, and what is the effect of those actions on nature? Without at least rough answers to these questions, economic analysis aimed at broadening the range of valued ecosystem services is fated to take place on weak ecological footing, thus undermining the results.

As a result, environmental economics has not delivered on capturing the majority of nature's benefits to inform public decision making. To date, the research aimed at providing missing prices has produced a set of case studies that demonstrate nature's value in particular, often superlative, places. They are valuable for those locations and provide clues to nature's value in other locations. But these studies are limited in their ability to assess trade-offs among competing uses for the majority of cases.

Ecological Endpoints that Capture Ecosystem Goods and Services

The concept of ecosystem services is best served by distinguishing between ecological features or processes and the ecological endpoints that are intended to represent or serve as proxies for the goods and services that people value (Wainger et al. 2001; Boyd and Banzhaf 2007; Kroeger and Casey 2007). Ecological endpoints can be measured through a variety of indicators (CENR 1999; CFEPTAP 2004; UNESCO 2006), some of which are more meaningful to nonscientists than others (Leschine et al. 2003; Newburn et al. 2005; Koontz and Bodine 2008).

Consider an example. What is the value of a natural process (nutrient cycling) or input (vegetative cover) that leads to cleaner water? This question can only be answered by placing a value on the clean water itself—the ecological endpoint. If you ask people, How valuable is nutrient cycling? they will first respond by asking, What is nutrient cycling? Once that is explained, they will answer, "The value of nutrient cycling is the clean water it makes available to me." Only the clean water has direct value to people. Thus the value of nutrient cycling must be derived from the value of the clean water provided. Clean water is the socially relevant endpoint of the underlying biophysical process.

When measuring ecological endpoints, the goal is to come as close as possible to the point at which people reveal virtual prices through choice. Another way of stating this is that the final biophysical units used in an ecosystem quantity index should be the ecological features, quantities, and qualities that are directly combined with other (nonecological) inputs to produce market and nonmarket benefits. Such indicators differ from monetary measures in that a system of indicators can be made flexible to allow inclusion of metrics that have a weight of evidence behind them, even if the exact relationship between the indicator and human well-being cannot be quantified. Assigning monetary values, in contrast, typically requires a tightly constrained relationship between a biophysical change and an environmental (and economic) outcome. Even when researchers allow individuals to draw their own conclusions about the values of environmental outcomes by using surveys or other stated preference techniques, it is considered good practice to ask respondents to associate a value with a quantitative outcome rather than a biophysical process, to promote rational choices on the part of respondents (e.g., Johnston et al. 2002).

Ecological endpoints, while potentially more flexible than monetary measures, must still be linked, by cause and effect, to actions we can take to improve or preserve those endpoints. The causal relationship between action and outcome (the "proof") may rely on no more than a conceptual model, but without some connection between action and outcome, we cannot even pretend to make a studied analysis of trade-offs and the "right decision."

Endpoints are the quantity and quality measures that are ultimately weighted or "valued" via economic assessment tools. In order to attach virtual prices to these endpoints, the endpoints should represent the same kinds of characteristics that consumers consider in a market setting: quality, scarcity or substitutability, and the reliability of the goods or services produced (see box 6.2). In some cases, ecological endpoints alone can act as rough proxies for the values of ecosystem services, such as water quantity measures (e.g., Postel and Thompson 2005). A purely biophysical measure tends to work well when the endpoint (water produced in an arid area) has clearly understood implications for conventionally valued outputs (crops, hydroelectric power, and drinking water). In other cases, biophysical metrics may represent more-complex measures or modeled outputs that reflect more information about scarcity of the service or irreversibility of decisions.

For example, "irreplaceability" is a metric favored by those aiming to protect biodiversity (Ferrier et al. 2000; Lawler et al. 2003). High irreplaceability is found where rare species, that are not found elsewhere, are present and where high species richness occurs. This metric therefore measures the quality of an area for promoting use and nonuse services derived from preserving biodiversity and includes an acknowledgment that such services are scarce. In general, the selection of indicators that are proxies for price is driven by theoretical and empirical economic research, even though such research is not replicable for every decision a resource manager must make.

Not every biophysical metric is an appropriate proxy for value, but well-designed indicator systems will select the most relevant for making trade-offs. Useful metrics include ones

Box 6.2. Analysis of Scarcity, Substitutability, and Complements

In order to determine what is most desirable and important about nature (noting again that *everything in nature is important*), it is imperative to examine the following:

- The scarcity of the natural feature or quality
- The ability to substitute other things for those natural features or qualities if they are not available in a particular place at a particular time
- The role of other, complementary social or natural features

The endpoints we have described empower this kind of analysis.

Consider two physically identical beaches eligible for cleanup and restoration and a budget that allows only one to be cleaned. Which should be cleaned up first? For the example, make the simplifying assumption that the only reason we care about the beaches is for their recreational potential. If the two beaches are physically identical, should we just flip a coin? No, for the following reasons: First, if one beach is the only one available to a population of residents (i.e., it is scarce), whereas the other beach is one of many beaches readily available to a similar population of residents, then cleaning up the beach where beach recreation is scarce will generate greater benefits. The importance and desirability of the scarce beach—all else being equal—will be greater than the desirability of the beach that is one of many.

Second, visitors may think of beaches as one of many recreation outlets. A day in the sun may also be enjoyed on a river or lake, for example. The more such substitute natural areas are available, the less important and desirable is the beach—all else being equal.

Finally, for some beach activities, complementary features affect the beach's desirability for recreational activities. Is there access to the beach, for example, in the form of roads, trails, or piers. Anglers will look for close offshore sportfish populations. For many people, the desirability of the beach is enhanced by the ability to get to it easily and enjoy other natural features while also enjoying the beach. Therefore, trade-offs are informed by considering ecological qualities (clean safe water, sportfish availability) as well as the preferences and needs of users as they choose where and how to recreate.

that reflect the quantity of goods and services produced, the ability of users to access the services, and the relative importance of different services to users when risk of service disruptions are taken into account. Such concepts of the need for goods in sufficient quantity, the need for complementary goods to allow use, and preferences that inform relative value are similar to those that might be assessed in an economic analysis.

Difficulties Likely to Be Encountered

The endpoints that are most desirable in theory may be among the most difficult to measure in practice. For example, it is relatively easy to measure atmospheric carbon. However, this is not a theoretically ideal endpoint, because the implications of atmospheric carbon levels to average people are not intuitive or directly meaningful. Desirable climate-related endpoints include location-specific temperature, species abundance, water availability, coastal damages avoided, and so on. Society uses atmospheric carbon as a rough proxy for these kinds of outcomes, however, because it is the cost-effective thing to measure.

Similarly, an indicator that may be applied to ecosystem-based management is an aquatic macroinvertebrate community index (Plafkin et al. 1989). But this type of indicator presents problems of communication and thus valuation. This index is considered highly useful for understanding a system, because it reflects food availability for fish and birds, indicates level of chronic stress on all aquatic organisms, and may be correlated with abundance of fish species or fish diversity. However, these relationships are not immediately obvious to non-ecologists when a macroinvertebrate index is reported, nor is it immediately apparent how this index supports decisions.

To fit within an economic trade-off analysis, the index must be interpretable as a measure of value. For example, we would ask, Can this indicator be linked to sportfish density or probability of a sportfish encounter in a recreational setting? (e.g., McConnell and Strand 1989; Freeman 1995) because this is an outcome more likely to be recognized as valuable. Or, if that particular connection cannot be made, the index could be linked to nonuse values, such as those that people hold for just knowing a certain assemblage of fish exists, even if they don't use that system as a source of recreation or food (Johnston et al. 2007). Although it may not be scientifically possible to link an indicator such as macroinvertebrate community structure to sportfish abundance, an indicator may still be useful, particularly if it helps to prescribe appropriate actions (Dale and Beyeler 2001).

To support decision making, the index scores would ideally be put into a context of demand, which includes consideration of scarcity and substitutability, so that managers could see what types of opportunities the indicator represents for management action. Managers might want to act to protect areas with high macroinvertebrate indices if they believe that such sites are increasingly scarce and serve as sources of biodiversity. For example, consider a threatened site that is one of only a few sites in the region with a high index score (indicating high macroinvertebrate diversity). Consider also that the site may be downstream of a large tract of protected land and therefore well situated to reliably produce this service into the future. In this case, preserving this high-scoring area might contribute to human well-being by preserving the option to restore habitat for sportfish and other aquatic species in the future, if other stressors can be managed. In this manner, managers may use indicators of ecological integrity or resilience to suggest appropriate trade-offs under certain conditions.

Proxy indices are important because they economize on information costs and can act as signals of a range of natural outcomes. However, the availability and use of proxies should not distract from the basic point: What society cares about is not macroinvertebrate community structure or atmospheric carbon, but rather the end results in intuitive terms of changing those conditions in specific places at specific times. We also note that statutes, regulations, and existing management practices may mandate the use of particular proxy-like endpoints that do not conform to the ideal. Here again, the point is not to jettison such proxy endpoints but to develop a better understanding of how those proxies relate to what society really cares about.

Guidance on Selecting Indicators

As described in the prior examples, ecological endpoints must be linkable to things people recognize as valuable. In addition, they must reflect quality, quantity, scarcity, and reliability of a resource in order that changes in indicators may eventually be related to value. Lastly, changes in the endpoints should be causally related to aspects of a system that can be managed.

To reflect quality, quantity, scarcity, or reliability requires that ecological endpoints be measured within appropriate space- and time-specific contexts (e.g., Popp et al. 2001). Available water matters to a farmer, but only if it is available in a particular place at a particular time. Thus, ecological endpoints must usually be highly disaggregated. From an economic perspective, it is meaningless to say that there are a billion acre-feet of subsurface water in the United States. We need to know where that water is and when it is there.

It is important to do more than document the decline of ecosystems. Instead, we must provide information that informs a management response. Our recommendations are generally similar to those of UNESCO (2006, citing Waltz 2000 and others) that suggested indicators should have the following properties (among others): (1) be meaningful to external audience, (2) have an agreed, scientifically sound meaning, (3) be representative of an environmental aspect of importance to society, and (4) assist decision making by being efficient and cost-effective to use.

Social values inevitably enter into the selection of ecological endpoints, but the assessment of how people value a service or the level of benefits they derive from that service is otherwise separate from the biophysical assessment. It is important to emphasize that economists aren't authorized to define endpoints and then turn around and demand that natural science cough up the answers. Rather, the natural and social science—with support and input from both science and government—should collectively debate and define these endpoints (see Kaufman et al., chap. 7 of this volume, for a related discussion).

Indicators developed to explore scientific questions may not be the ideal indicators to use to address management questions. In reviewing possible indicators used to judge environmental quality in coastal and estuarine systems, Salas and colleagues (2006) identify many of the indicators ecologists use to assess system condition. These indicators obviously have many valid scientific uses. However, from a management perspective, they can be difficult to apply, because of how they are scored. Indicators are frequently scored in reference to an ideal "pristine" condition: The pristine condition is "good," and the more intensively used systems are "poor." While such ratings are useful for characterizing habitat for species that avoid human contact or that are intolerant of human disturbance, such a classification is limited for evaluating other types of services, particularly those enhanced by having access by people. Thus, such scoring is not necessarily related to the ability of the system to produce a full range of ecosystem services.

Such indicators may also provide little guidance to those seeking to identify the systems most able to recover from disturbance (i.e., opportunities for restoration). If systems where people live never score well on an indicator, we cannot expect to improve that indicator without removing the people, which is not often a management option. Resilience indicators, on the other hand, potentially offer more appropriate information on which to base restoration decisions (see Leslie and Kinzig, chap. 4 of this volume).

Salas and colleagues (2006) provide an example of how indicators can be selected to represent functional quality and the biophysical conditions under which specific indicators are appropriate (e.g., NRC 2000; Niemi and McDonald 2004; Niemi et al. 2004). Such indicators are potentially useful for characterizing ecological endpoints related to nonuse and some use services. The determination of which indicators are appropriate for which locations must be effectively addressed and must include a determination of whether the indicators represent, and are sensitive to, the nature of the disturbance in the local ecosystem and whether they correct for natural environmental gradients (e.g., changes with latitude). Such "regionalization" of indicators, as it is sometimes called, presents a scientific question that reflects on how well the indicators accurately reflect proxies for services (Polasky and Dale 2007).

Linking Ecological Endpoints to Social Values

To develop a framework of indicators, it can be helpful to separate the selection of ecological endpoints and economic evaluation into two phases of assessments. In the first, the quality and quantity of a "service-producing" ecological endpoint is evaluated and reported using biophysical endpoints. In the second phase of assessment, the values associated with the ecological endpoints are assessed by considering the supply of and demand for a particular endpoint outcome. Value can be measured using either dollar values or nondollar metrics. Keeping the ecological evaluation distinct from the economic valuation allows the knowledge and theoretical foundations developed in ecology and economics to be properly coordinated.

The Production Function Approach

To create relationships that quantify services, and eventually benefits, two types of functional relationships are used. One or more functional relationships are needed to predict how features of the system (especially those that are potentially controllable by a resource manager, such as watershed land use) are related to the ability of the system to *supply* a meaningful ecological endpoint. Another functional relationship is required to link one or more ecological endpoints to *demand* for those services. The relationship between supply and demand, within a particular space–time context, allows benefits to be assessed.

For example, to relate the biophysical status of a fish population to the value of recreational fishing days, we need two things. First, we need to relate a range of environmental conditions (ideally manageable conditions), such as watershed land use, water pollution, or fishing effort, to the current and projected future fish population status. Second, we need relationships linking that population status and location characteristics to the values people hold for different services derived from that population, such as recreational fishing and existence values for species or ecosystems.

Therefore, the two types of functions needed to inform an analysis of ecosystem service value can be described as supply and demand functions:

Ecological production functions (supply of ecological goods) relate various biophysical features and qualities to ecological endpoints that serve as proxies for ecosystem services. They can be used to predict changes in the quality, quantity, or reliability of biophysical functions and outputs based on the quality and quantity of ecosystem features and biophysical inputs. In cases where endpoints are under threat or becoming increasingly scarce, models are most useful.

Economic demand functions relate characteristics of the user population and site context to the value people place on a service (e.g., willingness to pay), including the costs of accessing the service, availability of complements, and the regional scarcity and substitutability of the good or service.

The conceptual links between these functions and outputs are shown in figure 6.1.

To develop ecological production functions that measure the quantity and quality of services produced, we must relate measurable features of ecosystems (e.g., land cover) to the quality and quantity of subsequent ecological endpoints. The causal linkages between fundamental inputs and ecological outputs (the endpoints) are described by the ecological production function. The things we can readily measure about an ecosystem are its features. These are the attributes of systems that can be directly observed, described, and quantified, such as type of vegetation, distance to the stream, and type of surrounding land use. The features of the ecosystem produce functions of a given quality (rate of pollution assimilation) to produce a given quantity of an output (gallons of clean water).

Thus, ecological functions are the processes that generate desirable biophysical outcomes. Typically, a biophysical model is required to relate features to ecological outcomes (e.g., to relate a change in water quality to fish community changes), although processes may sometimes be measured directly. The quantity of the service outputs depends on both the areal extent of the system and the per-acre productivity. Of course, productivity can vary as a function of management and technological advances,

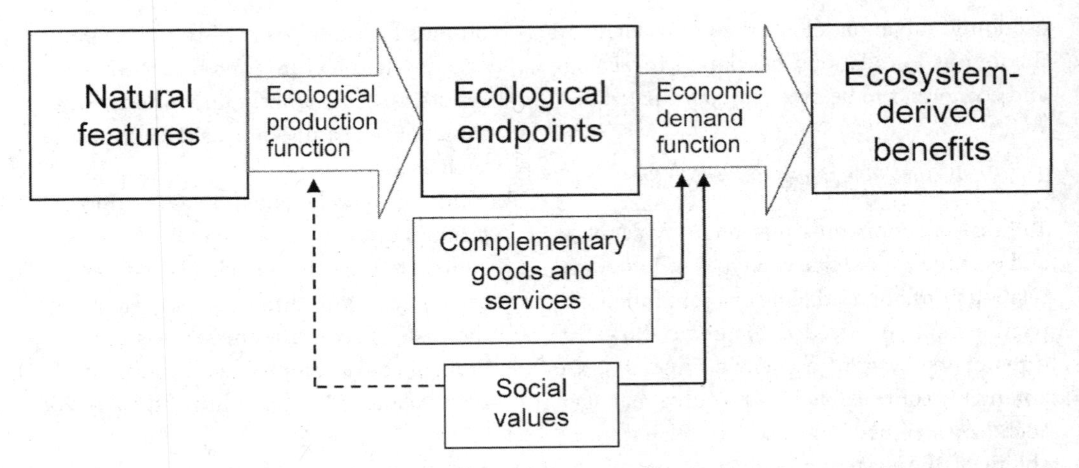

Figure 6.1 Analysis framework of benefits derived from natural systems. *Natural features* are readily observable characteristics of natural systems, such as type of vegetation and arrangement of land use. An *ecological production function* relates the type and quantity of natural features required to generate biophysical outputs, which we call *ecological endpoints* and which measure the ecological capacity to supply ecosystem goods and services. The ecological endpoints are process outputs or stocks of materials that may be valued directly (biodiversity) or are intermediate goods and services that have the potential to be valuable (groundwater storage). The *economic demand function* incorporates human preferences for the goods and services, considering costs of access, scarcity or substitutability, and reliability of those goods and services. Demand cannot be assessed directly when ecological endpoints do not represent the final good or service being demanded by users. In such cases, the demand analysis must include evaluation of a *technological production function* to relate ecological endpoints and the availability of complementary goods and services to the final ecosystem goods and services that people know they value. In the final step of the economic assessment, benefits can be valued in monetary terms or quantified with benefit indicators. Note that the dashed arrow from social values to the ecological production function reflects the role of human preferences in the choice of many ecological endpoints, while the solid arrow from social values to the economic demand function reflects the direct use of preferences in analyzing demand.

which requires making assumptions in order to anticipate changes. This distinction between features and functions makes explicit that we are using models (and embedded assumptions/hypotheses/theories) to estimate the quantity and quality of endpoints delivered.

For some ecosystem goods and services, the demand function will need to reflect output from a *technological production function* that relates the influence of complementary inputs (e.g., trails and piers) on demand for a particular environmental feature (e.g., open space). This function would also capture substitutability between ecological and nonecological inputs (e.g., the degree to which built water filtration is a substitute for natural filtration processes). For marketed goods, the costs of complementary and substitute goods would be a component of the demand function. A creative approach to demand analysis is required because we will not have prices for the complements and substitutes but can infer their availability from location context.

In the benefit assessment exercise, the ecological endpoints are related to demand or potential demand for the ecosystem goods and services. Demand is a function of who might be using the services and their preferences.

Consider a model for the value of recreational fishing. This model is likely to include aspects of accessibility (How long does it take most people to reach the site?), complementarity (Is the site near other amenities?), availability of substitutes (How many other fishing sites are available within the same travel distance?), congestion levels, and other characteristics that have been identified to contribute to preferences and values of sites (Smith et al. 1983; Bockstael et al. 1987; Bockstael et al. 1989; English et al. 1993). All these demand-side characteristics are measured to depict the benefits of fish abundance—the ecological endpoint. Ecological models can be used to assess how management improves fish abundance, but economic models are needed to assess how much people care—usually in a location-specific manner.

Table 6.1 shows the basic questions and associated metrics that could be developed as part of the production function approach. Each column in the table builds on results from the column to the left to advance an assessment of benefits. Note that the biophysical endpoints are related to benefits via both a technological production function (column 4) and a demand analysis (column 5). The change in social benefits then depends on how people respond or adapt to that change given characteristics of the service and region.

Can We Produce Ecological Production Functions?

The construction of ecological production functions is often stymied by a lack of comprehensive knowledge of biophysical relationships (De Leo and Levin 1997; Kremen 2005). This limitation is understandable. First, nature is a highly complex and nonuniform system (May 1973). Complexity means that causal relationships can frequently only be tested using rigorous, data-intensive empirical and scientific methods. Such methods are often difficult and costly. Nonuniformity means that even if you establish the causal relationship in one location, that relationship need not hold in any other location. Second, empirical analysis of causality requires collaboration between different disciplines (ecology and hydrology, for example). Cross-disciplinary collaboration in any scientific inquiry is always a practical barrier. Third, the biophysical sciences have many other things to study and have limited financial support to do all they are asked to do. If natural science has to-date failed to empirically determine these ecological production functions to our satisfaction, that failure should be viewed in the most sympathetic terms.

Ecologists often see a conflict with creating functional relationships that are both relevant and consistent. They might argue that many functions are highly site-specific and the variables used to measure the qualities of these functions would depend on the specific system being evaluated. For example, Wainger and colleagues (2007) sought to evaluate the equivalency of nutrient reduction strategies (i.e., best management practices [BMPs]) at different locations within a large watershed. The study sought relationships that would quantify the nutrients prevented from reaching the river mouth as a function of where a given BMP was installed or adopted. A functional relationship was needed to judge equivalency between management actions because nutrient trading was being considered as a means to manage costs of reducing nutrient runoff. It was important that each credit traded in the market represent an equivalent reduction in nutrients to the receiving water body.

In this case study, surprisingly little consensus among researchers was found regarding which characteristics could be measured (with available GIS [geographic information system] data) that would consistently indicate a site's ability to retain nutrients. Many factors had been shown to be important (Lowrance 1992; Litke 1999; Baker et al. 2006). However, capturing most of these variables

Table 6.1. Structure to trace changes in ecosystem service values resulting from land use or management change

Land use / management change	*Ecosystem feature change (via landscape distribution model)*	*Biophysical endpoint change (via ecological production function)*	*Ecosystem service change (via technological production function)*	*Social benefits change (via demand analysis)*
Cropping of marginal (previously fallow) farmland	Lower proportion of remnant natural vegetation; increased fragmentation	Fewer beneficial insects supported	Lower harvests and/or increased costs of crop production	Reduced producer profits
		Increased nutrient and sediment runoff to waterways linked to reduced fishery recruitment	Reduced fishery harvest	Loss of recreational benefits or producer harvests
		Reduced abundance of hunted species	Reduced hunting	Loss of hunting revenues; loss of recreational benefits
		Reduced abundance of common bird species	Reduced bird-watching	Loss of recreational benefits
		Diminished populations of rare species	Reduced population viability	Loss of existence/bequest benefits
Increased application of pesticides	Site-specific changes in water chemistry	Increased concentrations of toxins in drinking water sources	Increased health risks to humans	Lost worker productivity; lost individual benefits related to health outcomes

required detailed site measurements of hydrology and other characteristics. Yet, a trading system that required on-site monitoring of every site offering credits would likely be too expensive to implement. The study eventually relied upon statistical modeling that demonstrated that three variables explained a large proportion of the variance between watersheds (rather than between sites) and could reliably be used to compare nutrient concentrations in receiving streams (Jordan et al. 2003; Weller et al. 2003). However, it was unclear how well present-day nutrient runoff relationships used in the statistical models captured the effects of future land use or land management changes.

Much work remains to be done to facilitate the use of ecosystem service concepts in management decisions. There may be no absolute rules. Yet, if we cannot empirically detect and test functional relationships, environmental protection and intervention will remain a faith-based activity, relying on guesswork, rules of thumb, and hope that we get things right. What is needed is a scientific and social commitment to support that kind of difficult analysis and the type of monitoring needed to produce relevant analyses.

Toward Measuring Services: Ecological Endpoints

Given that nature is a highly complex and nonuniform system, if we want to depict the myriad ways in which nature benefits us, isn't the challenge we have presented too great to be taken seriously? In the short term, the challenge is immense. However, ignorance cannot be the preferred alternative.

So how should natural science focus and make the most constructive contribution to public policy? Our answer is that ecology should focus on the causal connections between practical social actions (preservation, management, regulation) and specific environmental outcomes that can be quantified with ecological endpoints (Boyd 2007a, b). To consistently communicate to nonscientists that what is being measured is important to their well-being, such endpoints are best described in terms of metrics that measure quality, quantity, scarcity, and/or reliability of endpoints that can be related to goods and services people value.

The complexity of the functional relationships necessary to inform production functions will vary greatly depending on the biophysical complexity of relationships and the proposed management actions, among other things. Bioeconomic models are tools that capture different levels of complexity, feedbacks, dynamics, and potential outcomes of proposed actions and incorporate ecological functions, processes, and economic actors (Barbier 1994; Popp et al. 2001; Sanchirico 2004; Finnoff et al. 2005; Wilson et al. 2006; Brookshire et al. 2007). In other cases, expert judgment, carefully elicited, may be used to provide guidance in lieu of complex models (Johnston et al. 2002). However, in the interests of cost-effectively evaluating management options at many diverse locations or across broad regions, static tools that apply indicators that can be measured and modeled with relative ease are often used (e.g., Binning et al. 2001).

Given that managers do not appear to be widely using and adapting existing economic models (Smith 1996; Newburn et al. 2005), the question becomes, How can tools be simplified or previous studies integrated to address decision makers' needs? A problem with many current models is that their ecological complexity may be either insufficient to capture unintended consequences of a proposed management action or too complex to allow for economic comparison of options. Finding an intermediate level of complexity that captures important dynamics, yet does not make management comparisons intractable, is an art as much as a science that must be further developed to make progress (Grimm et al. 2005).

Conclusions

Managers face tremendous challenges to improve management of marine ecosystems through ecosystem-based management. The state of coastal ecosystems reflects a complex interplay of outcomes resulting from decisions about fishing regulations, land use, and land management, among others. In particular, the challenges of coordinating the activities of so many actors that can influence aquatic ecosystems are daunting. The concept of ecosystem service accounting, while no panacea, is useful for bringing together both technical and social information. It has the potential to move beyond cost–benefit-type accounting and provides a context for disparate stakeholders to focus on shared goals and work at cross-jurisdictional scales to achieve ecosystem-based management.

Unfortunately, economic analysis of ecosystem services has a long way to go if it is to effectively—and practically—support managers. We must make progress both in understanding ecological production functions and in establishing acceptable methods for using alternative value metrics to reflect a wide range of services that cannot be monetized. Nevertheless, the measurement of endpoints that reflect human values is a useful first step toward informing decisions and the difficult trade-offs we must make in the use of our resources in political negotiations, regulations, legal decisions, and personal choices. If appropriately constructed, they will be helpful for making informed choices about risks of different actions.

We propose a framework that separates ecological and economic evaluation to ensure consistency with the theoretical underpinnings of both fields. The goal of maintaining a rigorous conceptual framework in which to measure ecosystem services is to (1) be clear about what we're trying to measure at different points in the analysis, (2) reveal assumptions and uncertainties, and (3) communicate effectively with nonscientists. Ecological endpoints can provide a reasonable proxy for economic value related to loss or degradation of natural resources if developed based on sound conceptual models. Such models must be made explicit and be open to scrutiny in order to effectively communicate the rationale for using proxies. To achieve a mutually acceptable framework will require improved collaboration among ecologists and economists, including willingness from both sides to be flexible in merging these two distinct fields.

How Can We Incorporate Concepts of Resilience?

Using endpoints that integrate ecological and economic concepts is one response to the lack of tools to monetize ecosystem benefits. They can be more cost-effective to measure than dollars in most cases, and they are practical for management. However, they may fail in an important respect; they may not prevent us from crossing important thresholds or destroying our life-support system through "death by a thousand cuts."

The perspective embedded in this approach largely mimics that of economic valuation, which assumes that we can make rational decisions by evaluating changes at the margin (e.g., the effect of converting a single parcel of land) and demonstrating that some services should be preserved at the expense of others. However, each individual decision to change land cover or land management will barely register an effect on public ecosystem services (vs. private gains) as measured through endpoints or dollars. Superlative sites are the exception. These are most easily characterized with indicators and monetary metrics as potentially valuable and worthy of trading off short-term use to preserve ecosystem services. More mundane and abundant sites are not as easily associated with high value. Therefore, this approach needs further development to allow responses to cumulative impacts to be evaluated, particularly once a system has become degraded.

However, the concept of resilience presents an opportunity to broaden the approach to embrace uncertainty and risk (see also Leslie and Kinzig, chap. 4 of this volume). Developing appropriate resource management techniques involves managing risk, and risk management offers a framework for integrating ecological endpoints with the value of risk reduction (Smith 1996; Shogren 2000). For a system to work, however, decisions are needed on how much risk groups are willing to accept for themselves and for future generations. Such risks will need to be debated within the context of both ecological and economic conditions, including the willingness and capacity of service users to accept risk.

This concept of valuing resilience for its ability to reduce the risk of future loss of ecosystem services, that is, "buying insurance," may be extraordinarily difficult to implement despite the fact that environmental economics has long recognized the concept through the use of option and quasi-option values (Arrow and Fisher 1974; Bishop 1982; Smith 1996).

Implementing option value or other risk management strategies has many impediments, not the least of which is coming up with a rational system for assessing risks. As Carpenter and colleagues (2006) summarize, "The most catastrophic changes in ecosystem services identified in the [Millennium Assessment] involved nonlinear or abrupt shifts. We lack the ability to predict thresholds for such changes, whether or not a change may be reversible, and how individuals and societies will respond." In other words, we cannot expect to have excellent information on the nature and extent of the risks we face from failing to preserve ecosystems. However, we may be able to assess our tolerance for uncertainty and catastrophic loss in order to improve the debate on how precautionary our policies should be.

Examples of frameworks where managers chose to manage risk rather than maximize benefits are widespread (Long and Fischhoff 2000; Morgan et al. 2000; Andrews et al. 2004; DeKay et al. 2004; Willis et al. 2004). What is less clear is whether frameworks that are not well-grounded in economic principles are successful at driving policy (Newburn et al. 2005). A framework to actively examine how management decisions are reducing the resilience of ecosystems would clearly need much work to evaluate trade-offs.

Resilience reflects the ability of the system to withstand perturbation and continue to provide services (Leslie and Kinzig, chap. 4 of this volume and references therein). However, maintaining resilience can appear expensive from the perspective of the current users of the system. Ensuring resilience of a relatively undisturbed ecosystem often involves making trade-offs among different user groups in the present, as well as between current and future generations. Such decisions are fraught with difficult choices about who will be the winners and who the losers. Resilience is often used as a proxy for sustainability, but what we are sustaining and for whom is not so easy to answer. Therefore, the degree of risk aversion and the capacity of affected groups to withstand risk must be assessed in order for resilience to be used in decision making.

Moving Forward

What gets measured get managed, goes the saying. Comprehensive measurement and tracking of gains and losses in appropriate ecological endpoints will allow us to capture aspects of the qualities and features of nature that are most directly important to society. Even if we stopped at the measurement of ecological endpoints, that would be a great advance for the fundamental aspiration of economic analysis—to have all nature's benefits count somehow when we make public policy decisions.

The proposed framework embodies a key concept of EBM, that we must manage multiple interconnected resources simultaneously, rather than trying to maximize a single good or service. The payoff of managing natural systems as producers of multiple services is that a higher level of goods and services can typically be generated for the same level of investment (Popp et al. 2001; Nelson et al. 2008). Many investments in natural assets will support a variety of services, but antagonistic services (ones that cannot be produced simultaneously in the same location) must be traded off at a particular location. However, antagonistic services can be accommodated by finding alternative sites best suited to produce them, because those sites minimize harm and the opportunity cost of foregoing other services.

Further work is needed to create a commonly accepted reusable framework in which economic assessments or trade-off analysis can be conducted cost-effectively. The approach of using indicators to reduce the time and effort needed to promote decisions that reflect societal goals is one approach to making use of past economic research. A great deal of useful theoretical and empirical economic analysis could be applied in decision making if it could be made more generalizable, accessible, and practical to use. Clearly, much work remains to achieve a better match between what managers use and what economics produces. A useful decision framework could use indicators or monetary values, but it needs to reduce the burdens on natural resource managers while simultaneously serving to improve decisions, set acceptable levels of risk, and clarify societal goals.

For all the debate over economics and ecology, it is the absence of ecological production functions that is often the limiting factor to public policy analysis of nature. Economics can help rank, weight, prioritize, and choose actions, but only if we know the consequences of those actions. We emphasize that the development of endpoints and the natural science necessary to depict causal connections between our choices and those endpoints is the single most important thing we can do.

References

Andrews, C. J., D. M. Hassenzahl, and B. B. Johnson. 2004. Accommodating uncertainty in comparative risk. *Risk Analysis* 24(5):1323–35.

Arrow, K. J., and A. C. Fisher. 1974. Environmental preservation, uncertainty and irreversibility. *Quarterly Journal of Economics* 88:312–19.

Baker, M. E., D. E. Weller, and T. E. Jordan. 2006. Improved methods for quantifying potential nutrient interception by riparian buffers. *Landscape Ecology* 21:1327–45.

Barbier, E. B., E. W. Koch, B. R. Silliman, S. D. Hackery, E. Wolanski, J. Primavera, E. F. Granek et al. 2008. Coastal ecosystem-based management with nonlinear ecological functions and values. *Science* 319:321–23.

Barbier, E. B. 1994. Valuing environmental functions: Tropical wetlands. *Land Economics* 70(2): 155–73.

Binning, C., S. Cork, R. Parry, and D. Shelton. 2001. *Natural assets: An inventory of ecosystem goods and services in the Goulburn broken catchment.* Report of the Ecosystem Services Project. Canberra, Australia: Commonwealth Scientific and Industrial Research Organisation (CSIRO) Sustainable Ecosystems.

Bishop, R. C. 1982. Option value: An exposition and extension. *Land Economics* 58(1):1–15.

Bockstael, N. E., K. E. McConnell, and I. E. Strand. 1989. Measuring the benefits of improvements in water quality: The Chesapeake Bay. *Marine Resource Economics* 6(1):1–18.

Bockstael, N. E., W. M. Hanemann, and C. L. Kling. 1987. Modeling recreational demand in a multiple site framework. *Water Resources Research* 23:951–60.

Boyd, J. 2007a. *Counting nonmarket, ecological public goods: The elements of a welfare-significant ecological quantity index.* Resources for the Future discussion paper. www.rff.org.

Boyd, J. 2007b. The endpoint problem. *Resources* Spring 2007.

Boyd, J. 2006. The nonmarket benefits of nature: What should be counted in green GDP? *Ecological Economics* 61:716–23.

Boyd, J., and S. Banzhaf. 2007. What are ecosystem services? *Ecological Economics* 63(2–3):616–26.

Bromley, D. W., ed. 1995. *Handbook of environmental economics.* Oxford: Blackwell.

Brookshire, D. S., L. A. Brand, J. Thacher, M. D. Dixon, K. Benedict, J. C. Stromberg, K. Lansey et al. 2007. Integrated modeling and ecological valuation: Applications in the semi arid southwest. In *Valuation for environmental policy: Ecological benefits.* Proceedings of a workshop sponsored by the US Environmental Protection Agency (EPA) National Center for Environmental Economics, 23–24 April 2007, Arlington, VA. http://yosemite.epa.gov/ee/epa/eermfile.nsf/vwAN/EE-0505-04.pdf/$File/EE-0505-04.pdf.

Carpenter, S. R., R. DeFries, T. Dietz, H. A. Mooney, S. Polasky, W. V. Reid, and R. J. Scholes. 2006. Millennium Ecosystem Assessment: Research needs. *Science* 314:257–58.

CENR (Committee on Environment and Natural Resources). 1999. *Ecological risk assessment in the federal government.* Washingon, DC: National Science and Technology Council, Executive Office of the President.

CFEPTAP (Chesapeake Fisheries Ecosystem Plan Technical Advisory Panel). 2004. *Fisheries ecosystem planning for Chesapeake Bay.* Annapolis, MD: NOAA Chesapeake Bay Office.

Chan, K. M. A., R. Shaw, D. Cameron, E. C. Underwood, and G. C. Daily. 2006. Conservation planning for ecosystem services. *PLoS Biology* 4:e379.

Daily, G. C., ed. 1997. *Nature's services: Societal dependence on natural ecosystems.* Washington, DC: Island Press.

Dale, V., and S. C. Beyeler. 2001. Challenges in the development and use of ecological indicators. *Ecological Indicators* 1:3–10.

DeKay, M. L., H. H. Willis, M. G. Morgan, H. K. Florig, and P. S. Fischbeck. 2004. Ecological risk ranking: Development and evaluation of a method for improving public participation in environmental decision making. *Risk Analysis* 24:363–78.

De Leo, G. A., and S. Levin. 1997. The multifaceted aspects of ecosystem integrity. *Conservation Ecology* 1(1):3. http://www.ecologyandsociety.org/vol1/iss1/art3/.

Desvousges, W. H., M. C. Naughton, and G. R. Parsons. 1992. Benefit transfer: Conceptual problems in estimating water quality benefits using existing studies. *Water Resources Research* 28(3):675–83.

English, D. B. K, C. J. Betz, M. Young, J. C. Bergstrom, and H. K. Cordell. 1993. *Regional demand and supply projections for outdoor recreation.* USDA Forest Service General Technical Report RM-230. Fort Collins, CO: Rocky Mountain Forest and Range Experiment Station.

Ferrier, S., R. L. Pressey, and T. W. Barrett. 2000. A new predictor of the irreplaceability of areas for achieving a conservation goal, its applicability to real-world planning, and a research agenda for further refinement. *Biological Conservation* 93:303–25.

Finnoff, D., J. F. Shogren, B. Leung, and D. Lodge. 2005. The importance of bioeconomic feedback in invasive species management. *Ecological Economics* 52:367–81.

Freeman, A. M., III. 1995. The benefits of water quality improvements for marine recreation: A review of the empirical evidence. *Marine Resource Economics* 10(4):385–406.

Grimm, V., E. Revilla, U. Berger, F. Jeltsch, W. M. Mooij, S. F. Railsback, H.-H. Thulke et al. 2005. Pattern-oriented modeling of agent-based complex systems: Lessons from ecology. *Science* 310(5750):987–91.

Heal, G., E. Barbier, K. Boyle, A. Covich, S. Gloss, C. Hershner, J. Hoehn et al. 2005. *Valuing ecosystem services: Toward better environmental decision making.* Washington, DC: National Academies Press.

Iovanna, R., and S. C. Newbold. 2007. Ecological sustainability in policy assessments: A wide-angle view and a close watch. *Ecological Economics* 63:639–48.

Johnston, R. J., E. T. Schultz, K. Segerson, E. Y. Besedin, J. Kukielka, and D. Joglekar. 2007. Development of bioindicator-based stated preference valuation for aquatic resources. In *Valuation for environmental policy: Ecological benefits.* Proceedings of a workshop sponsored by the US Environmental Protection Agency (EPA) National Center for Environmental Economics, 23–24 April 2007, Arlington, VA. http://yosemite.epa.gov/ee/epa/eermfile.nsf/vwAN/EE-0505-05.pdf/$File/EE-0505-05.pdf.

Johnston, R. J., G. Magnusson, M. J. Mazzotta, and J. J. Opaluch. 2002. The economics of wetland ecosystem restoration and mitigation: Combining economic and ecological indicators to prioritize salt marsh restoration actions. *American Journal of Agricultural Economics* 84(5):1362–70.

Jordan, T. E., D. E. Weller, and D. L. Correll. 2003. Sources of nutrient inputs to the Patuxent River Estuary. *Estuaries* 26(2A):226–43.

Koontz, T. M., and J. Bodine. 2008. Implementing ecosystem management in public agencies: Lessons from the US Bureau of Land Management and the Forest Service. *Conservation Biology* 22:60–69.

Kremen, C. 2005. Managing ecosystem services: What do we need to know about their ecology? *Ecology Letters* 8(5):468–79.

Kroeger, T., and F. Casey. 2007. An assessment of market-based approaches to providing ecosystem services on agricultural lands. *Ecological Economics* 64(2):321–32.

Lawler, J. J., D. White, and L. L. Master. 2003. Integrating representation and vulnerability: Two approaches for prioritizing areas for conservation. *Ecological Applications* 13(6):1762–72.

Leschine, T. M., B. E. Ferriss, K. P. Bell, K. K. Bartz, S. MacWilliams, M. Pico, and A. K. Bennett. 2003.

Challenges and strategies for better use of scientific information in the management of coastal estuaries. *Estuaries* 26(4):1189–1204.

Litke, D. W. 1999. *Review of phosphorus control measures in the United States and their effects on water quality.* US Geological Survey (USGS) Water-Resources Investigations Report 99-4007. Reston, VA: USGS.

Long, J., and B. Fischhoff. 2000. Setting risk priorities: A formal model. *Risk Analysis* 20(3):339–51.

Lowrance, R. 1992. Groundwater nitrate and denitrification in a coastal plain riparian forest. *Journal of Environmental Quality* 21:401–5.

May, R. M. 1973. *Stability and complexity in model ecosystems.* Monographs in Population Biology 6. Princeton, NJ: Princeton University Press.

McCauley, D. J. 2006. Selling out on nature. *Nature* 443:27–28.

McConnell, K. E., and I. E. Strand. 1989. Benefits from commercial fisheries when demand and supply depend on water quality. *Journal of Environmental Economics and Management* 17:284–92.

McLeod, K. L., J. Lubchenco, S. R. Palumbi, and A. A. Rosenberg. 2005. *Scientific consensus statement on marine ecosystem-based management.* The Communication Partnership for Science and the Sea (COMPASS). Signed by 221 academic scientists and policy experts with relevant expertise. http://www.compassonline.org/pdf_files/EBM_Consensus_Statement_v12.pdf.

Morgan, M. G., H. K. Florig, M. L. DeKay, and P. Fischbeck. 2000. Categorizing risks for risk ranking. *Risk Analysis* 20(1):49–58.

Nelson, E., S. Polasky, D. J. Lewis, A. J. Plantinga, E. Lonsdorf, D. White, D. Bael, and J. Lawler. 2008. Efficiency of incentives to jointly increase carbon sequestration and species conservation on a landscape. *Proceedings of the National Academy of Sciences* 105:9471–76.

Newburn, D., S. Reed, P. Berck, and A. Merenlender. 2005. Economics and land-use change in prioritizing private land conservation. *Conservation Biology* 19(55):1411–20.

Niemi, G. J., and M. E. McDonald. 2004. Application of ecological indicators. *Annual Review of Ecology, Evolution and Systematics* 35:89–111.

Niemi, G., D. Wardrop, R. Brooks, S. Anderson, V. Brady, H. Paerl, C. Rakocinski, M. Brouwer, B. Levinson, and M. McDonald. 2004. Rationale for a new generation of indicators for coastal waters. *Environmental Health Perspectives* 112(9):979–86.

NRC (National Research Council). 2000. *Ecological indicators for the nation.* Committee to Evaluate Indicators for Monitoring Aquatic and Terrestrial Environments. Washington, DC: National Academies Press.

Pagiola, S., K. von Ritter, and J. Bishop. 2004. *Assessing the economic value of ecosystem conservation.* World Bank Environment Department Paper no.101, October 2004. Washington, DC: World Bank.

Pearce, D. 1998. Cost-benefit analysis and environmental policy. *Oxford Review of Economic Policy* 14(4):84–100.

Plafkin, J. L., M. T. Barbour, K. D. Porter, S. K. Gross, and R. M. Hughes. 1989. *Rapid bioassessment protocols for use in streams and rivers: Benthic macroinvertebrates and fish.* US Environmental Protection Agency (EPA), Office of Water Regulations and Standards. Washington, DC: EPA. 440-4-89-001.

Plummer, M. L. 2009. Assessing benefit transfer for the valuation of ecosystem services. *Frontiers in Ecology and the Environment* 7(1):38–45.

Polasky, S., E. Nelson, E. Lonsdorf, P. Fackler, and A. Starfield. 2005. Conserving species in a working landscape: Land use with biological and economic objectives. *Ecological Applications* 15(4):1387–1401.

Polasky, S., and V. Dale. 2007. Measures of the effects of agricultural practices on ecosystem services. *Ecological Economics* 4:286–96.

Popp, J., D. Hoag, and D. Eric Hyatt. 2001. Sustainability indices with multiple objectives. *Ecological Indicators* 1(1):37–47.

Postel, S. L., and B. H. Thompson Jr. 2005. Watershed protection: Capturing the benefits of nature's water supply services. *Natural Resources Forum* 29:98–108.

Ready, R., and S. Navrud. 2005. Benefit transfer: The quick, the dirty, and the ugly? *Choices* 20(3):195–99. American Agricultural Economics Association.

Ricketts, T. H., G. C. Daily, P. R. Ehrlich, and C. D. Michener. 2004. Economic value of tropical forest to coffee production. *Proceedings of the National Academy of Sciences* 101:12579–82.

Rosenberger, R. S., and J. B. Loomis. 2001. *Benefit transfer of outdoor recreation use values: A technical document supporting the Forest Service Strategic Plan (2000 revision)*. Gen. Tech. Rep. RMRS-GTR-72. Fort Collins, CO: US Department of Agriculture, Forest Service, Rocky Mountain Research Station.

Salas, F., J. Patricio, and J. C. Marques. 2006. *Ecological indicators in coastal and estuarine environmental quality assessment*. Coimbra, Portugal: University of Coimbra Press.

Sanchirico, J. N. 2004. Designing a cost-effective marine reserve network: A bioeconomic metapopulation analysis. *Marine Resource Economics* 19(1):41–65.

Shabman, L., and K. Stephenson. 2000. Environmental valuation and its economic critics. *Journal of Water Resources Planning and Management* 126(6):382–88.

Shogren, J. 2000. Risk reductions strategies against the explosive invader In *The economics of biological invasions*, ed. C. Perrings, M. Williamson, and S. Dalmazzone. Cheltenham, UK: Edward Elgar.

Smith, V. K. 1996. Resource evaluation at a crossroads. In *Estimating economic value for nature*, ed. V. K. Smith. Cheltenham, UK: Edward Elgar.

Smith, V. K., W. H. Desvousges, and M. P. McGivney. 1983. The opportunity cost of travel time in recreation demand models. *Land Economics* 59(3): 259–78.

Spash, C. L., and A. Vatn. 2006. Transferring environmental value estimates: Issues and alternatives. *Ecological Economics* 6(2):379–88.

UNESCO. 2006. *A handbook for measuring the progress and outcomes of integrated coastal and ocean management*. IOC Manuals and Guides, 46; ICAM Dossier, 2. Paris: UNESCO (English).

Wainger, L. A., D. M. King, P. Hagan, T. Jordan, and D. Weller. 2007. *Establishing trading ratios for point–nonpoint source water quality trades: Can we capture environmental variability without breaking the bank?* Report to NOAA. University of Maryland Center for Environmental Science Technical Report no. TS-523-07. Cambridge, MD: University of Maryland.

Wainger, L. A., D. M. King, J. Salzman, and J. Boyd. 2001. Wetland value indicators for scoring mitigation trades. *Stanford Environmental Law Journal* 20(2):413–478.

Waltz, R. 2000. Development of environmental indicators systems: Experiences from Germany. *Environmental Management* 25(6):613–23.

Weller, D. E., T. E. Jordan, D. L. Correll, and Z.-J. Liu. 2003. Effects of land-use change on nutrient discharges from the Patuxent River watershed. *Estuaries* 26(2A):244–66.

Willis, H. H., M. L. DeKay, M. G. Morgan, H. K. Florig, and P. S. Fischbeck. 2004. Ecological risk ranking: Development and evaluation of a method for improving public participation in environmental decision making. *Risk Analysis* 24(2):363–78.

Wilson, K. A., M. F. McBride, M. Bode, and H. P. Possingham. 2006. Prioritizing global conservation efforts. *Nature* 440:337–440.

PART 3
Connecting Concepts to Practice

CHAPTER 7
Monitoring and Evaluation

Les Kaufman, Leah Bunce Karrer, and Charles H. Peterson

Up to this point this book has provided insights into the various components of marine ecosystem-based management (EBM), including recognizing the range of connections between people and ocean ecosystems, the implications of resilience thinking for marine EBM, and the role of economics. In this chapter we confront the pivotal question of how to *do* EBM. Currently there is no definitive prescription for moving marine EBM from planning to implementation. Here we highlight the relationships among EBM, adaptive management (AM), and system resilience and show how monitoring, research, and modeling provide the information necessary to enable management programs to move toward resilience and sustainability. We discuss challenges associated with translating scientific findings into usable information and tools to guide EBM and to improve management practices. We illustrate the process with a case study of the Marine Management Area Science Program, a project of Conservation International that is at the forefront of developing and applying these concepts to adaptive management of tropical nearshore ecosystems worldwide. A closing section offers thoughts about the role of scientists, both natural and social, in defining and implementing EBM.

Overview of Marine Ecosystem-Based Management, Adaptive Management, and Resilience

EBM emphasizes biological, social, and economic sustainability. However, it also encompasses a sharp divergence in worldview from the twentieth century paradigm of stability and tight control of the biosphere through human ingenuity. In its place is recognition of the living world as a nested and networked complex of systems that humanity can inhabit and adapt to through cycles of change, but cannot hold still (Levin 1999). Also new is the realization that, by virtue of their complexity, the precise effects of human behavior on ecosystems cannot be predicted well enough to operate solely on the basis of theory and computational models. Further reduction of uncertainty requires close observation and, wherever possible, manipulative experiments, to reveal the possible effects of our actions. The ultimate goal is no longer to keep an ecosystem in stasis, but rather to imbue social and ecological systems with sufficient resilience to emerge from new perturbations with basic components, structure, and services intact (Folke et al. 2005; Leslie and Kinzig, chap. 4 of this volume). For example, democratic government is argued to be more resilient than an authoritarian one because the ability to change rulers enables the citizenry's sense of things going amiss to be an important factor in the choices made by those rulers. Resilience has many different definitions, but all of them share the characteristic of capacity to resist or recover from perturbation without experiencing a fundamental change in state (Levin and Lubchenco 2008; Leslie and Kinzig, chap. 4 of this volume).

The evolving nature of EBM, therefore, dictates an adaptive approach. Management requires a process of learning and change that tracks the continuous transformations that define living, dynamic marine ecosystems. For the same reason, resilience in both the natural and social domains of a marine ecosystem is also an essential component of marine EBM. Investigators have focused on natural or social resilience individually, but the most important component of resilience may be the importance of recognizing the connections between human and natural systems (or what are often referred to as coupled social–ecological systems) and, in particular, the resilience of the adaptive policy process itself. Therefore, adaptive management and its relationship to resilience are the foundations of marine EBM, particularly given the increasing rate of global change and growing impacts of humans on ocean ecosystems (UNEP 2006).

EBM can proceed by one of two forms of adaptation: (1) passive adaptation, in which ecosystem change occurs spontaneously and managers react to it (e.g., fishery collapse results in a ban on fishing), and (2) active adaptation, in which managers proactively anticipate the need to change management practices, learn from experience, and adopt strategies accordingly (Walters 1986). From a scientific standpoint, active adaptive management is more conducive to scientific study because it allows for establishing pre-change baselines. The most scientifically rigorous form of active AM involves managers applying regulation in a form that can be treated as an experiment whose consequences can be analyzed. In other words, there are randomly assigned treatment and control areas, with independent replication of each type. This most extreme vision of AM is difficult to implement, because it likely will result in social inequities. Establishing different management actions in different regions almost always places unequal regulatory burdens on users, such as those who traditionally have used the treatment areas versus those who used the control areas. It is, therefore, critical during the conceptualization phase that researchers and other stakeholders identify and discuss the purpose of the regulations, what questions are to be answered, what uncertainties might emerge, and how compensation would be apportioned. This social contract can help manage expectations, minimize social inequities, and if necessary, compensate for them.

In contrast, passive adaptive management is analogous to what Connell (1974) has termed a natural experiment, using natural variation as the independent factor from which to infer consequences. In passive AM, managers use a combination of natural variation and inferred impacts of management actions to infer how management may be profitably modified. Unfortunately, such approaches are challenged in the way that any correlation is—by the presence of numerous confounding factors that render inference of causation tenuous at best. Consequently, it is difficult to draw definitive conclusions from passive adaptive management "experiments." In practice the AM that EBM embraces should involve management interventions that enable strong inferences and minimize opportunities for multiple conflicting explanations. But it also should be built upon a social contract that clearly articulates expectations between researchers and stakeholders. Monitoring, research, and modeling conducted in a synthetic, ecosystem context and encompassing all human activities that affect the ocean will enable scientists and other stakeholders to interpret the outcomes of AM experiments. Such interdisciplinary efforts require input and the use of approaches from both the social and natural sciences. In the following sections, we expand on each of these elements of marine EBM.

The Monitoring Imperative

In the relatively brief time that adaptive management has overlapped with experimental science, there have been few opportunities to conduct it at its fullest, as an iterative series of true experiments with policy refinements enlightened by outcomes (Lee 1999). Rigor aside, when people try something new, they tend to check back every once in a while to see how things are going. At its best, an AM experiment will be a repeated-measures BACI design: a before–after, control–impact comparison (Carr 2000). Before–after is actually from time point to time point indefinitely, because monitoring is done at regular intervals, just like health checkups. The control–impact is the comparison between a reference zone and any other management regime. The key concept here is that by having a nearby reference area for this comparison, all of the other confounding factors that normally plague such inferences are canceled out. The reference area and experimental regimes should, except for management action, be expected to follow similar trajectories as a consequence of seasonal, climatic, and other extrinsic influences.

Monitoring can range from minimalist to all encompassing. The choice of where to fall on this spectrum is determined by the objectives and available resources of a particular initiative. Since minimal monitoring provides limited information, it is most appropriate for situations where threats and their ecological and socioeconomic impacts are already well understood, so the monitoring can focus on the conditions expected to change. The advantage to this approach, of course, is that minimal monitoring is cheap. The major drawback, however, is that since limited information is provided, the variance in predictable impacts is wide, and therefore management needs to be more conservative and precautionary to ensure that impacts are minimized (Johannes 1998). And the consequences of such conservative, precautionary management strategies (e.g., no-take marine reserves) can be high social and economic costs (e.g., displaced fishermen).

The other end of the spectrum, the Cadillac of monitoring, is to strive to know everything about the possible states and trajectories of marine and social systems in a defined geographical area. These days it is possible to aspire to this, though at substantial cost. The advantage to knowing so much—were it possible also to know what it all meant—would be that resources could be extracted and environments altered up to, but just short of, those thresholds where irreparable harm would be inflicted. And were harm to happen anyway, we would have the best possible advance warning. The goal would be to derive as much benefit as possible from ocean ecosystems in a sustainable way, by approaching but not crossing thresholds of what the system can tolerate. The permissible proximity to these thresholds would be determined by the reaction time or resources needed to avert crossing over them. If marine EBM has a single objective, it is to keep us from crossing the thresholds that push the ecological and social systems of interest into alternate (and often, less valuable) states (see Leslie and Kinzig, chap. 4 of this volume). This is essential not only to avoid disaster, but also to ensure resilient ecosystems and human communities. However, prudence would dictate precautionary management and less need for a comprehensive monitoring effort (again, see Johannes' 1998 pithy thoughts on data-free management).

Effective Monitoring Depends on Appropriate Indicators

The good news is that we do not need to know everything about the system at all times. We only need to know the major ways that it can change and has changed (especially, but not only, for the worse), which can be revealed by monitoring indicators of these changes. Comparisons of conditions among management regimes (e.g., no-take marine reserves, other types of marine protected areas, and open access) and through time (BACI design) allow scientists to detect the effects of different management regimes and identify better ways to track these effects in the future. For example, Sandin and colleagues (2008) recently identified key indicators of reef health by assessing changes in reef condition along a gradient of human activities in the Line Islands, from heavily populated to pristine. Assessments of shifts in community composition revealed that top predator biomass, percent coral and coralline algae cover, and microbial composition were the strongest indicators of overall reef health (Johnston and Rohwer 2007; Sandin et al. 2008). These indicators may be relevant to assessments of reef health worldwide and provide a useful starting point for considering how to assess marine ecosystem health across other systems (see also box 7.1).

INDICATORS OF UNDERLYING PROCESSES AND CHANGE

Selecting EBM indicators requires understanding the underlying processes that drive change, and developing indicators that can cheaply and efficiently track these processes. Fisheries scientists made a great leap forward when they borrowed from population biology the insight that the level of removal for an exploited population that maximizes its rebound rate (and thus avoids system collapse) is equal to half of the population's carrying capacity (K). This point, $K/2$, is better known as maximum sustainable yield (MSY). MSY is one example of a reference point—that is, a specific value for an indicator that denotes the position and trajectory of human influence relative to what the system can tolerate. Consequently, yield (typically assessed as catch per unit effort) has become a critical indicator of fisheries' sustainability.

The most empowering approach to identifying these process indicators is to first seize upon a discrete process that is critical to marine ecosystem health and then isolate a parameter that is particularly responsive to the state of that process. For example, the biomass of apex predators is one indicator of trophic cascades, an ecological phenomenon that has garnered great attention through a series of papers calling attention to massive changes in the global ocean. These include recognition that the world's marine communities are being "fished down" (Pauly et al. 1998), that 90% (or at least some unacceptably large number) of large ocean predators have been eliminated (Myers and Worm 2003), that coral reefs are slip-sliding toward destruction (Pandolfi et al. 2003) on a "slippery slope to slime" (Pandolfi et al. 2005; see also Baird et al. 2004 re estuaries), and that the removal of large sharks can trigger trophic cascades that reorder ocean ecosystems and abruptly alter the health of some fisheries (Myers et al. 2007).

Indicators of ecosystem processes can also be revealed through advanced technologies (Hofmann and Gaines 2008). One example is the analysis of naturally occurring stable isotopes, useful in detecting shifts in individual feeding histories or overall food web structure, and thus potential indicators for the potential of a system to cross a threshold and exhibit a change in state (Michener and Kaufman 2007). Another example of a clever use of technology is the development of bioluminescence as an indicator for biological activity in the deep sea (Lapota et al. 2002).

WHOLE-ECOSYSTEM INDICATORS

Another approach to selecting EBM indicators is to identify indicators that reflect the entire ecosystem's overall health by synthesizing existing, individual-based data. Ecologists and fisheries biologists are experts at acquiring data on distribution, abundance, and community structure (basically, relative abundance patterns at any one site), which generate large volumes of individual-based data on age and growth, reproductive parameters, and feeding habits of key species. These fisheries and ecological survey data can be run through principal components or more-sophisticated factor analyses to visualize the state of a community and its trajectory through time. The resulting principal components can themselves serve as indicators. Multivariate analysis can also be used, perhaps more appropriately, as a histological stain to reveal underlying variables or the linking of variables into functional groups, which may better reflect community processes than the variables originally measured. We can then go back and measure those instead. When this works well, it eliminates a lot of extra effort and unwanted variance. For example, Link and Brodziak (2002) examined the northeastern US shelf ecosystem and combined species data into functional groups. This greatly reduced year-to-year variance (i.e., abundance of a functional group is robust against flux in the abundance of component species) and revealed compelling patterns otherwise lost in the complexity of the data, such as a shift from benthic to pelagic production.

SOCIAL AND ECONOMIC INDICATORS

The discussion of EBM indicators has naturally focused on ecological indicators; however, it is equally important to understand the status and drivers of the human populations who impact and are impacted by the marine ecosystem of interest (see also Shackeroff et al., chap. 3 of this volume). EBM recognizes that people are part of ecosystems, and thus we need to expand our view of indicators to include information about livelihoods, food security, nonmonetary benefits of healthy ecosystems (e.g., existence and option values), equity, local culture, environmental awareness, and social resilience. Social and economic monitoring enables marine managers and other practitioners to do the following:

- Estimate how coastal management is contributing to community development, including poverty alleviation and equitable

Box 7.1. Guidelines for EBM Indicators

In 2004, EBM visionaries gathered in Paris at a symposium cohosted by the Intergovernmental Oceanographic Commission and the Scientific Committee on Oceanic Research of the International Council for Science to share new ideas and data on indicators for EBM, specifically in a fisheries management context (Cury and Christensen 2005). The participants proposed the following guidelines, which clearly have relevance beyond fisheries:

- Ecosystem-based indicators need to have high inertia so that when they do change, it is in response to a major event, making them difficult to dismiss or ignore.
- A suite of indicators is needed, as opposed to just one or a few composite parameters.
- Both target reference points and limit reference points are needed. Target reference points correspond to the state of a fishery (or system) which is considered to be desirable; that is, management should aim to keep the system at this particular level. Limit reference points correspond to the limit beyond which the state of the fishery (or system) is not considered to be desirable. If this point is reached, management action should severely curtail fishing (or the appropriate activities).
- Predictive indicators predict changes in system state, whereas surveillance indicators identify changes that have already occurred, such as climate-driven regime shifts.
- Size-structure, aggregate, and spatial indicators are all important; however, aggregate indicators should be used in concert with a panel of more-specific variables so that information on the underlying processes is not lost.
- Information on physical processes and lower trophic levels is useful, as these are windows on environmental change and bottom-up processes, but they often do not directly reflect higher-trophic-level dynamics.
- The biomass of top predators and other indicators of trophic dynamics can indicate the presence of trophic cascades and functional shifts.

sharing of benefits. These data in turn are useful in determining management effectiveness.

- Value the marine resources in terms of cultural and economic significance, which is critical to lobbying for public and political support.
- Measure people's support for conservation efforts and changes in coastal activities.
- Facilitate stakeholder involvement by gaining a greater understanding of community perceptions and needs.
- Tailor management to the local situation, such as by developing education programs based on community members' understanding of resource conditions and threats.

As described above, EBM involves identifying indicators that reveal changes in critical processes. In line with this thinking, social and economic monitoring typically focuses on understanding the social, economic, and institutional impacts of management strategies by assessing indicators of these impacts. Demographic information and data on where and how people value and use coastal and marine resources are particularly useful for determining the degree and manner in which people are dependent on (and, potentially, threaten) the focal marine ecosystem. Data on environmental awareness and perceptions help establish level of awareness and support for management efforts. Information on the standard of living, built infrastructure, and human capital provide insight into household and community well-being, and governance (or institutional) indicators provide insight into the status and robustness of the management process, and thus, in part, its outcomes. A number of publications offer guidance on how to conduct social and economic monitoring in the context of marine protected areas and other marine ecosystem-based management strategies (e.g., Bunce et al. 2000; Pomeroy et al. 2004; McField and Kramer 2007).

In addition to monitoring the socioeconomic effects of management strategies, scientists are also monitoring processes of marine EBM design and implementation to determine the key factors that support or hinder positive ecological, social, and economic outcomes. In the case of marine protected areas (MPAs), important social factors include stakeholder participation in decision making, tangible economic benefits from MPA establishment (including alternative income projects), a perceived crisis in the resource base, strong leadership, a relatively small population, and enforcement capacity over the long term (Pollnac et al. 2001; Christie 2004; Pomeroy et al. 2004).

Integrating Social and Ecological Data: The Motivation for Models

Combining ecological, social, and economic indicators into integrated analyses is the next challenge. At one level, it could be as straightforward as putting all the data into a spreadsheet and running some statistics. However, the key is to figure out which social and economic indicators are influencing which ecological outcomes and vice versa. The obvious links are extractive activities, such as fishing, resulting in smaller populations of exploited species. But we need to delve further to investigate the feedback loops that drive such changes. For example, what are the long-term ecological changes associated with shifts in an economy from fishing to diving-related tourism, and what are the subsequent impacts on the economy? We might hypothesize that, at first, fishing pressure declines and the marine ecosystem rebounds but that as tourism increases, so do land-based sediment and nutrient inputs, eutrophication, and habitat destruction. Will fish populations suffer? Will fishing pressure increase again to supply the new demand from tourists? As the reef conditions degrade, will diving tourism decline as well, or will it switch to other forms of tourism? By answering these questions, we can better understand the fundamental relationships between people and the ecosystem of which they are a part to determine what additional socioeconomic and ecological indicators are critical to monitor.

Modeling for EBM

With models, scientists articulate current understanding of a system's dynamics in concise mathematical or graphical expressions. Models can help researchers organize and integrate data and provide frameworks that help us to design better experiments to zero in on the effects of management actions. Models guide the thinking of researchers and make thought experiments possible. Finally, when rendered user-friendly, as through a simple graphical interface or skillfully facilitated participatory modeling exercise, models can assist decision makers by helping them to foresee the consequences of specific policy options and thereby compare options by examining scenarios (e.g., Carpenter et al. 1999; Bousquet and LePage 2004; Boesch and Goldman, chap. 15 of this volume). The most important application of models is to elucidate processes and linkages that managers need to know about but that may otherwise go unnoticed.

Traditional fisheries management operates using overlapping demographic models for stocks of independent target species that occupy large, homogeneous spatial cells. In EBM, we are examining a complex, hierarchical, spatially distributed system, and thus a dynamic, spatially explicit model is needed to visualize system states, understand dynamics, and forecast outcomes for alternative policy scenarios (e.g., Babcock et al. 2005). Another requirement of EBM is that we model processes that are far more diverse and more sophisticated than population biology alone. What was formerly "surplus stock" is now seen as biomass flow in energetic pathways that may perform key ecosystem services (e.g., serve as prey for other species) in addition to supplying seafood for market. Traditional stock assessments can certainly contribute to EBM models, but now these data are joined by rich inputs from biophysical sensor arrays, targeted in situ monitoring, remote sensing, and data on human uses from questionnaires and other human survey instruments.

The minimum spatial scales for both monitoring and modeling are smaller than in most approaches of the last century. Statistically they must be adequate to detect and track divergent ecologies due to management strategy effects, and the spatial scale of some management regimes, such as marine reserves, can be very small indeed. The maximum resolution for computing scenarios must be on the order of a few square kilometers at most if the outputs are to be of practical use for decision making.

The tools needed for this kind of work are available now. One excellent approach now being utilized by The Nature Conservancy combines GIS (geographic information systems), the reserve-siting software MARXAN, and the dynamic modeling package ECOSPACE (TNC 2008). In our own Marine Management Area Science (MMAS) Program, we have used Java applets embedded in a GIS platform. Many prefer to program from the ground up, using MATLAB or primitive code. Practitioners have begun to broadly share technical advances and new ideas. For example, the EBM Tools Network brings together tool developers, practitioners, and training providers to develop EBM tools and support EBM implementation in coastal and marine environments (see http://www.ebmtools.org/). They are also initiating a training program to build the capacity of EBM practitioners to use tools effectively and appropriately. Several excellent models and utilities have recently proven very useful in EBM, including "Atlantis" out of Australia (Smith et al. 2007; Kaplan and Levin 2009). Atlantis is based upon a biogeochemical model of ecosystem processes that tracks flows through the major ecosystem components, including fished species and fishers, and also includes a sampling component that "observes" the model system as it progresses through a run. This yields new insights regarding monitoring design and appropriate system indicators. Models like Atlantis are invaluable for exploring multiple management options and as initial training and implementation tools. With time, some of these simulation models may prove transferable to diverse situations.

One very sticky problem remains. Making sense of human behavioral dynamics within an ecosystem context will require identifying how individual actors respond to changing environments. Valuation in the nonhuman components of a marine ecosystem is challenging enough to measure, map, and understand (see Wainger and Boyd, chap. 6 of this volume), but modeling the dynamics of the human actors within the system is much more daunting. Yet this integration must be achieved if the outputs are to be both interesting and useful.

The problem of capturing human behavior in marine ecosystems breaks naturally into three pieces: economic, sociocultural, and governance. The economic side is tractable. It relies on familiar statistical descriptions of wealth, welfare, and infrastructure and their distribution in space and by class (see Barbier, chap. 8 of this volume; Wainger and Boyd, chap. 6 of this volume). These relate in decipherable ways to the ecological landscape in both watershed and continental shelf environments. Sociocultural aspects are often more complex and less quantitative, as they involve understanding the diverse roles of marine ecosystem services in people's lives—for example, as religious symbols; as "last resort" occupations or occasional, supplemental means of livelihood (e.g., fishermen in Komodo, Indonesia, may go fishing when they need additional income for a special occasion such as a wedding); or as social infrastructure (e.g., fish buyers in rural Bahia, Brazil, provide emergency financial assistance to families in dire medical need). Traditional and local ecological knowledge can be very important in gleaning these types of information (Kliskey et al., chap. 9 of this volume). Finally, understanding the dynamics of governance requires knowledge of informal and formal rules and regulations of a community as well as process components, such as levels of stakeholder participation.

Understanding the decision-making process itself is also vital for marine EBM models. This is the dynamic by which people, alone and in groups, elect one option over another. Classical decision theory, particularly in regard to decision making under uncertainty, can help us to predict what a community might choose as the best among several policy alternatives. This helps us in EBM because what people choose as best for them is not necessarily the policy that maximizes system integrity or sustainability, and predicting this variance provides us a critical feedback loop in our marine system model. Another important area is group dynamics, including key emergent effects such as the "wisdom of the crowd" (Surowiecki 2004) and group behavior as a determinant of decision making (Durrett and Levin 2005). These considerations can help us to understand what causes decisions made by groups of people to range in quality from extremely wise to very unwise.

Research Beyond Monitoring and Modeling

Implementation of EBM will be much more feasible, acceptable, and effective if research targets the unanswered questions about the interplay between natural and social processes (Leslie and McLeod 2007). Developing new indicators is a vital need, but we should aspire to create indicators that are increasingly based on process, and thus hold predictive power, instead of merely measuring patterns. In particular, the development of new indicators of the mechanisms of change, as opposed to mere empirical measurements of change, is a key and challenging task for both natural and social scientists.

Applied research is needed to expand our understanding of the effectiveness of alternative ways to put EBM into practice. Here the use of BACI designs comes into play. Equally important is gaining insight into the complex dynamics and emergent patterns that result from processes and that shape system behavior in unexpected ways. If the mechanistic basis for why a particular EBM strategy fails remains unknown, then the ability to extrapolate from the test case is limited or nonexistent. Because ecosystems are complicated and because processes of interaction occur on multiple scales of time and space, applied research programs designed to evaluate particular management interventions will typically require substantial resources. Fortunately, this enhances basic understanding of ecosystem processes while at the same time providing practical answers to management questions.

One important nexus that needs much further attention is the interface between understanding ecosystem dynamics and managing human activities to facilitate resilience. Existing governance structures constrain the types of management intervention that are possible. For example, approaches to managing artisanal versus industrial fisheries differ dramatically, due to institutional differences among these two groups of fishers, the sociopolitical contexts in which they operate, and the level of information available about the marine ecosystems which they exploit (Castilla and Defeo 2005). Social acceptance is also critical to management implementation and effectiveness (e.g., Christie 2004). The fact that, ultimately, we are managing humans and their activities, rather than the ecosystems per se, means that natural science must be conducted in ways that engage stakeholders from the beginning. This entire arena of interplay between the natural and social systems is as yet insufficiently explored. It is not the resilience of the ecosystem alone that we seek to protect or restore, but rather the resilience of the coupled social–ecological system.

Translating Science into Action

The technical—and conventional—role of an EBM scientist is to visualize the full suite of ecosystem states that are possible, to provide a fix on current position and heading for the system within this space, and to predict the shift in ecosystem position and heading likely to be caused by a change in human behavior. The results are often groundbreaking discoveries with important EBM implications that are locked away in peer-reviewed publications or presented at obscure academic conferences that go unread or unheard of by stakeholders, policymakers, and managers. Stakeholders may be unaware of the publications or lack the time or capacity to determine their relevance to specific management issues. Meanwhile, scientists hesitate to become more engaged in policy and management discussions for fear of losing their impartiality. The result is a huge gap between the state of scientific knowledge and management practice. Fortunately, there has been an awakening in the scientific community to the fact that it is not only possible, but ethically responsible, to become engaged in management and policy issues and that it is possible to do so without compromising impartiality. As a result, EBM scientists are increasingly stepping out of their traditional roles and taking measures to ensure that their research contributes to management and policy decisions. In box 7.2, we propose guidelines for scientists interested in translating their research for EBM implementation.

Case Study In Translation: The Conservation International Marine Management Area Science Program

The Marine Management Area Science (MMAS) Program, which is within Conservation International's Center for Applied Biodiversity Science, was established in 2005 to provide scientific answers to questions confronting the management of marine multiple-use and protected areas. Bringing together expertise in both social and biological sciences, the MMAS Program was designed to play a leading role in improving the effectiveness of marine management area regimes at the international level. The research agenda focuses on the following questions:

- What are the costs and benefits of marine managed areas (MMAs)?
- What are they worth in cultural and economic terms?
- What economic incentives work? Why and how?
- Do MMAs enhance resilience? Why and how?
- What are the critical habitat and species linkages?
- What are the weak and strong links in the enforcement chain?

The program includes focal work at a network of sites, plus gap-filling science that is globally applicable. The four initial program sites are Abrolhos, Brazil; Belize; the Eastern Tropical Pacific Seascape (particularly Coiba Island, Panama); and Fiji. The MMAS Program is currently in its second year of implementation.

A central tenet of the program is to ensure that the resulting science leads to conservation action at the global, regional, national, and local levels. Building this principle into the program has been exciting and challenging as we (the authors) try to push the envelope on bridging the gap between conservation science and conservation action. We have found close collaboration between the researchers and members of the in-country conservation community to be essential to building this bridge. In each of our four priority locations there is an MMAS coordinator who is responsible for ensuring the research is relevant to and feeds into existing conservation efforts. This involves a great deal of communication and coordination throughout the research (from planning through data collection and analysis to dissemination of results), not only between the researchers and the coordinator, but also between the coordinator and the other stakeholders, including government agencies, policymakers, other environmental nongovernmental organizations, user groups, and community organizations. Typically, the coordinator arranges meetings between the principal investigator and the stakeholders early in the research process to discuss how the research can fit within their interests and how they would like to engage in the research. Consultations and updates continue throughout the project. In addition, we have been holding broader stakeholder meetings to discuss conservation goals, how the MMAS studies can help to address them, and in what form the MMAS results can best be translated and disseminated.

Perhaps of greatest practical importance, we strive for allocating 15% of each budget to the science-to-action component even if the budget needs are still being shaped. The research plans are developed between the researchers and the in-country MMAS coordinators and include not only hypothesis and methods, but also conservation impacts, involvement by target audiences, and deliverables (beyond peer-reviewed publications). Finally, an external assessment is conducted in parallel with the project, to determine how effective we are at translating science into action and how we can continue to improve.

Parting Thoughts on the Role of the Scientist in Marine EBM Implementation

We have examined a number of factors that contribute to the implementation of marine EBM. Marine EBM requires a shift in our perspective on the relationships between humans and the oceans. Adaptive management and active experiment-based learning, monitoring, modeling, and sharing of information as well as new indicators are needed, too. Indeed, a whole new clinical perspective on marine ecology is being developed. The medical analogy has been used a lot recently, to good effect. Usually the marine scientist is depicted as the doctor, and a marine ecosystem, as the patient. That may not be quite right. Perhaps it would be more productive to view the marine scientist as the doctor but consider coastal society to be the patient. The analogy is still less than perfect—one deals with individuals, not societies as a whole. Where do we practice? We still spend time in our laboratories and in our wet suits, but that is not clinical practice. Our examining rooms are anywhere that people gather to contemplate the sea and to think about their futures in it. This can mean committee meeting rooms, beachfront gatherings of stakeholders and community leaders, or furtive sessions in the dark recesses of stuffy hotels. Less than appetizing? Just think of it as hazardous duty.

The scary thing is that the people we are expecting to do some talking here are, after all, scientists first. Visualize fish geeks wearing wet suits, and fishing community groupies toting notepads into villages. Who are we, really? How many marine ecologists and social scientists count networking skills among their greatest strengths? The early 1990s saw a push for

Box 7.2. Guidelines for Translating Science into Action

1. *Address stakeholder priorities.* By asking stakeholders their priority information needs and developing research agendas accordingly, researchers are far more likely to produce findings that stakeholders will be interested in and will incorporate into their management decisions. In some places managers and researchers are coming together at a national or regional level to identify research priorities so that they can then solicit researchers to focus on these key topics. Asking before soliciting funding, setting an agenda, and planning the data collection together also builds trust and helps to keep all parties on the same page.
2. *Consult from conception to dissemination.* Holding regular consultations and sharing of findings to date with stakeholders ensures the research is tailored to the stakeholder needs and increases the likelihood that they will use the findings to shape EBM policy and programs.
3. *Build local capacity.* Engaging local scientists is important to build in-country expertise. However, engaging nonscientists (e.g., policymakers, fishermen, tour operators) can be even more strategic. When stakeholders are directly involved in the research, they are likely to validate the results and become advocates for the management implications that follow.
4. *Articulate a work plan that encompasses both the research and the implementation components of EBM.* Typically, research plans are developed with hypotheses, methods, anticipated results, and a budget. In addition, we propose the plan needs to articulate existing EBM efforts and how this research will contribute to them, the target audiences and how they will be involved in the research, and the ultimate deliverables beyond peer-reviewed publications (e.g., a white paper of policy recommendations, a poster, a guidebook, and/or a community meeting to discuss results). Project budgets also need to clearly include funding for this "science to EBM" component. Finally, it needs to be clear that this is a joint undertaking and that scientists and other stakeholders will need to work together to ensure the science feeds into EBM policies and programs.
5. *Translate results into accessible formats.* This will better convey them to the people whose behavior or policies influence the coupled systems of interest. Communications experts can be key allies in this process (e.g., in the United States, COMPASS [Communication Partnership for Science and the Sea] and SeaWeb were created to facilitate this translation of marine science into conservation action).

scientist-activists who were not afraid to speak out about the implications of their research results. A whole generation was encouraged in this mold by award, fellowship, and incentive programs to do good work and communicate far and wide. It worked, somewhat, until some of their own colleagues branded them (meaning us) as biased advocates, or worse. This did not scare anybody, but it does compromise the effectiveness of the activist-scientist. So if that is not the best paradigm, what is? Can we all become marine country doctors? There may still be plenty of room for studious social misfits who hide underwater or in their labs crunching numbers, but these individuals are not the advance guards of marine EBM implementation. We must find a voice, a heart, and a touch that will enable us to bring our fellow humans of all walks and all dispositions toward and into the sea on this wild, wild ride.

Acknowledgments

The authors are grateful to colleagues who helped through debate and discussion of the topics covered in this chapter: Burr Heneman, Suchi Gopal, Andy Rosenberg, Sarah Carr, Jason Link, David Bergeron, and Peter Auster. The authors are thankful for support from the Gordon and Betty Moore Foundation through the Marine Management Area Science Program; this paper is a contribution of that program. Les Kaufman also thanks the fishermen, organizations, and committees with whom he is engaged in early stages of EBM implementation in the waters of New England, California, Florida, and East Africa for innumerable fruitful discussions.

References

Babcock, E. A., E. K. Pikitch, M. K. McAllister, P. Apostolaki, and C. Santora. 2005. A perspective on the use of spatialized indicators for ecosystem-based fishery management through spatial zoning. *ICES Journal of Marine Science* 62: 469–76.

Baird, D., R. R. Christian, C. H. Peterson, and G. A. Johnson. 2004. Consequences of hypoxia on estuarine ecosystem function: Energy diversion from consumers to microbes. *Ecological Applications* 14:805–22.

Bousquet, F., and C. Le Page. 2004. Multi-agent simulations and ecosystem management: A review. *Ecological Modelling* 176(3–4):313–32.

Bunce, L., P. Townsley, R. Pomeroy, and R. Pollnac. 2000. *Socioeconomic manual for coral reef management*. Global Coral Reef Monitoring Network and Australian Institute of Marine Science. http://www.reefbase.org/socmon/.

Carpenter, S., W. Brock, and P. Hanson. 1999. Ecological and social dynamics in simple models of ecosystem management. *Conservation Ecology* 3(2):4.

Carr, M. 2000. Marine protected areas: Challenges and opportunities for understanding and conserving coastal marine ecosystems. *Environmental Conservation* 27:106–9.

Castilla, J. C., and O. Defeo. 2005. Paradigm shifts needed for world fisheries. *Science* 309(5739): 1324–25.

Christie, P. 2004. MPAs as biological successes and social failures in Southeast Asia. In *Aquatic protected areas as fisheries management tools: Design, use, and evaluation of these fully protected areas*, ed. J. B. Shipley, 155–64. Bethesda, MD: American Fisheries Society.

Connell, J. H. 1974. Field experiments in marine ecology. In *Experimental marine biology*, ed. R. Mariscal, 21–54. New York: Academic Press.

Cury, P. M., and V. Christensen. 2005. Quantitative ecosystem indicators for fishery management. *ICES Journal of Marine Science* 32:307–10.

Durrett, R., and S. A. Levin. 2005. Can stable social groups be maintained by homophilous imitation alone? *Journal of Economic Behavior and Organization* 57(3):267–86.

Folke, C., T. Hahn, P. Olsson, and J. Norberg. 2005. Adaptive governance of social–ecological systems. *Annual Review in Environment and Resources* 30:441–73.

Hofmann, G. E., and S. D. Gaines. 2008. New tools to meet new challenges: Emerging technologies for managing marine ecosystems for resilience. *BioScience* 58(1):43–52.

Holling, C. S., ed. 1978. *Adaptive environmental assessment and management.* Caldwell, NJ: Blackburn Press.

Johannes, R. E. 1998. The case for data-less marine resource management: Examples from tropical nearshore fisheries. *Trends in Ecology and Evolution* 13:243–46.

Johnston, I. S., and F. Rohwer. 2007. Microbial landscapes on the outer tissue surfaces of the reef-building coral Porites compressa. *Coral Reefs* 26(2):375–83.

Kaplan, I. C., and P. S. Levin. 2009. Ecosystem-based management of what? An emerging approach for balancing conflicting objectives in marine resource management. In, *The Future of Fisheries Science in North America,* ed. R. Beamish and B. Rothchild. Berlin: Springer.

Lapota, D., J. Andrews, S. Lieberman, and G. Anderson. 2002. Development of an autonomous bioluminescence buoy (bioBuoy) for long-term ocean measurements. *Oceans* '02 MTS/IEEE 1(29–31):396–401.

Lee, K. N. 1999. Appraising adaptive management. *Conservation Ecology* 3(2):3.

Leslie, H. M., and K. L. McLeod. 2007. Confronting the challenges of implementing marine ecosystem-based management. *Frontiers in Ecology and the Environment* 5(10):540–48.

Levin, S. A. 1999. *Fragile dominion: Complexity and the commons.* Reading, MA: Perseus Books.

Levin, S. A., and J. Lubchenco. 2008. Resilience, robustness, and marine ecosystem-based management. *BioScience* 58:27–32.

Link, J. S., and J. K. T. Brodziak. 2002. *Status of the northeast US continental shelf ecosystem: A report of the Northeast Fisheries Science Center (NEFSC) Ecosystem Status Working Group.* NEFSC Reference Document 02-11. Woods Hole, MA: NEFSC.

MacArthur, R.H. 1955. Fluctuations of animal populations and a measure of stability. *Ecology* 36: 533–36.

McField, M., and P. Kramer. 2007. *Healthy reefs for healthy people: A guide to indicators of reef health and social well-being in the Mesoamerican reef region with contributions by M. Gorrez and M. McPherson.* http://www.healthyreefs.org.

Michener, R. H., and L. S. Kaufman. 2007. Stable isotope ratios as tracers in marine food webs: An update. In *Stable isotopes in ecology and environmental science,* ed. K. Lajtha and R. H. Michener. Boston: Blackwell Scientific.

Myers, R. A., J. K. Baum, T. D. Shepherd, S. P. Powers, and C. H. Peterson. 2007. Cascading effects of the loss of apex predatory sharks from a coastal ocean. *Science* 315(5820):1846–50.

Myers, R. A., and B. Worm. 2003. Rapid worldwide depletion of predatory fish communities. *Nature* 423:280–83.

Pandolfi, J. M., J. B. C. Jackson, N. Baron, R. H. Bradbury, H. M. Guzman, T. P. Hughes, C. V. Kappel et al. 2005. Are US coral reefs on the slippery slope to slime? *Science* 307:1725–26.

Pandolfi, J. M., R. H. Bradbury, E. Sala, T. P. Hughes, K. A. Bjorndal, R. G. Cooke, D. McArdle et al. 2003. Global trajectories of the long-term decline of coral reef ecosystems. *Science* 301(5635): 955–58.

Pauly, D., V. Christensen, J. Dalsgaard, R. Froese, and F. Torres Jr. 1998. Fishing down marine food webs. *Science* 279(5352):860–63.

Pollnac, R., B. Crawford, and M. Gorospe. 2001. Discovering factors that influence the success of community-based marine protected areas in the Visayas, Philippines. *Ocean and Coastal Management* 44:683–710.

Pomeroy, R. S., J. E. Parks, and L. M. Watson. 2004. *How is your MPA doing? A guidebook of natural and social indicators for evaluating marine protected area management effectiveness.* Cambridge, UK: International Union for Conservation of Nature (IUCN).

Sandin, S. A., J. E. Smith, E. E. DeMartini, E. A. Dinsdale, S. D. Donner, A. M. Friedlander, T. Konotchick et al. 2008. Baselines and degradation of coral reefs in the Northern Line Islands. *PLoS ONE* 3(2):e1548.

Smith, A. D. M., E. J. Fulton, A. J. Hobday, D. C. Smith, and P. Shoulder. 2007. Scientific-tools to support the practical implementation of ecosystem-based fisheries management. *ICES Journal of Marine Science* 64(4):633–39.

Surowiecki, J. 2004. *The wisdom of crowds: Why the many are smarter than the few and how collective wisdom shapes business, economies, societies and nations.* New York: Anchor.

TNC (The Nature Conservancy). 2008. *E-BM Toolkit.* http://www.marineebm.org.

UNEP (United Nations Environment Programme). 2006. *Marine and coastal ecosystems and human wellbeing: A synthesis report based on the findings of the Millennium Ecosystem Assessment.* Nairobi: UNEP.

Walters, C. J. 1986. *Adaptive management of renewable resources.* New York: MacMillan.

CHAPTER 8
Ecosystem Service Trade-offs

Edward B. Barbier

A fundamental challenge to effective implementation of ecosystem-based management (EBM) for marine and coastal systems is that "ecosystem services are nearly always undervalued. . . . Examples of services that are at risk because they are undervalued include protection of shorelines from erosion, nutrient recycling, control of disease and pests, climate regulation, cultural heritage and spiritual benefits" (McLeod et al. 2005). Thus, valuation of marine and coastal ecosystem services is critically important to an EBM strategy that seeks to maintain and enhance the delivery of beneficial services.

An equally important role of valuing ecosystem services is to assist in the assessment of trade-offs. Coastal and marine ecosystems are some of the most heavily used natural systems globally; it is now recognized that the cumulative impacts from a range of human activities are threatening many of the world's remaining marine and coastal environments and the many benefits they provide (Halpern et al. 2008). To be effective, EBM has to consider the trade-offs between the gains from converting or exploiting coastal and marine ecosystems and any resulting loss in services. Economic valuation has a role in quantifying and analyzing these gains and losses, as well as determining the distributional impacts, that is, who wins and who loses in the economy.

Valuation can also assist an EBM strategy in identifying the key services that need protecting or supporting. As noted above, some ecological services of marine and coastal environments are frequently undervalued and thus are at risk of being ignored in policy decisions. But policymakers and managers also want to know which services provided by a given coastal or marine ecosystem in a specific location are the more valuable and most vital to human livelihoods. This is particularly important in considering EBM strategies for rehabilitating, restoring, or protecting natural environments. Does the storm-protection value of coastal wetlands justify the expansion of wetland restoration along the US Gulf Coast in the aftermath of Hurricanes Katrina and Rita? In expanding the coastal area devoted to wetlands, what human activities must be curtailed, and what will be the benefits forgone? Will other valuable services also be generated by the expanded wetlands restoration program, and how significant are they compared with storm protection? These are important questions that are directly relevant to EBM and which can be addressed through economic valuation.

As emphasized throughout this book, one of the goals of EBM is to ensure the resilience of marine and coastal ecosystems, that is, their capacity to avoid breaching thresholds that cause the systems to "flip" from one functioning state to another (see Leslie and Kinzig, chap. 4 of this volume). A key issue for managers is to know the consequences, in terms of the provision of vital services to humankind, of a loss of resilience in an ecosystem. For example, if a healthy coral reef system shifts from a state dominated by healthy coral cover and intact fish communities to an alternative state of scarce invertebrate consumers, excessive macroalgae, and depleted fish species, what are the implications for a variety of benefits, from commercial fishing to recreation and tourism?

What are the costs of maintaining the coral reef system within a healthy state to minimize breaching the thresholds that flip the system into the alternative state? Such questions are a challenge to economic valuation, as they raise additional problems of measuring values associated with irreversibility, uncertainty and discontinuous thresholds, which must be addressed if economic valuation is to prove useful for EBM.

The purpose of this chapter is to explore further how valuation of ecosystem services can inform trade-off decisions under EBM. First, I will describe economic approaches to valuing ecosystem services in general and then highlight the relative strengths and limitations of different approaches. Next, I will discuss shrimp aquaculture and mangrove loss in Thailand as an illustration of valuation approaches. I will conclude by examining further research issues in the valuation of ecological services that are relevant to EBM for coastal and marine systems.

Ecosystem Services and the Valuation Challenge

Regardless of how one defines and classifies ecosystem services, numerous challenges remain to adequately describe and assess linkages between ecosystem structure and functioning, services, and their subsequent values (Heal et al. 2005). Through the lens of ecosystem services, natural ecosystems are assets that produce a flow of beneficial goods and services over time. In this regard, they are no different from any other asset in an economy, and in principle, ecosystem services should be valued in a similar manner. That is, regardless of whether or not there exists a market for the goods and services produced by ecosystems, their social value must equal the discounted net present value (NPV) of these flows. For further discussion on value from an economic perspective, see Wainger and Boyd (chap. 6 of this volume).

For example, let's suppose that the flow of ecosystem services in any time period, *t*, can be quantified and that we can measure what each individual is "willing to pay" for the provision of these services. If we sum up, or aggregate, the willingness to pay by all the individuals benefiting in each period from the ecosystem services, we will have a monetary amount—call it B_t—which indicates the social benefits in the given time period *t* of those services. Hopefully, there will be a stream of such benefits generated by ecosystem services, from the present time and into the future. Because society is making a decision today about whether or not to preserve ecosystems, we want to consider the flow of benefits of these services, net of the costs of maintaining the natural ecosystems intact, in terms of their *present value*. To do this, any future net benefit flows are discounted into present value equivalents. In essence, we are treating natural ecosystems as a special type of capital asset—a kind of "natural wealth"—which, just like any other asset or investment in an economy, is capable of generating a current and future flow of "income" or "benefits."

However, what makes environmental assets special is that they give rise to particular measurement problems that are different from those for conventional economic or financial assets. This is especially the case for the beneficial services that are derived from the regulatory and habitat functions of natural ecosystems. For one, these assets and services are a special type of "natural capital" (Just et al. 2004). If an ecosystem is left intact, then the services from the ecosystem's regulatory and habitat functions are available in quantities that are not affected by the rate at which they are used. Like other assets in the economy, a natural ecosystem can be increased by investment, such as through restoration activities. Ecosystems can also be depleted or degraded, for example, through habitat destruction, land conversion, pollution impacts, and so forth. However, whereas the services from most assets in an economy are marketed, the benefits arising from the regulatory and habitat functions of natural ecosystems generally are not. If the aggregate willingness to pay for these benefits, B_t, is not revealed through market outcomes, then efficient management of such ecosystem services requires explicit methods to measure this social value (e.g., see Freeman 2003; Just et al. 2004). In fact, the failure to consider the "values" provided by key ecosystem services in current policy and management decisions is a major reason for the widespread disappearance of coastal and marine ecosystems and habitats across the globe (MA 2003). The failure to measure explicitly the aggregate willingness to pay for otherwise nonmarketed ecosystem services exacerbates problems caused by the global expansion of human populations and economic activity along the coasts, as the benefits of these services are "underpriced" and may lead to excessive land conversion, habitat fragmentation, and pollution.

Consider the example of the conversion of an area of the coast for commercial development (fig. 8.1). The marginal social benefits of ecological services at any time t are represented by the line MB_t for a natural coastal ecosystem of given area $\bar{S}$. For the purposes of illustration, this line is assumed to be sloping upward, which implies that for every additional square kilometer of coastal habitat land area, S, preserved in its natural state, more ecosystem service benefits will be generated at an increasing amount. Note that it is straightforward to determine the aggregate willingness to pay for the benefits of these services, B_t, from this line; it is simply the area under the MB_t line. If there is no other use for the land occupied by the ecosystem, then the opportunity costs of maintaining it are zero, and MB_t is at its maximum size when the entire coastal ecosystem is left intact at its original land area size, $\bar{S}$. The management decision is therefore simple; the coastal ecosystem should be completely preserved and allowed to provide its full flow of services in perpetuity.

However, population and economic development pressures in many areas of the world usually mean that the opportunity cost of maintaining the land for natural coastal ecosystems is not zero, due to increased demand for

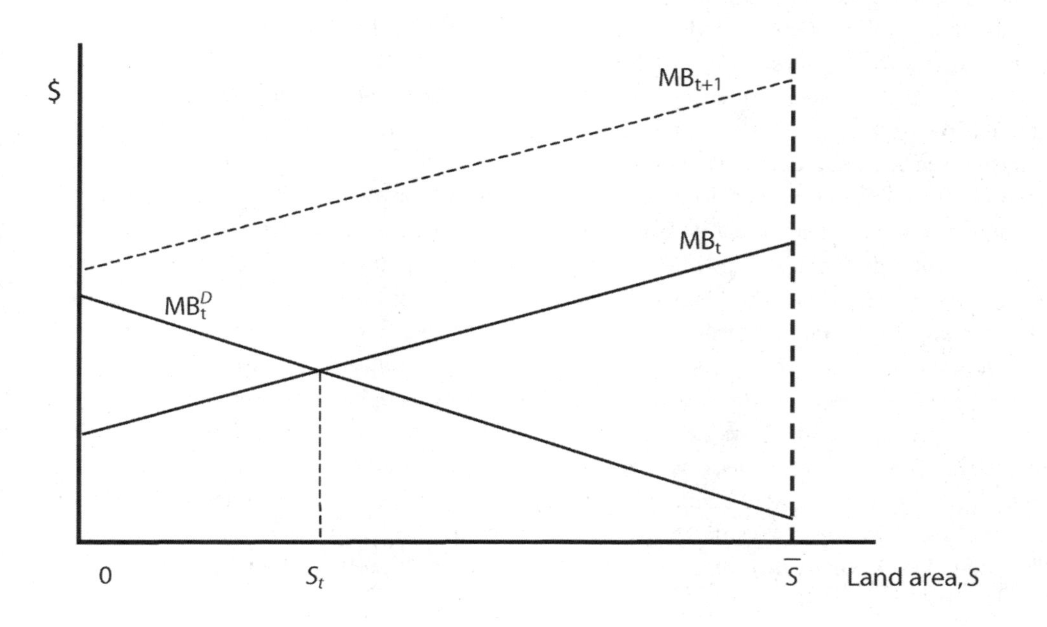

Figure 8.1 Optimal ecosystem conversion to coastal zone development.

land accompanying economic development in coastal zones. The ecosystem-based management decision needs to consider these alternative uses of coastal areas, and we therefore need to add these opportunity costs into figure 8.1. For example, suppose that the marginal social benefits of converting natural ecosystem land for these development options are now represented by a new line MB_t^D in the figure. Thus, efficient use of land would require that an amount S_t of coastal ecosystem area should be converted for development, leaving $\bar{S} - S_t$ of the natural ecosystem intact.

Both of these outcomes assume that the willingness to pay for the marginal benefits arising from coastal and surrounding marine ecosystem services, MB_t, is explicitly measured, or "valued". But if this is not the case, then these nonmarketed flows are likely to be ignored in the land use decision. Only the marginal benefits, MB_t^D, of the "marketed" outputs arising from coastal economic development activities will be taken into account, and as indicated in the figure, this implies that the entire ecosystem area $\bar{S}$ will be converted for development.

A further problem in valuing environmental assets is the uncertainty over their future values. It is possible, for example, that the benefits of natural ecosystem services will be larger in the future, as more scientific information becomes available over time (fig. 8.1). Recall that if the benefits of ecosystem services in the current period are correctly valued and incorporated in the development decision, then only S_t of ecosystem area should be converted for coastal zone development at time *t*. However, as we shall now see, even this may be too much development *today* if we are uncertain about *future* ecosystem service benefits. For example, suppose that in some future period $t+1$, it is discovered that the value of coastal and marine ecosystem services is actually much larger, so the marginal benefits of these services, MB_{t+1}, in present value terms (i.e., "discounted," or measured in terms of today's dollars) are now represented by the dotted line in the figure. If the present-value marginal benefits from coastal zone development in the future are largely unchanged, that is, $MB_{t+1}^D \approx MB_t^D$, then as figure 8.1 indicates, the future benefits of ecosystem services exceed these costs, and the coastal ecosystem area should be "restored" to its original area, $\bar{S}$. Unfortunately, in making development decisions, we often do not know today that the future value of ecosystem services will turn out to exceed future development benefits. Our simple example shows that if we have already made the decision today to convert S_t area of the coastal zone, then we will have to reverse this decision in the future period and "restore" the natural coastal ecosystem.

It should be apparent from this simple example that taking into account future ecosystem service values is further complicated by the problem of irreversibility. If development today leads to irreversible (or reversible only at high cost) loss of coastal and marine ecosystems, and the values of the services provided by these natural systems are uncertain, then this gives an addition reason for conserving ecosystems (Arrow and Fisher 1974; Henry 1974; Dixit and Pindyck 1994). As pointed out by Krutilla and Fisher (1985), if environmental assets are irreversibly depleted, their value will rise relative to the value of other reproducible and accumulating economic assets. This is a likely scenario for any coastal and marine ecosystem that is irreversibly converted or degraded as a result of expansion of coastal zone development or the cumulative generation of pollution by this activity. Because natural ecosystems are in fixed supply and are difficult to substitute for or restore, the beneficial services provided by their regulatory and habitat functions will decline as these assets are converted or degraded. The increasing relative scarcity of these services means that their value will rise relative to other goods and services in the economy. The additional measurement problem arriving from irreversible conversion of fixed ecosystem assets is illustrated in figure 8.2.

As in figure 8.1, if only the current benefits, MB_t, and opportunity costs, MB_t^D, of maintaining a natural ecosystem are considered, then an amount S_t of ecosystem area would be converted today. But suppose that the loss of coastal and marine ecosystem services arising from converting S_t causes the value of these services to rise. As a result, individuals benefiting from these services in a future time period t+1 would choose optimally to have less land converted to coastal zone development, that is, $S_{t+1}<S_t$. However, if ecosystem conversion is irreversible, then land development remains at S_t in time period t+1. But this additional development means that individuals in the future will be deprived of valuable coastal and marine services. This loss in welfare for individuals in the future is the "user cost" of irreversible loss of coastal and marine ecosystem services due to conversion today. In figure 8.2, the marginal user cost of development, measured in present-value terms, is represented as the straight line MUC_{t+1}, which rises as more coastal land is converted. The correct land use decision should take into account this additional cost of irreversible ecosystem conversion due to expansion of coastal zone development today. Deducting the marginal user cost from MB_t^D yields the net marginal benefits of the development option, MNB_t^D. The latter is the appropriate measure of the opportunity costs of maintaining coastal and marine ecosystems intact, and equating it with the marginal social benefits of ecosystem services determines the optimal land allocation. Only S_t^* of coastal ecosystem area should be converted for development, leaving $\bar{S}-S_t^*$ of the natural ecosystem intact.

Valuation of environmental assets under conditions of uncertainty and irreversibility clearly poses additional measurement problems. There is now a considerable literature advocating various methods for estimating environmental values by measuring the additional amount, or "premium," that individuals are willing to pay to avoid the uncertainty surrounding such values (see Ready 1995 for a review). Similar methods are also advocated for estimating the user costs associated with irreversible development, as this also amounts

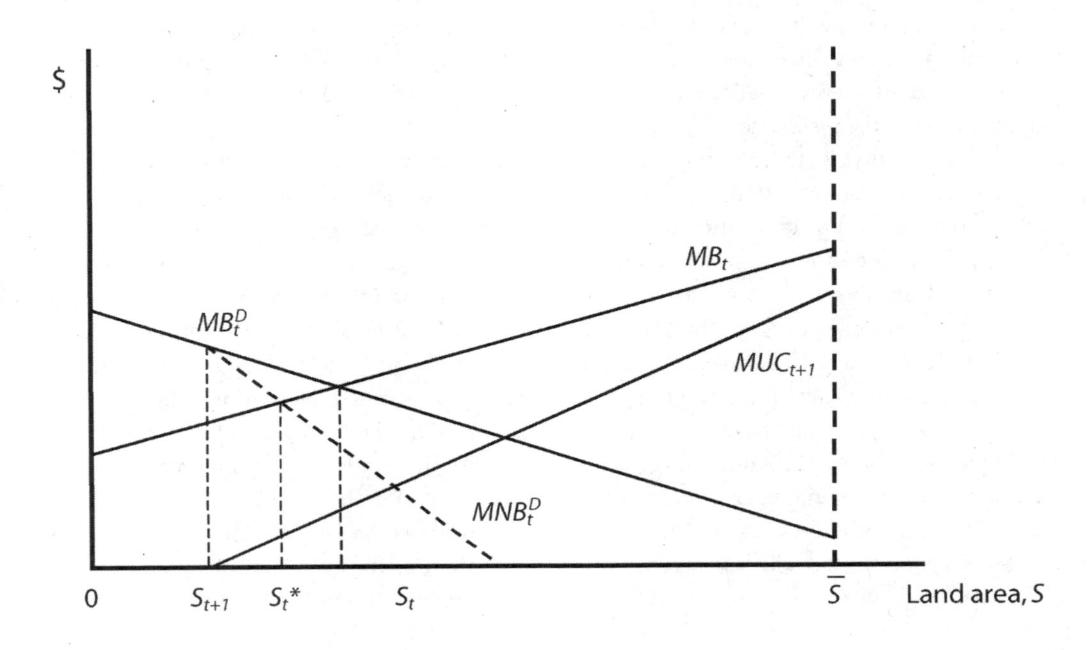

Figure 8.2 Irreversible conversion of ecosystems and uncertainty.

to valuing the "option" of avoiding reduced future choices for individuals (Just et al. 2004). However, the problem with such welfare measures is that they cannot be estimated from the observed behavior of individuals and are therefore difficult to implement empirically, particularly when there is uncertainty, not only about the future state of the environmental asset, but also over the future preferences and income of individuals. The general conclusion from the few empirical attempts to implement environmental valuation under uncertainty is that more empirical research is needed to decide under what conditions uncertainty can be ignored (Ready 1995).

As the chapters in this volume emphasize, ignoring uncertainty and irreversibility may not be wise for many marine and coastal ecosystems that are susceptible to system "flips" from one state system to another (Leslie and Kinzig, chap. 4 of this volume). In principle, the methods outlined here of incorporating the "user costs" of irreversible development are the same for taking into account the value of preserving the "resilience" of a natural ecosystem. For example, returning to figure 8.2, suppose that we know that if coastal development is limited to S_{t+1}, then the thresholds of the natural coastal and marine ecosystem are unlikely to be breached. However, as additional coastal area is converted, the resilience of the system is likely to decline, that is, there is an increasing risk that the natural ecosystem will "breach" its thresholds and "flip" to a different type of system that yields a different set of ecological services. If we are able to estimate the probability of the system's breaching its thresholds as land conversion proceeds beyond S_{t+1}, and if we can measure and value the changes to ecosystem services that result from the "altered" ecosystem state, then we might be able to construct a similar "marginal user cost" curve as MUC_{t+1} in figure 8.2. Once again, the correct land use decision should take into account this additional "resilience" cost of irreversible ecosystem conversion today.

In principle, therefore, taking into account the "resilience" cost of irreversible conversion of coastal and marine ecosystems can be handled through economic valuation. Actually measuring this cost, however, does impose a formidable challenge. As this simple example makes clear, the two key components of valuing the marginal user cost associated with reducing the resilience of ecosystems are determining (1) the probability of "breaching" thresholds as natural habitat area is converted or damaged and (2) the loss in valuable ecosystem services that result from any "flip" in ecosystems to an altered state.

Valuation Methods

Uncertainty, irreversible loss, and resilience costs are important issues to consider in valuing ecosystem service trade-offs. However, even before we tackle these "second-order" valuation issues, we are faced with a more basic problem. As emphasized by Heal and colleagues (2005), a more "fundamental challenge" in valuing these flows is that ecosystem services are largely not marketed. In other words, returning to figure 8.1, unless some attempt is made to value the aggregate willingness to pay for these services, MB_t, then it will be difficult to succeed at effective ecosystem-based management to balance development and conservation trade-offs.

Over the past decade, substantial progress has been made by economists working with ecologists and other natural scientists on this "fundamental challenge" to improve environmental valuation methodologies. A summary of methods that can be used to value nonmarketed services is shown in table 8.1. Below, I provide some brief observations regarding the advantages and shortcomings of these methods. More discussion of each of these and its application to the valuation of ecosystem services can be found in Freeman 2003, Pagiola et al. 2004, and Heal et al. 2005.

Table 8.1. Various valuation methods applied to ecosystem services

Valuation method	*Types of value estimated*[1]	*Common types of applications*	*Ecosystem services valued*
Travel cost	Direct use	Recreation	Maintenance of beneficial species, productive ecosystems, and biodiversity
Averting behavior	Direct use	Environmental impacts on human health	Pollution control and detoxification
Hedonic price	Direct and indirect use	Environmental impacts on residential property and human morbidity and mortality	Storm protection; flood mitigation; maintenance of air quality
Production function	Indirect use	Commercial and recreational fishing; agricultural systems; control of invasive species; watershed protection; damage costs avoided	Maintenance of beneficial species; maintenance of arable land and agricultural productivity; prevention of damage from erosion and siltation; groundwater recharge; drainage and natural irrigation; storm protection; flood mitigation
Replacement cost	Indirect use	Damage costs avoided; freshwater supply	Drainage and natural irrigation; storm protection; flood mitigation
Stated preference	Use and nonuse	Recreation; environmental impacts on human health and residential property; damage costs avoided; existence and bequest values of preserving ecosystems	All of the above

Source: Adapted from Heal et al. (2005), table 4-2.

[1]Typically, use values involve some human "interaction" with the environment, whereas nonuse values do not, as they represent an individual value of the pure "existence" of a natural habitat or ecosystem or wanting to "bequest" it to future generations. Direct use values refer to both consumptive and nonconsumptive uses that involve some form of direct physical interaction with ecosystem services, such as recreational activities, resource harvesting, drinking clean water, breathing unpolluted air, and so forth. Indirect use values refer to those ecosystem services whose values can only be measured indirectly, since they are derived from supporting and protecting activities that have directly measurable values.

First, the application of some valuation methods is often limited to specific types of ecosystem services. For example, the *travel cost* method is used principally for those environmental values that enhance individuals' enjoyment of recreation and tourism, *averting behavior* models are best applied to the health effects arising from environmental pollution, and *hedonic price* models are used primarily for assessing work-related environmental hazards and environmental impacts on property values.

In contrast, *stated preference* methods, which include *contingent valuation* methods, *conjoint analysis,* and *choice experiments,* have the potential to be used widely in valuing ecosystem services. These valuation methods share the common approach of surveying individuals who benefit from an ecosystem service or range of services, in the hope that analysis of these responses will provide an accurate measure of the individuals' willingness to pay for the service or services. In addition, stated-preference methods can go beyond estimating the value to individuals of single and even multiple benefits of ecosystems and in some cases elicit "nonuse values," such as the additional "existence" and "bequest" values that individuals attach to ensuring that a preserved and well-functioning system will be around for future generations to enjoy. For example, a study of mangrove-dependent coastal communities in Micronesia demonstrated through the use of contingent valuation techniques that the communities "place some value on the existence and ecosystem functions of mangroves over and above the value of mangroves' marketable products" (Naylor and Drew 1998). Similarly, choice experiments and conjoint analysis, which ask respondents to rank, rate, or choose among various environmental outcomes or scenarios, have the potential to elicit the relative values that individuals place on different ecosystem services. A study of wetland restoration in southern Sweden revealed through choice experiments that individuals' willingness to pay for the restoration increased if the result enhanced overall biodiversity but decreased if the restored wetlands were used mainly for the introduction of Swedish crayfish for recreational fishing (Carlsson et al. 2003).

However, as emphasized by Heal and colleagues (2005), a stated-preference study needs two key conditions: (1) the information must be available that allows description of the change in a natural ecosystem in terms of services that people care about, so that they can place a value on those services, and (2) the change in the natural ecosystem must be explained in the survey instrument in a manner that people will understand and that will cause them not to reject the valuation scenario. For many regulating and supporting services, one or both of these conditions may not hold. For instance, it has proven very difficult to describe accurately, through the hypothetical scenarios required by stated-preference surveys, how changes in ecosystem processes and components affect ecosystem regulatory and habitat functions—and thus the specific benefits arising from these functions that individuals value. If there is considerable scientific uncertainty surrounding these linkages, then not only is it difficult to construct such hypothetical scenarios, but also any responses elicited from individuals from stated-preference surveys are likely to yield inaccurate measures of their willingness to pay for ecosystem services.

In contrast to stated-preference methods, the advantage of *production function* (PF) approaches is that they depend on only the first condition, and not both conditions, holding. That is, for those regulatory and habitat functions where there is sufficient scientific knowledge of how these functions link to specific ecological services that support or protect economic activities, then it may be possible to employ the PF approach to value these services. The basic modeling approach underlying PF methods, also called "valuing the environment as input," is similar to determining the additional value of a change in the supply of any factor input (Barbier 1994, 2000, 2007; Freeman 2003). If changes in the regulatory and habitat functions of ecosystems affect the marketed production activities of an economy, then the effects of these changes will be transmitted to individuals through the price system via changes in the costs and prices of final goods and services.

An adaptation of the PF methodology is required for regulating services such as storm protection, flood mitigation, prevention of erosion and siltation, and pollution control. In such cases, the environment produces a nonmarketed service, such as "protection" of economic activity, property, and even human lives, which benefits individuals through limiting damages. Applying PF approaches requires modeling the "production" of this protection service and estimating its value as an environmental input in terms of the expected damages avoided by individuals (Barbier 2007).

However, PF methods have their own measurement issues and limitations. Applying the PF method raises questions about how changes in the ecosystem service should be measured, whether market distortions in the final goods market are significant, and whether current changes in ecological services may affect future productivity through biological "stock effects." A common approach in the literature is to assume that an estimate of ecosystem area may be included in the "production function" of marketed output as a proxy for the ecological service input. For example, this is the standard approach adopted in coastal habitat–fishery PF models, as wetland area is thought to be a suitable proxy for productivity that affects fish catch (Barbier 2000, 2007; Freeman 2003). For a different perspective on the production function approach using ecological endpoints, see Wainger and Boyd (chap. 6 of this volume).

In circumstances where an ecosystem service is unique to a specific ecosystem and is difficult to value, then economists have sometimes resorted to using the cost of replacing the service or treating the damages arising from the loss of the service as a valuation approach. This approach should be used with caution. For example, the few studies that have attempted to value the storm prevention and flood mitigation services of the "natural" storm barrier function of mangrove systems have employed the replacement cost method by simply estimating the costs of replacing mangroves by constructing physical barriers to perform the same services (Chong 2005). Shabman and Batie (1978) suggested that this method can provide a reliable valuation estimation for an ecological service, but only if the following conditions are met: (1) the alternative considered provides the same services, (2) the alternative compared for cost comparison is the least-cost alternative, and (3) there is substantial evidence that the service would be demanded by society if it were provided by that least-cost alternative. Unfortunately, very few replacement cost studies meet all three conditions.

In the absence of conducting reliable stated-preference surveys to elicit individuals' willingness to pay for ecosystem services, for some benefits an alternative to employing either the replacement cost or the cost-of-treatment method might be the expected-damage function (EDF) approach. The EDF approach is nominally straightforward; it assumes that the value of an asset that yields a benefit in terms of reducing the probability and severity of some economic damage is measured by the reduction in the expected damage. The essential step to implementing this approach, which is to estimate how changes in the asset affect the probability that the damaging event will occur, has been used routinely in risk analysis and health economics—for example, in cases involving airline safety performance, highway fatalities, and drug safety and in studies of the incidence of diseases and accident rates. I have shown previously (Barbier 2007) that the EDF approach can be applied, under certain circumstances, to value ecosystem services that also reduce the probability and severity of economic damages, such as the storm protection service of mangroves.

VALUING MANGROVE–AQUACULTURE TRADE-OFFS IN THAILAND

To illustrate the issues surrounding the nonmarket valuation of ecosystem services, and to show why such valuation is important to inform decisions regarding conservation versus development trade-offs, we will examine the specific case of aquaculture expansion and mangrove loss in Thailand. With the rapid growth of aquaculture worldwide, this is a case with broad applicability.

Over the past three decades, global output from aquaculture grew at an annual average rate of 9.1%, reaching 39.8 million metric tons in 2002 (FAO 2005). This growth rate was higher than any other animal-food-producing system, including livestock rearing for meat. By 2020, the baseline projection for global aquaculture production is 53.6 million metric tons but could be as high as 83.6 million metric tons (Delgado et al. 2003).

These trends in increased aquaculture production globally will require both land and water, placing additional stress on natural coastal and marine ecosystems. There is also concern about the environmental impacts of intensive systems, especially the large-scale production required for shrimp, salmon, and other high-valued species. For instance, aquaculture accounts for 52% of mangrove loss globally, with shrimp farming alone accounting for 38% of mangrove deforestation (Valiela et al. 2001). Intensive aquaculture systems can also lead to water shortages and pollution from effluent discharges, disrupting the functioning of coastal and marine ecosystems through nutrient overload (Goldburg and Naylor 2005). Another important ecological challenge for farming omnivorous and carnivorous species is their dependence on wild-caught marine fish, such as anchovies, sardines, capelin, and other lower-trophic species, for the fish meal and oils used in feeds. Increased growth in aquaculture may mean increasing reliance on capture fisheries supplying the input species used in feeds (Naylor et al. 2000; Delgado et al. 2003). Finally, there is growing concern that marine farming may increase the risk of invasive species problems in surrounding ecosystems through the increased number of escaped farm fish that interact with wild fish (Goldburg and Naylor 2005).

In Thailand, aquaculture expansion has been associated with mangrove wetlands destruction. Since 1961, Thailand has lost 1,500 to 2,000 km^2 of coastal mangroves, or about 50%–60% of the original area (FAO 2003). From 1975 to 1996, 50%–65% of Thailand's mangroves were lost to shrimp farm conversion alone (Aksornkoae and Tokrisna 2004). Mangrove deforestation in Thailand has focused attention on the two principle services provided by mangrove ecosystems: their roles as nursery and breeding habitats for offshore fisheries and as natural storm barriers to periodic coastal storm events, such as wind storms, tsunamis, storm surges, and typhoons. In addition, many coastal communities exploit mangroves directly for a variety of products, such as fuelwood, timber, raw materials, honey and resins, and crabs and shellfish. Various studies have suggested that these three benefits of mangroves are significant in Thailand (Sathirathai and Barbier 2001; Barbier 2003, 2007).

Valuation of the ecosystem services provided by mangroves is therefore important for two land use policy decisions in Thailand. First, conversion of remaining mangroves to shrimp farm ponds and other commercial coastal developments continues to be a major threat to Thailand's remaining mangrove areas. Second, since the December 2004 tsunami disaster, there is now considerable interest in rehabilitating and restoring mangrove ecosystems as natural barriers to future coastal storm events. Thus, valuing the goods and services of mangrove ecosystems can help to address two important policy questions: Do the net economic returns to shrimp farming justify further mangrove conversion to this economic activity? and, Is it worth investing in mangrove replanting and ecosystem rehabilitation in abandoned shrimp farm areas?

To illustrate how improved and more accurate valuation of ecosystems can help inform these policy decisions, table 8.2 compares the per hectare (ha) net returns to shrimp farming, the costs of mangrove rehabilitation, and the value of mangrove services. All land uses are assumed to be instigated over 1996–2004 and are valued in 1996 US$ per hectare.

Several analyses have demonstrated that the overall commercial profitability of shrimp aquaculture in Thailand provides a substantial incentive for private landowners to invest in such operations (Tokrisna 1998; Sathirathai and Barbier 2001; Barbier 2003). However, many of the conventional inputs used in shrimp pond operations are subsidized, thus increasing artificially the private returns to shrimp farming. In table 8.2 the net economic returns to shrimp farming, which are calculated once the estimated subsidies are removed, are based on nondeclining yields over a 5-year period of investment (Sathirathai and Barbier 2001). After this period, there tend to be problems of drastic yield decline and disease; shrimp farmers then usually abandon their ponds and find new locations. In table 8.2 the annual economic returns to shrimp aquaculture are estimated to be $322 per ha and, when discounted over the 5-year period at a 10%–15% rate, yield a net present value of $1,078 to $1,220 per ha.

There is also the problem of the highly degraded state of abandoned shrimp ponds after the 5-year period of their productive life. Across Thailand those areas with abandoned shrimp ponds degenerate rapidly into wasteland, since the soil becomes very acidic, compacted, and too poor in quality to be used for any other productive use, such as agriculture. Rehabilitation of the abandoned shrimp farm site requires treating and detoxifying the soil, replanting mangrove forests, and maintaining and protecting mangrove seedlings for several years. These restoration costs are considerable, $8,812 to $9,318 per ha in net present value terms (table 8.2). This reflects the fact that converting mangroves to establish shrimp farms is almost an "irreversible" land use, and without considerable additional investment in restoration, these areas do not regenerate into mangrove forests. Before the decision to allow shrimp farming to take place, the restoration costs could be treated as one measure of the "user cost" of converting mangroves irreversibly, and this cost should be deducted from the estimation of the net returns to shrimp aquaculture. As the restoration costs exceed the net economic returns per hectare, the decision should be to prevent the shrimp aquaculture operation from occurring.

Unfortunately, past land use policy in Thailand has ignored the user costs of shrimp farming, and as a result many coastal areas have been deforested of mangroves. Unfortunately, many short-lived shrimp farms in these areas have also long since fallen unproductive

Table 8.2. Comparison of land use values per ha, Thailand, 1996–2004 (US$)

Land use	*Net present value per ha (10%–15% discount rate)*
Shrimp farming	
Net economic returns[1]	1,078–1,220
Mangrove ecosystem rehabilitation	
Total cost[2]	8,812–9,318
Ecosystem goods and services	
Net income from collected forest products[3]	484–584
Habitat–fishery linkage[4]	708–987
Storm protection service[5]	8,966–10,821
Total	10,158–12,392

[1]Based on annual net average economic returns US$322 per ha for 5 years, from Sathirathai and Barbier 2001, updated to 1996 US$.

[2]Based on costs of rehabilitating abandoned shrimp farm site, replanting mangrove forests, and maintaining and protecting mangrove seedlings, from Sathirathai and Barbier 2001, updated to 1996 US$.

[3]Based on annual average value of $101 per ha over 1996–2004, from Sathirathai and Barbier 2001, updated to 1996 US$.

[4]Based on a dynamic analysis of mangrove–fishery linkages over 1996–2004 from and assuming the estimated Thailand deforestation rate of 3.44 km^2 per year (see Barbier 2007).

[5]Based on marginal value per ha of expected damage function approach of Barbier (2007).

and are now abandoned. Thus, an important issue today is whether it is worth restoring mangroves in these abandoned areas. Potential benefits from the restored mangroves include the net income from local mangrove forest products, habitat–fishery linkages, and storm protection (table 8.2).

Sathirathai and Barbier (2001) estimated the value to local communities of using mangrove resources in terms of the net income generated from the forests from various wood and nonwood products. If the extracted products were sold, then market prices were used to calculate the net income generated (gross income is minus the cost of extraction). If the products were used only for subsistence, the gross income was estimated based on surrogate prices, that is, the market prices of the closest substitute. Based on surveys of local villagers in Surat Thani Province, the major products collected by the households were various fishery products, honey, and wood for fishing gear and fuelwood. As shown in table 8.2, the net annual income from these products is $101 per ha, or a net present value of $484 to $584 per ha.

The coastal habitat–fishery linkage of mangroves in Thailand may be modeled through incorporating the change in wetland area within a multiperiod harvesting model of the fishery (Barbier 2007). The key to this approach is to model a coastal wetland that serves as a breeding and nursery habitat for fisheries as affecting the growth function of the fish stock. As a result, the value of a change in this habitat-support function is determined in terms of the impact of any change in mangrove area on the dynamic path of the returns earned from the fishery. The net present value of this service ranges from $708 to $987 per ha (table 8.2).

The value of the coastal protection service of mangroves in table 8.2 is derived by employing the EDF valuation methodology for estimating the expected damage costs avoided through increased provision of the storm protection service of coastal wetlands (Barbier 2007). As discussed previously, two components are critical to implementing the EDF approach to estimating the changes in expected storm damages: the influence of wetland area on the expected incidence of economically damaging natural disaster events, and some measure of the additional economic damage incurred per event. Both of these components can be estimated, provided that there are sufficient data on past storm events and preferably across different coastal areas, as well as estimates of the economic damages inflicted by each event. The estimated benefits from the storm protection service of mangroves in Thailand is $1,879 per ha, or $8,966 to $10,821 per ha in net present value terms (table 8.2).

The net present value of all three mangrove ecosystem benefits ranges from $10,158 to $12,392 per ha, clearly exceeding the net economic returns to shrimp farming (table 8.2). In fact, the net income to local coastal communities from collected forest products and the value of habitat–fishery linkages total from $1,192 to $1,571 per ha, which is greater than the net economic returns to shrimp farming. However, the value of the storm protection is critical to the decision as to whether or not to replant and rehabilitate mangrove ecosystems in abandoned pond areas. The storm protection benefit makes mangrove rehabilitation an economically feasible land use option.

In summary, this case study has shown the importance of valuing the ecosystem services that support coastal aquaculture. The irreversible conversion of mangroves for aquaculture results in the loss of ecosystem services that generate significantly large economic benefits. This loss of benefits must be taken into account in land use decisions that lead to the widespread conversion of mangroves. Finally, the largest economic benefits of mangroves appear to arise from the regulatory and habitat functions, reinforcing the importance of measuring the value of such services.

Implications for Other Coastal and Marine Systems

The Thailand case study illustrates the importance of estimating one particular benefit—the coastal protection service of mangroves—of any decision to restore these critical ecosystems in abandoned shrimp pond sites. A similar coastal land use decision is now being faced in the United States. Since the 2005 Hurricanes Katrina and Rita, attention in the United States has focused on restoring coastal wetlands, especially along the vulnerable Gulf Coast, to recover a valuable ecosystem "service"—the protection of low-lying coastal communities and property from tropical hurricanes and other storm events (Zinn 2005; USACE 2006; WGPPLC 2006). The rationale for these restoration schemes is to enhance the "natural barrier" function of coastal wetlands, such as salt marshes and mangroves. Before Hurricanes Katrina and Rita devastated the central Gulf Coast of the United States in 2005, the US Army Corps of Engineers had proposed a $1.1 billion multiyear program to slow the rate of wetland loss and restore some wetlands in coastal Louisiana. In the aftermath of these hurricanes, the US Congress is now expanding the program substantially to a $14 billion restoration effort (Zinn 2005). By estimating the storm protection value of coastal wetlands along the Gulf Coast and in other hurricane-prone areas, such as the South Atlantic, we could obtain a better idea of whether such restoration investments are justified—or perhaps even an underestimate of what areas need to be restored.

Conclusions

Important advances have been made recently in the economic valuation of ecosystem services, especially for certain coastal and marine systems. It is easy to focus on the difficulties and limitations of valuing ecosystem services, whereas a more critical challenge is that ecosystem services are largely not marketed, and unless some attempt is made to value the aggregate willingness to pay for these services, then management of coastal and marine ecosystems and their services will not be efficient (Heal et al. 2005). Often, coastal zone economic development occurs under the assumption that the services of natural coastal and marine ecosystems have little or no benefits, and the result may be an excessive, and potentially irreversible, loss of valuable systems and their services.

The Thailand case study in this chapter illustrates some of the trade-offs that must be considered when assessing a coastal development option such as shrimp farming. It is important to emphasize that the case study does not imply that all shrimp aquaculture should be halted in Thailand. It does suggest, however, the need for better policies to control excessive shrimp farm expansion and subsequent mangrove loss by accounting for the full range of services provided by the mangroves. To achieve this objective, there are clearly several steps that the government of Thailand could take to reduce the current perverse incentives for excessive mangrove conversion for shrimp farming. These include eliminating preferential subsidies for inputs of larvae, chemicals, and machinery used in shrimp farming; ending preferential commercial loans for clearing land and establishing shrimp ponds; employing land auctions and concession fees for the establishment of new farms in the "economic zones" of coastal areas; and finally, charging replanting fees for farms that convert mangroves (Barbier and Sathirathai 2004). Reducing the other environmental impacts of shrimp farming is also important, notably, problems of water pollution, the depletion of wild fish stocks used for feed, and disease outbreaks within ponds (Naylor et al. 2000; Goldburg and Naylor 2005).

In summary, this chapter has shown that nonmarket valuation of ecosystem services is increasingly important to informing decisions regarding trade-offs in an ecosystem-based management context. However, further progress in this endeavor faces several challenges. First, for valuation methods to be applied effectively, it is important that the key ecological and economic relationships be well understood. Unfortunately, our knowledge of the ecological functions underlying many services is still incomplete. Second, as described by Leslie and Kinzig (chap. 4 of this volume), ecosystems tend to display threshold effects and other nonlinearities that are difficult to predict, let alone model in terms of their economic impacts. Of particular importance are measuring the two costs associated with the loss of resilience of an ecosystem: the probability of crossing thresholds as natural habitat area is converted or damaged, and the loss in valuable ecological services that result from any irreversible alteration in the ecosystem state. In practice, estimating such costs for unknown ecological resilience effects is very difficult. These are challenges that must be addressed if economic valuation of ecosystem service trade-offs is to contribute to ecosystem-based management. As a review by the National Research Council concludes, "Given the imperfect knowledge of the way people value natural ecosystems and their goods and services, and our limited understanding of the underlying ecology and biogeochemistry of aquatic ecosystems, calculations of the value of the changes resulting from a policy intervention will always be approximate" (Heal et al. 2005, p. 218). Whether we can improve on our approximations so that valuation can better assist EBM approaches is the key challenge for the near future.

Finally, there have been attempts to extend valuation approaches to the ecosystem level through integrated ecological–economic modeling (for related discussion, see McLeod and Leslie, chap. 19 of this volume). This allows the ecosystem functioning and dynamics underlying the provision of ecosystem services to be modeled and can be used to value multiple rather than single services. For example, returning to the Thailand case study, it is well known that both coral reefs and sea grasses complement the role of mangroves in providing both the habitat–fishery and storm protection services. Thus full modeling of the integrated mangrove–coral reef–sea grass system could improve measurement of the benefits of both services. As we learn more about the important ecological and economic role played by such services, it may be relevant to develop multiservice production function modeling to understand more fully what values are lost when such integrated coastal and marine systems are disturbed or destroyed.

Acknowledgments

The author would like to thank the editors, Heather Leslie and Karen McLeod, as well as two anonymous referees, for their helpful comments and suggestions on this chapter.

References

Aksornkoae, S., and R. Tokrisna. 2004. Overview of shrimp farming and mangrove loss in Thailand. In *Shrimp farming and mangrove loss in Thailand*, ed. E. B. Barbier and S. Sathirathai. London: Edward Elgar.

Arrow, K. J., and A. C. Fisher. 1974. Environmental preservation, uncertainty and irreversibility. *Quarterly Journal of Economics* 88:312–19.

Barbier, E. B. 2003. Habitat–fishery linkages and mangrove loss in Thailand. *Contemporary Economic Policy* 21:59–77.

Barbier, E. B., and S. Sathirathai, ed. 2004. *Shrimp farming and mangrove loss in Thailand*. London: Edward Elgar.

Barbier, E. B., and G. R. Heal. 2006. Valuing ecosystem services. *The Economists' Voice* 3(3): article 2.

Barbier, E. B. 2007. Valuing ecosystem services as productive inputs. *Economic Policy* 22(49):177–229.

Barbier, E. B. 1994. Valuing environmental functions: Tropical wetlands. *Land Economics* 70(2):155–73.

Barbier, E. B. 2000. Valuing the environment as input: Applications to mangrove-fishery linkages. *Ecological Economics* 35:47–61.

Carlsson, F., P. Frykblom, and C. Lijenstolpe. 2003. Valuing wetland attributes: An application of choice experiments. *Ecological Economics* 47:95–103.

Chong, J. 2005. *Protective values of mangrove and coral ecosystems: A review of methods and evidence.* Gland, Switzerland: International Union for Conservation of Nature (IUCN).

Delgado, C. L., N. Wada, M. W. Rosegrant, S. Meijer, and M. Ahmed. 2003. *Fish to 2020: Supply and demand in a changing world.* Washington, DC: International Food Policy Research Institute.

Dixit, A., and R. Pindyck. 1994. *Investment under uncertainty.* Princeton, NJ: Princeton University Press.

FAO (Food and Agriculture Organization of the United Nations). 2003. *Status and trends in mangrove area extent worldwide,* ed. M. L.Wilkie and S. Fortuna. Forest Resources Assessment Working Paper no. 63. Forest Resources Division. Rome: FAO.

FAO (Food and Agriculture Organization of the United Nations). 2005. *The state of world fisheries and aquaculture 2004.* Rome: FAO.

Freeman, A. M., III. 2003. *The measurement of environmental and resource values: Theory and methods,* 2nd ed. Washington, DC: Resources for the Future.

Goldburg, R., and R. Naylor. 2005. Future seascapes, fishing and fish farming. *Frontiers in Ecology and the Environment* 3(1):21–28.

Halpern, B. S., S. Walbridge, K. A. Selkoe, C. V. Kappel, F. Micheli, C. D'Agrosa, J. F. Bruno et al. 2008. A global map of human impacts on marine ecosystems. *Science* 319:948–52.

Heal, G., E. Barbier, K. Boyle, A. Covich, S. Gloss, C. Hershner, J. Hoehn et al. 2005. *Valuing ecosystem services: Toward better environmental decision making.* Washington, DC: National Academies Press.

Henry, C. 1974. Investment decisions under uncertainty: The "irreversibility effect." *American Economic Review* 64(6):1006–12.

Just, R. E., D. L. Hueth, and A. Schmitz. 2004. *The welfare economics of public policy: A practical approach to project and policy evaluation.* Cheltenham, UK: Edward Elgar.

Krutilla, J. V., and A. C. Fisher. 1985. *The economics of natural environments: Studies in the valuation of commodity and amenity resources.* Washington, DC: Resources for the Future.

MA (Millennium Ecosystem Assessment). 2003. *Ecosystems and human well-being: A framework for assessment.* Washington, DC: Island Press.

McLeod, K. L., J. Lubchenco, S. R. Palumbi, and A. A. Rosenberg. 2005. *Scientific consensus statement on marine ecosystem-based management.* The Communication Partnership for Science and the Sea (COMPASS). Signed by 221 academic scientists and policy experts with relevant expertise. http://www.compassonline.org/pdf_files/EBM_Consensus_Statement_v12.pdf.

Naylor, R., R. Goldburg, J. Primavera, N. Kautsky, M. Beveridge, J. Clay, C. Folke, J. Lubchenco, H. Mooney, and M. Troll. 2000. Effect of aquaculture on world fish supplies. *Nature* 405:1017–24.

Naylor, R., and M. Drew. 1998. Valuing mangrove resources in Kosrae, Micronesia. *Environment and Development Economics* 3:471–90.

Pagiola, S., K. von Ritter, and J. Bishop. 2004. *How much is an ecosystem worth? Assessing the economic value of conservation.* Washington, DC: World Bank.

Ready, R. C. 1995. Environmental valuation under uncertainty. In *The handbook of environmental economics,* ed. Daniel W. Bromley, 568–93. Cambridge, MA: Blackwell.

Sathirathai, S., and E. Barbier. 2001. Valuing mangrove conservation, Southern Thailand. *Contemporary Economic Policy* 19:109–22.

Shabman, L. A., and S. S. Batie. 1978. Economic value of natural coastal wetlands: A critique. *Coastal Zone Management Journal* 4(3):231–47.

Tokrisna, R. 1998. *The use of economic analysis in support of development and investment decision in Thai aquaculture: With particular reference to marine shrimp culture.* A report to the Food and Agriculture Organization of the United Nations.

USACE (US Army Corps of Engineers). 2006. *Louisiana coastal protection and restoration*. Washington, DC: USACE.

Valiela, I., J. L. Bowen, and J. K. York. 2001. Mangrove forests: One of the world's threatened major tropical environments. *BioScience* 51:807–15.

WGPPLC (Working Group for Post-Hurricane Planning for the Louisiana Coast). 2006. *A new framework for planning the future of coastal Louisiana after the hurricanes of 2005*. Cambridge, MD: Center for Marine Science, University of Maryland.

Zinn, J. 2005. *Hurricanes Katrina and Rita and the coastal Louisiana ecosystem restoration. CRS Report for Congress*. Washington, DC: Congressional Research Service (CRS), Library of Congress, 26 September 2005.

CHAPTER 9
Integrating Local and Traditional Ecological Knowledge

Andrew (Anaru) Kliskey, Lilian (Naia) Alessa,* and Brad Barr*

The cumulative and transmitted knowledge, experience, and wisdom of human communities with a long-term attachment to the sea provide rich lessons for ocean and coastal ecosystem-based management (EBM). In this chapter we examine the extent to which such local and traditional knowledge intersect with ocean and coastal ecosystem-based management and the facets they contribute to resilience in marine management. The core principle of EBM for the oceans is the recognition and practice of an "entire-ecosystem view," in which humans are seen as part of, rather than separate from, coastal and marine ecosystems (McLeod and Leslie, chap. 1 of this volume; Shackeroff et al., chap. 3 of this volume). This view can be defined as including an integrated view across species, sectors, activities, and concerns; protection of ecosystem structure, function, and processes; recognition of interdependence among ecological, social, economic, and institutional perspectives; and acknowledgment of the need for community-based, adaptive management. The movement toward EBM in Western science-based management represents not only an adoption of holistic practices and principles, but also a change in philosophy, perhaps even a shift in worldview (Wallace et al. 1996; Moore and Russell, chap. 18 of this volume; Shackeroff et al., chap. 3 of this volume).

Alternatively, some aspects of EBM could be characterized as a return to a prior way of thinking rather than an actual move to a new way of thinking, or more cynically they could be characterized as the result of a collapsed paradigm (Kuhn 1970) and the embrace of a new paradigm to fill the void. Critics counter that the adoption of such approaches is less about a shift in worldview among a broad constituency than about a shift in the adoption of a more conservationist perspective by influential institutions (Dove 2006). This perspective views EBM as yet another paradigm that is not much different from past ones, such as integrated coastal management (Cicin-Sain 1993), but one that captured some institutions' interests due to funding opportunities, comparability in research interests, and similarity with well-established worldviews (Cicin-Sain and Belfiore 2006).

Traditional ecological knowledge (TEK) and local knowledge (LK) represent multiple cumulative perspectives that are very much in line with EBM. Several of the principles listed above are either explicit or implicit principles of one or more TEK systems. This chapter provides a short overview of what TEK and LK are and how they guide ways of knowing, reviews the key lessons to be learned from TEK and LK with relevance to EBM, discusses several critical challenges to synthesizing and integrating TEK and LK into Western science-based management frameworks such as EBM, and finally considers the outlook for integrating TEK and LK in EBM by examining case studies from two US marine protected areas.

Traditional Ecological Knowledge and Local Knowledge in Ocean and Coastal Ecosystems

Both TEK and LK have utility and importance for EBM, and yet they are fundamentally

*Equal first authors

different ways of knowing. Thus it is important to make the distinction between them. In this chapter we discuss the utility of each, although there is a greater emphasis on TEK, since by definition it has several key elements in common with EBM as well as a considerably longer and richer history.

TEK has numerous definitions, including "a cumulative body of knowledge, practice, and belief, evolving by adaptive processes and handed down through generations by cultural transmission, about the relationship of living beings (including humans) with one another and with their environment" (Berkes 1999, p. 8). This typically includes multigenerational indigenous knowledge, for example, Inupiaq knowledge, practices, and beliefs with respect to marine mammal harvesting. TEK is sometimes referred to as "indigenous knowledge" to emphasize the knowledge that has been acquired and transmitted over thousands of years by people with a sustained and direct connection with the land and sea. TEK is characterized in the above definition as a knowledge–practice–belief complex (Gadgil et al. 1993; Berkes 1999; Berkes et al. 2000) that comprises the local knowledge of species and environmental conditions, the local practices toward resource harvesting and protection, and the beliefs of the community regarding its role and identity in the environment. An example from marine systems of the first component of this complex, traditional knowledge is the Inupiaq knowledge of sea ice dynamics and marine mammal movements and population changes in Alaska (Huntington 2001; George et al. 2004) that contributes to Northern Eskimo management of the marine mammal harvest and resource—what some would characterize as largely effective comanagement, institutionalized into the dominant Western culture. The Chisasibi Cree fishing practices in James Bay, Canada, of rotating fishing areas over a 4-year cycle and of using nets of different mesh sizes based on proximity to a village (Berkes 1981, 1987, 1999) contributes to the productivity and sustainability of their traditionally managed fishery and highlights the second component of the knowledge–practice–belief complex. The third component is exemplified by the New Zealand Maori belief of *whanaungatanga*, that other species and phenomena in the environment are kin and that your behavior toward, and management of, the marine environment should be guided by this kinship (Patterson 1994). A similar belief is held and practiced by numerous other indigenous cultures around the world.

In contrast to TEK, local knowledge (or local ecological knowledge) refers to the knowledge of the local residents of a community, often the users of a marine resource. This may include multigenerational knowledge in historically settled communities, for example, that of watermen communities in Chesapeake Bay (see Boesch and Goldman, chap. 15 of this volume), but it also includes individuals and stakeholder groups with little or no historical attachment to the area, for example, the marine interests group in Morro Bay, Califorrnia (see Wendt et al., chap. 11 of this volume). LK may include the knowledge of only one or two generations or encompass accumulated knowledge of numerous generations, but typically it has a shorter time span than TEK. Also, whereas TEK is nested in the knowledge–practice–belief complex, LK primarily involves knowledge of local species and dynamics, may or may not be linked to a system of practice, and is typically not embedded within an explicit belief system. LK encompasses a wide range of knowledge, from that of a local stakeholder with less than a lifetime of experience in the local area to the accumulated wisdom of an entire community comprising multiple generations of families with several hundred years of occupancy and use in a locale.

Local residents who are not indigenous are often assumed to have little knowledge of their environment, despite the fact that their livelihoods and well-being often depend on their knowledge of their local natural systems, for example, fish harvesters from coastal communities in Newfoundland and Labrador, Canada (Neis et al. 1999; Neis and Felt 2000). It is reasonable to assume that LK and TEK would converge in knowledge per se but diverge in beliefs, and culture. Thus, local knowledge and traditional knowledge are two different knowledge systems because they are embedded in their respective cultures, and they cannot always be treated together. Thus TEK and LK are inherently different, and each offers valuable lessons for managers.

Lessons for Ecosystem-Based Management from TEK and LK

In many ways TEK contrasts with scientific knowledge: TEK is predominantly qualitative rather than quantitative, intuitive rather than purely rational, holistic rather than reductionist, spiritual rather than mechanistic, based on diachronic data (long-term in one locality) rather than synchronic data (short-term in a broad area), and moral rather than objective (Berkes 1993). While TEK and Western scientific knowledge represent different ways of knowing, the emergence of EBM suggests a convergence between TEK and science-based understanding. A primary element of this convergence is the holistic or ecosystem viewpoint that is the core principle of EBM and is an underlying perspective for many TEK systems. A second lesson from some LK systems as well as a number of TEK systems is the adoption of adaptive approaches or strategies that correspond with adaptive learning in EBM. These elements are discussed below.

Ecosystem View

One of the clearest marine examples of the adoption and practice of an entire-ecosystem view is the ancient Hawaiian managed watershed, or *ahupua'a* (Johannes 1981; Costa-Pierce 1987; Berkes 1999), a TEK practice that is being revitalized on some Hawaiian islands. *Ahupua'a* typically extended from the headwaters and mountains of an island, down to the nearshore lagoon, and out beyond the reef to the sea (fig. 9.1). Often all the necessary resources for life could be found in the *ahupua'a,* and management was by a *konohiki* (custodian) who coordinated the community's efforts to sustain a productive *ahupua'a.* Flooded taro terraces (lo'i), dryland terraces (mala), and fishponds were key components in many *ahupua'a* (fig. 9.2). Lo'i provided the staple of the Hawaiian diet, taro. A productive lo'i system required sufficient quantities of cool, clean water. As such, the health of the whole community was very obviously tied to maintenance of a healthy watershed and ecosystem. The lack of extensive lo'i systems in modern-day Hawai'i is probably the most visual indicator of land and water use change from traditional to modern times. This all-encompassing view extended to the relationship of people with their watersheds.

Similar connections between people and the watersheds in which they live and ways of managing land–marine environments include the Fijian *vanua,* the Solomon Islands' *puava,* and the Yap Island *tabinau* (Ruddle 1991; Ruddle et al. 1992; Berkes 1999). The Dene (a First Nation people of Northern Canada) term *nde* is a holistic term for land, the living environment, relations of the living and nonliving, and everything in the environment having life and spirit (Cobb et al. 2005). Similarly, the James Bay, Canada, Cree term *aschkii* refers to the physical environment, including plants, animals, and humans (Cobb et al. 2005). These ecosystem views, practiced in TEK, are more than a perspective or understanding that encompasses connectivity and the interaction between biophysical components of the marine ecosystem—they are also belief systems. For some indigenous peoples, an ecosystem view

Figure 9.1 Orthographic view of Hā'ena on the island of Kaua'i, Hawai'i, USA. From Carlos Andrade, University of Hawai'i at Manoa.

treats all parts of the ecosystem as kin to oneself: for example, the Maori *whanaungatanga* that carries forward to management practices (Ramstad et al. 2007). While the knowledge and practice components of TEK may be readily incorporated into marine EBM, the belief component represents a way of knowing that is more difficult to simply integrate. Incorporating TEK *into* a science-founded management framework also has many political and ethical dimensions (e.g., Nazarea 2006), frequently involving complex cultural and spiritual aspects based on nonscientific principles (Gadgil et al. 1993). These are discussed below in the Challenges to Integrating TEK and LK with EBM section.

TEK places humans in the ecosystem through a combined, holistic view of the world and the individual's place in it. This approach is exemplified by Dasmann's (1976) distinction between "ecosystem people" and "biosphere people," where ecosystem people live in one or two ecosystems, rely on those systems for sustenance, and suffer the consequences if they overexploit or "break the rules." Biosphere people live in the globalized world and serially exploit many ecosystems, so if a resource of value is lost from one, they simply move on to another source in the biosphere—they are not immediately subject to the perils of "breaking the rules." Berkes and colleagues (2006) refer to such serial exploiters as "roving bandits" and cite the example of the sequential depletion of global sea urchin stocks from 1945 to 2000 as illustrative of this effect. This distinction between local and global perspectives helps to explain the dichotomy between communities whose practices are rooted in TEK and those societies that rely on Western management systems. For practitioners attempting to meld TEK and EBM, there is the real challenge of whether these contrasting and arguably

(A)

(B)

Figure 9.2
(A) Lo'i system of traditional taro terraces in Ha'ena ahupua'a, Kaua'i, Hawai'i. (B) Traditional Hawaiian fishponds in He'eia ahupua'a, Oahu, Hawai'i. A and B from Kaeo Duarte, Kamehameha Schools, Hawai'i.

dichotomous approaches can be integrated—and even if this were possible, whether holders of TEK would want this.

We believe that our inability to effectively find coherent, just, and democratically mandated solutions to the human dimensions of ocean and coastal ecosystem change is a systemic one that centers on the culture of those most invested in ignoring change (Nazarea 2006). In this manner, it is a question of understanding culture, respecting the relationship of culture with the marine ecosystem, and acknowledging the ability of TEK-based systems to respond to changes in ways that promise the most success for equitable and sustainable livelihoods. For further discussion of equity in EBM, see Moore and Russell (chap. 18 of this volume).

A key component of the holistic ecosystem view is the recognition of the dynamic, constantly changing nature of the world. TEK also develops and explains processes that occur outside the temporal and spatial scales of a person through a rich and diverse associated culture (e.g., stories, myths, dances) and emphasizes the status quo (i.e., the present moment in any given place) as sacred, thus avoiding many of the pitfalls associated with managing toward a specific "endpoint" or equilibrium, which is unlikely to exist in any social–ecological systems. These features result in a resilient management system that is place-based but not bounded by a specific result within a short time frame. Rather, it focuses on continued generation of a product or outcome that will be, in the framework of the users, available to subsequent generations indefinitely. At the same time, TEK accommodates uncertainty. It is ultimately this component that is most attractive conceptually and most challenging in reality: the precautionary principle. TEK acquires knowledge and transmits it to individuals across generations with a central caveat that all resources are provided almost marginally to the user and that small variations in social and biophysical parameters can upset this balance. Examples of TEK-based management strategies include limited entry methods, marine protected areas, closed areas, closed seasons, size limits, and fishing gear restrictions (Johannes 2002b). This is not to say that TEK-based efforts always work (Johannes 2002a; Guzman et al 2003). Some suggest that TEK and associated management systems were not particularly good at maintaining sustainable relations between people and ecosystems (e.g., Diamond 1986; Jackson et al. 2001) and that building contemporary marine management on traditional resource management will fail because of the fundamentally different worldviews on which they rest (Dwyer 1994). Johannes (2002a, b) rebuts these suggestions as overgeneralizations and lacking extensive evidence. We can conclude that some TEK systems have worked and some have not and that important lessons on recognizing resource limits and developing tools to live within those limits can be discerned in both cases.

Adaptive Management

Cod fishers in Fogo Island, Newfoundland, developed several adaptive strategies in response to the unpredictable seasonal variation in cod abundance and spatial variability in abundance at different fishing grounds from their long-term LK and experience of the local seas (McCay 1978b; Davidson-Hunt and Berkes 2003). Strategies such as fishing several locales, maintaining a wide variety of fishing gear to allow fishing under different habitat conditions, maintaining equipment and personnel to support several seasonal fisheries, diversification into fishing other species, and diversification into alternative livelihoods (e.g., farming) enabled the Fogo Island community to adapt to a fishery exhibiting wide temporal and spatial fluctuation. This example demonstrates adaptation in the face of changing and uncertain environments, but adaptive management is more than that. It involves designing interventions in ways that not only are appropriate in the face of uncertainty, but also have built-in ways of evaluating and adaptively responding to new information and change in the system (Walters 1986; Lee 1999). The cooperative management of New Jersey's hard clam spawner sanctuaries highlights an "incrementalist" or "muddling through" adaptive management strategy incorporating LK (McCay 1978a), as do the Chisasibi Cree TEK-based fishing practices described above (Berkes 1999). While adaptive management is a widely advocated and important element of EBM, there are numerous challenges with implementing adaptive management and few successful examples in Western environmental management (Lee 1999; Gregory et al. 2006). Adaptive management is particularly difficult to do in largely risk-averse management agencies; the lessons from adaptive management in TEK frequently apply to local-scale systems, whereas Western agencies that might consider adopting EBM frequently have regional- and national-scale foci.

Indigenous and local populations' predictions of marine ecosystem changes and development of response strategies suggest a possible cognition of forthcoming conditions and recognition of appropriate mitigation and adaptation responses. Perception of climate change is often structured by knowledge of resource–environment interactions and by different outcomes associated with the changed conditions (Vedwan and Rhoades 2001). For example, by integrating LK of the biology and behavior of a variety of fish species; the different fishing grounds in the Gulf of California; awareness of the irregularities of weather conditions, tides, and currents; and the ability to choose among a variety of fishing gear, small-scale fishers in the Gulf of California, Mexico, have developed an adaptation strategy to respond to climate change impacts on their marine resource (Vasquez-Leon 2002). See Ezcurra et al. (chap. 13 of this volume) for a more detailed treatment of ecosystem-based management in the Gulf of California, as well.

The incorporation of TEK and LK can provide information on adaptive behavior and limits to adaptation (Rayner and Malone 1998; Neis and Felt 2000; Smit and Pilifosova 2001). Western science-based management, and its culture of "monitoring ecosystems," often fails to keep pace with observations and expectations in understanding the implications of change, particularly in marine systems (Mantua et al. 2002). However, observations and expectations derived from accumulated experiences taken over a wide perspective, including TEK and LK, can help people identify emergent properties, thresholds, and synergies between systems (Chave and Levin 2002; Kinzig et al. 2006; Leslie and Kinzig, chap. 4 of this volume). One task in ecosystem-based management is the articulation and understanding of measurable variables and criteria that can be used to identify and describe repeating patterns and emergent properties of a specific marine system. TEK and LK may serve not only to illuminate potential ecosystem indicators but to articulate the intimate relationships that cultures have with their marine environment (e.g., Neis et al. 1999; Cobb et al. 2005; Kaufman et al., chap. 7 of this volume; Shackeroff et al., chap. 3 of this volume). Such relationships include not only rules and guidelines for the use and conservation of resources, but also those for social interactions, traditions, and community building (Pinkerton 1989; Dyer and McGoodwin 1994).

Challenges to Integrating TEK and LK with EBM

The interaction between TEK and Western/science-based management models, such as EBM, can be problematic for several reasons. Traditional people do not necessarily want their knowledge incorporated in these frameworks. They may feel EBM privileges scientific knowledge even more than other models of management, such as integrated coastal management (Dove 2006). Associated with this critique are the ethical issues surrounding the erosion of linguistic and cultural knowledge (e.g., Drew and Henne 2006) and the assimilation of TEK following colonization, industrialization, urbanization, and globalization (Nazarea 1999; Brodnig and Mayer-Schonberger 2000. Focusing on Oceania, Johannes (1978) describes the impacts of Westernization on traditional marine conservation methods, the breakdown of traditional authority, and the establishment of colonial laws and practices. A central mechanism in the demise of traditional methods in this case was the abrogation of traditional nearshore tenure and fishing rights (Johannes 1978; Gadgil and Berkes 1991). These changes may reduce, in turn, social resilience in marine social–ecological systems (Folke et al. 2005). More recently, Johannes (2002b) has pointed to a reversal in this trend and a renaissance of community-based marine resource management. He identified cooperative management, in which decision making and implementation are locally and community based, as a major factor involved in this shift. Other key issues for indigenous or local communities considering comanagement approaches that integrate TEK and Western management models are the intellectual and cultural property rights associated with "their" knowledge (Moran et al. 2001; Dove 2006).

The contemporary application of TEK in North America often pits local indigenous approaches against the dominant Western science-based approach, or at least sees TEK as subordinate to a dominant paradigm. The implementation of TEK with respect to Western environmental management has been characterized as being motivated either by attempts to complement general science-based knowledge with site-specific, contextualized knowledge or by attempts to challenge science-based knowledge and approaches (Gadgil et al. 2003). Thus, TEK is viewed by some as complementing Western science-based resource management through community-based monitoring programs (Gadgil et al. 2003; Cobb et al. 2005) that incorporate traditional knowledge into marine ecosystem monitoring. Other authors have explicitly examined the difficulties of incorporating TEK within Western systems of environmental management and/or assessment, including issues of formatting and framing TEK (Usher 2000; Sherry and Myers 2002; Drew 2005). For Western science-based agencies, a key issue is the admissibility of local observations (Gadgil et al. 2003), which can lead to the teasing apart of local knowledge from traditional beliefs, in other words, disaggregating the knowledge–practice–belief complex, and assimilation of that knowledge which is empirically valid (Colding and Folke 2001). This process of identifying useful local or traditional knowledge, scientific validation, and subsequent abstraction of useful elements, and its generalization for use in a more global context, has been termed "scientization" by Agrawal (2002). While this approach preserves the rules of science, it can be considered anathema to the integrity of TEK, with its tight mesh between knowledge and belief.

Another critical issue in melding TEK and EBM is the potential mismatch between place-based local and traditional knowledge and the broad regional and national scales that EBM is typically focused on. While EBM and large marine ecosystems may align with the "ecological thinking" of indigenous and local people, there is a risk that TEK and local control of resources are likely to be overlooked at larger geographic scales. However, local observations based on TEK can be aggregated from the local to regional scale, as shown by monitoring in Hudson Bay (Gadgil et al. 2003) and in the Bering Sea (Gill 2007). Involving traditional peoples and fisheries in the evolution, design, funding, and evaluation of EBM can help mitigate this challenge (e.g., Neis and Felt 2000).

Recognizing change or the inevitability of change is fundamental to applying timely mitigation responses within the social–ecological system before a perturbation or loss of resilience that could contribute to a shift in ecological state (see also Leslie and Kinzig, chap. 4 of this volume). TK and LEK can provide information that can advance this analysis.

Local knowledge provides longitudinal data incorporating perspective and experience, and that local knowledge is consensual, replicable, and to some extent experimental and predictive, and it can be generalized (Denny 1986; Bielawski 1995; Kuzyk et al. 1999; Cruikshank 2001). However, in modernity, local knowledge on its own is poorly equipped to perceive, let alone integrate, change at larger scales, due to the complexities and rates of change. Together with Western science, however, TEK and LK provide a temporal series of references which may be applied to modern management challenges, providing adequate attention is given to the social context in which management frameworks are applied (Cinner 2007).

Outlook for Integrating TEK and LK with EBM

The steady transition from a nature- to a human-dominated epoch of environmental change (Vitousek et al. 1997; Messerli et al. 2000) has exacerbated the vulnerability of many contemporary maritime cultures and the marine environments on which they rely. Recognition that ecosystem health is as much about sustaining human communities, cultural diversity, economic opportunity, and habitat as it is about sustaining the biological functions of the ecosystem provides compelling impetus for the inclusion of TEK and LK with Western science-based management systems (Lamb 1982; Messerli et al. 2000; Turner et al. 2000; ICSU 2002). Increasingly, attempts are being made to proactively incorporate TEK and LK into marine EBM. However, effectively integrating these forms of knowledge is neither simple nor always easily embraced by resource management agencies, which almost uniformly suffer the burdens of being understaffed and underfunded to meet the demands of the responsibilities they are expected to address.

What seems to be driving this shift toward seeking out local and traditional ecological knowledge is the requirement in EBM to acquire a deeper understanding of the place being managed. Resource managers are trained as generalists, studying theoretical models and illustrative case studies and acquiring a general understanding of ecosystem structure and function (Meffe 1998; Kainer et al. 2006). Many resource agencies have a culture and policies of moving personnel around from site to site so that they can gain the widest possible exposure to the social–ecological systems in which their sites are located (Sellers 1997). While these experiences are valuable, and understanding the theoretical framework that supports ecosystem structure and function is useful and appropriate, EBM requires site-specific information gathered over long periods of time to give perspective to those observations—getting beyond snapshots to video, recording the variability, the cycles of life, and the extreme events in that place. It is only through living and working in a place day to day, generation to generation, that one can achieve this depth of knowledge. Therefore, LK and TEK are increasingly recognized as ways to fill this knowledge gap, to partner with and learn from others who are deeply connected with a specific place (Slocombe 1998).

Over the past few decades, there has been a shift—in intent if not always in reality—in the practice of resource management away from top-down "command and control" approaches to more bottom-up civic engagement (Crosby et al. 1986; Dalton 2005). Certainly, descriptions of how EBM should be conducted, including the US Commission on Ocean Policy report (USCOP 2004), structure the institutionalization of EBM around citizen participation. Citizen advisory councils are now the rule rather than the exception in most resource management agencies, particularly in place-based management of protected areas. Where EBM is being implemented, advisory councils abound, involving a broad array of partner government agencies, constituent and user groups, nongovernmental organizations, and the public (see O'Boyle and Worcester, chap. 14 of this volume, and Wendt et al., chap. 11 of this volume, as examples).

In terms of public involvement in EBM, there are two key roles for those outside management agencies. The first is to offer advice on management decisions and matters related to the development of policy and regulation. Most of the advisory councils provide this sort of advice. Resource agencies are becoming increasingly reliant on the advice of advisory councils when they try to reach broader public consensus on controversial issues; they seek the councils' involvement early and have them facilitate coordination with the groups they represent, well before any decisions are reached. In the National Marine Sanctuary Program, each of the thirteen sanctuaries has an advisory council that meets regularly and is involved in most major issues related to sanctuary management, including playing a central role in the periodic review of sanctuary management plans. Council members are generally leaders of various constituencies and user groups in the sanctuary gateway communities and not only possess considerable local knowledge, but also are leaders within the community. One of the reasons that the views of advisory councils are given great weight is the recognition of the LK these constituent representatives bring to the table.

The other key role is specifically related to the knowledge base that guides and informs EBM. The management agency can collect physical, chemical, and biological oceanographic information; characterize ecosystem structure and function; and attempt to illuminate the underlying processes that create the observed structure and function. They can also begin to characterize the socioeconomic, and perhaps even the historical, elements of the ecosystem. At best, however, this information offers benchmarks in a highly complex and dynamic ecosystem. LK and TEK represent the deeper knowledge of place cast in a considerably longer time horizon, offering context for the much shorter-term socioecological data and information collected by the resource management agency. Developing institutions that can facilitate the collection and integration of these two types of knowledge is challenging, but a number of innovative approaches have been attempted for both LK and TEK. For example, the Fishermen and Scientists Research Society (FSRS) was established in eastern Canada in 1994 with funding from the Government of Canada, including Fisheries and Oceans Canada, to facilitate the communication and cooperation among agency scientists and managers and fishermen in this region (FSRS 2008). While the original scope of FSRS was to collect fisheries and oceanographic information related to groundfish, it has expanded to include a broader suite of species (e.g., lobster, shrimp, pelagic fish, and turtles) and research methods (e.g., mark and capture, physiology, and habitat mapping). The FSRS has begun to see some successes and is considered a valuable, nonconfrontational forum for coordination and communication among these two groups (CEMSMF 1999). Another example comes from FishResearch.org, a clearinghouse for collaborative fisheries research opportunities for fishermen and scientists on the US West Coast and in New England (www.fishresearch.org). The group is engaged in more than 125 research projects, most of which generate EBM-relevant information.

Just as the acquisition of LK can take generations, and TEK perhaps centuries, the integration of local and traditional ecological knowledge will also take time and dedicated effort on the part of both the holders of that knowledge and those responsible for implementing EBM. Trust and respect take time to build, particularly between different cultures and especially when cultures have a history of being exploited, mistreated, and disrespected. A common element in several promising attempts to integrate TEK and science knowledge is the adoption of interdisciplinary approaches combining social science and marine science (Aswani and Hamilton 2004; Moller et al. 2004; Drew and Henne 2006). It remains for practitioners of EBM to ensure that TEK is treated justly in EBM policy processes and during any integration of TEK and the science process. Genuine comanagement approaches, for example, using participatory research techniques (e.g., Freire 1982; Fals-Borda 1987; Gaventa 1988), are critical, and they will require not just taking knowledge, but sharing control of research and policymaking.

TEK is something of a special case. It is likely to be held by some group, usually an indigenous culture that may have some legal claim to rights of access and/or ownership of marine and coastal areas. This right may be conferred through treaty or be asserted by that group and recognized in law or policy through nontreaty mechanisms. Therefore, new institutions need to be created to provide appropriate access to decision making that recognize these special rights conferred to a group by law or policy. Two of these new innovative institutions have been developed by the US National Marine Sanctuary Program and partners from states, indigenous groups, and other federal agencies, and they provide formal acknowledgment of the special status of the indigenous groups involved. We share these examples to illustrate how institutions can formalize collaborative use of TEK (or LK) in EBM.

Papahānaumokuākea Marine National Monument

In 2006, the largest fully protected marine area on Earth, encompassing 362,061 km^2 of the Pacific Ocean (NOAA 2006), was established in the Northwestern Hawaiian Islands (fig. 9.3). A focal element of the Proclamation of the Papahānaumokuākea Marine National Monument (Presidential Proclamation 8031, 2006), as it also was in the Executive Orders designating the Coral Reef Ecosystem Reserve (EO 13178, 2000; EO 13196, 2001) that preceded the monument, was honoring and respecting the "great cultural significance" of the Northwestern Hawaiian Islands to Native Hawaiians.

While the monument has only recently been established, and therefore the impact of Native Hawaiian TEK has not been fully realized, the intent to effectively integrate TEK is clearly both desired and required. The Native Hawaiian community is actively participating in the management of the monument, both as monument staff and as advisors helping to guide monument decision making, and the important role of Native Hawaiians is being institutionalized into the management structure of the monument. A Native Hawaiian working group has been created, made up of cultural experts, *kū puna* (elders), and practitioners, and the State Office of Hawaiian Affairs is a member of the monument management board, representing and advocating for Native Hawaiian culture.

Native Hawaiian TEK has already played a role in establishing the regulations for the monument, explicitly permitting Native Hawaiian cultural practices that might otherwise be prohibited, such as allowing sustenance take of fish and other monument resources for personal consumption as part of Native Hawaiian traditional and cultural use. The monument staff has been working and continues to work with the Native Hawaiian community and other cultural experts to identify how TEK and associated practices may be integrated into monument management and research activities (Moani Pai, personal communication, 17 April 2007). It should be stressed that this is a case in progress and that some players remain cautious of the interplay between TEK and EBM:

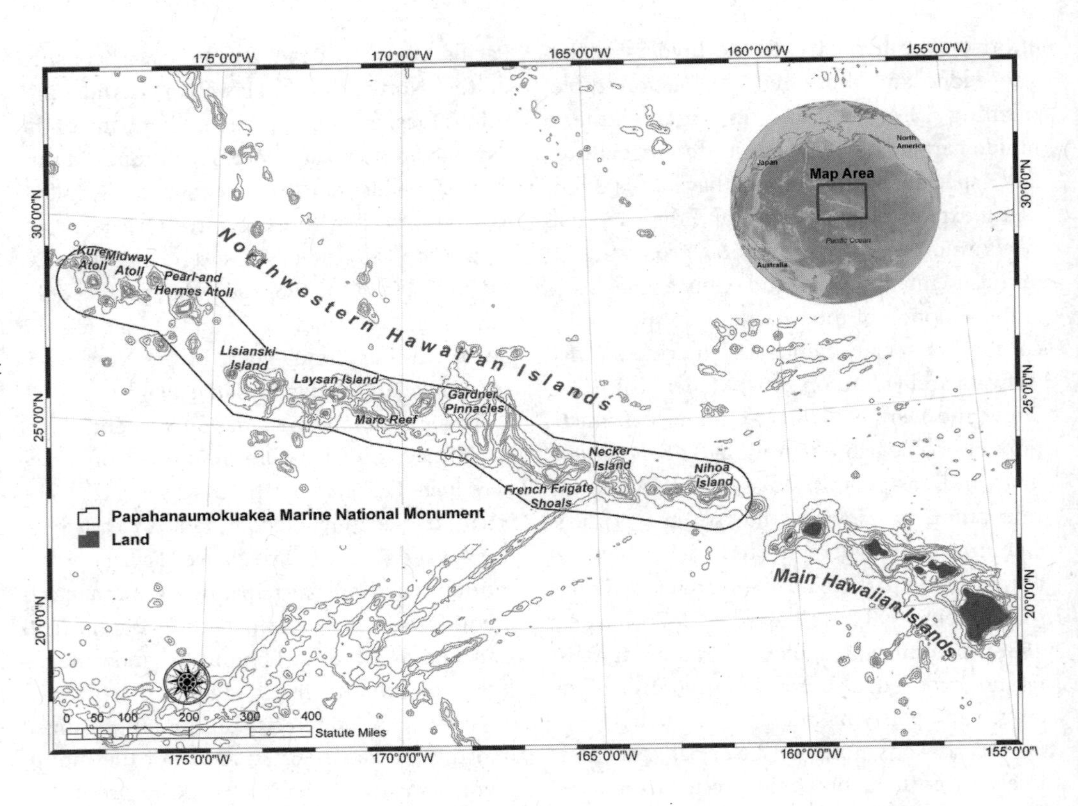

Figure 9.3 Map of Papahānaumokuākea Marine National Monument, Northwest Hawaiian Islands, USA. From the National Oceanic and Atmospheric Administration.

We just helped them organize and facilitate a meeting of the Northwest Hawaii working group, but there is a long way to go before they can talk about integrating TEK into monument management. Don't forget that any access to the monument requires a permit. All permits are issued with some major conditions and requirements. Some permits will require NEPA documentation, for example, the recent shark culling permit. Surely the intent is there to recognize the significance of Native Hawaiian cultural heritage in the area; how and to what degree TEK is integrated, remains to be seen. (Anonymous)

While it is a difficult and complex task to balance the desires and rights of a traditional culture and a dominant culture's insistence on viewing ocean and coastal waters as a commons, institutions such as those created as part of the monument are a step toward achieving some appropriate balance between these two perspectives. Some may view the monument designation itself as a heavy-handed blow to the Native Hawaiian culture, as it was established by unilateral decree of the US president, while others anticipate that this ecosystem will be managed more effectively with the Native Hawaiian perspective helping to guide the management of the monument.

Olympic Coast National Marine Sanctuary

Another promising sign of expanding opportunities to more effectively integrate traditional knowledge into ecosystem-based management is the establishment in 2006 of the Intergovernmental Policy Council (IPC) at the Olympic Coast National Marine Sanctuary in the Pacific Northwest (fig. 9.4). The IPC is a formal collaborative body composed of representatives of the sovereign governments of the four treaty groups (Hoh, Makah, and Quileute tribes and the Quinault Indian Nation), the State of Washington, and NOAA's Olympic Coast National Marine Sanctuary. These three governments have comanagement responsibilities for resources within the boundaries of the national marine sanctuary. The IPC provides an appropriate forum for coordination among these sovereign governments. It has been described by Jim Woods of the Makah tribe as integrating the tribes "into the management and decision-making process about resources that we co-manage with the state within the sanctuary . . . an important step in improving federally-mandated government-to-government communication between the tribes, the state and the sanctuary on coastal marine matters" (NMS 2008). But a more cautious view is reflected in this statement:

> I was left with the impression that this IPC is seen as providing an important platform on which to build. These same parties are nearing completion of a "charter," and that road has not been easy. Nor should one expect that there will not be differences of view among each tribe and other parties to the agreement on how to plan and what policies to set in relationship to the revision of the GMP. This likelihood of continued difference does not detract from the significance of the IPC, which is a big step forward, but it is not likely the first step in a Sousa band march to the sea together. (Anonymous)

The effective integration of TEK in the Olympic Coast National Marine Sanctuary must begin with some institution, some mechanism crafted specifically to offer the opportunity to share that knowledge, built on a firm foundation of respect and trust. Forums like the Intergovernmental Policy Council can provide such a foundation, building relationships of trust, recognizing the value of TEK in the effective collaborative management of natural and cultural resources, and creating institutional arrangements that can be sustained over time.

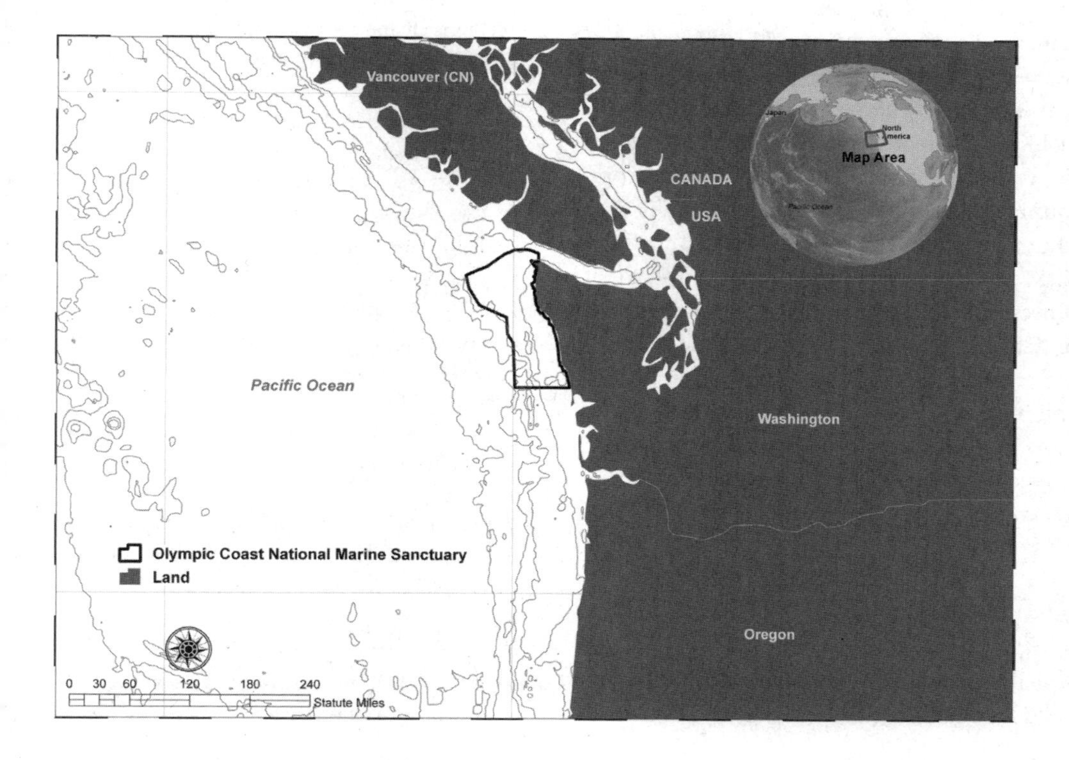

Figure 9.4 Map of Olympic Coast National Marine Sanctuary, Washington, USA. From the National Oceanic and Atmospheric Administration.

Conclusion

The integration of TEK with oceans and coastal EBM remains dependent on bridging the gap between skepticism among Western science-based management toward the belief component of TEK and mistrust among local and indigenous communities toward Western science-based agencies' ability to protect and respect the integrity of TEK. The examples of the comanagement of Papahānaumokuākea Marine National Monument and the Intergovernmental Policy Council at the Olympic Coast National Marine Sanctuary point to both the difficulties and the opportunities in attempting the marriage of TEK and EBM. Part of the reason TEK systems and practice have been successful in some communities is that they are (or at least were) functioning, often in isolated communities with social systems that operate effectively because they have processes led by elders whose decisions are respected and because members of the community are held accountable to strategies adopted by the community. Thus recognizing TEK in terms of Berkes' (1999) knowledge–practice–belief complex is vital. Any successful integration of TEK or LK in ocean and coastal EBM will hinge on the respective capacity of managers and communities: Managers must understand TEK or LK, ensure that TEK or LK is treated justly in EBM policy processes, and share control of that process; communities must have leaders who understand the scientific process and are willing and able to share their knowledge. Shared and just integration of the two knowledge systems has the potential to strengthen the social and ecological resilience of coastal and marine systems.

Acknowledgments

Mahalo (thank you) to Moani Pai, the Native Hawaiian Cultural Liaison with the Papahānaumokuākea Marine National Monument, for her insights and information on the comanagement process in Hawai'i. This chapter also benefited from insightful comments by the other contributors of this volume and two anonymous external reviewers.

References

Agrawal, A. 2002. Indigenous knowledge and the politics of classification. *International Social Science Journal* 173:287–97.

Aswani, S., and R. J. Hamilton. 2004. Integrating indigenous ecological knowledge and customary sea tenure with marine and social science for conservation of bumphead parrotfish (*Bolbometopon muricatum*) in the Roviana Lagoon, Solomon Islands. *Environmental Conservation* 31:69–83.

Berkes, F. 1987. Common property resource management and Cree Indian fisheries in subarctic Canada. In *The question of the commons*, ed. B. J. McKay and J. M. Acheson, 66–91. Tucson: University of Arizona Press.

Berkes, F. 1981. Fisheries of the James Bay area and northern Quebec: A case study in resource management. In *Renewable resources and the economy of the north*, ed. M. M. R. Freeman, 143–60. Ottawa: Association of Canadian Universities for Northern Studies.

Berkes, F., T. P. Hughes, R. S. Steneck, J. A. Wilson, D. R. Bellwood, B. Crona, C. Folke et al. 2006. Globalization, roving bandits, and marine resources. *Science* 311:1557–58.

Berkes, F., J. Colding, and C. Folke. 2000. Rediscovery of traditional ecological knowledge as adaptive management. *Ecological Applications* 10(5):1251–62.

Berkes, F. 1999. *Sacred ecology: Traditional ecological knowledge and resource management.* Philadelphia: Taylor & Francis.

Berkes, F. 1993. Traditional ecological knowledge in perspective. In *Traditional ecological knowledge concepts and cases*, ed. J. T. Inglis, 1–10. Ottawa, Canada: International Development Research Centre.

Bielawski, E. 1995. Inuit indigenous knowledge and science in the Arctic. In *Human ecology and climate*

change: People and resources in the Far North, ed. D. Peterson and D. Johnson, 219–28. Washington, DC: Taylor & Francis.

Brodnig, G., and V. Mayer-Schonberger. 2000. Bridging the gap: The role of spatial information technologies in the integration of traditional ecological knowledge and Western science. *Electronic Journal of Information Systems in Developing Countries* 1:1–15.

CEMSMF (Committee on Ecosystem Management for Sustainable Marine Fisheries). 1999. *Sustaining marine fisheries*. Ocean Studies Board, Commission on Geosciences, Environment and Resources, National Research Council. Washington, DC: National Academies Press.

Chave, J., and S. Levin. 2002. *Scale and scaling in ecological and economic systems*. Beijer Institute Discussion Paper no. 154. Beijer Institute of Ecological Economics. Stockholm: Royal Swedish Academy of Sciences.

Cicin-Sain, B., and S. Belfiore. 2006. Linking marine protected areas to integrated coastal and ocean management: A review of theory and practice. *Ocean and Coastal Management* 48:847–68.

Cicin-Sain, B. 1993. Sustainable development and integrated coastal management. *Ocean and Coastal Management* 21:11–43.

Cinner, J. E. 2007. Designing marine reserves to reflect local socioeconomic conditions: Lessons from long-enduring customary management systems. *Coral Reefs* 26:1035–1045.

Cobb, D., M. K. Berkes, and F. Berkes. 2005. Ecosystem-based management and marine environmental quality indicators in Northern Canada. In *Breaking ice: Renewable resource and ocean management in the Canadian North*, ed. F. Berkes, 71–94. Calgary, Alberta: University of Calgary Press.

Colding, J., and C. Folke. 2001. Social taboos: Invisible systems of local resource management and biological conservation. *Ecological Applications* 11:584–600.

Costa-Pierce, B. A. 1987. Aquaculture in ancient Hawaii. *BioScience* 37:320–30.

Crosby, N., J. Kelly, and P. Shaefer. 1986. Citizens panels: A new approach to citizen participation. *Public Administration Review* 170–78.

Cruikshank, J. 2001. Glaciers and climate change: Perspectives from oral tradition. *Arctic* 54:377–93.

Dasmann, R. F. 1976. Toward a dynamic balance of man and nature. *Ecologist* 6:2–5.

Dalton, T. M. 2005. Beyond biogeography: A framework for involving the public in planning of US marine protected areas. *Conservation Biology* 19: 1392–1401.

Davidson-Hunt, I., and F. Berkes. 2003. Learning as you journey: Anishinaabe perception of social–ecological environments and adaptive learning. *Conservation Ecology* 81:5–25.

Denny, J. P. 1986. Cultural ecology of mathematics: Ojibway and Inuit hunters. In *Native American mathematics*, ed. M. P. Closs, 129–80. Austin: University of Texas Press.

Diamond, J. 1986. The environmentalist myth. *Nature* 324:19–20.

Dove, M. R. 2006. Indigenous people and environmental politics. *Annual Review of Anthropology* 35:191–208.

Drew, J. A., and A. P. Henne. 2006. Conservation biology and traditional ecological knowledge: Integrating academic disciplines for better conservation practice. *Ecology and Society* 11(2): 34.

Drew, J. A. 2005. Use of traditional ecological knowledge in marine conservation. *Conservation Biology* 19:1286–93.

Dwyer, P. D. 1994. Modern conservation and indigenous peoples: In search of wisdom. *Pacific Conservation Biology* 1:91–97.

Dyer, C. L., and J. R. McGoodwin. 1994. *Folk management in the world's fisheries: Lessons for modern fisheries management*. Boulder: University of Colorado Press.

Fals-Borda, O. 1987. The application of participatory action research in Latin America. *International Sociology* 2: 329–47.

Folke, C., T. Hahn, P. Olsson, and J. Norberg. 2005. Adaptive governance of social–ecological systems. *Annual Review in Environment and Resources* 30:441–73.

Freire, P. 1982. Creating alternative research methods: Learning to do it by doing it. In *Creating knowledge: A monopoly?*, ed. B. Hall, A. Gillette, and R. Tandon, 29–37. New Delhi: Society for Participatory Research in Asia.

FSRS (Fishermen and Scientists Research Society). 2008. http://www.fsrs.ns.ca/.

Gadgil, M., P. Olsson, F. Berkes, and C. Folke. 2003. Exploring the role of local ecological knowledge in ecosystem management: Three case studies. In *Navigating social–ecological systems: Building resilience for complexity and change,* ed. F. Berkes, J. Colding, and C. Folke, 189–209. Cambridge, UK: Cambridge University Press.

Gadgil, M., F. Berkes, and C. Folke. 1993. Indigenous knowledge for biodiversity conservation. *Ambio* 22:151–56.

Gadgil, M., and F. Berkes. 1991. Traditional resource management systems. *Resource Management and Optimization* 8:127–41.

Gaventa, J. 1988. Participatory research in North America. *Convergence* 21:9–28.

George, J. C., H. P. Huntington, K. Brewster, H. Eicken, D. W. Norton, and R. Glenn. 2004. Observations on shorefast ice dynamics in arctic Alaska and the responses of the Inupiat hunting community. *Arctic* 57:363–74.

Gill, M., ed. 2007. *Developing an implementation plan for community-based monitoring within the circumpolar biodiversity monitoring program.* Workshop proceedings, 27–28 November 2006, Anchorage. Whitehorse, Yukon, Canada: Circumpolar Biodiversity Monitoring Program.

Gregory, R., L. Failing, and P. Higgins. 2006. Adaptive management and environmental decision making: A case study application to water use planning. *Ecological Economics* 58:434–47.

Guzman, H. M., C. Guevara, and A. Castillo. 2003. Natural disturbances and mining of Panamanian coral reefs by indigenous people. *Conservation Biology* 17:1396–1401.

Huntington, H. 2001. Using traditional ecological knowledge in science: Methods and applications. *Ecological Applications* 10:1270–74.

ICSU (International Council for Science). 2002. *Report of the Scientific and Technological Community to the World Summit on Sustainable Development (WSSD).* ICSU Series on Science for Sustainable Development. Report no. 1. Paris: ICSU.

Jackson, J. B. C., and 18 coauthors. 2001. Historical overfishing and the recent collapse of coastal ecosystems. *Science* 293:629–38.

Johannes, R. E. 2002a. Did indigenous conservation ethics exist? *SPC Traditional Marine Resource Management and Knowledge Information Bulletin* 14:3–7.

Johannes, R. E. 2002b. The renaissance of community-based marine resource management in Oceania. *Annual Review of Ecology and Systematics* 33:317–40.

Johannes, R. E. 1978. Traditional marine conservation methods in Oceania and their demise. *Annual Review of Ecology and Systematics* 9:349–64.

Johannes, R. E. 1981. *Words from the lagoon: Fishing and marine lore in the Palau district of Micronesia.* Berkeley: University of California Press.

Kainer, K. A., M. Schmink, H. Covert, J. R. Steppe, E. M. Bruna, J. L. Dain, S. Espinosa, and S. Humphries. 2006. A graduate education framework for tropical conservation and development. *Conservation Biology* 20:3–13.

Kinzig, A. P., P. Ryan, M. Etienne, H. Allison, T. Elmqvist, and B. H. Walker. 2006. Resilience and regime shifts: Assessing cascading effects. *Ecology and Society* 11(1):20.

Kuhn, T. S. 1970.*The structure of scientific revolutions,* 2nd ed. Chicago: University of Chicago Press.

Kuzyk, G., D. E. Russell, R. S. Farnell, R. M. Gotthardt, P. G. Hare, and E. Blake. 1999. In pursuit of prehistoric caribou on Thandlät, southern Yukon. *Arctic* 52:214–19.

Lamb, H. H. 1982. *Climate, history, and the modern world.* London: Methuen.

Lee, K. N. 1999. Appraising adaptive management. *Conservation Ecology* 3(2):3.

Mantua, N., D. Haidvogel, Y. Kushnir, and N. Bond. 2002. Making the climate connections: Bridging scales of space and time in the US GLOBEC program. *Oceanography* 15:75–86.

McCay, B. J. 1978a. Muddling through the clam beds: Cooperative management of New Jersey's hard clam spawner sanctuaries. *Journal of Shellfish Research* 79:327–40.

McCay, B. J. 1978b. Systems ecology, people ecology, and the anthropology of fishing communities. *Human Ecology* 6: 397–422.

Meffe, G. R. 1998. Conservation scientists and the policy process. *Conservation Biology* 12:741–42.

Messerli, B., M. Grosjean, T. Hofer, L. Núñez, and C. Pfister. 2000. From nature-dominated to human-dominated environmental changes. *Quaternary Science Reviews* 19:459–79.

Moller, H., F. Berkes, P. Lyver, and M. Kislalioglu. 2004. Combining science and traditional ecological knowledge: Monitoring populations for comanagement. *Ecology and Society* 9(3):2.

Moran, K., S. R. King, and T. J. Carlson. 2001.

Biodiversity prospecting. *Annual Review of Anthropology* 30:505–26.

Nazarea, V. D. 1999. *Ethnoecology: Situated knowledge/located lives.* Tucson: University of Arizona Press.

Nazarea, V. D. 2006. Local knowledge and memory in biodiversity conservation. *Annual Review of Anthropology* 35:317–35.

Neis, B., and L. Felt, ed. 2000. *Finding our sea legs: Linking fishery people and their knowledge with science and management.* St. John's, Newfoundland, Canada: ISER Books.

Neis, B., D. C. Schneider, L. Felt, R. L. Haedrich, J. Fischer, and J. A. Hutching. 1999. Fisheries assessment: What can be learned from interviewing resource users? *Canadian Journal of Fisheries and Aquatic Science* 56:1949–63.

NMS (National Marine Sanctuaries). 2008. http://sanctuaries.noaa.gov/news/features/0107_octribes.html.

NOAA (National Oceanic and Atmospheric Administration). 2006. *Northwestern Hawaiian Islands proposed National Marine Sanctuary draft environmental impact statement and management plan.* Draft Management Plan, vol. 2 of 2. Honolulu, Hawaii. http://hawaiireef.noaa.gov/PDFs/NWHI_OLD_MP1_042006.pdf.

Patterson, J. 1994. Maori environmental values. *Environmental Ethics* 16:397–411.

Pinkerton, E. 1989. *Co-operative management of local fisheries.* Vancouver: University of British Columbia Press.

Ramstad, K. M., N. J. Nelson, G. Paine, D. Beech, A. Paul, P. Paul, F. W. Allendorf, and C. H. Daugherty. 2007. Species and cultural conservation in New Zealand: Maori traditional ecological knowledge of tuatara. *Conservation Biology* 21:455–64.

Rayner, S., and E. Malone, ed. 1998. *Human choice and climate change*, vol. 3, *The tools for policy analysis.* Columbus, OH: Battelle Press.

Ruddle, K. 1991. A research framework for the comparative analysis of the traditional sole property rights fisheries management systems in the Pacific Basin. *Resource Management and Optimization* 8:143–54.

Ruddle, K., E. Hviding, and R. E. Johannes. 1992. Marine resources management in the context of customary tenure. *Marine Resource Economics* 7:249–73.

Sellers, R.W. 1997. *Preserving nature in the national parks: A history.* New Haven, CT: Yale University Press.

Sherry, E., and N. Myers. 2002. Traditional environmental knowledge in practice. *Society and Natural Resources* 15:345–58.

Slocombe, D. S. 1998. Lessons from experience with ecosystem-based management. *Landscape and Urban Planning* 40:31–39.

Smit, B., and O. Pilifosova. 2001. Adaptation to climate change in the context of sustainable development and equity. In *Climate change 2001: Impacts, adaptation, and vulnerability*, ed. J. J. McCarthy, O. F. Canziani, N. A. Leary, D. J. Dokken, and K. S. White. Cambridge, UK: Cambridge University Press.

Turner, N. J., M. B. Ignace, and R. Ignace. 2000. Traditional ecological knowledge and wisdom of aboriginal peoples in British Columbia. *Ecological Applications* 10:1275–87.

USCOP (US Commission on Ocean Policy). 2004. *An ocean blueprint for the twenty-first century.* Final report. Washington, DC: USCOP. ISBN 0 9759462 0 X.

Usher, P. J. 2000. Traditional ecological knowledge in environmental assessment and management. *Arctic* 53:183–93.

Vasquez-Leon, M. 2002. Assessing vulnerability to climate change risk: The case of small-scale fishing in the Gulf of California, Mexico.*Investigaciones Marinas* 30. ISSN 0717-7178. Online.

Vedwan, N., and R. E. Rhoades. 2001. Climate change in the western Himalayas of India: A study of local perception and response. *Climate Research* 19:109–17.

Vitousek, P. M., H. A. Mooney, J. Lubchenco, and J. M. Melillo. 1997. Human domination in Earth's ecosystems. *Science* 277:494–99.

Wallace, M. G., H. J. Cortner, M. A. Moote, and S. Burke. 1996. Moving toward ecosystem management: Examining a change in philosophy for resource management. *Journal of Political Ecology* 3:1–36.

Walters, C. J. 1986. *Adaptive management of renewable resources.* New York: MacMillan.

CHAPTER 10
Building the Legal and Institutional Framework

Janis Searles Jones and Steve Ganey

It is now widely recognized that the trajectory of degradation in the world's oceans is extensive (Shackeroff et al., chap. 3 of this volume), has negative consequences for both ecosystems and associated human communities, and thus is undesirable (McLeod and Leslie, chap. 1 of this volume). The mandates and structures of current governance and management systems are partially to blame (Rosenberg and Sandifer, chap. 2 of this volume). Consequently, numerous national and international bodies have urged movement toward ecosystem-based management (EBM) of coasts and oceans (UNCED 1992; POC 2003; USCOP 2004; MA 2005; UNEP 2006).

Calls for EBM, in turn, have raised a host of operational questions, including the potential that resilience theory may provide a useful conceptual framework for implementing EBM. The relationship between the resilience conceptual framework and marine EBM is addressed in detail earlier in this volume by Leslie and Kinzig, but at its core, resilience science recognizes that social and natural systems are inextricably linked and respond dynamically to changes within and outside themselves (Walker and Salt 2006). Managing for resilient ecological and human systems has many implications for marine management and policy. Our focus here is on ocean governance and how a resilience-informed approach would change how we implement existing laws within existing governance systems, and ultimately require changes to legal frameworks and systems.

This chapter provides a brief overview of the existing domestic statutory regime and identifies opportunities for and obstacles to ecosystem-based management and efforts to maintain or restore ecological resilience under the current regime. We then discuss key reforms that could bring resilience science into marine law and policy and thus advance ecosystem-based management.

We suggest that ample discretion exists under current laws to implement EBM but that a new mandate and governance changes based on resilience theory would provide significant advantages. The chapter closes with a proposal to create transitional strategies to move existing activities to an ecosystem-based management approach by a time certain, while holding new activities or programs to a higher ecosystem-based standard. That standard would be embodied in a new national oceans policy, overseen by a single federal agency.

Current US Oceans Management

In the United States, oceans are the largest component of the public domain, spanning nearly 4.5 million square miles, an area about 23% larger than the nation's overall land area. We know enough today about marine systems and their functions to understand that our oceans are not only a series of exploitable resources (e.g., fish, pharmaceuticals, oil and gas), but rather a dynamic and connected ecosystem that includes people. We understand that what we take out of and put into the oceans can profoundly affect the health, functioning, and resilience of entire marine ecosystems

(Shackeroff et al., chap. 3 of this volume; Leslie and Kinzig, chap. 4 of this volume).

Our current system of oceans management and governance is largely a historical artifact and reflects dated understanding of how to sustain ecosystem services through time—particularly seafood production—and the misguided appeal of optimization-oriented, command-and-control resource management approaches (Holling and Meffe 1996; Leslie and Kinzig, chap. 4 of this volume). And our current system reflects our dualistic regime of federal and state powers and authorities (Crowder et al. 2006). What we know about the oceans today suggests the need for a regulatory and governance regime to address our oceans as an integrated whole. Unfortunately, at present, there is no single federal oceans policy that guides management of living and nonliving marine resources, and there is no single oceans agency responsible for prudent management of our oceans public domain. Rather, jurisdiction and management of our oceans and marine resources is fractured, defined both by geographical divisions and by resource category within those divisions. The result is a system organized around political divisions largely unrelated to ecological features, where many are in charge of something but no one is in charge of everything.

Geographical Divisions

It is only in recent decades that nations have asserted jurisdiction over the ocean. Historically, coastal countries had authority over the territorial seas off their coasts, generally recognized as a 3-mile-wide area. Customary international law considered the area beyond those territorial seas to be international waters, which were largely unregulated. In the post–World War II era, advances in technology led to areas of intense offshore fishing by foreign vessels and growing interest by many nations in asserting broader jurisdiction over marine resources off their coasts. Internationally, the 1982 United Nations Convention on the Law of the Sea is the culmination of many years of effort. The convention permits coastal nations to assert jurisdiction out to 200 nautical miles from their shorelines, now known as exclusive economic zones (EEZs).

Domestically, jurisdiction over marine waters off the US coast is divided between the federal government and the states. In 1953, the US Congress granted control over the 0-to-3-nautical-mile band of ocean, with few exceptions, to the states through the Submerged Lands Act. In 1976 Congress passed the Fishery Conservation and Management Act, which asserted US jurisdiction over the EEZ, with the ultimate goal of replacing the foreign fishing fleet with a domestic fleet.

Management of the waters off the US coast is therefore bifurcated between nearshore waters managed by the states and offshore waters managed by the federal government, with the nearshore waters being further divided among the states. Depending on the resource at issue, tribal and local governments also may have jurisdiction and regulatory authority in nearshore waters. While this geographic division is relatively simple to depict on a map, it does not reflect the interconnectedness, interrelatedness, and lack of fixity of many ocean resources. Management decisions in the nearshore made by one state may affect resources and management options in adjacent state and federal waters, and visa versa. For example, a decision by state A to authorize the catch of unsustainable levels of a particular fish stock may affect resources in adjacent state B or in federal waters in multiple ways. State A's decision may sufficiently depress the stock so that neither state B nor the federal government can responsibly authorize fishing on that same stock. State A's decision may have adverse habitat impacts that affect other target and nontarget species that live in or transit through state B's and federal waters. State A's decision may result in significant catches of nontargeted species, the removal of which has adverse effects beyond state A's boundaries. State A's decision may affect predator–prey dynamics in a manner that extends beyond state A's boundaries. The effects of state A's actions will rarely be contained within state A's jurisdictional boundaries. Our current system does not, however, effectively foster collaborative and integrated decision making, and with very rare exceptions it does not provide managers authority to prevent a harmful action from going forward in another jurisdiction.

Resource Category Divisions

The geographically based jurisdictional divisions create a complex regulatory environment. The complexity is compounded by the numerous legal divisions of state and federal management authority based on resource categories. At the federal level, there are more than 140 federal laws governing various aspects of our oceans and coasts (POC 2003; USCOP 2004). These laws delegate regulatory authority widely throughout the federal government, ultimately involving dozens of federal and state agencies in oceans management issues (Christie and Hildreth 2007). For an appendix and short description of the primary federal oceans and coastal laws, see the US Commission on Ocean Policy final report (USCOP 2004), appendix D. Regulatory regimes vary wildly at the state level. In Massachusetts, for example, different state agencies are responsible for fisheries management, coastal wetlands restoration, and water quality, whereas in the neighboring state of Rhode Island, one agency oversees all of these functions. In summary, while numerous agencies are responsible for conserving or exploiting (or both) some aspect of the marine environment, no agency is responsible for ensuring that the combined and cumulative effects of that conservation and exploitation leave us with a marine environment that is intact, functioning, and resilient.

Even where federal statutes do include jurisdictional considerations, they primarily focus on resource categories. The Marine Mammal Protection Act is a federal statute that regulates most marine mammals. The Magnuson–Stevens Fishery Conservation and Management Act (MSA) is a federal statute that authorizes conservation and management of fish, where necessary. For many marine mammal species, direct or indirect interactions with fisheries are a significant source of mortality and injury. And yet, our federal statutory regime does not explicitly integrate the fisheries and marine mammal regulatory systems, and it provides little guidance for what to do in cases of conflict.

Existing Authority for Ecosystem-Based Management

Here we highlight opportunities under the current governance structure to apply resilience science in the implementation of ecosystem-based management, and we suggest governance changes that would move EBM forward in a more systemic manner. The best expressions of resilience concepts in legal instruments tend to be clustered at the highest scale: international instruments that deal with irreversible outcomes and the precautionary approach (see also Rosenberg and Sandifer, chap. 2 of this volume; Rosenberg et al., chap. 16 of this volume). For example, the Rio Declaration on Environment and Development explicitly states: "In order to protect the environment, the precautionary approach shall be widely applied by States according to their capabilities. Where there are threats of serious or irreversible damage, lack of full scientific certainty shall not be used as a reason for postponing cost-effective measures to prevent environmental degradation" (UNCED 1992).

Similar statements may be found in several other international agreements and codes of conduct relevant to ocean governance (UNCED 1992; UNFCCC 1994; UN 1995). Important as they are, these instruments tend not to control at the level of managing specific marine ecosystems. The precautionary intent is diluted through either decision-making institutions or processes or more-specific instruments that do not incorporate resilience concepts. One has yet to see the precautionary approach called for in the 1995 Fish Stocks Agreement applied to bluefin tuna management, though the trajectory of that fishery is approaching, or has perhaps already passed, an important threshold (Safina and Klinger 2008).

At the national level, statutes such as the National Environmental Policy Act and the Endangered Species Act include several ideas from resilience theory, if incompletely, in their design. These provisions help managers coordinate decision making among multiple authorities, think more broadly about habitat and ecosystems, and avoid irreversible environmental harm. Even some of the principal ocean resource statutes, including the Magnuson–Stevens Act, contain language about avoiding irreversible adverse effects on the marine environment. However, this broad intent in the MSA does not control when balanced against statutory imperatives to efficiently optimize fishery production. Beyond calls for consultation and to avoid irreversible effects, there is little else in existing federal statutes to facilitate efforts to maintain or restore ecological resilience or advance the principles of resilience science discussed by Leslie and Kinzig (chap. 4 of this volume).

This section will highlight examples of statutes that do provide discretion for managers to employ some resilience concepts that could advance ecosystem-based management. The section will then discuss various obstacles to applying resilience concepts and realizing ecosystem-based management through existing tools (see also Rosenberg and Sandifer, chap. 2 of this volume).

Fisheries Management—The Magnuson–Stevens Act

Congress passed the original version of the Magnuson–Stevens Fishery Conservation and Management Act in 1976, in large part to assert jurisdiction over the exclusive economic zone and to domesticate the foreign fisheries occurring off the US coasts. Like many other federal environmental statutes, the act does two things, broadly speaking: (1) establishes a policy and standards by which federal fisheries are regulated in the United States and (2) establishes a system and structure by which those standards are applied.

The original Magnuson–Stevens Act focused on commercially viable fish exploitation, on a species-by-species basis. Its purpose was to foster and facilitate fisheries. In its policy approach, the original act is arguably antithetical to EBM concepts. The act specifically calls for optimization and efficiency of commercial fishery production, a reasonable but narrowly defined marine management goal (contrast with the POC 2003 and USCOP 2004 ocean management goals) that has been shown to lead to severe declines in system resilience and ecosystem functioning (Holling and Meffe 1996; Worm et al. 2006; Leslie and Kinzig, chap. 4 of this volume).

To execute the act's policies, Congress created a relatively unique two-level management structure. In an effort to emphasize regional, collaborative management, Congress created eight regional fishery management councils composed of federal and state agency personnel and a number of politically appointed knowledgeable individuals. The councils are dominated by the commercial fishing industry (Eagle et al. 2003). Within their respective jurisdictions, these councils are largely responsible, in the first instance, for identifying fisheries for which conservation and management are necessary and for drafting fishery management plans, amendments, and proposed regulations to manage those fisheries. The councils then forward these plans, amendments, and regulations to the National Marine Fisheries Service, the federal agency ultimately responsible for federal fisheries management. The agency is charged with determining whether the management measures are consistent for Congressionally specified standards. While the councils are technically advisory, the agency's ability to manage fisheries is constrained: Except in extreme cases, the agency may only approve, disapprove, or partially approve and partially disapprove fishery management plans and amendments. Should the agency disapprove a plan, the plan goes back to the council. In the absence of a plan, fisheries may continue under preexisting regulations, if any.

With the council system, Congress created a unique approach to public resources management, which has been widely criticized for its conflicts of interest, lack of accountability, and management failures (e.g., POC 2003; USCOP 2004). In employing the two-tier governance structure, Congress deliberately invested economically interested industries with quasi-regulatory authority, creating a set of institutions and relationships that have proven damaging to public resources and highly resistant to change. In terms of the effects of governance structures on marine resources, 20 years after its enactment, it became clear that the original intent of the Magnuson–Stevens Act had perhaps been accomplished far too well. Within a few short decades fisheries off the US coast had been domesticated, and the domestic fleet was overcapitalized, catching far too many fish far too quickly and having significant adverse effects on the fish populations themselves and on the marine environment (NRC 1999).

In response, Congress amended the Magnuson–Stevens Act with the Sustainable Fisheries Act of 1996. Congress welded three conservation provisions onto what had been predominantly an exploitation statute: the duty to end overfishing and rebuild fisheries, the duty to protect essential fish habitat to the extent practicable, and the duty to reduce bycatch and bycatch mortality to the extent practicable. Congress did not, however, fundamentally alter the two-tier governance structure that had led to the need for the conservation-minded amendments.

We focus here on the essential fish habitat (EFH) provisions, as an example of an existing statutory and regulatory mechanism for incorporation of some ecosystem-based management principles and processes into what has traditionally been a single-species decision-making process. The EFH obligations provide both a means to consider linkages between exploited fish and their environment and a mechanism to address some of the jurisdictional limitations of our current management system. We note, however, that the EFH provisions do not establish thresholds or require monitoring or management changes in response to habitat degradation.

ESSENTIAL FISH HABITAT

Historically, MSA management efforts had been focused on achieving high fisheries yields on a species-by-species basis. With the imposition of the EFH provision, Congress broadened the fisheries managers' mandate to include protection of important components of the marine environment. The amendment reflected the simple reality that fish need habitat to exist. Congress defined "essential fish habitat" as "those waters and substrate necessary to fish for spawning, breeding, feeding or growth to maturity" [16 U.S.C. § 1802(10)]. Congress further required that fishery management plans describe and identify essential fish habitat and minimize to the extent practicable the adverse effects of fishing on that habitat [16 U.S.C. § 1853(a)(7)].

NOAA Fisheries, charged with implementing the EFH mandate, further defined key terms. "Waters" in essential fish habitat include "aquatic areas and their associated physical, chemical, and biological properties that are used by fish and may include aquatic areas historically used by fish where appropriate" [50 C.F.R. § 600.10]. "Substrate" in essential fish habitat includes "sediment, hard bottom, structures underlying the waters, and associated biological communities" [50 C.F.R. § 600.10]. In the identification and designation of essential fish habitat, fisheries managers are charged with thinking broadly about the current and historical habitat needs of exploited species, with an expansive definition of habitat, which includes such things as "associated biological communities." Injecting such considerations into fisheries management represents an incremental step forward in the march toward ecosystem-based management.

Once essential fish habitats have been defined, fisheries managers must consider whether fishing is causing adverse effects to that habitat. The adverse effects evaluation is a broad one, requiring consideration of not only changes of waters or substrate, but also "loss of, or injury to benthic organisms, prey species and their habitat, and other ecosystem components, if such modifications reduce the quality and/or quantity of EFH" [50 C.F.R. § 810(a)].

The addition of habitat considerations, particularly those as broadly defined as essential fish habitat, requires managers, and provides the public the opportunity, to incorporate ecosystem considerations not previously considered in fisheries management decisions. Essential fish habitat provides the avenue to consider not only the physical habitat requirements of target species, but also their prey–field habitat requirements. These new considerations are tempered, however, by a relatively weak statutory mandate. While the work must be done to identify essential fish habitat, the duty to minimize the effects of fishing on that habitat is only "to the extent practicable." While the essential fish habitat mandate inserted nontraditional ecosystem considerations into the management arena, Congress continued to permit consideration of traditional economic concerns. Thus, in decisions to protect habitat, councils and the agency have the authority to act in a manner that is considerate of the ecosystem, but they may consider the economic consequences of doing do. In the first round of EFH amendments to fishery management plans, no council recommended that fisheries be modified to protect newly defined essential fish habitat.

Allowing the economic cost of habitat protection to trump protection is a reflection of the limits of existing mandates. The EFH provision represents a significant opportunity to broaden fisheries management decisions beyond single-species maximum sustained yield exploitation to protection of fish habitat, but it is significantly limited by the influence of the current socioeconomic system that relies on fisheries. Despite widespread evidence of habitat destruction and the commonsense need to protect habitat in order to continue to be able to fish, subsequent amendments to the statute have not strengthened the mandate. So we are left with an opportunity to make progress, but not a mandate to manage for marine resilience.

CONSULTATION REQUIREMENT

In addition to inserting habitat considerations into traditional fisheries management decisions, in framing the essential fish habitat provisions, Congress also provided a tool to address the fractured jurisdiction described above. It is clear that activities other than fishing may adversely affect fish habitat. Through the Magnuson–Stevens Act, the councils and

NOAA Fisheries have authority only over fishing activity. In recognition of the fact that nonfishing action may harm fish habitat, Congress included a consultation requirement in the 1996 amendments to the act.

The essential fish habitat consultation process requires any federal agency that proposes to authorize, fund, or undertake an action that may "adversely affect any essential fish habitat" to consult with NOAA Fisheries [16 U.S.C. § 1855(b)(2)]. Should NOAA Fisheries conclude that "an action authorized, funded, or undertaken, or proposed to be authorized, funded or undertaken, by any State or Federal agency would adversely affect any essential fish habitat," Congress requires NOAA Fisheries to "recommend to such agency measures that can be taken by such agency to conserve such habitat" [16 U.S.C. § 1855(b)(4)(A)]. Furthermore, Congress required federal agencies that receive recommendations from NOAA Fisheries to respond in writing, including "a description of measures proposed by the agency for avoiding, mitigating, or offsetting the impacts of the activity on such habitat" [16 U.S.C. § 1855(b)(4)(B)]. While Congress did not require other federal agencies to comply with NOAA Fisheries recommendations, the reasoning behind any deviations from the recommended measures must be explained [16 U.S.C. § 1855(b)(4)(B)].

This interagency consultation process also allows the regional fishery management councils to provide comments and recommendations to NOAA Fisheries and federal or state agencies concerning actions that may effect essential fish habitat [16 U.S.C. § 1855(b)(3)(A)]. The essential fish habitat process, while not mandating a particular outcome for habitat protection, does serve to cross jurisdictional boundaries by involving federal and state agencies and the regional fishery management councils in a dialogue about the effects of nonfishing activities on important fish habitats.

This consultation process, while not a perfect example of an ecosystem-based management process, does align itself with an ecosystem ethic by recognizing linkages between essential fish habitat and nonfishing activities and by facilitating cooperation with other agencies that may have jurisdiction over these nonfishing activities (Macpherson 2004). Consultation also gives other federal agencies the ability to think through and adopt mitigation strategies to actions that may affect essential fish habitat or, at the very least, to explain their reasons for rejecting measures that could lessen the effect on habitat.

As with the essential fish habitat mitigation provision, the ecosystem-based management opportunity provided by the consultation requirement is not absolute. Fisheries managers ultimately lack the statutory authority to stop nonfishing activities that may harm fish habitat. Nevertheless, consultation requirements in the Magnuson–Stevens Act serve as an example of existing provisions in statutes governing marine resources that may be used by managers to advance ecosystem-based management principles, processes, and measures. Other sections of the MSA provide additional authority to consider linkages between managed species and the marine environment in making fisheries management decisions. For example, the statute authorizes fishery management plans to impose restrictions necessary for the conservation and management of the fishery [16 U.S.C. § 1853(b)(12)]. "Conservation and management" is a key term of art in the MSA, defined by Congress to include measures "required to rebuild, restore, or maintain . . . any fishery resources and the marine environment and . . . which are designed to assure that . . . irreversible or long-term adverse effects on fishery resources and the marine environment are avoided" [15 U.S.C. § 1802(5)]. The EFH requirement has not, however, resulted in incorporation of thresholds related to ecological resilience into fisheries management, and it is unlikely to do so without significant statutory direction.

Protected Species Management—The Endangered Species Act

Where the Magnuson–Stevens Act is directed primarily at exploitation, with conservation provisions added, there are several statutes affecting oceans resources that are directed primarily at conservation, with exploitation exceptions. Chief among these is the Endangered Species Act (ESA) [16 U.S.C. 1531 et seq]. Congress enacted the ESA in recognition of the increasing pace of habitat destruction, which was leading to species extirpation. The purpose of the ESA is "to provide a means whereby the ecosystems upon which endangered species and threatened species depend may be conserved" and to develop programs for conservation and recovery of those species [16 U.S.C. § 1531(b)]. Unlike agency-specific resource exploitation statutes, the ESA's conservation requirements apply to all federal agencies.

The ESA establishes criteria for listing plants and animals as threatened or endangered, and it generally requires designation of critical habitat for listed plants and animals. The act prohibits the unpermitted take of listed species and requires interagency consultation and cooperation to ensure that no federal activity is likely to jeopardize the continued existence of a listed species or adversely modify its critical habitat.

The ESA contains many key features that offer opportunities to advance ecosystem-based management of marine species, both by raising ecosystem concepts and by injecting them across jurisdictions into resource management decisions. While this section will focus on the critical habitat and interagency consultation provisions, a few other features of the act are worth noting. First, the ESA generally prohibits the unpermitted taking of any listed species by any person [16 U.S.C. § 1583(a)(1)(A)]. Second, Congress included civil and criminal penalties for violations of the act [16 U.S.C. § 1540(a)–(b)]. And third, Congress authorized citizen enforcement of the law [16 U.S.C. § 1540(g)]. On the one hand, this has served to make the ESA one of the most effective conservation statutes in the world. On the other hand, this has served to make the ESA a lightning rod of controversy.

Ecological resilience is addressed by the ESA to some degree, for example, via the requirement to establish criteria for endangered or threatened or recovery status. "Recovery" under the ESA, however, does not mean recovery in the traditional sense of the word. Species are considered recovered when they may be de-listed—when they are no long likely to go extinct in the foreseeable future. While powerful, the ESA becomes relevant only at the margins of existence and therefore provides a limited opportunity to develop proactive policies to maintain and/or restore the ecological resilience of coastal and ocean areas.

CRITICAL HABITAT

Critical habitat for a threatened or endangered species is broadly defined as specific areas in the geographic range of the species that contain "those physical or biological features . . . essential to the conservation of the species and . . . which may require special management considerations or protection" [16 U.S.C. § 1532(5)(A)(i)]. Critical habitat may also include areas outside the existing geographical range occupied by a listed species upon a determination that those areas are essential for the conservation of that species [16 U.S.C. § 1532(5)(A)(ii)]. Congress did prohibit designation of "the entire geographical area which can be occupied" by the listed species as critical habitat unless it is specifically found that circumstances require such a broad designation [16 U.S.C. § 1532(5)(C)].

Once critical habitat is designated, federal agencies must ensure that their actions are not likely to destroy or adversely modify critical habitat [16 U.S.C. § 1536(a)(2)]. "Destruction or adverse modification means a direct or indirect alteration that appreciably diminishes the value of critical habitat for both the survival and recovery of a listed species. Such alterations include, but are not limited to, alterations adversely modifying any of those physical or biological features that were the basis for determining the habitat to be critical" [50 C.F.R. § 402.02]. This duty to ensure against harm to critical habitat has proved to be a powerful conservation tool. Activities that might be lawful under resource extraction statutes, like the Magnuson–Stevens Act, must be modified to avoid harm to critical habitat before they can go forward. While the ESA is fairly criticized for its focus on preventing the extinction of particular imperiled species, curtailing activities that harm the habitat designated as critical for one species may have ancillary benefits for other species, simply by reducing anthropogenic effects on that habitat.

The critical habitat concept does have limits. The ESA requires agencies to designate critical habitat concurrently with species listing "to the maximum extent prudent and determinable" [16 U.S.C. § 1533(a)(3)(A)]. Agencies have relied heavily on this qualification to avoid critical habitat designations. Congress commanded that agencies make critical habitat designations based on "the best scientific data available." Congress then tempered that command by requiring agencies to consider the economic impact of designating an area as critical habitat [16 U.S.C. § 1533(b)(2)]. Finally, the ESA permits exclusion of an area from critical habitat designation upon a determination that the benefits of excluding the area will outweigh the benefits of including the area [16 U.S.C. § 1533(b)(2)]. This discretion is not permitted if failure to designate an area should "result in the extinction of the species concerned" [16 U.S.C. § 1533(b)(2)]. Nevertheless, even with these limits, where critical habitat exists, all federal agencies are forced to consider the habitat needs of listed species in their management decisions. For example, NOAA Fisheries designated critical habitat for the western population of endangered Steller sea lions in 1993. Since that time, the agency has had to evaluate whether fisheries that target sea lion prey in sea lion critical habitat adversely modify that habitat. In the late 1990s, after unsuccessfully defending itself in court, NOAA Fisheries made new analyses that indicated that some fisheries were likely competing with sea lions in a manner that diminishes the mammals' foraging effectiveness, adversely modifying critical habitat. Accordingly, the agency has modified fisheries regulations for sea lion prey species to reduce competition in both time and space.

INTERAGENCY CONSULTATION

In addition to assurance against destruction or modification of critical habitat under the ESA is the requirement that all federal agencies "insure that any action authorized, funded, or carried out" by that agency "is not likely to jeopardize the continued existence of any [listed] species" [16 U.S.C. 1535(a)(2)]. Because of the widespread federal permitting requirements and funding opportunities, many nonfederal projects are brought within the ambit of the ESA through this provision. Interagency consultation is the mechanism for these assurances.

Depending on the species, either the US Fish and Wildlife Service or the National Marine Fisheries Service is responsible for the conservation and recovery of that species [16 U.S.C. § 1532(15)]. Any federal agency that wants to authorize an action that may affect a listed species or its designated critical habitat must consult with the appropriate service. The appropriate service prepares a biological opinion, concluding whether or not the proposed activity is likely to jeopardize the listed species or adversely modify its critical habitat [16 U.S.C. § 1536(b)]. Should the service determine that the action is not likely to cross those thresholds, the action may go forward as proposed. Should the service conclude that jeopardy or adverse modification is likely, the action may not go forward. Instead, it must be modified in such a way that the jeopardy and adverse modification thresholds are not crossed [16 U.S.C. § 1536(b)

(3)(A)]. Modifications to some North Pacific groundfish fisheries to account for Steller sea lion prey needs, as described above, are an example of this process.

COOPERATION WITH STATES

In the ESA, in order to advance species conservation, Congress enacted specific provisions requiring cooperation with the states and authorizing state–federal management and cooperative agreements. Congress further authorized the services to provide financial assistance to states that enter into cooperative agreements, providing incentives for cross-jurisdictional collaboration [16 U.S.C. 1535].

The ESA provides important tools for advancing ecosystem considerations within the current regime—a clear conservation mandate with broadly applicable substantive standards, combined with formalized lateral and horizontal cross-jurisdictional consultation and collaboration and incentives. The ESA also requires establishment of various listing and de-listing thresholds and attendant management requirements. The ESA's biggest limitation is that its mandates generally do not come into play until a species is approaching extinction, and in its implementation it has served primarily to avoid extinction, not to conserve ecosystems. The species protected by the ESA's provisions may or may not be a species whose health or abundance has a disproportionate effect on its ecosystem, and concerted actions to conserve that species may or may not be useful to conservation of the resilience of the broader system. ESA protections and considerations often amount to emergency room triage, rather than a systematic approach to ecosystem conservation.

The National Environmental Policy Act

Perhaps the most promising available tool to advance ecosystem-based management in the current statutory and regulatory environment and to manage for the resilience of coupled social–ecological systems is the much lauded and much maligned National Environmental Policy Act (NEPA). In many ways, NEPA is a simple statute. It requires that federal agencies understand the consequences of their actions before they take them. It does not require federal agencies to reach a particular outcome in their decisions about public resources. NEPA is often referred to as a procedural rather than a substantive statute. This is due in part to US Supreme Court decisions holding that NEPA cannot "mandate particular results but only prescribe the necessary process" [Robertson v. Methow Valley Citizen's Council, 490 U.S. 332, 350 (1989)]. See also Kleppe v. Sierra Club [427 U.S. 390, 410 (1976)], holding that once an agency has made a NEPA-based decision, "the only role for a court is to ensure that the agency has taken a 'hard look' at the environmental consequences; it cannot 'interject itself within the area of discretion of the executive as to the choice of the action to be taken.'" But Congress believed that better information would lead to better decisions. Congress thereby established a particular analytical process and a particular set of considerations that have significantly changed conventional public resource management thinking by requiring rigorous analysis of the environmental consequences of resource exploitation.

Congress passed NEPA in 1969 in part in reaction to the growing awareness of the great damage being done to our nation's natural resources by government action (and inaction) due to ignorance. At the time, Congress's objectives were both revolutionary and entirely sensible. In NEPA, Congress established a new national environmental policy: "It is the continuing policy of the Federal Government . . . to use all practicable means and measures . . . in a manner calculated to foster and promote the general welfare, to create and maintain conditions under which man and nature can exist in productive harmony, and fulfill the social, economic, and other requirements of present and future generations of Americans" [42 U.S.C. § 4331(a)].

In the absence of an overarching federal ecosystem-based management mandate, NEPA is the most broadly applicable tool available to incorporate ecosystem considerations into marine resource decision making. NEPA applies to all federal agencies. Rather than modify the statutes creating various agencies or regulatory programs, Congress changed all federal agencies' approaches to their statutory missions by requiring all federal agencies to consider the environment in all of their programs and in all of their major decisions.

Many marine-related activities occur in or affect public waters, therefore requiring federal involvement that triggers NEPA jurisdiction. For those that affect only state waters, there is often a federal permit that provides the federal nexus required for NEPA to apply. In addition, many states have enacted so-called little NEPAs, state laws that largely track the requirements and procedures of the federal law. Some state statutes are even more stringent, imposing substantive in addition to procedural requirements (Sive et al. 2005). As EBM approaches become more widespread, NEPA provides a vehicle to at least force these concepts into the analysis and consideration of managing fisheries, offshore energy development, ocean dumping, and military activities, among other activities.

THE ENVIRONMENTAL IMPACT STATEMENT

For all major federal actions that may affect the quality of the human environment, NEPA requires the preparation of an environmental impact statement (EIS) [42 U.S.C. § 4332(C)]. Congress mandated that agencies consider the environmental impact of a proposed action, alternatives to that action, the relationship between short-term uses and long-term productivity, and any irreversible and irretrievable commitments of resources that would occur should the action go forward [42 U.S.C. § 4332(C)(i)–(v)]. In preparing the EIS, agencies are required by Congress to use "a systematic, interdisciplinary approach which will insure the integrated use of the natural and social sciences and the environmental design arts" [42 U.S.C. § 4332(A)].

Agencies must consider the direct and indirect effects of the proposed action [40 C.F.R. § 1502.16] and, importantly, must consider the cumulative impacts "on the environment which results from the incremental impact of the action when added to other past, present, and reasonably foreseeable future actions regardless of what agency (Federal or non-Federal) or person undertakes such other actions. . . . Cumulative impacts can result from individually minor but collectively significant actions taking place over time" [40 C.F.R. § 1508.7]. This obligation to consider indirect and cumulative effects, regardless of actor, is a significant one with the potential to reframe marine resource management decisions.

Another notable feature of NEPA is the specific procedure an agency must follow when it is assessing reasonably foreseeable significant adverse effects and the needed information is incomplete or unavailable [40 C.F.R. § 1502.22]. If the information is required to make a reasoned choice and the cost of getting the information is not exorbitant, the agency must get the information [40 C.F.R. § 1502.22(a)]. If the cost of the information is exorbitant or the means to get the information is unknown, the agency must do the following in its EIS: (1) explain that the information is incomplete or unavailable, (2) explain the relevance of the information, (3) summarize the existing credible and relevant scientific information, and (4) evaluate the information based on theoretical or research approaches that are generally accepted [40 C.F.R. § 1502.22(b)(1)–(4)]. Given our still significant gaps in information about many marine resources, processes, and functions, this procedure is highly relevant. As with NEPA overall, while it does not require an agency to reach a certain decision in light of uncertainty, it does require a relatively rigorous approach to uncertainty.

While NEPA does not mandate that an agency take the most environmentally friendly course of action identified by its analysis of effects, it was Congress's hope that an interdisciplinary approach to analysis of options would result in better, more informed decisions. In the terrestrial context, NEPA has certainly interjected broad principles of conservation biology and ecology into public lands resource management. NEPA is beginning to serve the same function in the marine arena (e.g., NOAA 2004, 2005a, 2005b).

Ultimately, NEPA's primary limitation is its lack of a substantive mandate. NEPA can provide the analytical tools needed to consider the effects of actions on temporal and spatial scales that are relevant to ecosystem-based management. NEPA can force consideration of the broader ecological effects of proposed actions. But NEPA cannot require the selection of an alternative that is considerate of the ecosystem. As such, NEPA is an important and useful analytical tool in the quest for ecosystem-based management of marine resources, but it is not a substitute for a more far-reaching ocean policy mandate. NEPA's lack of a substantive mandate also limits its utility in advancing resilience concepts. While NEPA's obligation to supplement analyses in the face of changed circumstances has the potential to lend itself as a tool for structuring adaptive management, NEPA can force only the analysis, not the management adjustments the analysis counsels. While NEPA forces consideration of uncertainty, it does not force management changes in the face of uncertainty. While NEPA explicitly considers not only the biophysical, but also the socioeconomic consequences of action or inaction, it does not require any particular result based on those impacts. NEPA may open the door to relevant considerations, but it will not require managers to cross the threshold.

Existing Institutional Structures

In addition to legal authority, understanding the institutional structure and decision-making process by which laws are implemented also is vital to moving EBM forward and other efforts to apply resilience science. At present, the federal ocean governance regime largely adheres to a balkanized paradigm that does not easily lend itself to implementing EBM. There are exceptions, of course, but the majority of institutional innovation relevant to resilience and ecosystem-based management is happening on the regional and state levels where place-based management efforts are on the rise (Young et al. 2007; see also part 4 case studies, this volume).

On the US Pacific coast, the 2006 West Coast Governors' Agreement on Ocean Health is intended to chart a new regional course to restore and maintain the health of the California Current ecosystem. Individual states (e.g., New York, New Jersey, Massachusetts, Florida, and California) are advancing ocean governance reforms in response in part to the reports of the two ocean commissions (POC 2003; USCOP 2004). Massachusetts passed its state Oceans Act of 2008 to embark on a comprehensive ocean zoning plan (MOCZM 2008). Local communities from the Gulf of Maine to Morro Bay, California, and to Port Orford, Oregon, have stepped forward with place-based management proposals (see McLeod and Leslie, chap. 17 of this volume, for more details). The breadth and pace of these emerging initiatives stand in stark contrast to the lack of institutional innovation federally.

Summary: Existing Authority for Ecosystem-Based Management

There is agreement among many scientists, practitioners, and policymakers that the knowledge and tools to move toward ecosystem-based management of marine resources exist within our current set of rules (e.g., McLeod et al. 2005; Leslie et al. 2008; Rosenberg and Sandifer, chap. 2 of this volume). The statutory provisions described above are but a sample of the numerous provisions in state and federal laws that permit or require consideration of ecosystem principles and allow ecosystem-based management decisions. The institutional structure of the current regime, however, tends to obstruct more-widespread innovation and application of these opportunities. The statutory tools that do exist are scattered throughout various statutes, applied only to certain agencies or to certain resources or only under certain circumstances. Due to NEPA's function as an umbrella over most federal action, proper and rigorous application of its requirements to decisions affecting marine resources would be a useful bridge from traditional management approaches to new ways of making decisions.

The current regime provides even less potential to manage for ecological and social resilience. The result is a trajectory characterized by increasing losses of ecological, social, and economic diversity and resilience. Sadly, it is only the undesirable, fractured jurisdictional and institutional elements of the existing governance regime—elements which contribute greatly to the current unsustainable trajectory—that can be characterized as persistent and resistant to change. A comparison of key elements from resilience thinking with current ocean governance illustrates how the regime principally obstructs efforts to maintain or restore ecological and social resilience (table 10.1).

While managing oceans as ecosystems rather than as sources of isolated resources is possible within our current statutory context, there is also general agreement that (1) those opportunities have yet to be fully realized and (2) incremental gains, while necessary, are not sufficient to tackle our marine problems (POC 2003; USCOP 2004; Crowder et al. 2006; Leslie and McLeod 2007). The challenge for this generation is to recognize the emerging window of opportunity for change and act decisively to breach thresholds needed to shift the existing institutions and regimes into a new sustainable state that enhances ecological, social, and economic diversity and resilience (Anderies et al. 2006; Lebel et al. 2006; Leslie and Kinzig, chap. 4 of this volume). The obstacles to breaching these thresholds are significant: Social, political, and institutional change generally does not come easily.

Reforms to Advance Ecosystem-Based Management

While there are opportunities to advance ecosystem-based management within the existing jumble of laws, institutions, and jurisdictions, new policy and regulatory frameworks are needed. The need to incorporate new understanding of the coupling between social and ecological systems (Shackeroff et al., chap. 3 of this volume) and resilience science (Leslie and Kinzig, chap. 4 of this volume) also suggests the need for a revitalized governance approach. We suggest a two-pronged approach, creating an overarching federal ocean policy and allowing for adjustments to the structure and procedures of governance to better implement that policy (POC 2003; USCOP 2004; Rosenberg and Sandifer, chap. 2 of this volume). This approach is in keeping with the resilience science tenet that there is no "one size fits all" set of governance recommendations and allows for considerable regional and subregional innovation (e.g., Wilson 2006).

Given the current degraded state of many of our oceans resources, any move toward ecosystem-based management, whether incremental or paradigm shifting, will most likely require increased restrictions on human activities. Successful implementation of marine ecosystem-based management will entail trade-offs of short-term costs for longer-term benefits

Table 10.1. Comparison of key elements of resilience thinking with existing ocean governance

Resilience thinking[1]	*Existing ocean governance*[2]
Embraces change; recognizes the dynamic, constantly changing nature of social–ecological systems and their drivers of change	Embraces status quo; suffers from the sliding-baseline syndrome; poorly recognizes, understands, and responds to the drivers of slow system change
Accounts for uncertainty and manages risk; understands that abrupt, unanticipated changes are possible	Exhibits a historical tendency for risk-prone decision making; poorly accounts for and even manipulates uncertainty to maximize delivery of certain short-term benefits; identification and monitoring of controlling variables and regime shift thresholds is rare
Recognizes that narrowly managing for maximum sustainable yield can reduce system resilience; focuses on the journey rather than the destination (i.e., does not manage for specific states or endpoints)	Specifically manages for maximum sustainable yield in fisheries; emphasizes optimizing the delivery of particular products or endpoints across a variety of resource categories; attempts to maintain specific system states that deliver maximum benefits from one or a few system components
Calls for adaptive management approaches	Tends to implement policies without plans for monitoring and learning from the experiment; rarely employs adaptive management
Emphasizes linkages between social and ecological systems; understands resilience characteristics are influenced by multiple domains	Emphasizes a more mechanical, reductionism approach; systems are broken down into constituent parts; scientists and managers specialize in disciplines to understand the properties of the different elements alone
Recognizes that management must include a multiscale perspective (i.e., system dynamics are determined by what's happening at both smaller and larger scales than the focal scale)	Emphasizes analysis and management at the focal scale; consideration of how system dynamics may be influenced by smaller- and larger-scale forces is relatively new and rare
Emphasizes longer time frames; considers both past and future system states	Emphasizes shorter management and planning time frames (3–5 years); consideration of the evolution of system states and the trajectory of systems is rare
Emphasizes consideration and management of a suite of disturbances (cumulative effects)	Tends toward isolated consideration of activities and threats; has historically not emphasized coordinated management of cumulative effects
Maintains diversity	Tends to promote, maintain, and optimize more narrowly defined values, sets of interests, and system components
Establishes redundant, overlapping institutions and governance processes	Emphasizes area and resource category specialization and administrative efficiency in institutions and governance processes

[1]Adapted from Anderies et al. 2006; Walker and Salt 2006; Leslie and Kinzig, chap. 4 of this volume.

[2]Adapted from POC 2003; USCOP 2004; McLeod et al. 2005; Rosenberg and Sandifer, chap. 2 of this volume.

and will therefore face significant opposition (see Rosenberg and Sandifer, chap. 2 of this volume). Accordingly, we also propose that existing activities would continue under the current regime, which permits but generally does not require the type of holistic ecosystem-based management envisioned, for a reasonable period of time, with incentives to transition to the new ecosystem-based standards by a time certain. New activities would not be permitted to go forward unless they were compliant with the new overarching federal mandate.

Overarching Federal Policy

Both the US Commission on Ocean Policy and the Pew Oceans Commission concluded after years of study and solicitation of public input that the nation would benefit from enactment of a national ocean policy act to establish a coherent framework for managing US ocean territory and resources. The Joint Ocean Commission Initiative, composed of leaders from both commissions, continues to educate the public and policymakers about the critical need for a new mandate. The three most important elements of such a policy must be (1) a new overarching and unequivocal mandate accompanied by (2) decision rules and delegation of clear authority to resolve conflicts and manage trade-offs and (3) new coordinated, inclusive planning and management decision-making structures appropriately scaled to reflect ecosystem rather than political boundaries (box 10.1).

The preference for a mandate in lieu of guidance or voluntary policy directives is largely a response to the level of resistance a shift to ecosystem-based management will likely attract. Given the nature of the changes EBM and resilience imply, capturing the ecosystem-based management goal in a new policy mandate is necessary to provide the motivation, legal mechanisms, and incentives needed to coordinate management at levels capable of achieving the goal.

A federal mandate should also establish an independent committee of scientists to provide technical guidance and advice on how we might maintain and evaluate concepts such as ecosystem diversity, structure, functioning, productivity, and resilience. The National Forest Management Act, for example, authorizes the secretary of the interior to appoint a committee of scientists when revising planning regulations. The secretary convened such a committee in the late 1990s after two attempts to revise regulations were unsuccessful, in large part due to controversy over requirements for species viability. The committee of scientists' recommendations included making sustainability the overarching goal of national forest management. Predictably, that recommendation was not avidly embraced by the Forest Service. Including a requirement that agencies adopt the recommendations of the committee of scientists would be advisable.

Implementing the overarching federal policy will require new regional governance structures composed of federal, state, tribal, and local authorities. The essence of these structures will be to provide the cross-jurisdictional interface needed to foster cooperation. Enabling cooperation should encourage and empower local, bottom-up approaches to ecosystem-based management while balancing such efforts with a top-down federal approach that requires coordination and the satisfaction of minimum standards. Complementary partnerships and creative tension are the intent, and while the policy should require action to create partnerships, it should allow flexibility as to how those must be structured. It is likely that multiple types of these regional arrangements could be facilitated based on the particulars of different ecosystems, activities, and so forth, and allowing for this diversity should be considered a benefit. To further diversify and build resilience into these decision-making structures, each new regional ecosystem-based management entity should also establish and empower advisory councils to include a broad array of perspectives and interests.

The principal function of these regional partnerships should be the development of enforceable plans that are based on an active, adaptive management approach. Such an approach accepts change as a natural feature of social and ecological systems and treats management as a series of experiments from which we can learn (Kaufman et al., chap. 7 of this volume). Presently, management is an experiment from which we learn little because we fail to embed active monitoring and interdisciplinary scientific investigations within management and governance. Reversing that problem will make regional ecosystem-based management governance more resilient by creating incentives for participation through establishing cooperative research and monitoring approaches and incorporating gathered knowledge into a decision-making process that cultivates learning and is more flexible to adaptation.

Another logical imperative of the new mandate to maintain ecosystems in a healthy, productive, and resilient condition is the need to maintain large management areas that protect a range of species and habitat diversity, so species can move to more suitable areas and otherwise adapt in response to climate change and its effects (e.g., Elmqvist et al. 2003). Sustaining diversity is simply essential to managing for ecological resilience (Leslie and Kinzig, chap. 4 of this volume). Given tensions swirling around protected areas today, developing appropriate criteria for marine protected area (MPA) designation and indicators of MPA success will clearly be among the most controversial new EBM requirements, and as such they should be clearly mandated as part of the national oceans policy act. The mandate can be tempered by the fact that a diversity of approaches to meet this objective exist and clearly would emerge from a diversity of unique regional governance processes.

Box 10.1. Incorporating Elements of Resilience Science in a New Ocean Governance Mandate

Developing a national ocean policy act around the following ideas would codify resilience concepts in US ocean governance and advance EBM:

- **Mandate** one overarching and unequivocal policy goal—to protect, maintain, and restore ecosystems in a healthy, productive, and resilient condition so they may continue to provide the services humans want and need.
- **Establish decision rules** to manage trade-offs and resolve conflicts or delegate authority to one entity to create such rules (see Rosenberg and Sandifer, chap. 2 of this volume).
- **Establish the largest scale of EBM jurisdictional units** (i.e., large marine ecosystems [LMEs]), but also require and create incentives for a diversity of smaller-scale units to work within the statutory unit to create tighter feedbacks and overlapping institutions with shared authority.
- **Require active adaptive management** of each system to either (1) avoid thresholds that will shift a system into a regime where the system is no longer capable of providing the breadth of goods and services desired or (2) drive an undesirable system into a different regime capable of providing those goods and services.
- **Require the development of a model for each social–ecological system** to be managed that does the following:
    - **Describe the history** of the system, alternate system regimes, adaptive cycles, and the current trajectory of the system.
    - **Define key system attributes** ecosystem services, including nonconsumptive services such as nutrient cycling; slow variables; thresholds; intervention points; management triggers; penalties; and consequences of failure to act.

Concluding Remarks

The current US ocean governance regime is contributing to severe and unacceptable losses of ecological, social, and economic diversity. While ecosystem-based approaches can be advanced within the context of existing authorities, the political and institutional structure of the current system thwarts such innovation and is itself quite persistent and resistant to change. A window of opportunity is now opening to shift the institutional structure of US ocean governance into a more sustainable regime. The time has come to craft an overarching oceans mandate which establishes new governance structures compatible with the latest science and practice of ecosystem-based management. At this unprecedented moment in human history, where our capacity to change the Earth's systems has set in motion sweeping ecological and social changes, we must focus on this new reality and realign our laws and governance systems to be capable of addressing the challenge of our time: securing sustainability.

References

Anderies, J. M., B. H. Walker, and A. P. Kinzig. 2006. Fifteen weddings and a funeral: Case studies and resilience-based management. *Ecology and Society* 11(1):21.

Christie, D. R., and R. G. Hildreth. 2007. *Law in a nutshell: Coastal and ocean management.* St. Paul, MN: West Publishing Co.

Crowder, L. B., G. Osherenko, O. R. Young, S. Airame, E. A. Norse, N. Baron, and J. C. Day et al. 2006. Resolving mismatches in US oceans governance. *Science* 313:617–18.

Eagle, J., S. Newkirk, and B. H. Thompson Jr. 2003. *Taking stock of the regional fishery management councils.* Washington, DC: Island Press.

Elmqvist, T., C. Folke, M. Nystrom, G. Peterson, J. Bengtsson, B. Walker, and J. Norberg. 2003. Response diversity, ecosystem change, and resilience. *Frontiers in Ecology and the Environment* 1:488–94.

Holling, C. S., and G. K. Meffe. 1996. Command and control and the pathology of natural resource management. *Conservation Biology* 10:328–37.

Lebel, L., J. Anderies, B. Campbell, C. Folke, S. Hatfield-Dodds, T. Hughes, and J. Wilson. 2006. Governance and the capacity to manage resilience in regional social–ecological systems. *Ecology and Society* 11(1):19.

Leslie, H. M., and K. L. McLeod. 2007. Confronting the challenges of implementing marine ecosystem-based management. *Frontiers in Ecology and the Environment* 5(10):540–48.

Leslie, H. M., A. R. Rosenberg, and J. Eagle. 2008. Is a new mandate needed for marine ecosystem-based management? *Frontiers in Ecology and the Environment* 6(1):43–48.

MA (Millennium Ecosystem Assessment). 2005. *Ecosystems and human well-being: Current state and trends.* Washington, DC: Island Press.

Macpherson, M. 2004. Integration of ecosystem management approaches into federal fishery management through the Magnuson–Stevens Fishery Conservation and Management Act. *Ocean and Coastal Law Journal* 6:1–32.

McLeod, K. L., J. Lubchenco, S. R. Palumbi, and A. A. Rosenberg. 2005. *Scientific consensus statement on marine ecosystem-based management.* The Communication Partnership for Science and the Sea (COMPASS). Signed by 221 academic scientists and policy experts with relevant expertise. http://www.compassonline.org/pdf_files/EBM_Consensus_Statement_v12.pdf.

MOCZM (Massachusetts Office of Coastal Zone Management). 2008. http://www.mass.gov/czm/oceanmanagement/oceans_act/index.htm.

NOAA (National Oceanic and Atmospheric Administration). 2004. *Alaska groundfish fisheries final programmatic supplemental environmental impact statement.* July 2004. http://www.fakr.noaa.gov/sustainablefisheries/seis/intro.htm.

NOAA (National Oceanic and Atmospheric Administration). 2005a. *Final environmental impact statement for essential fish habitat identification and conservation in Alaska.* April 2005. http://www.fakr.noaa.gov/habitat/seis/efheis.htm.

NOAA (National Oceanic and Atmospheric Administration). 2005b. *Pacific coast groundfish fishery management plan, essential fish habitat designation and minimization of adverse impacts, final environmental impact statement.* December 2005. http://www.nwr.noaa.gov/Groundfish-Halibut/Groundfish-Fishery-Management/NEPA-Documents/EFH-Final-EIS.cfm.

NRC (National Research Council). 1999. *Sustaining*

marine fisheries. Washington, DC: National Academies Press.

POC (Pew Oceans Commission). 2003. *America's living ocean: Charting a course for sea change. A report to the nation*. Washington, DC: Pew Trusts.

Safina, C., and D. Klinger. 2008. Collapse of bluefin tuna in the western Atlantic. *Conservation Biology* 20(2):243–46.

Sive, D., et al. 2005. "Little NEPAs" and the environmental impact assessment process. In *Environmental litigation*, 1175. SK094. Philadelphia: ALI-ABA .

UN (United Nations). 1995. *UN Conference on Straddling Fish Stocks and Highly Migratory Stocks*. 1995. Agreement for the implementation of the provisions of the United Nations Convention of the Law of the Sea of 10 December 1982 relating to the conservation and management of straddling fish stocks and highly migratory fish stocks. New York: UN.

UNCED. 1992. *Report of the United Nations Conference on Environment and Development*. Rio de Janeiro: United Nations.

UNEP (United Nations Environment Programme). 2006. *Marine and coastal ecosystems and human wellbeing: A synthesis report based on the findings of the Millennium Ecosystem Assessment*. Nairobi: UNEP.

UNFCCC (United Nations Framework Convention on Climate Change). 1994. 1771 UNTS 107. Bonn: UN.

USCOP (US Commission on Ocean Policy). 2004. *An ocean blueprint for the twenty-first century*. Final report. Washington, DC: USCOP. ISBN 0 9759462 0 X.

Walker, B., and D. Salt. 2006. *Resilience thinking: Sustaining ecosystems and people in a changing world*. Washington, DC: Island Press.

Wilson, J. A. 2006. Matching social and ecological systems in complex ocean fisheries. *Ecology and Society* 11(1):9.

Worm, B., E. B. Barbier, N. Beaumont, J. E. Duffy, C. Folke, B. S. Halpern, J. B. C. Jackson et al. 2006. Impacts of biodiversity loss on ocean ecosystem services. *Science* 314:787–90.

Young, O. R., G. Osherenko, J. Ekstrom, L. B. Crowder, J. Ogden, J. A. Wilson, J. C. Day et al. 2007. Solving the crisis in ocean governance: Place-based management of marine ecosystems. *Environment* 49(4):20–32.

PART 4

Marine Ecosystem-Based Management in Practice

CHAPTER 11
Morro Bay, California, USA

Dean E. Wendt, Linwood Pendleton, and Don Maruska

In response to threats of offshore oil development and the potential dumping of selenium-tainted agricultural wastewater, our community created a new organization, the San Luis Obispo Science and Ecosystem Alliance (SLOSEA). This integrated group of scientists, resource managers, and stakeholders is studying and supporting marine resource management in Morro Bay on the central California coast. The following chapter describes the creation of SLOSEA, illustrates parallels with resilience concepts, and details how SLOSEA has applied marine ecosystem-based management (EBM) science to important management questions.

Morro Bay Estuary and the adjacent coastal ocean (hereafter, the Morro Bay ecosystem) is one of only two significant wetland systems on California's central coast (the other being Elkhorn Slough in Monterey Bay). The Morro Bay ecosystem juxtaposes a globally significant hotspot for terrestrial biodiversity with a rich and productive coastal marine ecosystem. The ecosystem serves as a link for many migratory species (e.g., birds, steelhead trout) and as a permanent home for a variety of fishes, mammals, invertebrates, and plants, including sixteen federally threatened or endangered species, six of which are endemic to the area (MBNEP 2000). The watershed for the estuary covers 48,000 acres (75 square miles) with two primary tributaries, the Los Osos and Chorro creeks (fig. 11.1). The watershed includes riparian corridors, agricultural lands, oak grassland, coastal chaparral, and relatively limited urbanization. About 68% of the watershed is agricultural land, urbanized residential areas make up approximately 11% of the area, and remaining land uses include open space, areas for public facilities, and recreation areas. The majority of the land in the watershed is privately owned, with some ownership by federal, state, and local municipalities.

The estuary is a 2,300-acre semienclosed body of water, which empties into the larger Estero Bay, an open coastal embayment. The distribution of habitats within the bay include eelgrass beds, mudflats, salt marsh, sandy beaches, and to a lesser degree, emergent rocky substrata. The ecosystem supports migratory birds such as Brandt geese, a suite of infaunal animals only found in estuaries, and plants such as eelgrass, which serves as nursery areas for fish. The area outside the estuary is dominated by both sandy and rocky intertidal and subtidal benthic habitats with extensive kelp beds north and south of Estero Bay. Within central California, the Morro Bay ecosystem is one of the most likely places to have strong land-to-sea linkages, as most watersheds in this area empty directly into the ocean, rather than a semienclosed embayment like this one.

The long-term sustainability of the services that this ecosystem provides is threatened by increased sedimentation, increased bacterial loading, and increased nutrient inputs, which have acted cumulatively to alter ecosystem productivity (MBNEP 2000). Recent studies show additional major shifts in the bay over the last 30 years from a system dominated by mussels to one dominated by invasive bryozoan species (Needles 2007). Point source and nonpoint source pollution remain a problem during certain times of the year, directly impacting

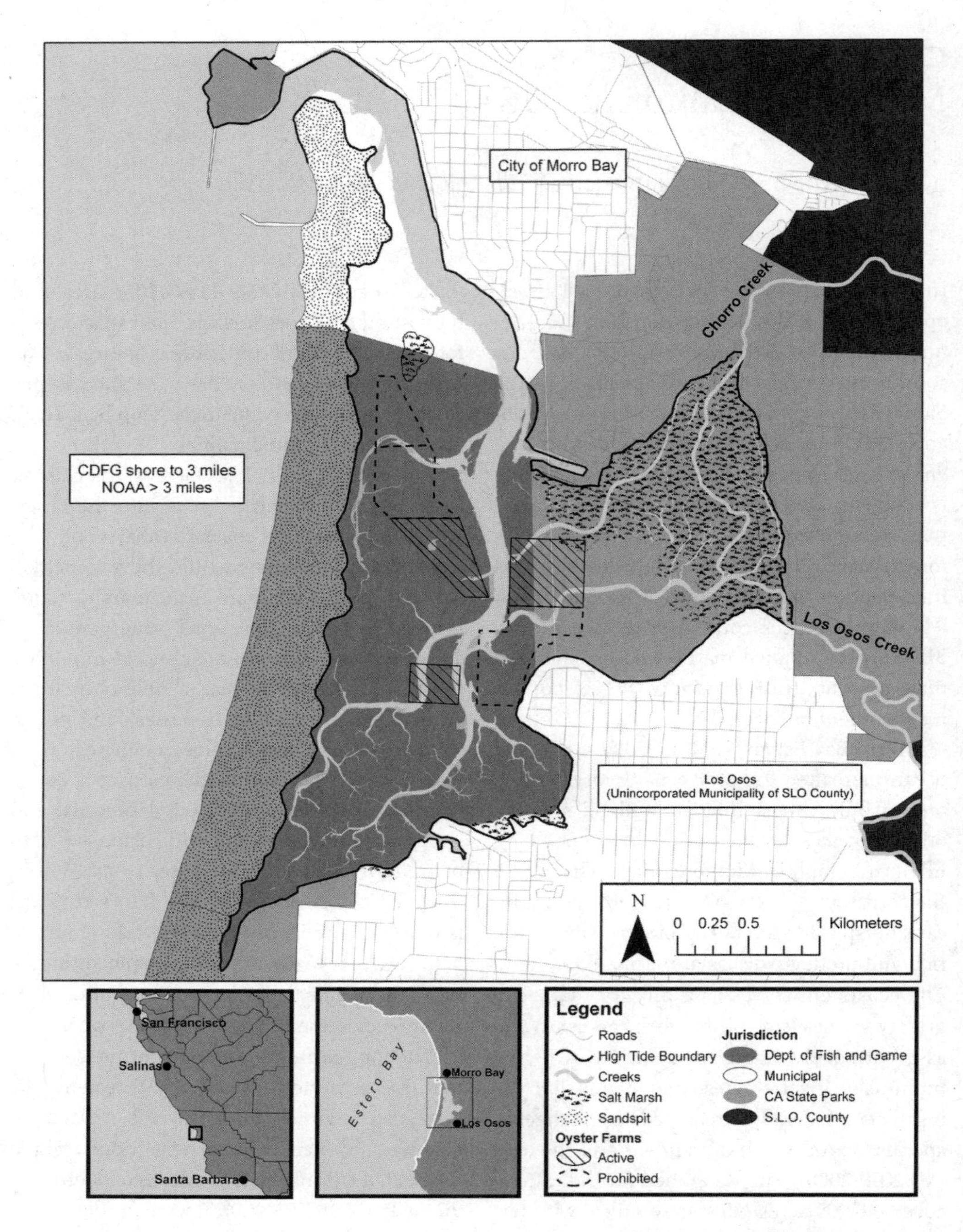

Figure 11.1 Morro Bay is located on the central coast of California approximately 320 km north of Los Angeles and 370 km south of San Francisco. Shown are the current jurisdictional boundaries, the locations of oyster leases, and SLOSEA water quality arrays. The towns of Morro Bay and Los Osos have populations of about ten thousand and fifteen thousand people, respectively.

ecosystem services such as shellfish harvest and recreational activities. Fishing practices and harvest levels have contributed to depletion of some of the nearshore fish species in waters north (e.g., Mason 1998) and south (e.g., Love et al. 1998) of the Morro Bay ecosystem, but a recent analysis suggests that local nearshore fish may not have suffered the same declines (Stephens et al. 2006).

The towns of Morro Bay, Los Osos, and Cayucos have a combined population of approximately thirty thousand people. In addition to an active tourist industry, the ecosystem supports both recreational and commercial fishing fleets (although to a lesser degree than historically), recreational boating, land-based agriculture including farming and grazing, and an active oyster and abalone aquaculture industry. Another iconic feature of the Morro Bay ecosystem is the 200-foot-tall stacks of the natural gas–fired power plant. When running at full capacity, the power plant pumps out of Morro Bay about 450,000 gallons of water for cooling per minute, which is then discharged in the near coastal environment along Morro Rock. Another service delivered by the ecosystem is the processing of nutrients from approximately 4,300 septic systems from the town of Los Osos, which surrounds the southern and eastern portions of the bay, and the sewage outfall 0.5 miles offshore in Estero Bay for the towns of Morro Bay and Cayucos. In fact, the towns of Morro Bay and Cayucos still receive one of the last remaining federal waivers (301-h modified discharge permit) that allows them to discharge partially treated sewage into the coastal environment. Significant benthic algal blooms and ongoing restrictions on oyster farming because of high levels of fecal coliform bacteria suggest that these inputs are having an impact. The ecosystem is also home to two golf courses and several state parks, including Montaña de Oro and Morro Bay State Park.

As with most coastal areas, the Morro Bay ecosystem is a tapestry of jurisdictional boundaries, government agencies, municipalities, and private land ownership (fig. 11.1). Within this relatively small area, regulatory authority is held by the California Department of Fish and Game (CDFG), US Fish and Wildlife Service, NOAA Fisheries, Bureau of Land Management, National Forest Service, California Coastal Commission, California Water Quality Control Board, California Department of State Parks and Recreation, and several local governing bodies, including San Luis Obispo County, the cities of San Luis Obispo and Morro Bay, and the Community Services District of Los Osos. There are no existing mandates or memoranda of understanding dictating that the regulatory agencies or municipalities interact with one another. Consequently, one of the primary functions of SLOSEA (as described below) is to increase interagency communication and management coordination.

The SLOSEA Organization

SLOSEA has developed as an organization in the wake of ongoing efforts both locally and at the state and federal levels. The most significant state-level development was the enactment of the California Ocean Protection Act in 2004. The act established the California Ocean Protection Council, whose general mission is to improve the protection and management of California's ocean and coastal resources and to implement the governor's ocean action plan released in October 2004; among many actions, the plan calls for ecosystem approaches to management.

At a local level, there is a significant history of activism associated with the estuary that began in the late 1960s. These grassroots efforts culminated in 1994, with the designation of Morro Bay as California's first State Estuary and acceptance into the National Estuary Program in 1995. The Morro Bay National Estuary Program (MBNEP) is a collaborative watershed organization that works with stakeholders and resource managers to restore the health of the watershed and estuary. Another group central to the existence of SLOSEA is the Marine Interests Group of San Luis Obispo County (MIG). The MIG formed in 2002 as a consensus stakeholders' group to discuss the extension of the Monterey Bay National Marine Sanctuary, which resulted from fears of offshore oil development and potential selenium dumping from California's central valley into Estero Bay (a management option being considered by the US Bureau of Reclamation). The existence of SLOSEA rests squarely on the earlier work and collaboration of both the Morro Bay National Estuary Program and the MIG. The SLOSEA program is working with the MIG and the MBNEP to build on their existing strengths and to overcome previous limitations by conducting research and monitoring over spatial scales that complement what the other organizations do (table 11.1) and by establishing an integrated, cross-jurisdictional management community for the entire Morro Bay ecosystem (land, estuary, and coastal ocean).

The Functional Unit of Marine EBM: An Integrated Ecosystem Group

To understand the process we used to develop SLOSEA and its focal areas of research and monitoring, it is useful to first understand the organizational structure. Three core entities exist within the program: (1) a leadership team, consisting of the SLOSEA program director, the director of the Morro Bay National Estuary Program, and the independent facilitator of the MIG; (2) a science team, which consists of about ten academic scientists and many research staff; and (3) an advisory committee, including representatives of all organizations with jurisdictional authority and management responsibilities in the ecosystem, stakeholders that live and work in the ecosystem, and three individuals from the science team. Taken together, these three entities form an "integrated ecosystem group," that is, the functional unit for implementing EBM. In many respects, SLOSEA's integrated ecosystem group is a modern representation of the traditional ecological knowledge management systems discussed by Kliskey and colleagues in chapter 9 of this volume. This organizational structure provides a place-based program that draws on a broad array of knowledge from both scientists and stakeholders to make sound management decisions across jurisdictional boundaries. Indeed, SLOSEA functions because all of these groups work together to improve information, understanding, and management. As Kaufman and colleagues (chap. 7 of this volume) suggest, scientists need to move beyond their narrowly circumscribed roles as academics or technicians; we suggest here that managers and stakeholders must also undertake such a frame shift in order to implement marine EBM. SLOSEA's integrated ecosystem group strives to accomplish, at least in part, this kind of wholesale change.

Developing a Conceptual Model of the Ecosystem

One of the first products from the advisory committee and the science team was a conceptual model of the Morro Bay ecosystem. This model helped us identify potential key linkages within the ecosystem and define ecosystem boundaries. Although not quantitative, the conceptual model enabled managers and stakeholders to articulate critical questions and further develop initiatives to test specific hypotheses of interest. The SLOSEA model represents the ecosystem as three integrally connected compartments, including the watershed, the estuary, and the coastal ocean (fig. 11.2). Each of the compartments of the

Table 11.1. Activities of MIG, MBNEP, and SLOSEA (primary foci in uppercase)

	Watershed	*Bay/Estuary*	*Coastal ocean*
Marine Interests Group (MIG) The MIG seeks to (1) promote understanding of the marine resources off the coast of San Luis Obispo County and the needs and interests of the stakeholders involved with their use and enjoyment, (2) openly examine potential ways to sustain and enhance the resources, and (3) recommend desirable courses of action to support the resources and their sustainable use.			• OUTREACH & EDUCATION • Basic research • Monitoring • Linking science to management
Morro Bay National Estuary Program (MBNEP) The MBNEP is a collaborative organization that brings local citizens, local government, nonprofits, agencies, and landowners together to protect and restore the physical, biological, economic, and recreational values of the Morro Bay Estuary.	• RESTORATION • CONNECTING TO RESOURCE MANAGERS • OUTREACH & EDUCATION	• RESTORATION • MONITORING • CONNECTING TO RESOURCE MANAGERS • OUTREACH & EDUCATION • Basic research	
San Luis Obispo Science and Ecosystem Alliance (SLOSEA) SLOSEA is an integrated group of scientists, resource managers, and stakeholders focused on creating sustainability and resilience of the marine resources within San Luis Obispo County. Their mission is to improve management by facilitating transfer of data and knowledge from academic scientists and the community to resource managers.		• BASIC RESEARCH • MONITORING • LINKING SCIENCE TO MANAGEMENT • Outreach & education	• BASIC RESEARCH • MONITORING • LINKING SCIENCE TO MANAGEMENT • Outreach & education

ecosystem is connected by species movements (most notably, the use of the bay as a nursery ground for fish and invertebrates) and freshwater input and tidal exchange. One of the fundamental goals of SLOSEA is to increase our understanding of the linkages and dynamics of the ecosystem, including its human components, so as to provide practical information to resource managers for more cost-effective and ecologically effective management.

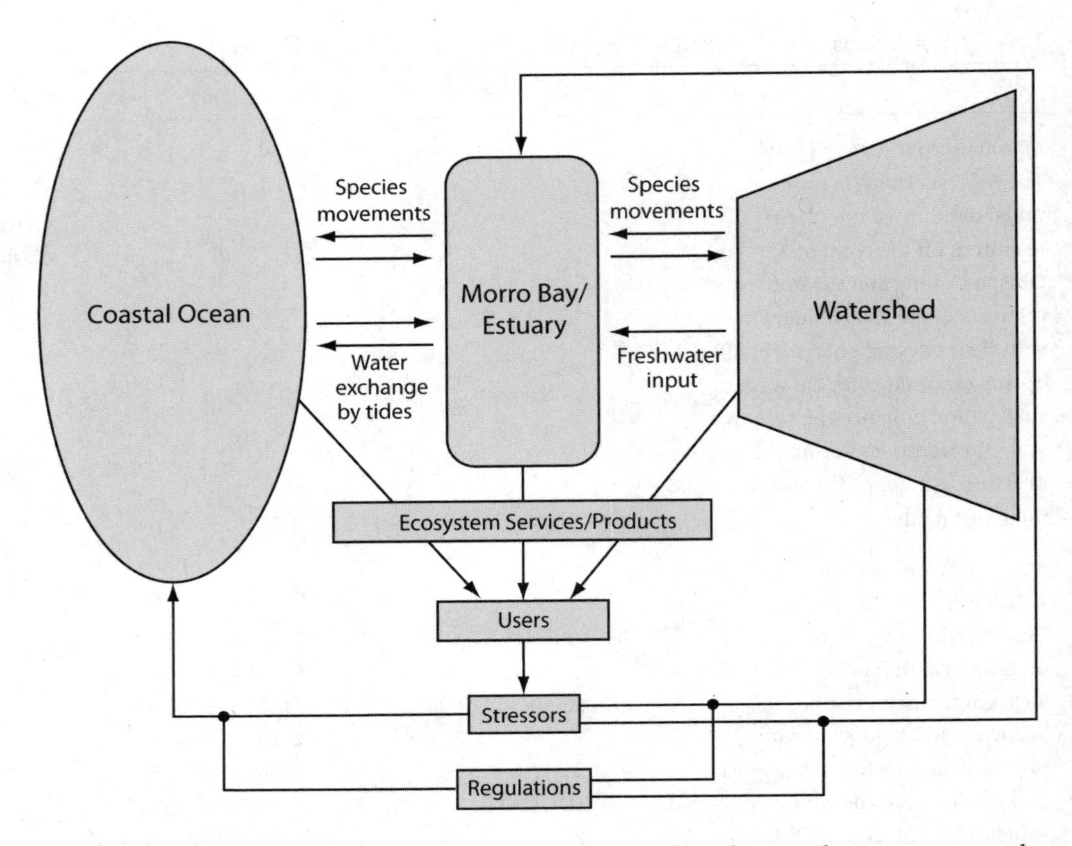

Figure 11.2 Conceptual diagram of the Morro Bay ecosystem, linking the coastal ocean, estuary, and upland areas. This was developed by the SLOSEA advisory committee and science team.

Stakeholder Engagement

In our experience a central aspect of EBM is developing and encouraging discussion about science and management among diverse players. SLOSEA tries to capture knowledge about the ecosystem from multiple sources and groups—scientists, fishermen, resource managers, businesspeople, environmentalists, and all others who live and work in the ecosystem. We believe that doing so provides a clearer, more holistic understanding of the area and develops a place-based identity, a critical element in managing for resilience and sustainability (McLeod et al. 2005; Kliskey et al., chap. 9 of this volume; Leslie and Kinzig, chap. 4 of this volume). We have engaged diverse players in SLOSEA through a balance of science-centered stakeholder consultation and informed collaboration (Webler and Tuler 2006). Once everyone is "at the table," we are also challenged regularly with trying to convey (and receive critical input on) complex scientific methodologies and ecological concepts to nonscientists and nontechnical audiences. This challenge has been overcome through professional facilitation of advisory committee meetings, minimizing the use of scientific jargon, relying on conceptual diagrams and graphic representation of information where possible, and utilizing the SLOSEA Web site as an interactive forum to display results and post observations about the ecosystem. We use this suite of tools to engage stakeholders in substantive ways, tap their knowledge about the area, and build ecosystem identity among SLOSEA participants and the larger scientific, management, and stakeholder communities.

A Collective Vision for the Morro Bay Ecosystem

A key step in developing SLOSEA was creating a collective vision for the Morro Bay ecosystem from the advisory committee. A set of shared hopes for SLOSEA formed the basis for this vision and guides the continuing evolution and development of SLOSEA's activities (see box 11.1).

Research and Monitoring

One of the critical first steps in developing SLOSEA was defining the areas within which to focus research and monitoring activities of relevance and interest to the management community. This included investigating connections to larger ecosystem scales, given the recognition that to affect change we need to differentiate between factors over which we have management control versus those operating at broader scales, such as climate change (i.e., understanding cross-scale linkages; see Guichard and Peterson, chap. 5 of this volume; Leslie and Kinzig, chap. 4 of this volume). Active research projects are now underway in six areas:

- Water quality
- Biological indicators
- Economic indicators
- Critical spawning and nursery environments
- Human access
- Collaborative fisheries research

One of the central objectives of the SLOSEA program is to provide managers with data that can inform their management decisions. This occurs as the advisory committee develops the

Box 11.1. Shared vision for SLOSEA (San Luis Obispo Science and Ecosystem Alliance)

Clean and Healthy Ecosystem

- Clean and healthy water and marine life so that people enjoy the water and eating from it
- Protection of the watershed, estuary, and nearshore area as a nursery
- Public shoreline access with resource protection that enables people to access the lands they helped acquire and earns sustained public support and stewardship
- Flexible systems and solutions that work in a dynamic environment so that we can avoid firefighting or crises

Lasting Insights, Tools, and Support

- Cost-effective tools to monitor the ecosystem that become part of ongoing management
- Long-term time-series data that can be maintained beyond the term of the project
- Successful collaborations and partnerships among resource managers and stakeholders
- Method for extending the approach to a broader area to encompass the larger biogeographic region
- Expanded public interest and support for the marine environment
- Model of successful ecosystem-based management to share with other ecosystems

Deeper Understanding of the Resources

- Accurate and reliable data for understanding the ecosystem resources and dynamics
- Identification of thresholds and tipping points to sustain critical resources
- Clear voice for the silent flora and fauna of the ecosystem to preserve these natural treasures
- Better understanding of the values and socioeconomic interests of people within the ecosystem in order to improve resource management

Improved Decision Making

- Enhanced scientific understanding leading to better decisions by resource managers
- Good working relationships among public, users, scientists, and regulators to develop common ground
- Improved sociopolitical understanding and decision making to extend these successes elsewhere

areas of scientific inquiry, as well as through regular and sustained interaction among the diverse members of this group to hone details of methodologies, discuss recent results, and tailor data products to management needs. This approach to marine EBM requires that scientists respond to feedback from resource managers and stakeholders about their research and be willing to undergo "course corrections." Scientists working with SLOSEA have also needed to become communicators beyond their academic circles (see Kaufman et al., chap. 7 of this volume). At the same time, it is essential that resource managers and stakeholders understand the process of scientific inquiry, including the time and logistical constraints associated with such endeavors.

AN EXAMPLE: WATER QUALITY RESEARCH AND MONITORING

To address questions surrounding water quality in the estuary and particularly the sources of nutrients to the bay, SLOSEA designed and deployed four water quality arrays within the estuary at the creek mouths, the back bay, and the entrance of the mouth of the estuary (fig. 11.1).

Each array continuously measures salinity, temperature, turbidity, dissolved oxygen, nitrate concentrations, water velocity, and fluorescence (as a proxy for phytoplankton biomass) in order to provide baseline information about physical, chemical, and biological processes within the bay and changes through time.

One issue of critical interest to SLOSEA is the effect of excess nutrients on the ecosystem services delivered by the estuary, for instance, its role as a nursery for fishes and habitat for oyster farming. Based on the conceptual model developed earlier (fig. 11.2), the advisory committee is investigating all possible sources of nutrients, including the watershed, the open coast, and groundwater input (recall that Los Osos is on septic systems). Are the major sources of nutrient input to the bay local and thus controllable (i.e., watershed or septic systems) or regional and potentially not human-caused (i.e., upwelling events) and thus not controllable? The water quality monitoring data will help us to address this question, as well as enable us to track changes in the bay ecosystem on a range of temporal cycles—from the daily tidal fluctuations to seasonal, annual, and longer-term cycles.

Connecting Natural Science with Economics: Biological and Economic Indicators

EBM activities, and indeed all types of coastal ecological restoration, are expensive, time-consuming, and often politically fractious. EBM projects compete with other public projects for sustained funding, and considerable public debate often exists about whether changes in coastal economies and cultures are tied to ecosystem decline or other factors beyond the control of coastal planners, managers, and conservation professionals. Unfortunately, the debate often is based more on anecdote than science because as scientists, we have failed to show that noncatastrophic ecosystem changes (i.e., changes that fall within the realm of past experience) have an impact on economic activity. Further, while we have developed a number of indicators to monitor ecosystem health (e.g., fecal coliform bacteria levels, dissolved oxygen, and other water quality parameters such as those described in the previous section), we know little about how measurable changes in ecosystem health are linked to measurable changes in economic activity at the local level. As a result, even if we know that ecosystem indicators reveal that EBM has led to improvements in ecosystem quality (e.g., declining levels of fecal indicator bacteria, more eelgrass, or fewer hypoxic events), we are often unable to show that these improvements have yielded measurable improvements in the way people use, enjoy, and benefit economically from these ecosystems.

In most cases, communities are unaware of the economic value of ecosystem-dependent activities. Economic data are rarely collected at a level that corresponds well with the spatial extent of a marine ecosystem, whether that ecosystem is a large marine ecosystem, a major estuary (e.g., San Francisco Bay), or a smaller embayment like Morro Bay. As a result, it is difficult for planners to put the costs of restoration and management in the context of the economic activities that might benefit from such policies. Making the matter even more complicated is the fact that restoration and coastal development often represent large one-time values, while the economic values of ecosystem goods and services usually are in the form of smaller values distributed over long periods of time. Without a good understanding of the annual, and potentially sustainable, values of ecosystem-dependent economic activities, the costs of management and the coastal development often seem disproportionately large.

In Morro Bay, the public discussion about the economic value of implementing EBM strategies often focuses on commercial and charter recreational fishing, ignoring the potential economic value of other types of ecosystem-dependent economic activity (especially tourism, recreation, and housing values) that depend on ecosystem conditions (see Kildow 2007 for a recent review of the economic effects of water quality on home values). Commercial fishing activity is concentrated among a small, and often shrinking, cadre of commercial fishers and captains, and charter recreational fishing operations are similarly concentrated at a limited number of locations. Recreational activities, on the other hand, are much more diffuse. Kayaking, recreational fishing, bird watching, and diving often can be enjoyed by individuals without the assistance of commercial businesses. To date, we do not know the economic contribution of marine-based recreation or tourism to the Morro Bay economy. Even more importantly, we do not know how these activities have changed over time. As a result, the economic contributions of ecosystem- and marine-dependent recreational and tourism activities are left out of public discussion, policy, planning, and management.

Linking EBM to Economic Activity

There is little empirical evidence to show how ecological decline has affected local marine-based economies, and even less evidence we can use to predict how EBM will affect local economic activity. In the late 1980s, a number of studies demonstrated the economic impacts on fisheries of large-scale losses of mangroves and wetlands (Lynne et al. 1981; Kahn and Kemp 1985; Ellis and Fisher 1987). Similarly, many bioeconomic studies have attempted to link environmental conditions to fisheries output, but the measures of environmental quality are often not available or appropriate at the scale of most EBM studies (see Knowler 2002 for a review). Further, the empirical studies tend to be static; the economic impact of changes in ecosystems must be predicted based on static relationships between ecosystem quality and economic activity (Knowler 2002). Time series that demonstrate the empirical effects that ecosystem changes have on nonextractive economic activities are even rarer.

Developing a Baseline of Economic Activity

Recognizing the need to develop baselines and monitoring data for ecosystem-dependent human activities, a number of scholars and organizations are attempting to collect data on integrated ecological and socioeconomic indicators of ecosystem health. The Organisation for Economic Co-operation and Development (OECD), the European Union, and Environment Canada all have developed frameworks for the collection of integrated coastal and marine indicators (see Bowen and Riley 2003 for a review). More recently, the NOAA Coastal Restoration Center has developed a framework for measuring the human dimensions of coastal restoration (Salz and Loomis 2005).

Here, we describe the basic foundation we established in Morro Bay to begin the collection of economic indicators of marine ecosystem-dependent activities. As with other SLOSEA activities, we began developing economic indicators by collaborating with our stakeholder partners from the working waterfront. It is along the waterfront that changes in ecosystem-dependent economic activity in Morro Bay have been felt most acutely. The fishery in Morro Bay has changed substantially over the last several decades; many vessels have been sold, and fishing-related businesses (e.g., chandleries, marine ice vendors, and sportfishing operations) have closed their doors. The waterfront also has enjoyed a renaissance of tourism and recreation, with retail stores, restaurants, and hotels competing for waterfront space.

Initially, the list of candidate economic activities was a long one, reflecting the business interests of many stakeholders. Many waterfront businesses and activities, however, are not directly dependent upon ecosystem health (e.g., curio and saltwater taffy shops). We worked with our stakeholders to pare the list of candidate indicators by asking a simple question: How does each activity depend on the ecological condition of the bay? To make the question even more direct, we turned it on its head and started with the list of ecosystem goods and services developed by the SLOSEA team. For each ecosystem good and service, we asked, What economic activity might change if this ecosystem good or service changed? By taking this approach, we created a shorter list of economic activities (table 11.2).

To properly understand the economic consequences of ecosystem change requires economic indicator data that reflect both the output and value of activities. Output measures for economic activities in Morro Bay fall into two basic categories: measures of physical output (e.g., landings of fish, volume of sediments removed) and measures of human activity (e.g., recreational visit days, park attendance). Value measurements also fall into two categories: measures of economic impact (usually measured as gross revenues or expenditures) and estimates of economic value (usually measured as consumer and producer surplus or the willingness of the user to pay to participate in an activity beyond the costs of participation).

Except for commercial fishing, for which state and federal agencies report gross revenues for landed catch, the collection of onsite data about gross revenues and consumer surplus data is too difficult for such data to be considered as indicators (indicators should be easily collected and readily available). The problem is twofold. First, private firms are reluctant to reveal gross or net revenue data. Second, original studies to estimate consumer and producer surplus values are costly and difficult to apply to a repeated time series. Because of the difficulty of collecting economic impact and value data, we focus primarily on measures of output for our economic indicators. However, we use estimates of economic impact and value from supplementary surveys and the literature to place individual indicators in an economic context that helps us to weigh the relative economic importance of changes in specific indicators.

As of this writing, we have collected useful economic indicator data from a number of sources, including the California Department of Fish and Game, the Pacific Fisheries Management Council, Duke Energy (the power plant), the Army Corps of Engineers, and the Morro Bay Harbor District. To better understand these data, we have reached out to economic interests whose livelihood or activity is reflected by these data (e.g., fishermen, charter boat captains, the local chamber of commerce). In doing so, we have been able to engage many

Table 11.2. Economic activities and indicators in Morro Bay, CA

Ecosystem service	*Economic activities*	*Economic indicator*
Fish	Commercial fishing	Landings (by species), value (by species)
Fish	Commercial passenger fishing vessels (CPFV)	Passenger trips, landings (by species)
Fish	Marine ice sales	Tons, gross sales
Fish	Recreational fishing	Trips
Water quality	Oyster production	Pounds, gross value
Water quality, wildlife	Kayaking	Charters, trips, gross revenues
Water quality, wildlife	Park attendance	Attendance days
Water quality	Beach attendance	Attendance days
Wildlife	Hunting	Activity days
Water quality, wildlife	Tourism	Hotel occupancy
Sediments	Boating/navigation	Volume of sediments removed, dredging costs
Sediments	Electric power plant cooling	Maintenance costs for cooling intake systems

stakeholders who may be affected by EBM. From these discussions, we have developed interactive, online exercises to help stakeholders explore the links between economic activity (e.g., commercial fish landings) and a variety of factors that are mentioned in public discussion as causes of fisheries change but which in fact may or may not contribute to these changes in the ways expressed by conventional or public wisdom (box 11.2).

The Need for Economic Indicator Data: Analysis, Monitoring, and Dialogue

Economic indicator data are essential to definitively demonstrate the link between changing ecological conditions and the economic outcomes of management, restoration, and protection. Economic indicators should be part of monitoring programs, just as biological indicators are. As with all environmental policies, public support is required to secure investment and cooperation with EBM initiatives. The public needs to understand the baseline of economic activities that may benefit from EBM. Even more importantly, the public needs to be shown that EBM benefits people as well as flora and fauna. The need for accountability—to show that the economic benefits of EBM are real and substantial—can only be achieved by more fully incorporating economic indicators into monitoring systems that already include environmental indicators. Finally, economic indicator data can engage members of the public in the process of scientific exploration that is at the heart of EBM. Economic indicator data deal with measurable information about activities that are generally well known

Box 11.2. Exploring Biological and Economic Data for Morro Bay

After explaining what the data are and where we got them, our online exercise shows commercial fish landings for rockfish (*Sebastes* spp.) in Morro Bay.

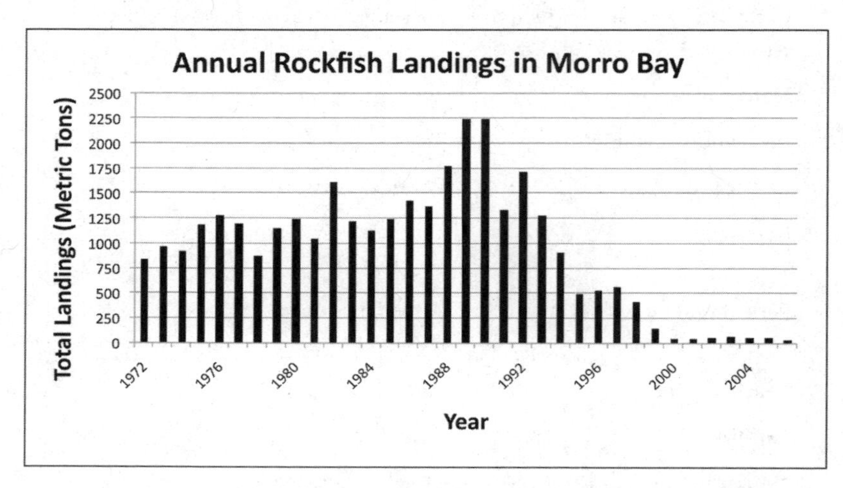

Two features of the time series stand out: a dramatic increase in rockfish landings in Morro Bay between 1972 and 1990 and a dramatic decline immediately after 1991. These data, especially the decline in landings after 1991, reflect the conventional wisdom that commercial catch of rockfish in Morro Bay has declined tremendously. Rather than provide direct analysis about the causes of this decline, we invite our stakeholders to examine the data for themselves. Through an online survey we ask the following:

What is going on here?

Rockfish landings increased substantially in the late 1980s and then declined in the early 1990s. Let's take a look at the possible causes of these changes. Where would you like to start?

Increased landings in the 1980s—or—Decreased landings in the 1990s

After choosing which event to consider, the user is allowed to vote on what he or she believes are the factors most important to the cause of the increase or decrease in landings.

What do you think?

What factors do you believe were important in contributing to the decrease in rockfish catch in Morro Bay during the late 1980s? (Choose all that apply.)

1. *Decreased prices for rockfish ($/lb)*
2. *Change in the number of fishing vessels in Morro Bay*
3. *Increases in catch of*
   a. flatfish; b. salmon; c. lingcod; d. urchins; e. crabs; f. prawns
4. *Changes in sea surface temperature*
5. *Other (Please tell us if you think other factors are at work here.)*

The survey lets us gauge opinion among our users and also, through the "other" choice, points us to factors about which we may be unaware (say, the closure of a major fish processor or fuel prices). We then begin an iterative process in which the user chooses a single factor he or she believes is most important in the increase/decrease in landings.

If you had to choose just one ***single factor****, which factor do you believe was the* ***most important*** *in contributing to the decline in rockfish catch in Morro Bay after 1990?*

1. *Decreased prices for rockfish*
2. *Fewer fishing vessels in Morro Bay*
3. *Increases in catch of*
   a. flatfish; b. salmon; c. lingcod; d. urchin; e. crabs; f. prawns
4. *Changes in sea surface temperature*
5. *Overfishing*
6. *Stricter fishing regulations*

Each choice is linked to a time series of data. For instance, if falling prices are to blame, we let the user examine the regional prices for rockfish for those years.

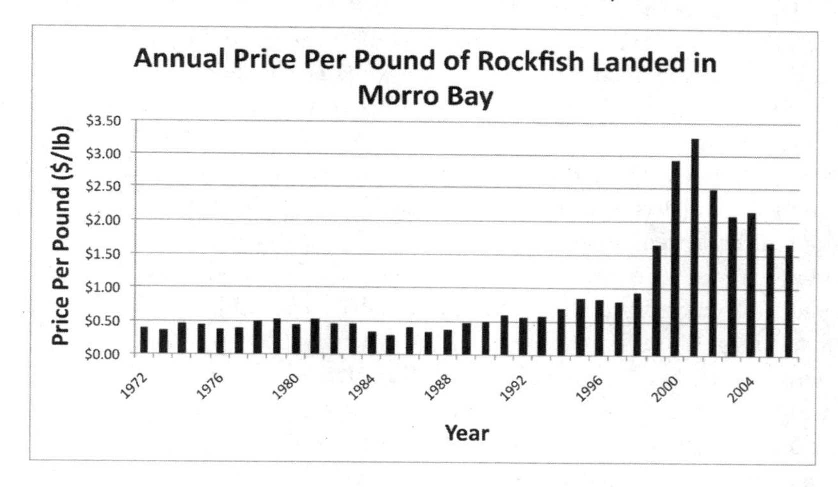

And then we ask about effect.

Does this new information change your opinions about the factors that contributed to the decreased catch of rockfish in the early 1990s?

If the answer is yes, we take the user back to the beginning to explore other factors. Throughout, we give our users an opportunity to tell us about other factors that they believe are important contributors to change. As a result, we can incorporate this "local knowledge" directly into our statistical analyses of economic effects of changing ecological and environmental conditions. We also ask the users about their personal experiences, demographic information, and occupations. This allows us to match responses to different types of stakeholders. We will develop these types of interactive exercises for a variety of economic activities.

to most members of the public. In Morro Bay, we are using economic indicator data to engage the public in a scientific discussion of the links between ecosystem and people in a way that allows stakeholders to come to their own conclusions.

How Does Resilience Science Inform EBM in Morro Bay?

In general, SLOSEA is approaching the management of human–environment interactions within the resilience conceptual framework as described by Leslie and Kinzig (chap. 4 of this volume). Specifically, we are (1) increasing collaboration and communication among agencies through active participation of managers, scientists, and stakeholders on an advisory committee (i.e., increasing cross-jurisdictional communication and coordination); (2) actively developing an integrated set of science and monitoring activities to better understand forcing factors, cumulative impacts, and cross-scale linkages within the ecosystem; (3) exploring the connections between economic and ecological well-being; (4) ensuring that data are used by resource managers to affect change within their respective agencies; and (5) providing a forum for engagement of the community about trade-offs among different sectors. Below, we demonstrate how resilience concepts have informed the work of SLOSEA.

Cumulative Impacts on Oyster Production

Oysters have been produced in Morro Bay since the late 1930s. In 1935, the California Department of Fish and Game allotted 1,677 of the 2,300 total acres of the bay for oyster farming (Barrett 1963). Morro Bay was the leading producer of oysters in California during World War II and continued to have high levels of production on average through the 1970s. Over the last 25 years, production in Morro Bay has declined steadily, reaching very low levels during the last decade (fig. 11.3). Some of this decline is attributable to changes in the state of the ecosystem over the past 60 years. Three main factors impact the available growing area for oysters in the bay: (1) presence of fecal coliform bacteria, most likely from sewage and agricultural practices; (2) sedimentation; and (3) the return of eelgrass to the bay and its status as a protected species. We next describe each of these factors in more detail.

Water quality has had substantial effects on oyster production. In 1985 the California Department of Health Services (CDHS) closed the entire bay to oyster culture, due to the presence of nondisinfected sewage that entered the mouth of the bay from the wastewater treatment plant of Morro Bay and Cayucos (Baltan 2007). In 1988 the city of Morro Bay upgraded its facility and began chlorinating the effluent, and the bay was again conditionally approved for 760 acres of lease area. Since then, the lease area has steadily declined as the result of continued water quality problems. More than half of the 760 acres currently zoned for oyster culture is prohibited because of sporadic and unpredictable spikes in fecal coliform bacteria (fig. 11.1; Baltan 2007).

Sedimentation is another factor that has influenced oyster culture. A technical report produced for the MBNEP calculated that the bay has lost 25% of its volume in the last 100 years and that all open water areas should fill within the next 300 years, based on current rates of sedimentation (Haltiner 1998). In fact, Barrett's (1963) map of Morro Bay oyster culture areas designates as active regions some eastern portions of the bay that today are tidal marsh and clearly unavailable for oyster culture (fig. 11.1, portions of the bay east of the current leased areas). To corroborate this observation, the current oyster-farming companies report that they are having trouble utilizing the eastern portions of the open lease areas because of lack of water depth.

A third impact that has been an issue, although to a lesser extent thus far, is the distribution of eelgrass, *Zostera marina*. Although the distribution of eelgrass fluctuates from year to year, when new beds appear, they legally cannot be disturbed either through shading or by activities that affect the benthic environment (e.g., oyster culture). Although clearly a trade-off, the reestablishment of eelgrass beds in Morro Bay is seen as a positive shift in the state of the ecosystem, because eelgrass is an important nursery environment and habitat for multiple resident species of fishes and invertebrates, and it is a food source for migrating birds such as Brandt geese.

In addition to shrinking available areas for culture, social factors have influenced the decline in oyster production, namely, the death in 1998 of the single person responsible for production of oysters in Morro Bay during the 1990s. His family tried to maintain production after his death but was unable to do so because of "lack of expertise" producing oysters. In 2004 another oyster-farming company purchased half of the available lease area in Morro Bay and is making a significant effort to restore production. However, it is clear that production still remains very low compared with historic values (fig. 11.3), despite efforts to increase production; the low production is at least in part due to the cumulative impacts of the factors described above. Despite active management to improve water quality and slow sedimentation in Morro Bay (e.g., effluent chlorination described above and watershed restoration projects of MBNEP), poor water quality still poses significant problems. Actions underway to improve water quality include an upgrade to tertiary treatment for Morro Bay, the construction of a sewer and treatment plant for Los Osos, and multiple restoration projects in the watershed. Importantly, the response time of the ecosystem is likely fairly long, as evidenced by continued closure of available lease areas due to water quality despite the active efforts to reduce pollution. Moreover, in terms of sedimentation, the newly established salt marsh is such a stable ecosystem state that complete reduction in new sediment alone will not likely return the ecosystem to an alternate state. It would seem that to do so would require dredging. This option will likely not be taken, because of high costs and "collateral" ecological damage. In the end, despite the community's desire to have oyster production in Morro Bay, the ecosystem may no longer be able to support it. Even if we could implement instantaneous management actions to reduce the cumulative impacts of pollution and sedimentation in the bay, the response time of the system might be too slow to support an active industry in the interim.

Exploring Resilience through Experimentation

In another example of the application of resilience thinking to advance marine EBM, SLOSEA scientists initiated a study in 2007 to examine the effects of removing a dominant invasive species of bryozoan in the bay. We were interested in exploring the potential of the fouling community—that is, the community of plants and animals that grow on floating docks, boats, pier pilings, and hard substrata—to breach a threshold and return to a previous, mussel-dominated state. Through a comparison of current surveys with historical data sets over the last 30 years, Needles (2007) documented important shifts in the fouling community in the bay. Although the number of native and exotic species did not change significantly in the past 30 years, species composition has shifted drastically. Morro Bay historically had a mussel-dominated fouling community, and this community is now dominated almost exclusively by the invasive bryozoan *Watersipora subtorquata*. The bryozoan covers 86% of available substrata on average (Needles 2007). We performed a short-term (8-month) "gardening" experiment to investigate the likelihood of

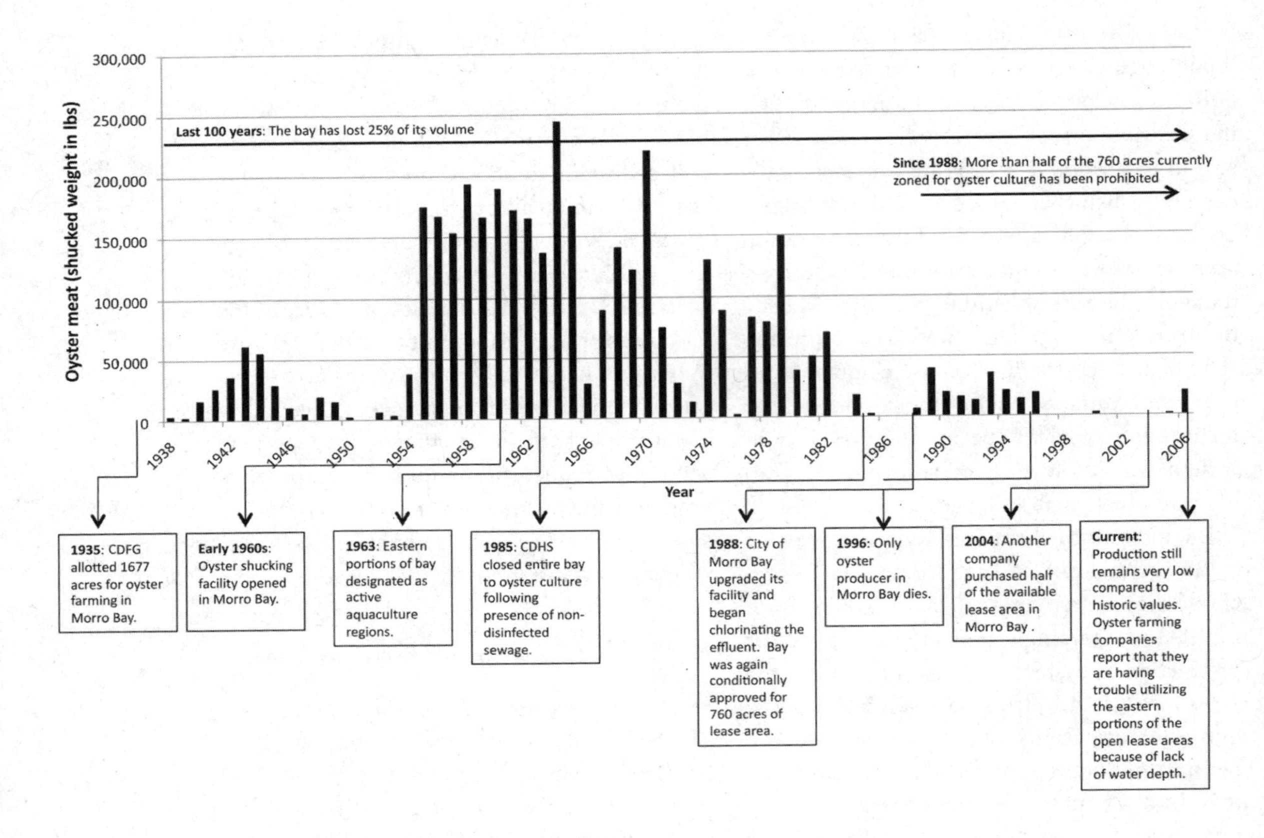

Figure 11.3 Pacific oyster production in Morro Bay from 1938 to 2006.

shifting the community back to the a mussel-dominated state. After removing new recruits of the invasive bryozoan from uncolonized surfaces through time, we found that the only significant change in species composition compared with control surfaces (i.e., not gardened) was the dramatic increase in another invasive bryozoan species, *Schizoporella unicornis*. Thus the altered ecosystem state has tremendous resilience because of functional diversity (see Leslie and Kinzig, chap. 4 of this volume). Thus, removing a dominant invasive may not have the intended consequence of returning an ecosystem to an earlier state, and costly management action should be considered carefully.

What Are the Prospects for EBM in Morro Bay?

The prospects for continued implementation of EBM in Morro Bay are good. The concept of ecosystem services and the notion of managing those services according to people's needs resonates strongly in the scientific, environmental, and marine-dependent businesses and communities. SLOSEA has the good fortune of being surrounded and supported by local organizations such as the MIG and the Morro Bay National Estuary Program and working within the context of larger policy changes at the state level that call for more holistic approaches to resource management, such as the California Ocean Protection Act (COPA 2004).

In order to optimize the impact of EBM activities in this region, a few additional elements need to be implemented over the next several years. First, since EBM necessarily focuses on management of the cumulative impacts of multiple activities on ecosystem services, implementation of this approach requires a governance structure that effectively integrates agencies and jurisdictions. Institutional fragmentation in Morro Bay (and elsewhere) makes this integration challenging. SLOSEA can work across agencies to some extent due to the representation of key institutions on the advisory committee. However, each of the agencies continues to operate in an individualistic fashion. In order to enhance SLOSEA's effectiveness, we see a need for state agencies to more formally support an integrated, ecosystem-based perspective. Enhancing the viability of the Morro Bay ecosystem and securing its long-term value will benefit from strengthened commitments to specify scope, engagement, and protocols for action. We envision the creation of a regional stewardship council to integrate the efforts of resource managers and stakeholders for greater impact.

Second, to be effective, this regional ecosystem-based management program needs to have the authority to craft policies and regulations within the ecosystem region. This is not to suggest that higher-level management structures (at the state and federal levels) are not needed, but a strong local authority also is required, as suggested by Ostrom (2005) in her discussion of polycentric, or many-centered, institutions. Such nested institutional structures can facilitate resilience of both management institutions and the ecosystems of interest.

Finally, as a relatively localized EBM effort, SLOSEA needs to connect to management efforts at a broader geographic scale, such as the south central coast (e.g., Año Nuevo to Pt. Conception) and the California Current as a whole (stretching from British Columbia to Baja). Otherwise, EBM efforts on the local scale in Morro Bay may be overwhelmed by ecological and institutional changes at larger geographic scales.

Key Messages

1. Engagement of robust local communities is a necessary precursor to broader-scale ecosystem-based management. SLOSEA has flourished by creating a strong network of resource managers, stakeholders, and scientists, who share information, learn together, deliberate important trade-offs, and engage in collective action.
2. Visual models of ecosystem dynamics can help frame discussions and formulate research questions. "Ecosystem-based management" is a blizzard of words for many stakeholders. Diagrams and linkages with concrete examples bring the concept to life.
3. Resource management issues can guide rigorous scientific inquiry. In Morro Bay, we identify the critical issues being raised by resource managers and stakeholders, search for relevant science to address the issues, and ensure that resource managers share ownership of the scientific direction from the start.
4. Creative responses to new issues, information, and opportunities are vital to EBM success, in addition to the developing body of experience and good practices.

Acknowledgments

The authors acknowledge the SLOSEA advisory committee for committing time and energy to our program. Without their valuable interaction and direction, marine EBM would not be happening in Morro Bay and San Luis Obispo County. We thank Leslie Rebik for editorial assistance and preparation of the manuscript. We greatly appreciate the interactions and feedback on our chapter from the editors and other authors in this volume. We thank Ben Halpern and two anonymous reviewers for critical and constructive criticism of the chapter.

References

Baltan, J. 2007. *Twelve-year sanitary survey report: Shellfish growing area classification for Morro Bay, California*. Prepared for the California Department of Health Services Division of Drinking Water and Environmental Management.

Barrett, E. M. 1963. The California oyster industry. Prepared for the California Department of Fish and Game. *Fish Bulletin* no. 123.

Bowen, R. E., and C. Riley. 2003. Socio-economic indicators and integrated coastal management. *Ocean & Coastal Management* 46:299–312.

COPA. 2004. California Ocean Protection Act. US Senate Bill no. 1319. Ch. 719. http://resources.ca.gov/copc/docs/OPC_Code_Div_26.5.pdf.

Ellis, G. M., and A. C. Fisher. 1987. Valuing the environment as input. *Journal of Environmental Management* 25:149–56.

Haltiner, J. 1998. *Sedimentation processes in Morro Bay, California*. Prepared by Phillip Williams & Assoc., Ltd., for the Coastal San Luis Resource Conservation District, with funding by the California Coastal Conservancy.

Kahn, J. R., and W. M. Kemp. 1985. Economic losses associated with the decline of an ecosystem: The case of submerged aquatic vegetation in the Chesapeake Bay. *Journal of Environmental Economics and Management* 12:246–63.

Kildow, J. 2007. The influence of coastal preservation and restoration on coastal real estate values. In *The economics of estuary and coastal restoration: What's at stake*, ed. L. Pendleton. Arlington, VA: Restore America's Estuaries.

Knowler, D. 2002. A review of selected bioeconomic models with environmental influences in fisheries. *Journal of Bioeconomics* 4:163–81.

Love, M. S., J. E. Caselle, and W. Van Buskirk. 1998. A severe decline in the commercial passenger fishing vessel rockfish (*Sebastes* spp.) catch in the Southern California Bight, 1980–1996. *CalCOFI Reports* vol. 39.

Lynne, G. D., P. Conroy, and F. J. Prochaska. 1981. Economic valuation of marsh areas for marine production processes. *Journal of Environmental Economics and Management* 8:175–86.

Mason, J. E. 1998. Declining rockfish lengths in the Monterey Bay, California, recreational fishery, 1959–94. *Marine Fisheries Review* 60(3):15–28.

MBNEP (Morro Bay National Estuary Program). 2000. *Comprehensive conservation management plan: Executive summary*. http://www.mbnep.org/files/pdfs/execsum.pdf.

McLeod, K. L., J. Lubchenco, S. R. Palumbi, and A. A. Rosenberg. 2005. *Scientific consensus statement on marine ecosystem-based management*. The Communication Partnership for Science and the Sea (COMPASS). Signed by 221 academic scientists and policy experts with relevant expertise. http://www.compassonline.org/pdf_files/EBM_Consensus_Statement_v12.pdf.

Needles, L. A. 2007. *Big changes to a small bay: Exotic species in the Morro Bay fouling community over thirty years*. Master's thesis, California Polytechnic State University (San Luis Obispo).

Salz, R., and D. Loomis. 2005. The human dimensions of coastal restoration. In *Science-based restoration monitoring of coastal habitats*, ch. 14. http://coastalscience.noaa.gov/ecosystems/estuaries/restoration_monitoring.html.

Stephens, J., D. Wendt, D. Wilson-Vandenberg, J. Carroll, R. Nakamura, E. Nakada, S. Rienkeke, and J. Wilson. 2006. Rockfish resources of the south central California coast: Analysis of the resource from partyboat data, 1980–2005. *CalCOFI Reports* vol. 47.

Webler, T., and S. Tuler. 2006. Four perspectives on public participation process in environmental assessment and decision making: Combined results from 10 case studies. *Policy Studies Journal* 34(4):699–722.

CHAPTER 12
Puget Sound, Washington, USA

Mary Ruckelshaus, Timothy Essington, and Phil Levin

Observable, widespread declines in the status of species, habitats, and ecosystem functions—combined with studies illustrating the complex interactions characterizing many coastal systems—have led to high-level policy and scientific support for ecosystem-based management (EBM) as an approach to restore watersheds and coastal oceans (POC 2003; USCOP 2004; McLeod et al. 2005; NSTC 2006). The "management" portion of EBM involves decisions that weigh political and social consequences of activities; analysis of ecological outcomes typically does not heavily influence the choice of strategies to restore ecosystems, at least in the United States (e.g., GAO 2001, 2005). Why the cold shoulder to science? In our view, scientific information is underused in EBM for two primary reasons: (1) management decision structures are not explicitly designed to incorporate rigorous scientific input at relevant stages, and (2) the scientific community has not adequately considered in advance how best to conduct, synthesize, and then communicate their scientific analyses so that they can inform key decisions.

It is clear to us as scientists that we know a lot about how marine systems function, and about how those functions can be altered by human activities (e.g., Lenihan and Peterson 1998; Duffy 2003; James et al. 2008; Lotze et al. 2006). It is just as clear to us that there is much about marine ecosystem structure and function that we do not understand. Thus, an additional challenge is to clearly express these uncertainties in a constructive way when communicating with decision makers (Ludwig et al. 1993; Rosenberg and Sandifer, chap. 2 of this volume). In this chapter, we use the large-scale EBM process under way in the Puget Sound region of Washington State to illustrate successes and failures in using science to inform management. The successes we highlight in infusing science into the planning stages of this management effort include, first, encouraging adoption of clearly stated human and natural system goals for the ecosystem that can be measured using indicators of its status and function. Second, conceptual and quantitative models of food webs and ecosystems can highlight important linkages among ecosystem components and point toward strategies likely to achieve goals. We illustrate the promise of even simple conceptual and quantitative models to suggest likely ecological outcomes in Puget Sound under alternative management strategies. Throughout the examples, we discuss how inevitable uncertainty can be communicated and can inform difficult policy decisions, rather than causing delay while people wait for the elusive state of perfect knowledge.

Vision for a "Thriving Natural System" in Puget Sound

The Puget Sound region encompasses 41,500 km^2 of upland, freshwater, estuarine, and marine habitats, and it currently supports a large and increasingly urban population from Vancouver, British Columbia, to Olympia, Washington (fig. 12.1). Population projections suggest that human numbers in the greater Puget Sound region will increase by three million people in the next 20 years (Sound Science 2007). A number of indicators show that the processes supporting the Puget Sound ecosystem have been disrupted or impaired. The American Fisheries Society has listed Puget Sound as one of the five national "hotspots" of marine fishes at risk of extinction (Musick et al. 2000). Over forty species of birds, mammals, fishes, plants, and invertebrates currently are listed as threatened, as endangered, or as candidates for state and federal endangered species lists (PSAT 2007; Sound Science 2007). Moreover, some of these threatened species, such as Puget Sound chinook salmon and orca whales, are icons of the Pacific Northwest that have been celebrated in art, culture, and tradition for centuries. As a recent illustration of the state of the sound, in 2006 the first-ever salmon consumption advisory from the Washington Department of Health made headlines, cautioning against eating chinook salmon more than once a week because of their PCB and mercury levels (Black 2006; Cornwall 2006; Judd 2006).

In response to signs of trouble, in late 2005 Washington Governor Christine Gregoire appointed a public–private partnership group (hereafter, the partnership) to chart a course forward to "ensure that Puget Sound forever will be a thriving natural system, with clean marine and freshwaters, healthy and abundant native species, natural shorelines and places for public enjoyment, and a vibrant economy that prospers in productive harmony with a healthy Sound." The governor went on to say that her vision for an approach to restoring Puget Sound is to "preserve the health, goods and services needed by the year 2020 to ensure that the Puget Sound's marine and freshwaters will be able to support healthy populations of native species, as well as water quality and quantity to support both human needs and ecosystem functions" (PSP 2006). Fulfilling this charge involves articulating clear and measurable goals for the Puget Sound ecosystem and its elements and describing solutions—strategies and actions that are consistent with achieving those goals.

The partnership worked with their diverse members, a science working group, and the general public to articulate their final recommendations to the governor on how to achieve her vision (PSP 2006). The partnership recommendations include clearly stated ecosystem goals and a list of initial actions designed to move Puget Sound toward recovery. The ecosystem-wide goals adopted by the partnership capture the natural and human components of the Puget Sound region and encompass five major themes: (1) species and food webs, (2) habitats, (3) water quantity, (4) water quality, and (5) human health and well-being, including economic, social, and cultural values (table 12.1). The principles underpinning the partnership goals are summarized in their report:

> A healthy ecosystem has three key properties: (1) it is *resilient* to changes in natural- and human-caused changes in environmental conditions; (2) [it] has built-in *redundancy in its parts* so that not all members of a species or habitat type are limited to a single location. Spreading the risk of catastrophic losses of species or habitats improves the ability of the ecosystem to withstand localized losses of key components; and (3) [it] has a *representative* sample of the diversity of species and habitat types that characterized its historical state (PSP 2006).

The partnership report presents a novel approach to large-scale ecosystem restoration by adopting a systemwide framework that includes humans as part of the natural ecosystem. The approach laid out uses the concept of ecosystem services to help frame the challenge of achieving multiple, likely competing human and natural system goals. The process through which the ecosystem goals were adopted was

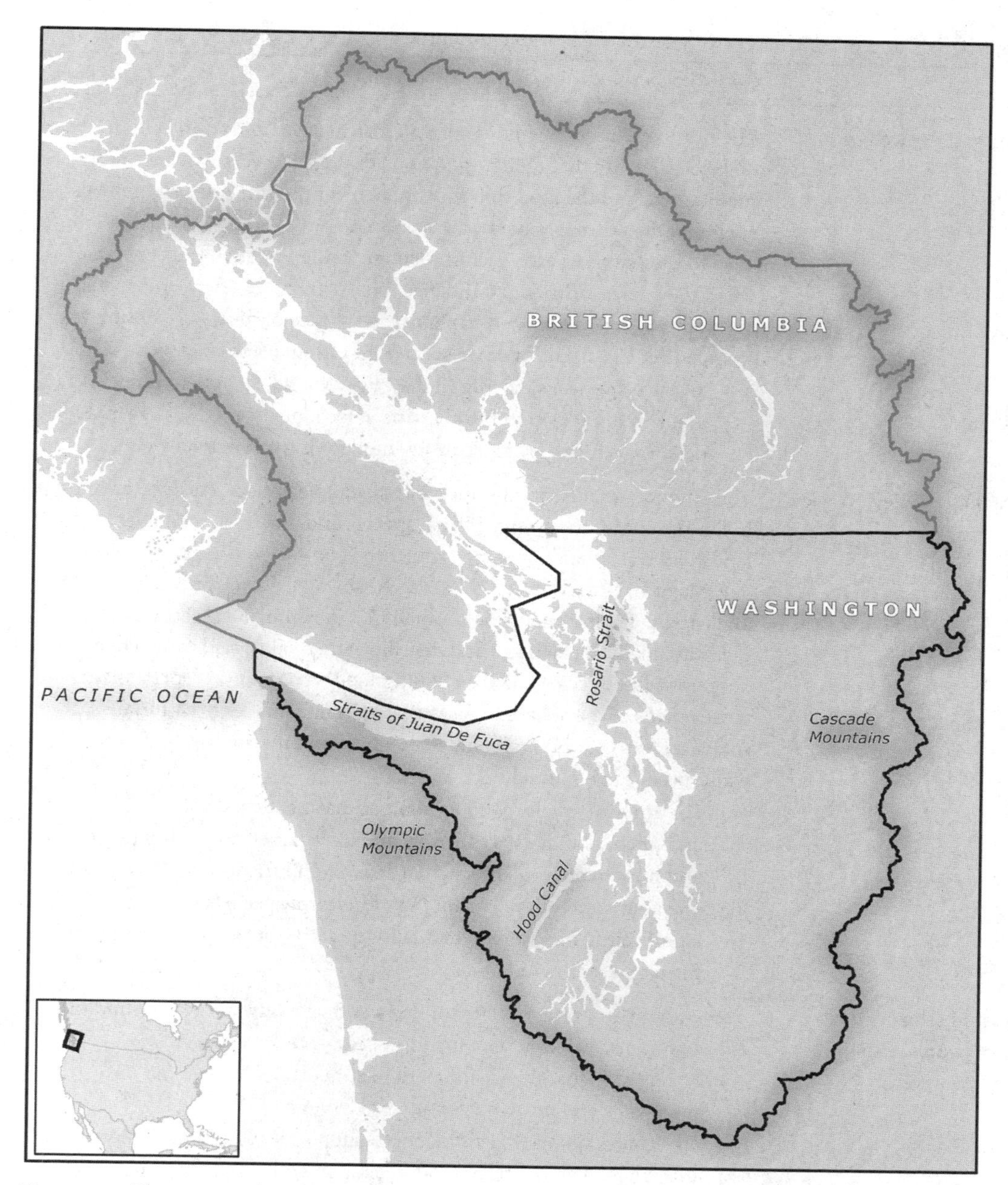

Figure 12.1 The geographic setting of the Puget Sound–Georgia Basin ecosystem. The Puget Sound Partnership is focusing on ecosystem-based management of the marine waters and lands within Washington State, while recognizing that the entire ecosystem spans Washington State and British Columbia and thus transboundary coordination is needed to restore the ecosystem. The Cascade and Olympic mountain ranges form the boundaries of the greater Puget Sound ecosystem on its eastern and western sides, and the San Juan Islands and Strait of Juan de Fuca bound it on its northern portions within Washington State.

Table 12.1. Ecosystem goals and narrative descriptions of measurable outcomes for the Puget Sound ecosystem

Ecosystem goal	*Narrative outcomes*
Puget Sound species and the web of life thrive.	The numbers, productivity, condition, distribution, and composition of species ensure that their populations thrive, biodiversity is naturally maintained, and the food web functions to support the ecosystem needs such that the outcomes below are achieved. • Species support ecosystem functions in the Puget Sound ecosystem when they have sufficient abundance, productivity, diversity, and spatial distribution so that the species persist in the face of future environmental changes and the needs of other species and humans are met. • Invasive species occurring in the Puget Sound ecosystem do not occur in high enough numbers or in locations such that they impede native species persistence or the functioning of the food web.
Puget Sound habitat is protected and restored.	The amount, quality, and location of marine, nearshore, freshwater, and upland habitats sustain the diverse species and food webs of Puget Sound lands and waters such that the outcomes below are achieved. • Major habitat types in terrestrial, freshwater, estuary, nearshore, and marine environments occur such that they provide (1) sufficient quantity (including acreage and distribution), quality, and connectivity to support viable species and functioning food webs and (2) representative occurrences of those habitat types that existed historically within the major terrestrial and freshwater ecoregions and marine subbasins and estuaries of Puget Sound. • The amount, quality, and location of marine, nearshore, freshwater, and upland habitats are formed and maintained by natural processes and human stewardship so that ecosystem functions are sustained. • The abundance and distribution of invasive species do not significantly impair habitat quality, quantity, or the processes that form and maintain habitats.
Puget Sound rivers and streams flow at levels that support people, fish and wildlife, and the environment.	The amount and range of stream flows and groundwater levels support the ecosystem such that the outcomes below are achieved. • Freshwater quantity is sufficient to support freshwater and terrestrial food webs and human uses and enjoyment. • Freshwater quantity is sufficient to support estuarine, nearshore, and marine food webs and the habitats upon which they depend.
Puget Sound marine and fresh waters are clean.	The quality of water and sediments support the ecosystem such that the outcomes below are achieved. • Toxics and pathogen levels in marine mammals, fish, birds, shellfish, and plants do not harm the persistence and health of these species. • Loadings of toxics, nutrients, and pathogens do not exceed levels consistent with healthy ecosystem functions. • The waters in the Puget Sound region are safe for drinking, swimming, and other human uses and enjoyment.

Table 12.1. *(continued)*

Ecosystem goal	*Narrative outcomes*
A healthy human population is supported by a healthy Puget Sound.	Human health is not threatened by changes in the ecosystem such that the outcomes below are achieved. • Fish and shellfish are safe for people to eat. • Marine and fresh waters are clean for swimming, fishing, and other human uses and enjoyment.
The quality of human life is sustained by a healthy Puget Sound.	A functioning ecosystem supports social and cultural well-being and economic vitality such that the outcomes below are achieved. • Aesthetic values, opportunities for recreation, and access for the enjoyment of Puget Sound are continued and preserved. • Terrestrial and marine resources are adequate to sustain the treaty rights, as well as the cultural, spiritual, subsistence, ceremonial, and medicinal needs and economic endeavors, of the tribal communities of Puget Sound. • The Puget Sound ecosystem supports thriving natural resource and marine industry uses such as agriculture, aquaculture, fisheries, forestry, and tourism. • The Puget Sound's economic prosperity is supported by and compatible with the protection and restoration of the ecosystem.

Source: PSP 2006.

invaluable—the detailed discussions between scientists and policy leaders about what the diverse partnership group wanted from the Puget Sound ecosystem, and what functioning ecosystem goals would look like, helped to clarify and solidify their stated multiple objectives. In 2007 the Washington State legislature codified the recommendations of the initial partnership and created a new state agency and a public–private entity to oversee the partnership's mission. The difficult work ahead is to make decisions about how to achieve the goals adopted by the partnership and, in cases where trade-offs occur, to determine acceptable decision-making rules for competing objectives. Work is ongoing to produce the first of many iterations of an overarching ecosystem strategy for Puget Sound by the end of 2008—below we outline how science has informed (see box 12.1), and can continue to support, this ongoing, large-scale ecosystem effort.

Identifying and Tracking Measurable Outcomes in Puget Sound

The number of measurable outcomes needed to describe the status of the five ecosystem goals adopted in the Puget Sound system quickly adds up. Thus, in addition to developing metrics for each goal, there is strong political support in Puget Sound for choosing a smaller subset of indicators for the ecosystem that could be used in a more public-friendly scorecard to track progress over time (see also Kaufman et al., chap. 7 of this volume).

What can science add to the discussion of which ecosystem elements are most indicative of overall ecosystem functioning? The primary recommendation in reviews of ecosystem indicators is that they should be based on a well-understood and accepted conceptual model of the system of interest (NRC 2000; Heinz Center 2002, 2006). In addition, it is important to

Box 12.1. What Do We Know about Puget Sound: Summarizing the "State of the Science"

In anticipation of the growing political interest in ecosystem-based management in the Puget Sound region, scientists from state, federal, tribal, academic, NGO, and private institutions produced a document summarizing what is known about the Puget Sound ecosystem, key threats to its functioning, and major gaps in scientific understanding (Sound Science 2007). This collaborative effort, involving over a hundred natural and social scientists, was intended primarily to synthesize what is known about the ecosystem to give decision makers an overview of the entire natural and social system, from mountain summits to the depths of Puget Sound. *Sound Science* explains the concept of ecosystem services and provides examples of these services within Puget Sound and interactions among the natural and human parts of the ecosystem. The concrete examples help managers and policy leaders see how their decisions could affect other components of the system and the delivery of ecosystem services. For example, shoreline hardening allows bank stability for agricultural lands, homes, and commercial and industrial buildings; however, it also reduces the quality of beach spawning habitat for surf smelt, an important prey species for recreationally and commercially fished salmon (Rice 2005).

The primary human factors influencing the function of the Puget Sound ecosystem include human population growth and concomitant land use and development practices; industrial and transportation activities; commercial and recreational harvest; and waste disposal. The primary natural factors affecting ecosystem function include climate and global warming impacts on ocean, terrestrial, and freshwater conditions, and topographic and bathymetric diversity effects on freshwater flows, terrestrial vegetation, and oceanographic features. Key findings include the following: (1) activities and conditions on land affect the quality of freshwater, estuarine, and marine environments within the ecosystem; (2) human health is inextricably linked to freshwater and marine system health; (3) marine, terrestrial, and freshwater food webs have been substantially altered, but the causes and consequences of these changes are not well understood; and (4) changes in global climate already, and will continue to, significantly impact the ecosystem, making attainment of some ecosystem goals more challenging. While these findings seem self-evident to scientists, policy leaders, and the general public, they are important messages to express as strategies to achieve ecosystem goals are developed (PSP 2006; McClure and Ruckelshaus 2007).

include indicators that capture elements of broadly agreed-upon ecosystem goals—especially often-neglected human well-being goals. Other important scientific considerations for identifying a set of indicators are listed in table 12.2. The choice of indicators may also be based on social or political, rather than ecological, criteria (Heinz 2006). For example, the status of an iconic species such as the orca whale in Puget Sound is likely to resonate with the public. Useful indicators may also relate to the delivery of ecosystem services. For this region, these might include harvest of salmon, shellfish, or commercial timber species; wildlife viewing; and clean water for drinking and swimming.

At this stage, outcomes for each ecosystem goal in Puget Sound represent qualitative statements of what a "healthy" condition would be for human health and well-being, species, habitats, water flows, and water quality (table 12.1). Developing quantitative targets that more explicitly state "how much" of each outcome is needed to meet a healthy condition is an important and difficult next step. Viability or conservation targets have been established for a handful of federally listed and state-listed species and some habitat types within the Puget Sound ecosystem (Floberg et al. 2004; NMFS 2007a, b; SSPS 2007). Setting targets for single species or habitat types is itself a complex task, yet these targets often do not account for interactions with other components of the ecosystem. For example, population viability modeling in support of setting Endangered Species Act (ESA) de-listing targets for chinook and Hood Canal summer chum salmon in Puget Sound included alternative harvest levels in persistence models (Ruckelshaus et al. 2002a; Sands et al. 2007). The analyses accounted for one ecosystem service delivered by these salmon (providing food to humans), but there were no estimates of the other ecosystem roles of salmon, namely, providing marine-derived nutrients in support of

Table 12.2. Desirable attributes of ecosystem indicators and some example species and habitats from Puget Sound

Purpose of indicators	*Example species, habitats, and associated processes*
Show response (or vulnerability) to human activities	Habitats susceptible to impaired freshwater flows (e.g., wetlands, marshes) and species such as abalone, Olympia oysters, sea pens, Pacific cod, hake, salmon, rockfish, tufted puffins, marbled murrelets, western grebes, southern resident killer whales
Reflect functioning of other processes	Phytoplankton, sea pens, eelgrass (impaired sediment, nutrient processes); zooplankton (impaired sediment, nutrient and phytoplankton processes); temperate corals (physical disturbance of sea floor); freshwater sedges (impaired water flows, floodplain dynamics); orca, predatory birds (bioaccumulation of toxics, altered food web)
Span a range of time scales over which responses to environmental changes are expected	*Very short term*: dynamic species such as phytoplankton or zooplankton communities, aquatic invertebrates, and annual plants that may be leading indicators of ecological change but suffer from huge baseline variability *Long term*: species such as orcas, geoducks, and Douglas fir that integrate over very long temporal and spatial scales but are not good as early warnings signals
Are resident species within the ecosystem	Great blue herons, pigeon guillemots, crows, and terrestrial, estuarine, and marine plants
Are "overabundant"species that may signal ecosystem degradation	*Sargassum muticum, Spartina* spp., some jellyfish and ctenophores, terrestrial species on Washington's noxious weed list
Are species that are important for other species (e.g., "ecosystem engineers" or species with key food web roles)	Eelgrass, kelp, forage fish (i.e., small schooling fish, including herring, eulachon, prickleback, surf smelt, and sand lance), canopy tree species in terrestrial forests (e.g., Douglas fir, cedar, hemlock)
Measure general food web or ecosystem function	Primary and secondary productivity, proportion of biomass produced and consumed by different species groups and trophic levels, ratios of predators to prey to scavenging species (particularly within communities), species diversity or richness within communities, numbers of species listed as threatened or endangered (WA State or federal listing), changes in ratios of biologically active nutrients or elements and their rates of uptake or cycling

Note: Choosing among these and lists of similar indicators for other ecosystem goals (e.g., water quality, human well-being) within a system such as Puget Sound needs to occur so that progress toward goals can be tracked.

terrestrial ecosystems (Scheuerell et al. 2005) and providing food for resident orcas (Ford et al. 1998).

Setting numerical targets for each of the ecosystem goals in Puget Sound will be informed by ecosystem and food web models that are under development. Such models can illustrate likely alternative future states with varying risk levels for each of the water, species, habitat, and human goals and changes in the overall dynamics and resilience of the system (e.g., Brand et al. 2008). For a number of indicators, we can explore how they reflect changing ecosystem structure and function. In some instances, we may see clear tipping points when the ecosystem indicator changes sharply, thus providing us with a clear management target. In other cases, the complexity of the system will not produce an obvious threshold, and thus a single scientifically based answer will not emerge (Kaplan and Levin 2009). Challenging next steps, namely, determining trade-offs among the myriad hopes for the ecosystem and acceptable levels of risk for ecosystem components, fall to policy leaders within the region.

As for a smaller subset of indicators that could be tracked in Puget Sound, a report card on some elements of the Puget Sound ecosystem has been summarized biannually since 1998; the most recent report was released early in 2007 (PSAT 2007). Indicators in the existing Puget Sound report card focus primarily on water quality, the status of major habitat types (lowland forests and eelgrass), toxic contamination or disease (in marine sediments, mussels, English sole, harbor seals, and salmon), and the status of a handful of marine species in the region (i.e., orcas, salmon, groundfish, herring, pinto abalone, and marine birds). The monitoring that supports the Puget Sound report card is important for beginning to describe the overall functioning of the ecosystem. However, without an analysis of key drivers and determinants of ecosystem functioning in Puget Sound, it is difficult to know the extent to which the monitoring and reporting adequately assess ecosystem structure and function. Additionally, significant parts of the ecosystem are missing from the report card, including the functioning of most upland and freshwater habitats and species; the role and functioning of physical and ecological processes in watershed, nearshore, and marine habitats; the sources and flows of ecosystem services such as tourism and other businesses dependent on the natural system; and provision of freshwater flows for food webs and direct human uses, to name a few.

What is needed next from the scientific community is an ecosystem model that encompasses interactions among the major ecosystem elements and their drivers so that changes in both key components of the ecosystem and its emergent dynamics can be projected into the future and compared against systemwide goals (NRC 2000; PSP 2006). Although complex marine ecosystem models are available for adjacent regions (Pauly et al. 2006; Brand et al. 2008), such models are not yet available for Puget Sound. Until quantitative models are developed, conceptual models that describe how changes in a food web can lead to changes in multiple human uses and other values can be powerful communication tools. As an example, figure 12.2 illustrates a few of the many possible ways that human activities can change food web function in eelgrass meadows (*Zostera marina* and *Z. japonica*) and thus affect the services humans can derive from the system through shoreline development, nonconsumptive uses such as ecotourism, and harvest. In this example, indicators of human well-being include elements of both the human (i.e., spatial and temporal patterns of harvest, development) and the ecological (i.e., changes in eelgrass, herring) mechanisms that affect desired ecosystem services. Including both human and ecological processes allows estimates of whole-system response, more fully informing desired corrective actions. A full conceptual model for this part of the ecosystem would also include interactions among human activities. For example, shoreline development affects public access for consumptive and nonconsumptive uses and thus may affect the intensity or locations of activities such as kayaking and recreational harvest. Further work is needed to identify which of these metrics are the best indicators of human well-being.

Developing Priority Strategies

Consideration of the cumulative and interacting effects of human activities on system goals is at the heart of ecosystem-based management. Without an understanding of physical effects on food web linkages and the spatial and temporal scales over which those effects occur, unintended consequences of human activities are likely (e.g., Aydin et al. 2005; Beamish et al. 2005; Whitney et al. 2005). Once the threats

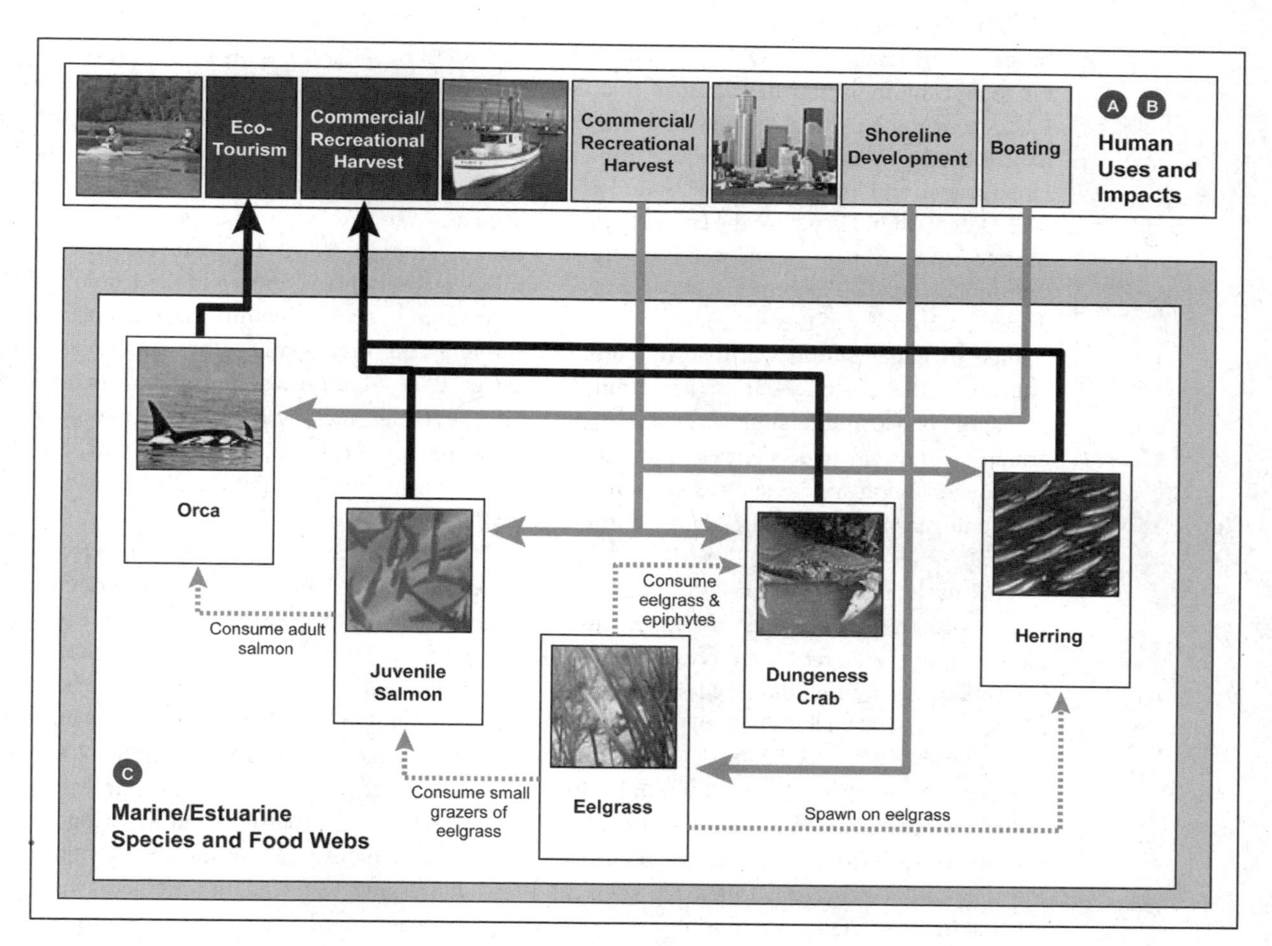

Figure 12.2 Example of the interrelationships between human impacts to, and uses of, species in a simplified eelgrass-based food web in Puget Sound. Gray arrows indicate pathways through which human activities can affect species; black arrows indicate the potential changes to ecosystem services humans can receive from those species. Some of the many possible interactions between species in this food web are depicted as dashed arrows. From PSP 2006.

with the greatest impact on ecosystem status are identified, strategies can be prioritized to focus on actions with likely positive changes to ecosystem elements and the system's overall resilience. A formal ecosystem assessment such as an integrated ecosystem assessment (IEA) can be a crucial element of an EBM strategy (see box 12.2).

Protection and Restoration Strategies

The Puget Sound ecosystem is defined as the lands and waters from the Cascade mountain range in the east to the Olympic mountain range and the mouth of the Strait of Juan de Fuca, from the San Juan Islands and the Canadian border in the north, and south through Hood Canal and southern Puget Sound (fig. 12.1). Puget Sound is part of the larger Georgia Basin ecosystem that extends well into British Columbia, Canada (Floberg et al. 2004), and addressing a number of the drivers of ecosystem change in Puget Sound will involve working outside its defined boundaries. For example, managing or accommodating climate impacts, transportation and trade policies, and harvest on widely ranging species such as salmon, herring, and marine birds will require coordination with other states and countries.

Threats to portions of the ecosystem that Pacific salmon inhabit and a number of subregions within the Puget Sound ecosystem have been relatively well studied and summarized (Floberg et al. 2004; HCDOP 2007; PSAT 2007; Sound Science 2007; SSPS 2007), yet a full ecosystem assessment of the status, trends, and cumulative threats to natural and human systems in Puget Sound is not yet available. The Puget Sound Partnership called for such an ecosystem assessment as a priority next step in their recommendations to the governor (PSP 2006).

The science group advising the partnership provided some general guidance for identifying strategies until the formal ecosystem assessment is completed. Strategies that are set up as testable hypotheses that can be refined over time with information from monitoring and modeling are most likely to succeed. General guidance from a wealth of experience in achieving conservation goals in ecosystems suggests that protection strategies are the most certain to maintain structure and function (NRC 1992, 1996). Protecting ecosystem processes—such as stream flows, trophic interactions, beach nourishment from sediment transport, and estuarine circulation—can go a long way toward maintaining habitats that are important for supporting food web and human uses (Paine 1980; Beechie et al. 2003; Fresh et al. 2004). In contrast, although it will be a necessary approach in achieving ecosystem goals in most systems, restoration of already damaged ecosystem components is a much less certain strategy (NRC 1992, 1996). Furthermore, recovery plans for two of the iconic and wide-ranging species within the ecosystem—salmon and orcas—and other assessments on portions of the ecosystem identified specific strategies such as habitat protection and reducing toxic contaminants, nutrients, and pathogens in the sound (NMFS 2007a, b; PSAT 2007; SSPS 2007).

The partnership highlighted a handful of immediate strategies and specific actions deemed high priority based on a mix of scientific guidance, logistical, and political reasons. One of the highest-priority strategies is to build support among the public so that a sustained funding source will be available (PSP 2006). Other early actions are aimed at addressing clear threats (e.g., protecting and restoring key species and habitats, cleaning up toxic sites, and addressing water quality problems from septic systems and storm water runoff) and resolving a governance structure for the Puget Sound ecosystem.

The EBM effort in Puget Sound is very young, and until an integrated ecosystem assessment is completed, it is difficult to estimate

Box 12.2. Integrated Ecosystem Assessment: A Framework for Conducting Ecosystem-Based Management

An integrated ecosystem assessment (IEA) is an organizing framework for designing and evaluating ecosystem-scale approaches to management. The basic IEA approach (similar to that included in formal decision analytical tools for fisheries assessments) uses quantitative analyses of cumulative impacts and ecosystem modeling to estimate how key ecosystem components (e.g., species, habitats, and human uses) change under alternative management options. These assessments provide scientific support for policy and decision making for managers and stakeholders; they also offer a systematic and transparent way for scientists to enhance and communicate their understanding of ecosystem dynamics. NOAA has adopted the IEA framework to help guide its assessment and management of marine resources and their ecosystems and currently is piloting a series of IEAs throughout US coastal environments.

An IEA consists of five steps that are iterated over a regular assessment cycle so that management strategies can be adapted over time as understanding of the system increases. The steps in an IEA briefly are described below.

1. **Define Scope by Articulating Objectives and Scale of Management**
The deceptively simple step of identifying a set of common objectives among multiple interests and the geographic scale of management often is overlooked. In the simplest case, identifying objectives for an ecosystem could involve combining individual interests and guiding a group of several management sectors toward agreement on a collective set of aims. In more-complex cases where stakeholder groups include governments, private businesses, NGOs, and citizens, identifying common objectives can be a more time-consuming process of analysis and deliberation about desired alternative future states of the ecosystem. In most cases, trade-offs in desired ecosystem services must be confronted when objectives are set that involve multiple demands on ecosystems. In the end, the outcome of this important step is a clear shared vision among interest groups of the desired future condition of the ecosystem and the services it will provide. Those involved in EBM efforts skip this step at their peril. Without broad agreement on objectives among all parties (including managers and those whose future activities will be affected by management decisions), it is very difficult to agree on the management changes necessary to achieve desired ecosystem functions.

2. **Develop Indicators and Thresholds**
To measure progress toward ecosystem objectives, the next step is to develop indicators of ecosystem state and function. Often, the desired state of an ecosystem cannot easily be directly monitored, thus indicators must be identified that provide reliable measures of state or function. This can initially be accomplished using first principles and lessons learned from similar systems. Ultimately, food web and ecosystem models combined with monitoring and field testing are the best approaches for choosing indicators that reliably gauge ecosystem health.

The other crucial aspect of this step is to identify thresholds for each indicator that can be used to evaluate status. These thresholds, or levels corresponding to some change in the biophysical state of the indicator, can be used as "red flags" to trigger changes in management actions if

continued

Box 12.2. *continued*

indicators fall below desired levels. Thresholds can be difficult to establish, because indicator responses are often system-specific and thus require a lot of local data. A practical approach in the usual case where local information is sparse is to establish interim "benchmarks," or landmarks of indicator status that are improvements upon the existing condition. With more information, formal thresholds can be established that describe levels of each indicator that are sufficient to achieve ecosystem objectives.

3. **Conduct Ecosystem Risk Analysis**
An ecosystem risk analysis gauges the status of individual indicators relative to thresholds, their vulnerability to perturbations, and the overall functioning of the ecosystem. Approaches for conducting these analyses range from single-species population viability analyses to more-complex food web or ecosystem models used to simulate the dynamics of multiple indicators and assess changes in their status relative to thresholds. Results from a risk analysis can inform policy decisions about which indicators are highest priority for action (e.g., possibly those which are most vulnerable) and what actions are most likely to improve an indicator's status (e.g., those actions addressing the biggest threats).

4. **Evaluate Management Strategies**
A critical step in an IEA is to develop management scenarios representing suites of proposed strategies designed to achieve ecosystem objectives. Models of parts or the whole system can then be used to quantitatively evaluate likely changes in ecosystem indicators resulting from changes in management actions. Models are especially useful as strategic management tools, revealing trade-offs among management goals and effects of different management policies on ecosystem services. More-sophisticated models can be used to simulate the dynamic effects of management scenarios on the complex components of the ecosystem. The ecosystem model can thus serve as a filter to identify both promising and suboptimal policies and management approaches.

5. **Monitor, Evaluate Strategies and Progress, and Adapt as Needed**
The final and critical step in an IEA is to monitor indicators and effects of implemented strategies, evaluate their impacts, and change management approaches as necessary. Monitoring of ecosystem indicators is notoriously difficult to sustain, even for single-sector management, yet incentives can be provided if funding for continuing valued activities or human uses is made contingent upon observing real progress in implementing strategies or improvement in indicators. The more quantitative and model-based approaches to IEA can include simulated monitoring of an ecosystem in the virtual world of a model; lessons learned from such approaches can help to design monitoring programs that are likely to detect changes in indicators of interest. Depending on simulated or real-world monitoring results, changes in management strategies might be advised. It is then the prerogative of the appropriate governing body to determine whether objectives need to be reevaluated and how frequently IEA should be iterated.

the relative magnitude of impact that will result from alternative strategies. The explicit consideration of human well-being and natural ecosystem function in the ecosystem goals for Puget Sound will force trade-offs to be openly evaluated. There are likely to be several alternative approaches to achieving the goals, and although ecosystem outcomes may countervail in one place (e.g., maintaining more freshwater in-stream flows in some areas may mean parts of the watershed can support fewer housing developments), across the whole ecosystem, it is more likely that desired levels of the human and natural system goals will be achievable under some scenarios and not others.

We offer three concrete and fairly simple examples below to illustrate the benefits of gaining a better understanding of the scientific linkages among ecosystem elements in Puget Sound. Each of the examples also points out the importance of taking the extra time needed to translate complex scientific information for policymakers and managers so that the nuances of what is—and is not—known can inform better decisions. Over time, it is our hope that the list of priority actions identified in the partnership's annual ecosystem plan will be increasingly informed by scientific understanding of how the system functions and how those functions relate to the ecosystem goals.

Role of Species Interactions: Potential Food Web Effects of Marine Protected Areas

Area-based management tools such as marine reserves—regions of the ocean where fishing and other extractive activities are prohibited—have become a flagship tool of EBM. By eliminating take from within an area, marine reserves protect many components of the ecosystem occurring within their boundaries, although they cannot provide protection from all impacts, such as pollution or climate change (Lubchenco et al. 2003). Although marine reserves are a management tool that addresses entire ecosystems, the vast majority of theoretical and empirical work on marine reserves has emphasized single-species dynamics. How community structure or function responds to a marine reserve depends fundamentally on the properties of the individual populations that comprise the community, as well as the interactions among populations (e.g., Mangel and Levin 2005). Indeed, it is the interactions among species that make communities more than the sum of their parts. Unfortunately, in Puget Sound (and in most marine systems) we know very little about the species interactions that structure marine communities. Direct interactions such as competitive and predator–prey relationships are poorly understood. And, studies of such pair-wise interactions may be misleading. For example, although we might expect the removal of a predator to result in an increase of its prey, this may not occur if predation reduces the biomass of competitors (Mangel and Levin 2005).

We thus are left with key uncertainties about how well reserves and other types of marine protected areas (MPAs) might achieve particular conservation or fisheries objectives. Effective management should confront these uncertainties (Ludwig et al. 1993). Models can play an important role in understanding of and coping with uncertainty, because they allow us to examine the consequences of our assumptions and determine the sensitivity of our management guidance to uncertainty (Mangel 2000).

Currently, Washington Department of Fish and Wildlife manages twenty-four MPAs in Puget Sound, covering a total of about 700 subtidal hectares. Most of these MPAs (75%) are fully protected, no-take marine reserves. Evidence suggests depleted stocks of lingcod and some rockfishes have responded positively within marine reserve boundaries. For instance, Eisenhardt (2002) found dramatic recovery of lingcod within reserves and significant, though less impressive, rebuilding of copper rockfish. On the other hand, he also noted reduced numbers of Puget Sound rockfish and quillback rockfish.

Because lingcod appear to be important predators of small rockfish (Beaudreau and Essington 2007), it is possible that the accumulation of predator biomass in Puget Sound marine reserves prevents or slows recovery of their rockfish prey. However, because fishing in Puget Sound targets adult rockfish (Stout et al. 2001) and lingcod appear to eat only juvenile rockfish, predation does not simply replace fishing as an agent of mortality—it changes which age class suffers the mortality. Analysis of rockfish populations reveals that changes in mortality rates of smaller individuals will have a greater effect on the population's growth rate than the same change in the mortality rates of larger adults (e.g., Mangel et al. 2006). Thus, shifting the source of rockfish mortality from fishing to predation by lingcod in marine reserves could negatively affect rockfish populations.

We used simple computer models to explore how different assumptions about the importance of predation alter our expectations about the effectiveness of marine reserves. Our goal with this example is to show how simple models can provide intuition about the roles of ecological interactions in marine reserves and thus potentially improve our ability to choose management strategies that are consistent with ecosystem-level goals.

We followed the approach of Tolimieri and Levin (2005), in which we divided the rockfish population into four stages: (1) juveniles (egg, larvae, and young-of-the-year); (2) pre-productive fish; (3) young adults; and (4) old adults. We estimated total egg production by adults and survival rates for each of these stages following the approaches of Methot and Piner (2001) and Tolimieri and Levin (2005).

In our models we first envisioned a scenario in which 20% of the region was in a marine reserve and a 3% fishing mortality rate occurred outside the marine reserve. We then estimated the time to recovery of a depleted rockfish population, assuming that there was no change of the predation rate inside the reserve. In this case, recovery time for the rockfish population was about 67 years, which was 30% faster than if there was the same average fishing mortality but no reserve. This occurs because the population within the reserve contains more, older adults with higher fecundity than in regions where fishing occurs. Thus, if we ignore the potential for an increase in predation within the reserve, reserves are clearly the preferred management option for rockfish because they allow for more rapid rebuilding with a given harvest rate.

We next repeated the modeling exercise with the added impacts that might accompany an increase in abundance of a predator such as lingcod. Because we do not know the additional predation rate on rockfish juveniles that would accompany an increase in predator numbers, we explored a range of predation rates. Our goal was to estimate the predation rate on rockfish within the marine reserve that would negate the benefits of the reserve for rockfish relative to simply regulating harvest rates. As expected, recovery times of the hypothetical rockfish population increased as the predation-related mortality within reserves increased (fig. 12.3). Our models indicate that even very small increases in predation mortality within reserves (i.e., 1.2%) are sufficient to negate the benefit of reduced fishing pressure. Thus, in this example, a small increase in predation within marine reserve boundaries renders them ineffective as an alternative to conventional harvest management for rockfish.

In this case, the effectiveness of marine reserves as a strategy (for a specific objective such as recovering rockfish populations) depends

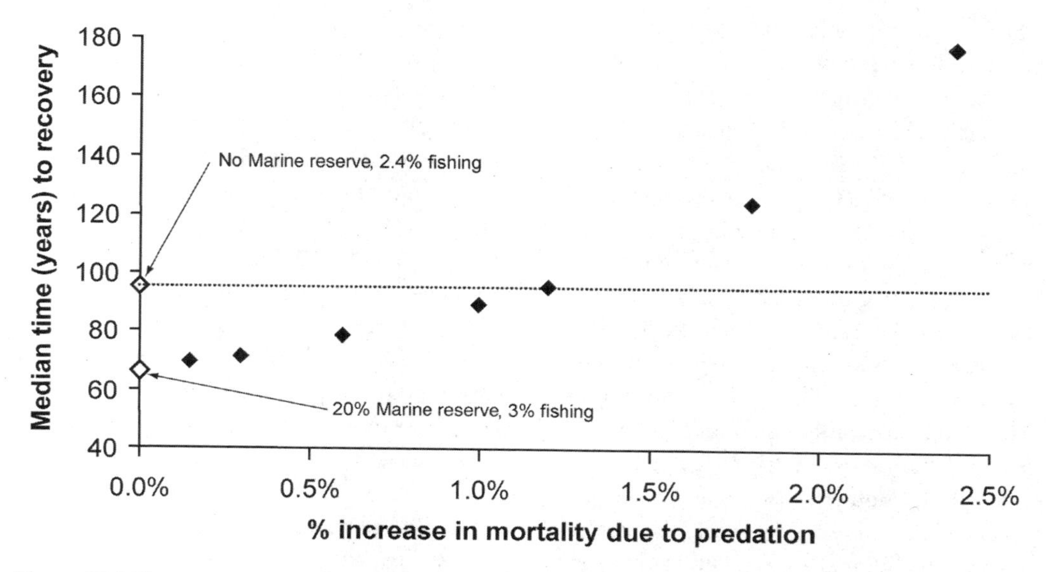

Figure 12.3 Time to recovery of a modeled Puget Sound rockfish population under different assumptions about harvest rates and the percentage of the metapopulation that is protected within a no-take marine reserve. Under the scenario where 20% of the metapopulation is protected within the reserve, a 3% fishing rate occurred outside the reserve. The x-axis depicts the changes in predator-induced mortality under the different numerical experiments with the metapopulation model. In this example, an increase of 1.2% in mortality due to predation within the reserve negates any benefits of the reserve for rockfish recovery.

critically on the role of predation within the reserve. These simple models suggest that a manager who considered data on predation rates within and outside of marine reserves would be better informed of the likely consequences of specific policy choices. Even before definitive data are available, this exercise illuminates the importance of considering species interactions when designing strategies for their recovery. For example, in this case both predator and prey are targeted by fishers. Thus, understanding this predator–prey relationship reveals a potential trade-off between goals that seek to maximize services provided by the prey and goals that seek to maximize those provided by the predator. Strategies that employ marine reserves and those that do not can both produce recovery of the rockfish but at different rates. Thus, the manager must weigh the benefits of enhancing the predator population through marine reserves versus the costs of potentially longer recovery times for their rockfish prey. Uncertainty about this predator–prey relationship is only the beginning of unraveling critical community-level interactions and the likely effectiveness of alternative management strategies (e.g., Sale et al. 2005).

The Importance of Translating Science: Hypoxia in Hood Canal

One important barrier in providing scientific guidance to policymakers in support of EBM is describing what is known, and not known, about the underlying causes of a particular ecosystem problem. Though this issue is not unique to Puget Sound, it is particularly noteworthy here because there is a striking disconnect between public perception of human impacts and the scientific evidence supporting that perception (PSP 2006).

This disconnect is well illustrated by the public perception concerning the consequences of low dissolved oxygen and fish kill events in Hood Canal. Hood Canal is a narrow, deep, fjord-like estuary within Puget Sound that supports lucrative commercial and recreational fisheries for Dungeness crab, geoduck clams, and anadromous salmon (fig. 12.1). Because of the limited tidal exchange of ocean water in the southern reaches and subsequent long residence time of deep water, the southern areas of Hood Canal are regularly subjected to bouts of reduced dissolved-oxygen (DO) conditions, also known as hypoxia. These conditions date back at least to the mid twentieth century, when early monitoring detected 2 to 3 months per year of DO reduced below 3 mg/L, where biological impacts of hypoxia are typically observed (HCDOP 2007). Over the past decade, the frequency, intensity, and spatial extent of hypoxic events has markedly increased. For instance, in 2003, the two most southern areas of Hood Canal included in monitoring had bottom DO levels that rarely exceeded 3 mg/L and that frequently dropped below 1 mg/L (HCDOP 2007). Many marine fish and invertebrates cannot withstand extended periods of time in water that has less than 2 mg/L DO.

The public within the region began to recognize the extent of this problem, following a fish kill event that occurred in October 2003. Scuba divers reported anomalous behavior of rockfish aggregating in shallow water, and scores of dead rockfish and shiner perch washed up on Hood Canal shores (WDFW 2006). A second fish kill event occurred in September 2006. Both of these events prompted widespread public concern and calls for scientific study to diagnose the causes and to identify corrective actions.

Despite the disturbing images of dead fish littering the shoreline, there is little scientific information for estimating the impact of these events on the Hood Canal ecosystem. Scientific monitoring of the Hood Canal biota has been conducted by the Washington State Department of Ecology and the Department of Fish and Wildlife (WDFW). The WDFW has conducted bottom trawl surveys in Hood Canal once every 3 years, beginning in the late 1980s. Department of Ecology has collected bottom grab samples since 1999, but only recently have more than twenty samples per year been collected.

Analysis of the bottom trawl data has revealed few statistically significant impacts of the 2003 hypoxic event on demersal fish or invertebrates (T. Essington, unpublished data). Only three fish species show a significant reduction in abundance following the event (pile perch, shiner perch, and spiny dogfish), and eight species show increased abundance. When the impacted and nonimpacted areas are compared spatially, six species show reduced abundance in impacted areas, and three show increased abundance. But the real question is whether population abundances were reduced more drastically in impacted areas compared with those in areas unaffected by hypoxia. Using this more rigorous standard of proof for detecting local impacts of hypoxia, we are unable to conclude that any species' abundance had been clearly impacted by hypoxia. Our inability to detect these ecological effects does not mean that they do not exist; we may simply lack the precision in our survey data to detect them. Indeed, an analysis of the "power" of our data suggests that it is quite low. Even if there was a 50% reduction in the abundance of each species in the impacted areas, we would only be able to detect it about 7% of the time.

Analysis of mobile and sessile invertebrates also revealed inconclusive, and in some cases counterintuitive, results (T. Essington, unpublished data). Of the thirty-three species of crabs, shrimps, cnidaria, and echinoderms regularly captured in the bottom trawl gear, only three showed significant declines in abundance over time, and four showed a significantly reduced abundance within the impacted area. Five species' population trends differed between impacted and unimpacted sites, but only two of these species' trends were in the expected direction (i.e., a reduction over time in the hypoxic zone with stable or increasing trend over time in unimpacted areas). The other three species demonstrated a regional *increase* in abundance after the 2003 hypoxia event. Clearly, the impacts of hypoxic events on the Hood Canal biota are far more complex than those reported in popular literature and the media, which describe vast "dead zones" in Hood Canal and elsewhere in Puget Sound (McClure and Stiffler 2006).

Despite widespread public recognition of a "crisis" in Hood Canal, the scientific evidence necessary to characterize the extent and nature of this crisis for biota is extremely limited. In contrast, data on physical and chemical attributes are relatively good due to recent investments in oceanographic mooring buoys that continuously monitor water conditions. The essential problem is the limited investment in biological monitoring needed to identify baselines that form the basis of any scientific comparison. This presents a difficult challenge: How, as scientists, can we explain the fundamental uncertainty about the impacts of hypoxic events and provide guidance on whether specific actions are advisable? Presumably, the majority of stakeholders and policymakers have little appreciation for the term "statistical power" and are rightly skeptical of scientists' claims of "no statistically significant effect," given that they've seen the media coverage of beaches littered with dead sea life and videos depicting gasping fish slowly suffocating underwater. In our experience, time spent in the trenches with the managers, stakeholders. and policymakers discussing results and what they do and do not show is the best hope for informing public opinion and management decisions.

Benefits of Ecosystem Approaches: Holistic versus Myopic Views of Ecosystems

Lessard and colleagues (2005) suggest that holistic approaches to natural resource policy are of greater benefit because they identify a larger suite of policy options than those that do not consider species interactions or other ecological processes. Here we demonstrate a related point, namely, that the expected outcome of a policy decision depends largely on which ecological processes you consider. Moreover, we also suggest that science can provide guidance to policymakers by identifying different outcomes expected from alternative ecological phenomena, by testing and eliminating nonviable hypotheses, and by translating the remaining uncertainty in a manner that can be readily appreciated.

We demonstrate these alternative expectations by drawing upon well-described ecological concepts and theory on the nature and magnitude of species interactions. Specifically, here we suggest that one of the more contentious policy issues in Puget Sound, salmon stock enhancement via hatcheries, can be viewed as beneficial or destructive, depending on which ecological processes one considers. Recovery plans designed to protect and restore ESA-listed salmon species in Puget Sound involve actions that cover a large fraction of upland, freshwater, estuarine, and marine environments within the ecosystem (NMFS 2007a; SSPS 2007). One of the strategies to recover salmon for human and natural ecosystem needs is through production of salmon in hatcheries. Hatchery salmon have been reared and released into Puget Sound since early in the nineteenth century, with the original expectation that more salmon produced meant more salmon that could be harvested. This policy was essentially predicated on the most narrow view of the natural world, that a hatchery salmon is equivalent to a wild salmon and that there is no ecological constraint on the total number of salmon in the ecosystem (fig. 12.4A) (Naish et al. 2007). Clearly, this particular worldview suggests hatchery production is a prudent "techno-fix" to the problem of declining salmon stocks (Meffe 1992). However, a broader, ecological view recognizes that hatchery and wild fish are distinct, yet that they share critical environmental resources (fig. 12.4B). If those shared resources are limiting, then hatchery fish production is likely to lead not to more total salmon, but instead to the replacement of wild fish with hatchery fish.

A more holistic view consists of a food web module that identifies some of the main predators and competitors of salmon, along with an explicit consideration of the prey species that constitute potentially limiting resources (fig. 12.4C). Now insights from ecological theory help us identify and consider quite a few more questions in evaluating the policy question of how many hatchery salmon should be released to achieve a target. For instance, given recent declines in herring abundance due to habitat degradation (Gustafson et al. 2006), one of the prey resources shared between hatchery and wild salmon could be becoming more limiting each year. So, hatchery production might have been a good technological solution 30 years ago, but it may not be now that the ecological context has changed. We also can consider orcas in this policy question because chinook salmon are a primary prey species for orcas in Puget Sound (Ford et al. 1998; NMFS 2003; Krahn et al. 2004). Orcas within Puget Sound have been listed under the US Endangered Species Act since 2005 (NMFS 2005), and future policies will aim to enhance their population status (NMFS 2007b). Assuming this recovery effort is successful, how could more orcas affect federally listed wild salmon, and might hatchery salmon releases exacerbate or lessen those impacts? Again, drawing on ecological theory, it might make sense to use hatchery salmon releases as a temporary measure to mitigate against the increased predation mortality that a recovered orca population would produce, especially if orca feeding is limited by physiological or other constraints that make the amount of salmon eaten per orca relatively insensitive to the number of salmon available. If this constraint exists, then the addition of hatchery salmon would essentially "dilute" the wild salmon component that orcas feed upon and would result in a lower fraction of wild salmon being consumed. On the other hand, if orcas are attracted to larger schools of fish, then this benefit might be reduced or eliminated.

Finally, we consider the alternative food web responses mitigated by dogfish predation on salmon and their prey. Dogfish abundance in Puget Sound has been declining over the past decade, and if this decline continues, undoubtedly there will be calls to "save the dogfish," given the unique role they play in this ecosystem. In the face of a dynamic dogfish population feeding on salmon and herring, what might be the effect of hatchery salmon releases on wild salmon? Do hatchery releases cause an increase in dogfish predation on wild salmon? Ecological theory tells us this can happen if wild and hatchery fish overlap in time and space and dogfish are attracted to these areas. Dogfish populations might also benefit from being "fed" hatchery salmon and become more abundant, thus resulting in an increase in predation rates on wild salmon. Alternatively, if wild and hatchery fish do not overlap in space and time, hatchery salmon releases might attract dogfish away from schools of wild salmon, thereby reducing dogfish predation rates on wild fish.

The lesson behind these examples is that the more holistic view of the ecosystem produced a much richer range of policy options and outcomes than any of the simpler views, while also highlighting potential trade-offs

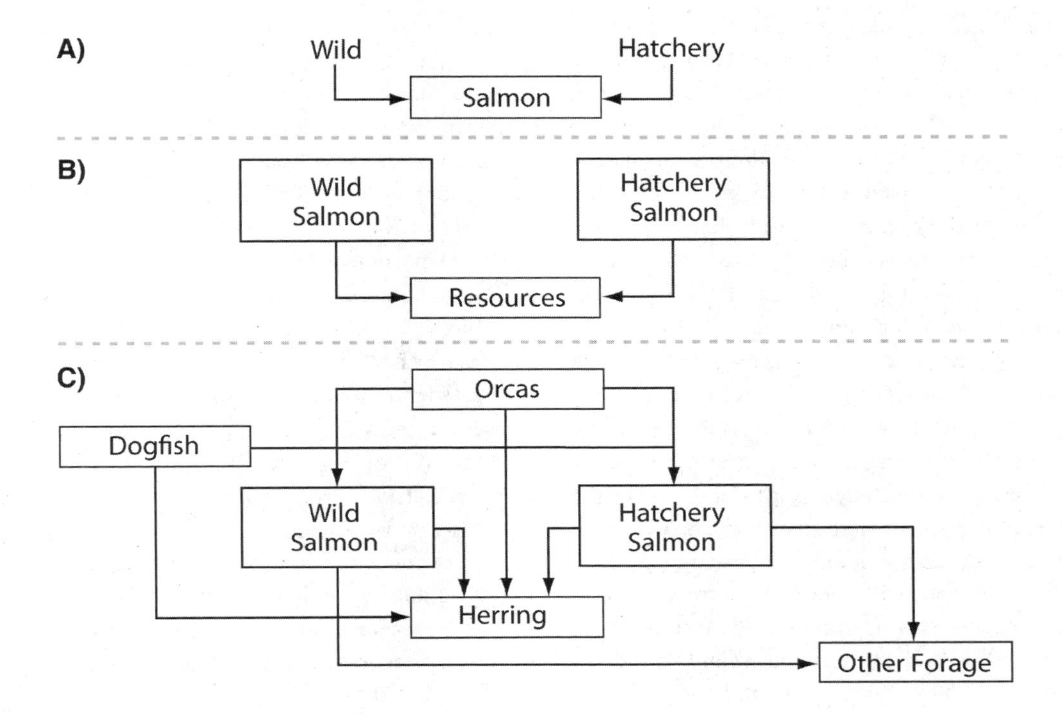

Figure 12.4 Sample food web diagrams for considering the impacts of hatchery-origin salmon on wild salmon. (A) Naive view of wild and hatchery fish both contributing equivalently to salmon stocks. (B–C) Arrows depict ecological interactions expressed through feeding linkages.

between conflicting conservation mandates. These policy options include a more strategic approach to hatchery releases, which may include time, area, and sizes of release that will minimize negative impacts while maximizing potential benefits. Such considerations of "strategy" allow the debate about hatchery releases to ascend beyond the "hatchery fish are bad" versus "no they're not" debate and instead try to explore when they have positive, negative, or unknown impacts on wild salmon targets. Also, these hypotheses are all supported by and generated from ecological theory. To turn these into more useful scientific guidance, we need data and analyses to discriminate among the alternative hypotheses and then translate their implications for those faced with making policy decisions. Both conceptual and quantitative models can be used as rapid screening tools to help us refine the hypotheses and also to rule out hypotheses because they require biologically implausible parameters to hold true. Models also can be used to engage policymakers through "gaming," where models with easy-to-use and attractive interfaces can capture the attention and imagination of nonscientists.

No matter how many tools, theories, or direct observations scientists have to inform a policy decision, it is critical for them to clearly articulate that such decisions almost always need to be made in the absence of a full understanding of system behavior. Indeed, even the most complex food web module described in figure 12.4 is incomplete. Other marine mammals in Puget Sound consume salmon. There are several species of salmon to consider. Other species rely on herring as prey. And "other forage fish" is a broad category that represents numerous species (Sound Science 2007). Dogfish not only feed in the pelagic regions, but also are important predators in the demersal food web. Our job as scientists in informing management is therefore not necessarily to point to the "right" answer, but to convey ecologically plausible possibilities, the nature and extent of uncertainty, and potential strategies that are robust to uncertainties. In almost all cases these potential strategies will be precautionary and accompanied by monitoring that will reveal the key species interactions that are important for dictating food web dynamics. This view of the role of science—where an explicit set of alternative hypotheses is identified, management actions are made with these alternatives in mind, and ecosystems are monitored closely—is, of course, the adaptive management paradigm (Walters 1986). Successful implementation of the adaptive management approach is not easy, however, and requires institutional structures that are flexible and expectations of the roles and capabilities of scientific advice that are realistic (Walters 1997; Kaufman et al., chap. 7 of this volume).

Moving Forward in Puget Sound: Ongoing Role of Science

In many ways, Puget Sound is like many other marine ecosystems around the world. The ecosystem is degraded, and there are calls from many sectors to employ the tenets of ecosystem-based management. Unlike in most regions, political leaders, managers, and scientists in the Puget Sound region are in an enviable position. Broad ecosystem goals, early actions, and a framework for designing a full ecosystem strategy have been adopted by the diverse Puget Sound Partnership group, representing public and private institutions, tribes, and citizen groups. The partnership's mission, governance structure, and accountability system are mandated in state law. In particular, the partnership is designed to build upon existing and well-seasoned science–policy fora that are working well in the region to achieve goals for parts of the ecosystem. For example, the extensive portion of the ecosystem traversed by Pacific salmon is being managed under a scientifically well-vetted ecosystem planning approach. The planning and implementation of the federal salmon recovery plans involve "top-down" guidance on recovery goals and coordination of regional strategies for harvest and hatchery management, coupled with "bottom-up" work from multistakeholder watershed councils to adopt habitat management strategies to achieve their salmon goals. A science team works closely with the policy leaders and managers to conduct analyses that are timely and critical to the process, evaluate local and regional strategies against objectives, and translate its findings for watershed and regional councils to consider in implementation (PSTRT 2007; SSPS 2007). The science underpinning the strategy development is not only useful for the watershed groups restoring salmon, but also peer reviewed and thus accessible to the broader scientific community (e.g., Ruckelshaus et al. 2002b; Bartz et al. 2006; Scheuerell et al. 2006; Battin et al. 2007). Similar science–policy interactions producing applied science on smaller scales in Puget Sound are well established for harvest and hatchery management (e.g., Brodeur et al. 2003; Kope 2006) and forest practices (e.g., Hayes et al. 2006; Quinn et al. 2007).

The ongoing work of the partnership under its new science–policy governance structure will be to integrate the existing science–policy forums and to fill knowledge gaps so that a whole-ecosystem perspective is provided. A step in this direction has occurred within the scientific community in Puget Sound. An unanticipated but positive benefit of developing the *Sound Science* document (box 12.1; Sound Science 2007) was the discussions that were initiated within the diverse community of scientists in the region who mostly work on small pieces of the overall ecosystem (see also Rosenberg and Sandifer, chap. 2 of this volume). Better coordination of existing research efforts and identifying ways to fill key gaps in our understanding will advance as those discussions continue.

Governor Gregoire and the partnership highlighted the primary importance of educating the public about the risks facing their treasured natural resources, as well as the important role that scientific information has and will continue to play in illuminating those risks and potential solutions for the ecosystem (PSP 2006). The two early challenges put before the scientific community are daunting: determining what is the best set of measurable ecosystem indicators for tracking progress toward goals, and estimating the likely and integrative effects of actions on ecosystem elements. The difficult job of choosing ecosystem indicators is not a new problem, and excellent reviews of the subject agree about the value of using ecosystem models to help make those choices (NRC 2000; Fulton et al. 2005). Since an ecosystem model is one that typically will be developed and improved upon over time, it would be wise to include the indicators as part of the adaptively revised EBM plan. Now that ecosystem goals exist for Puget Sound, important decision feedbacks linked to indicators for each goal can be built into the strategy evaluation.

The next offerings from science to support smarter ecosystem-scale actions in Puget Sound will be (1) an integrated ecosystem assessment and (2) guidance on near-term approaches to addressing some of the high-priority threats identified by the partnership. These threats include understanding sources and transport of toxics, nutrients, and pathogens; abating storm water runoff impacts; identifying priority protection and restoration sites along nearshore habitats of the ecosystem; and evaluating the effectiveness of alternative habitat protection approaches such as acquisition, easements, and regulation. The ecosystem assessment will provide a means by which the impacts of human actions on parts of the ecosystem can be estimated. This, in turn, will allow estimates of the resulting changes in the services provided from the ecosystem as its components change. Finally, it will allow us to generate a set of explicit hypotheses mechanistically linking strategies to the threats they are aimed at abating, and then to predicted outcomes in ecosystem elements. The hypothesized links between the status of an ecosystem goal and the actions implemented to achieve it can be tested and the strategies refined as needed through adaptive management.

The policy leaders in Puget Sound have made significant strides in moving toward an EBM approach for their ecosystem. The important details of how scientific information will feed into management decisions within the ecosystem governance structure have been worked out in practice for the first iteration of the ecosystem plan—called the Action Agenda—which was completed in December 2008 (PSP 2008). The onus is now on the scientific community to deliver on key data collection, analysis, and modeling needs so that the Action Agenda can be adapted as new scientific information becomes available. Just as important as conducting relevant, peer-reviewed science is for scientists to translate the nuances of their results so that managers, policy leaders, and the public can make informed decisions.

Key Messages

1. The Puget Sound Partnership was created to recover the sound's marine and

freshwater ecosystems and associated human prosperity by 2020 and has adopted measurable social and ecological goals to meet this overall objective. Powered by a state legislative mandate, the group represents public and private institutions, tribes, and citizen groups.

2 Inclusion of a range of ecological processes in management scenarios and models can change forecasted policy outcomes; more holistic views of the ecosystem—especially including land–sea linkages and their impacts on indicators—produce a richer range of policy options and outcomes than do simpler views.
3. The role of science is to inform management and to convey ecologically plausible possibilities, the nature and extent of uncertainty, and potential strategies that are robust to uncertainties.
4. Next steps to connect science to policy in Puget Sound are to undertake an integrated ecosystem assessment and provide guidance to identify indicators of ecosystem condition and address high-priority threats. Scientists must deliver on key data collection, analysis, and modeling needs and work to better communicate their results to managers, policy leaders, and the public.

Acknowledgments

Ideas in this chapter were improved with input from Paul McElhany and Nick Tolimieri. Comments from the editors and two anonymous reviewers helped improve earlier versions of this chapter. We thank the partners and staff of the Puget Sound Partnership for challenging us to think carefully about how science can help move the ecosystem effort forward in Puget Sound. Spirited discussions within the science working group of the partnership have helped to frame these issues in a scientifically rigorous and tractable way.

References

Aydin, K. Y., G. A. McFarlane, J. R. King, B. A. Megrey, and K. W. Myers. 2005. Linking ocean food webs to coastal production and growth rates of Pacific salmon (Oncorhynchus spp.), using models on three scales. *Deep Sea Research II* 52:757–80.

Bartz, K., K. Lagueux, M. D. Scheuerell, T. Beechie, and M. Ruckelshaus. 2006. Translating alternative land use restoration scenarios into changes in stream conditions: A first step in evaluating salmon recovery strategies. *Canadian Journal of Fisheries and Aquatic Sciences* 63:1578–95.

Battin, J., M. Wiley, M. Ruckelshaus, R. Palmer, E. Korb, K. Bartz, and H. Imaki. 2007. Projected impacts of future climate change on salmon habitat restoration actions in a Puget Sound river. *Proceedings of the National Academy of Sciences* 104:6720–25.

Beamish, R. J., G. A. McFarlane, and J. R. King. 2005. Migratory patterns of pelagic fishes and possible linkages between open ocean and coastal ecosystems off the Pacific coast of North America. *Deep Sea Research II* 52:739–55.

Beaudreau, A. H., and T. E. Essington. 2007. Spatial, temporal and ontogenetic patterns of predation on rockfishes by lingcod. *Transactions of the American Fisheries Society* 136:1438–52. Seattle, WA: National Oceanographic and Atmospheric Administration.

Beechie, T. J., P. Roni, E. A. Steel, E. Quimby, ed. 2003. *Ecosystem recovery planning for listed salmon: An integrated assessment approach for salmon habitat.* US Dept. of Commerce, NOAA Tech Memo NMFNWFSC-58.

Black, C. 2006. New alert on eating local salmon: State warns to limit meals of Puget Sound chinook. *Seattle Post-Intelligencer* 27 October 2006.

Brand, E. J., I. C. Kaplan, E. A. Fulton, A. J. Herman, J. C. Field, and P. S. Levin. 2008. *A spatially explicit ecosystem model of the California Current's food web and oceanography.* NOAA Fisheries, NOAA Tech Memo. Seattle, WA: National Oceanographic and Atmospheric Administration.

Brodeur, R. D., K. W. Myers, and J. H. Helle. 2003. Research conducted by the United States on the early ocean life history of Pacific salmon. *North Pacific Anadromous Fish Community Bulletin* 3:89–131.

Cornwall, W. 2006. State issues warning on eating Sound salmon. *Seattle Times* 27 October 2006.

Duffy, J. E. 2003. Biodiversity loss, trophic skew and ecosystem functioning. *Ecology Letters* 6:680–87.

Eisenhardt, E. 2002. A marine preserve network in San Juan Channel: Is it working for nearshore rocky reef fish? *2001 Puget Sound Research Conference*. Olympia, WA: Puget Sound Action Team.

Floberg, J., M. Goering, G. Wilhere, C. Macdonald, C. Chappell, C. Rumsey, Z. Ferdana, A. Holt, P. Skidmore, and T. Horsman. 2004. *Willamette Valley–Puget Trough–Georgia Basin ecoregional assessment*. Arlington, VA: The Nature Conservancy. http://www.ecotrust.org/placematters/assessment.html.

Ford, J. K. B., G. M. Ellis, L. G. Barrett-Lennard, A. B. Morton, R. S. Palm, and K. C. Balcomb III. 1998. Dietary specialization in two sympatric populations of killer whales (*Orcinus orca*) in coastal British Columbia and adjacent waters. *Canadian Journal of Zoology* 76:1456–71.

Fresh, K., C. Simenstad, J. Brennan, M. Dethier, G. Gelfenbaum, F. Goetz, M. Logsdon et al. 2004. *Guidance for protection and restoration of the nearshore ecosystems of Puget Sound*. Puget Sound Nearshore partnership report no. 2004-02. Seattle: Washington Sea Grant Program, University of Washington.

Fulton, E. A., A. D. M. Smith, and A. E. Punt. 2005. Which ecological indicators can robustly detect effects of fishing? *ICES Journal of Marine Science* 62:540–51.

GAO (US Government Accountability Office). 2005. *Chesapeake Bay Program*. Report to the Subcommittee on Interior and Related Agencies, Committee on Appropriations, US Senate. Report no. GAO-06-96 (2005). Washington, DC: GAO.

GAO (US Government Accountability Office). 2001. *South Florida ecosystem restoration*. Report to the Subcommittee on Interior and Related Agencies, Committee on Appropriations, US Senate. Report no. GAO-01-361 (2001). Washington, DC: GAO.

Gustafson, R. G., J. Drake, M. J. Ford, J. M. Myers, E. E. Holmes, and R. S. Waples. 2006. *Status review of Cherry Point Pacific herring* (Clupea pallasii) *and updated status review of the Georgia Basin Pacific herring distinct population segment under the Endangered Species Act*. US Dept. of Commerce, NOAA Tech Memo NMFS-NWFSC-76. Seattle, WA: National Oceanic and Atmospheric Administration.

Hayes, M. P., T. Quinn, D. J. Dugger, T. L. Hicks, M. A. Melchiors, and D. E. Runde. 2006. Dispersion of coastal tailed frog (*Ascaphus truei*): A hypothesis relating occurrence of frogs in non-fish-bearing headwater basins to their seasonal movements. *Journal of Herpetology* 40:533–45.

HCDOP (Hood Canal Dissolved Oxygen Program). 2007. http://www.hoodcanal.washington.edu/index.jsp.

Heinz Center. 2006. *Filling the gaps: Priority data needs and key management challenges for national reporting on ecosystem condition*. A report of the Heinz Center's State of the Nation's Ecosystems Project. Washington, DC. www.heinzctr.org/ecosystems.

Heinz Center. 2002. *The state of the nation's ecosystems: Measuring the lands, waters, and living resources of the United States*. Cambridge, UK: Cambridge University Press.

James, M. C., C. A. Ottensmeyer, and R. A. Myers. 2005. Identification of high-use habitat and threats to leatherback sea turtles in northern waters: New directions for conservation. *Ecology Letters* 8:195–201.

Judd, R. 2006. Puget Sound not exactly a super model of purity. *Seattle Times* 29 October 2006.

Kaplan, I. C., and P. S. Levin. 2009. Ecosystem-based management of what? An emerging approach for balancing conflicting objectives in marine resource management. In *The Future of Fisheries Science in North America*, ed. R. Beamish and B. Rothschild. Berlin: Springer.

Kope, R. G. 2006. Cumulative effects of multiple sources of bias in estimating spawner-recruit parameters with application to harvested stocks of chinook salmon (*Oncorhynchus tshawytscha*). *Fisheries Research* 82:101–10.

Krahn, M. M., M. J. Ford, W. F. Perrin, P. R. Wade, R. P. Angliss, M. B. Hanson, B. L. Taylor et al. 2004. *Status review of southern resident killer whales* (Orcinus orca) *under the Endangered Species Act*. US Dept. of Commerce, NOAA Tech Memo NMFS-NWFSC-62. Seattle, WA: National Oceanic and Atmospheric Administration.

Lenihan, H. S., and C. H. Peterson. 1998. How habitat degradation through fishery disturbance enhances impacts of hypoxia on oyster reefs. *Ecological Applications* 8:128–40.

Lessard, R. B., S. F. D. Martell, C. J. Walters, T. E. Essington, and J. F. Kitchell. 2005. Should ecosystem management involve active control of species abundances? *Ecology and Society* 10(2): article 1.

Lotze, H. K., H. S. Lenihan, B. J. Bourque, R. H. Bradbury, R. G. Cooke, M. C. Kay, S. M. Kidwell et al. 2006. Depletion, degradation and recovery potential of estuaries and coastal seas. *Science* 312:1806–09.

Lubchenco, J., S. R. Palumbi, S. D. Gaines, and S. Andelman. 2003. Plugging a hole in the ocean: The emerging science of marine reserves. *Ecological Applications* 13:S3–S7.

Ludwig, D., R. Hilborn, and C. Walters. 1993. Uncertainty, resource exploitation, and conservation: Lessons from history. *Science* 260:17–36.

Mangel, M. 2000. Irreducible uncertainities, sustainable fisheries and marine reserves. *Evolutionary Ecology Research* 2:547–57.

Mangel, M., and P. S. Levin. 2005. Regime shifts, phase shifts and paradigm shifts: Making community ecology the basic science for fisheries. *Philosophical Transactions of the Royal Society B* 360:95–105.

Mangel, M., P. Levin, and A. Patil. 2006. Using life history and persistence criteria to prioritize habitats for management and conservation. *Ecological Applications* 16:797–806.

McClure, M., and M. Ruckelshaus. 2007. Collaborative science: Moving ecosystem-based management forward in Puget Sound. *Fisheries* 32:458–61.

McClure, R., and L. Stiffler. 2006. Marine life is disappearing from Puget Sound, and fast. *Seattle Post-Intelligencer* 9 October 2006.

McLeod, K. L., J. Lubchenco, S. R. Palumbi, and A. A. Rosenberg. 2005. *Scientific consensus statement on marine ecosystem-based management.* The Communication Partnership for Science and the Sea (COMPASS). Signed by 221 academic scientists and policy experts with relevant expertise. http://www.compassonline.org/pdf_files/EBM_Consensus_Statement_v12.pdf.

Meffe, G. K. 1992. Techno-arrogance and halfway technologies: Salmon hatcheries on the Pacific coast of North America. *Conservation Biology* 6:350–54.

Methot, R, and K. Piner. 2001. *Status of the canary rockfish resource off California, Oregon and Washington.* Document submitted to Pacific Fishery Management Council, Portland, Oregon. http://www.pcouncil.org/groundfish/gfstocks.html.

Musick, J. A., M. M. Harbin, S. A. Berkeley, G. H. Burgess, A. M. Eklund, L. Findley, R. G. Gilmore et al. 2000. Marine, estuarine, and diadromous fish stocks at risk of extinction in North America (exclusive of Pacific salmonids). *Fisheries* 25:6–30.

Naish, K. A., J. Taylor, P. S. Levin, S. S. Crawford, A. M. Muir, T. P. Quinn, D. Huppert, and R. W. Hilborn. 2007. A review of salmonid enhancement activities. *Advances in Marine Biology* 53:61–194.

NMFS (National Marine Fisheries Service). 2007a. *Endangered and threatened species: Recovery plans.* National Oceanic and Atmospheric Administration. Federal Register 72(12):2493–95.

NMFS (National Marine Fisheries Service). 2005. *Endangered and threatened wildlife and plants: Endangered status for southern resident killer whales.* National Oceanic and Atmospheric Administration. 50 CFR Part 224. Federal Register 70(222):69903–12.

NMFS (National Marine Fisheries Service). 2007b. *Proposed recovery plan for southern resident killer whales* (Orcinus orca). National Oceanic and Atmospheric Administration. Northwest Region Protected Resources Division. Seattle, WA. http://www.nwr.noaa.gov/Marine-Mammals/Whales-Dolphins-Porpoise/Killer-Whales/ESA-Status/upload/SRKW-Prop-Recov-Plan.

NMFS (National Marine Fisheries Service). 2003. *Summary: May 2003 Southern Resident Killer Whale Prey Relationships Workshop.* Seattle, WA: National Oceanic and Atmospheric Administration. http://www.nwfsc.noaa.gov/research/divisions/cbd/marine_mammal/kwworkshops/index.cfm.

NRC (National Research Council). 2000. *Ecological indicators for the nation.* Committee to Evaluate Indicators for Monitoring Aquatic and Terrestrial Environments. Washington, DC: National Academies Press.

NRC (National Research Council). 1992. *Restoration of aquatic ecosystems: Science, technology, and public technology.* Washington, DC: National Academies Press.

NRC (National Research Council). 1996. *Upstream: Salmon and society in the Pacific Northwest.* Washington, DC: National Academies Press.

NSTC (National Science and Technology Council). 2006. *Charting the course for ocean science in the United States for the next decade.* Joint Subcommittee on Ocean Science and Technology. 7 December 2006. http://www.ostp.gov/nstc/.

Paine, R. T. 1980. Food webs: Linkage, interaction strength and community infrastructure. *Journal of Animal Ecology* 49:667–85.

Pauly, D., T. Pitcher, D. Preikshot, and J. Hearne, ed. 2006. Back to the future: Reconstructing the Strait of Georgia ecosystem. *Fisheries Centre Research Reports* 6(5):1–89.

POC (Pew Oceans Commission). 2003. *America's living ocean: Charting a course for sea change. A report to the nation.* Washington, DC: Pew Trusts.

PSAT (Puget Sound Action Team). 2007. *State of the Sound 2007.* Office of the Governor, State of Washington. Publication no. PSAT 07-01. http://www.psat.wa.gov/Publications/state_sound07/sos.htm.

PSP (Puget Sound Partnership). 2006. *Sound health, sound future: Protecting and restoring Puget Sound.* Olympia, WA. http://www.psp.wa.gov/.

PSP (Puget Sound Partnership). 2008. Action Agenda. Olympia, WA. http://www.psp.wa.gov/.

PSTRT (Puget Sound Technical Recovery Team). 2007. *Northwest salmon recovery.* Seattle, WA: Northwest Fisheries Science Center. http://www.nwfsc.noaa.gov/trt/index.cfm.

Quinn, T., M. Hayes, D. J. Dugger, T. L. Hicks, and A. Hoffman. 2007. Comparison of two techniques for surveying headwater stream amphibians. *Journal of Wildlife Management* 71:282–88.

Rice, C. A. 2005. Effects of shoreline modification on a northern Puget Sound beach: Microclimate and embryo mortality in summer spawning surf smelt (*Hypomesus pretiosus*). *Estuaries and Coasts* 29:63–71.

Ruckelshaus, M., K. Currens, R. Fuerstenberg, W. Graeber, K. Rawson, N. Sands, and J. Scott. 2002a. *Planning ranges and preliminary guidelines for the delisting and recovery of the Puget Sound chinook salmon.* Evolutionarily Significant Unit. Seattle, WA: NOAA Fisheries, Northwest Fisheries Science Center. http://www.nwr.noaa.gov/Salmon-Recovery-Planning/Recovery-Domains/Puget-Sound/upload/trtpopesu.pdf.

Ruckelshaus, M. H., P. Levin, J. B. Johnson, and P. M. Kareiva. 2002b. The Pacific salmon wars: What science brings to the challenge of recovering species. *Annual Review of Ecology and Systematics* 33:665–706.

Sale, P. F., R. K. Cowen, B. S. Danilowicz, G. P. Jones, J. P. Kritzer, K. C. Lindeman, S. Planes, N. V. C. Polunin, G. R. Russ, and Y. J. Sadovy. 2005. Critical science gaps impede use of no-take fishery reserves. *Trends in Ecology and Evolution* 20:74–80.

Sands, N. J., K. Rawson, K. Currens, B. Graeber, M. Ruckelshaus, R. Fuerstenberg, and J. Scott. 2007. *Dogz in the 'Hood: The Hood Canal summer chum salmon ESU.* Seattle, WA: NOAA Fisheries, Northwest Fisheries Science Center. http://www.nwfsc.noaa.gov/trt/pubs.html.

Scheuerell, M. D., P. S. Levin, R. W. Zabel, J. G. Williams, and B. L. Sanderson. 2005. A new perspective on the importance of marine-derived nutrients to threatened stocks of Pacific salmon (*Oncorhynchus* spp.). *Canadian Journal of Fisheries and Aquatic Sciences* 62(5):961–64.

Scheuerell, M., R. Hilborn, M. Ruckelshaus, K. L. Bartz, K. Lagueux, A. Haas, and K. Rawson. 2006. The Shiraz model: A tool for incorporating anthropogenic effects and fish–habitat relationships in conservation planning. *Canadian Journal of Fisheries and Aquatic Sciences* 63:1596–1607.

Sound Science. 2007. *Sound Science: Synthesizing ecological and socioeconomic information about the Puget Sound ecosystem.* M. Ruckelshaus and M. McClure, coordinators. Seattle, WA: National Oceanic and Atmospheric Administration (NMFS), Northwest Fisheries Science Center.

Stout, H. A., B. B. McCain, R. D. Vetter, T. L. Builder, W. H. Lenarz, L. L. Johnson, and R. D. Methot. 2001. *Status review of copper rockfish, quillback rockfish, and brown rockfish in Puget Sound, Washington.* US Dept. of Commerce, NOAA Tech Memo NMFS-NWFSC-46. Seattle, WA: National Oceanic and Atmospheric Administration.

SSPS (Shared Strategy for Puget Sound). 2007. *Puget Sound salmon recovery plan.* Seattle, WA: SSPS. http://www.nwr.noaa.gov/Salmon-Recovery-Planning/Recovery-Domains/Puget-Sound/.

Tolimieri, N., and P. S. Levin. 2005. The roles of fishing and climate in the population dynamics of bocaccio rockfish. *Ecological Applications* 15:458–68.

USCOP (US Commission on Ocean Policy). 2004. *An ocean blueprint for the twenty-first century.* Final report. Washington, DC: USCOP. ISBN 0 9759462 0 X.

Walters, C. J. 1986. *Adaptive management of renewable resources.* New York: MacMillan.

Walters, C. J. 1997. Challenges in adaptive management of riparian and coastal ecosystems. *Conservation Ecology* 1(2):1.

WDFW (Washington Department of Fish and Wildlife). 2006. *Underwater video of the September 2006 low-dissolved-oxygen event in Hood Canal, Puget Sound.* WDFW. http://www.pugetsoundnearshore.org/.

Whitney, F. A., W. R. Crawford, and P. J. Harrison. 2005. Physical processes that enhance nutrient transport and primary productivity in the coastal and open ocean of the subarctic NE Pacific. *Deep Sea Research II* 52:681–706.

CHAPTER 13

Gulf of California, Mexico

Exequiel Ezcurra, Octavio Aburto-Oropeza, María de los Ángeles Carvajal, Richard Cudney-Bueno, and Jorge Torre

The Gulf of California in northwest Mexico, also known as the Sea of Cortés, together with its area of direct influence in the Pacific Ocean, covers an area of 375,000 km^2 (fig. 13.1). The region includes not only one of the most important large marine ecosystems on the planet, but also the Sonoran and Baja Californian deserts and significant portions of two terrestrial hotspots: the southern part of the California Biotic Region and the northern portion of the Mesoamerican dry tropical forest. In the gulf itself, over one hundred islands surrounded by powerful upwellings of cold, nutrient-rich waters house a unique biodiversity that faces increasing pressures.

The gulf's subtidal habitats, together with 6,000 km^2 of coastal lagoons and 2,560 km^2 of mangrove forests, serve as reproductive, nesting, and nursing sites for hundreds of resident and migratory species. A large proportion of the world's marine phyla are represented in the gulf. The gulf is home to 891 fish species, 181 marine birds, 34 marine mammals, and 4,853 known marine macroinvertebrates (some authors estimate that more than 4,000 invertebrate species remain undescribed). Of these species, 831 are endemic to the region, including the *totoaba* (a giant sea bass) and the *vaquita* (the Gulf of California's harbor porpoise).

Not only is the gulf biologically important, it also provides the socioeconomic sustenance of the inhabitants of the region who have developed systems of natural resource use, access, and appropriation, often threatening the long-term sustainability of the resources. The most important threats to biodiversity are driven by the growth of economic activities that cause the deterioration of coastal areas due to decreasing freshwater flows, pollution by agrochemicals and urban waste, sedimentation, and the use of inappropriate fishing technologies such as bottom-trawling nets (Lavín et al. 1998; Carriquiry and Sánchez 1999; Lavín and Sánchez 1999; Steller et al. 2003; Beman et al. 2005). This rate of biodiversity loss is substantially higher in many coastal lagoons around the gulf, where shrimp farms have been developed during the last decade (Glenn et al. 2006). Critical mangrove forest habitat is being lost in Mexico at an annual rate of 1.0%–2.5% because of sedimentation, eutrophication, and changes in water flows caused by the construction of shrimp ponds, marinas, inland channels, and deforestation (INE 2005).

Significance of Marine Resources in the Gulf's Regional History

For thousands of years, the population of the Gulf of California has benefited from the marine resources of the region, using the bays, coastal lagoons, river mouths, salt marshes, and estuaries to find food. In isolation from the rest of Mesoamerica, the early inhabitants of Baja California were fishers and coastal hunters and developed one of the most incredible assemblages of cave paintings in the American continent that bear witness of their close association to the sea and its resources. Other coastal indigenous groups from Sonora and Sinaloa, such as the Cucapá, the Seri, and the Yaqui, also developed unique lifestyles as fishermen and sailors, with cultures finely adapted to the sea and its coastal resources.

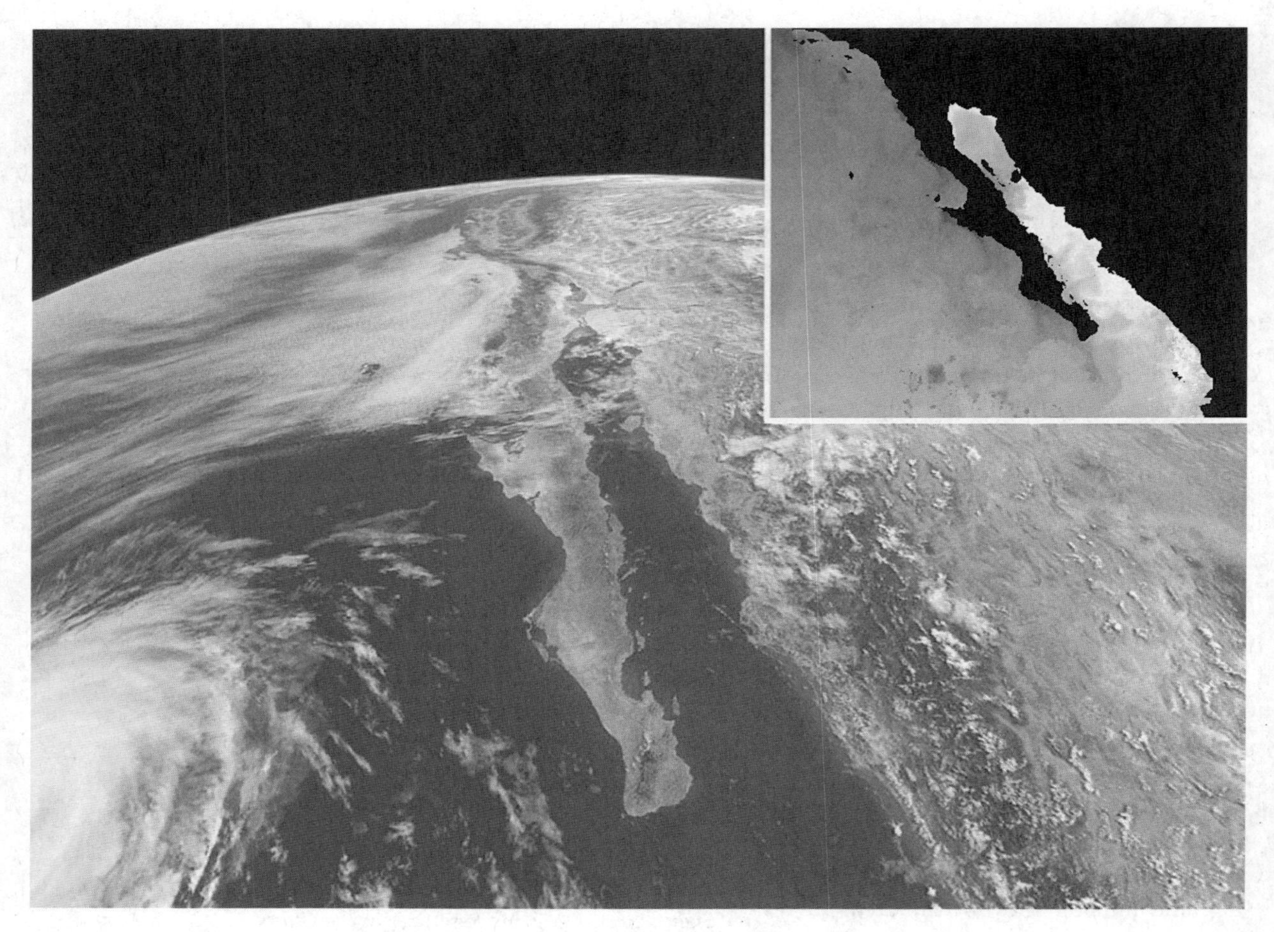

Figure 13.1 Satellite view of the Gulf of California. Coastal fog can be seen in the northern-Pacific side of the peninsula, as the prevailing northwesterly winds cross over the cold California Current. In the southern-Pacific side, a typical fall hurricane storm is seen forming. These storms often enter through the mouth of the gulf and bring torrential autumnal storms to the region. In the insert, a sea surface temperature scan is shown for the same month; lighter colors indicate warmer temperatures. The thermal difference between the gulf and the Pacific coastal upwelling is clearly visible, as well as the tide-induced upwelling in the gulf's Midriff Islands region. Derived from NASA satellite images 2007.

After the disappearance of the missions, the discovery of pearls led to the exploitation of one of the most valuable resources and export items the gulf had to offer. The largest pearl beds in the world were located on the east coast of Baja California Sur, and in the nineteenth century the city of La Paz was recognized worldwide for its pearls (Cariño 1996). Nevertheless, by 1892 pearl diving had declined significantly, possibly as a result of overharvesting (Cariño and Monteforte 1999).

Until the early twentieth century, fishing was only practiced in inshore areas, with small vessels powered by oars and sail and using lines and hooks as the main fishing gear (Robles and Carvajal 2001). A new era in the history of fishing came in the 1930s when outboard motors and gill nets came into use. Inshore fisheries in estuaries and lagoons started to increase steadily. Valuable species began to decrease, like the *totoaba*—an endemic fish to the Sea of Cortés and highly appreciated for export—that suffered significant population declines in the region (Bahre et al. 2000). In 1933, the shrimp fishery began to operate trawler boats over soft seabeds. Since then, sandy seabeds in the Gulf of California are being swept by shrimp trawls every year, with a significant impact to the ecosystem through the bycatch of other species—fish, octopus, conchs, sponges, and sea stars. As the twentieth century progressed, the shrimp fishery became one of the most important activities within the fishing sector. In 1997, the five states surrounding the Gulf of California produced 57,000 tons of shrimp, representing approximately 70% of the national shrimp production and 90% of the Pacific landings (Robles et al. 1999).

Parallel to changes seen in marine and coastal habitats, river ecosystems were also being subjected to changes caused by human activities. The Colorado River was until the 1930s the largest river flowing into the Gulf of California, with its vast delta covering 300 km^2 of wetlands (Sykes 1937; Fradkin 1984; Ezcurra et al. 1988). The dense cottonwood and mesquite forests on the riverbanks were a major source of charcoal and lumber, allowing the development of steamboat traffic on the Colorado River from the Sea of Cortés into the Yuma trail during the nineteenth and early twentieth centuries. In 1935, however, the Colorado River was dammed after an international water treaty was signed between the United States and Mexico, triggering the development of the Imperial and the Mexicali agricultural valleys but bringing the demise of the great delta, which dried up in less than a decade (Bergman 2002; Arias et al. 2004 and references therein). Similarly, in the 1940s dams and channel works were initiated in the Fuerte, Mayo, and Yaqui rivers, and farming was encouraged through government subsidies, such as cheap electrical tariffs for water pumping, and low-interest credit. These irrigation projects resulted in reduction of the freshwater supply to river estuaries and coastal lagoons, causing degradation and loss of native riparian ecosystems (Robles et al. 1999).

By the 1950s, fishing had become a major driving force of regional development (Hernández and Kempton 2003; Sala et al. 2004). The success of this industry attracted financial resources and led to the establishment of freezing and packing companies, as well as shipyards. This period was characterized by an apparent inexhaustible abundance of marine resources. In reality, however, the decline of the fishing sector was already under way, driven by unsustainable harvests, overcapitalization of the fishing fleet, and wasteful fishing technologies (Robles and Carvajal 2001). In the 1960s, a purse seine fishery for sardines developed in the region. This technology, a particularly selective fishing gear, is still in use today. The Monterey sardine fishery is the most productive one with a mean of approximately 200,000 metric tons per year (DOF 2004a). The yellowfin tuna fishery also developed during this time, focusing its efforts in the mouth of the gulf (Torres-Orozco et al. 2006). In the 1980s, the Humboldt squid fishery began and is currently the second most important fishery in the region in terms of biomass, with annual catches of 50,000–120,000 metric tons between 1996 and 2001 (de la Cruz-González et al. 2007).

Socioeconomic Dynamics

The gulf region is a large, sparsely populated area with human densities of only one-third of Mexico's national average. Indicators of economic development (education, housing, and human fertility) suggest a relatively high level of economic development compared with the rest of Mexico (Ezcurra 2003). It is also a relatively wealthy region, with the per capita contribution to the country's GDP being 5% above the national average. This productive advantage is even higher in the Baja California peninsula and the State of Sonora, where the per capita income is about 22% higher than the national average. In particular, the region is a major contributor to the national fisheries, producing approximately 50% of the landings and 70% of the value of all fisheries in Mexico (Robadue 2002; Enríquez-Andrade et al. 2005).

In spite of low fertility rates, the success of the regional economy has brought a large demographic increase, chiefly derived from immigration. While the demographic growth rate in Mexico has declined considerably in the last decades, from a national average of more than 3% to less than 2%, growth rates around the gulf—and especially in the Baja California peninsula—still remain high. The cities with the most dynamic and active economies grew especially rapidly: Tijuana, fueled by the immigration magnet of the maquiladora industry, grew at a rate of 6.5%, while the population of Los Cabos, under the impulse of a tourism boom, grew at the extraordinary rate of 9.7% (López-López et al. 2006). There is, however, considerable differentiation among the five states, and only some specific zones seem to be driving the region's economic growth. Furthermore, most of the regional population is concentrated in the coastal zone, where growth is fastest and where pressure on natural resources is greatest.

Until a few decades ago, the economy of the gulf region was mainly based on agriculture, fisheries, and mining. In fisheries, overcapitalization of the shrimp-trawling fleet and other fishing fleets that share limited marine resources has caused profits to plummet (Cisneros-Mata 2004). Apart from a series of studies that have shown marked signs of overfishing in the gulf, it is interesting to note that, according to Mexico's own *Carta Nacional Pesquera* published by the National Fisheries Commission, all of the important fisheries in the gulf have been showing a downward trend in catches and profits during the last three to four decades. This has resulted in the classification of all large fisheries, with the exception of the Humboldt squid, as overfished (DOF 2004a).

Recently, there has been a major shift in the economic structure. Macroeconomic policy changes, trade agreements, and globalization have created new opportunities in export-oriented manufacturing and in services such as tourism. This economic environment results in increasing demands for natural resources derived from population growth. In recent years, the agriculture, livestock, and fisheries sector has been by far the one showing the lowest growth rates. While the annual growth rate of this sector between 1980 and 2000 was 0.7%, the rest of the regional economy grew at an average combined rate of 4.7%. Furthermore, official statistics published in the last 5 years clearly show that most of the aquifers in Mexico's northwest are "overdrafted," and in some cases rapidly dwindling (DOF 2003), while fisheries have reached a "maximum sustainable threshold" (DOF 2004a). In essence, resource management government agencies have publicly accepted the collapse of the resources they administer, making it clear that further economic growth has reached a limit. This overexploitation of resources may also explain why the areas within the region showing a high dependence on natural resource use are lagging behind in economic development.

As natural resources reach an exploitation limit and new economic opportunities stimulate demand for alternative uses of these resources and the ecosystem services they provide, new conflicts have arisen. This is especially true in the case of common resources, which lack clear property rights and enforceable regulations for access. For example, sportfishing operators argue with industrial fishing boats over the impact of longlines on billfish; artisanal fishers quarrel with shrimp trawlers over access to fishing grounds (Meltzer and Chang 2006), and open-sea shrimp fishers have criticized the impact of coastal shrimp farms on postlarval recruitment. The current clash between sectors contrasts with the regulated access and collaborative atmosphere that sound ecosystem-based management demands at a regional level.

In short, extractive use of natural resources seems to have reached its limit, while the rapid growth of the manufacturing and service sectors is placing additional strain on regional resources. Rapid demographic growth results in increasing pressures on resources such as water, which is preciously scarce in the arid coasts of the gulf, and in increasing amounts of pollutants generated by unbridled growth. Thus, this population growth comes at the expense of depleting underground aquifers and damaging the natural ecosystems and watersheds that surround large urban conglomerates. It is extremely difficult to meet the demand for services such as running water and sewage in cities that double in size every 6 to 10 years.

Ecosystem Resilience and Natural Resource Management

A number of efforts within the last 15 years have underscored the need to understand the gulf as a single large marine ecosystem (LME). The gulf is one of the world's sixty-four LMEs, relatively large regions characterized by distinct bathymetry, hydrography (tides, currents, and physical conditions of the water), biological productivity, and trophically linked populations (Sherman 1994). A global effort by the United Nations' Intergovernmental Oceanographic Commission, the US National Oceanic and Atmospheric Administration, and the International Union for Conservation of Nature (IUCN) is currently under way to improve the long-term sustainability of resources within LMEs and their associated watersheds, with a particular focus on ecosystem-based approaches for fisheries. Several efforts by regional NGOs and scientists (including the Coalition for the Sustainability of the Gulf of California) have articulated conservation priorities for the region as a whole (Enríquez-Andrade et al. 2005). More recently, following intensive public consultation, the Mexican federal government developed a territorial use plan (*Ordenamiento Ecológico*) for the gulf using the LME approach (DOF 2006a, b).

Resilience, State Shifts, and Thresholds

Resilience science may provide a useful context in which to understand this marine ecosystem from a regional perspective and offer the conceptual framework needed to understand current ecological and social changes and implications for the future of the region. The concept of resilience highlights the nonlinear dynamics observed in many ecosystems, as well as their capacity to shift between multiple possible states (Leslie and Kinzig, chap. 4 of this volume). Under a relatively low degree of perturbation, a system will tend to regress to its original state, but when the perturbation is strong, many ecosystems will shift to new stability domains instead of reverting to their original states (Walker and Salt 2006). Transitions among stable states have been described for many ecosystem components in the Gulf of California, and understanding the causes of these shifts may help inform future management actions.

Perhaps the clearest example of a transition toward a new steady state is the shrimp fishery. It is well known that shrimp trawling severely damages the seafloor (Young and Romero 1979; Dayton et al. 2002). The report presented by Steinbeck and Ricketts (1941) predicted that "a very short time will see the end of the shrimp industry in Mexico," due to bottom trawling. Although the fishery has declined, as described in Mexico's own *Carta Nacional Pesquera*, the predicted collapse has not occurred as once imagined. Rather, it seems that the repeated combing of the seafloor by the trawlers has shifted the original ecosystem to a new steady state, similar in some ways to the dynamics of a tropical forest when converted to pasture. Unfortunately, there are no good descriptions of the original system, and understanding of the state shifts has been derived primarily from indirect evidence, such as interviews with fishermen and local residents, and a systematic description of the shift in baselines in fisheries descriptions from old and young fishermen in the gulf (Sáenz-Arroyo et al. 2005a, b, 2006).

In other fisheries, transitions between ecosystem states and the subsequent failure to return to the original system also have been documented. Velarde and colleagues (2004), for example, found two readily distinguishable states in the gulf's sardine fishery. Prior to 1989, there was a clear linear relationship between fishing effort and sardine landings, with fishing effort growing exponentially. In 1989, the sardine fishery collapsed, forcing the reorganization of the fleet to decrease their effort in order to allow the stocks to recover. However, the system did not regress to its previous domain, but rather persisted in a new state in which stocks are lower, and catch is mostly regulated by oceanographic conditions (fig. 13.2). The sardine fishery shifted from a mature phase (pre-1989, where landings were basically predicted by fishing effort), to a disturbance phase (1989–1990, when the fishery collapsed), and then into a reorganization phase in which the system adjusted to the new high-extraction conditions and likely stabilized in a new domain (mirroring the adaptive cycle, as explained by Walker and Salt 2006). The disappearance of top predators, such as that documented in the gulf by Sala and colleagues (2004) due to unsustainable fishing practices, may be destabilizing the entire system and making it more vulnerable to these types of irreversible changes.

Together with fisheries, quite possibly the most rapidly changing components of the gulf ecosystems are the coastal lagoons, especially the mangrove forests. These habitats are rapidly disappearing under the growing pressures of coastal development, mostly for tourism infrastructure and shrimp farms (fig. 13.3; Páez-Osuna 2001; Páez-Osuna et al. 2003; Whitmore et al. 2005). Mexican regulations state that mangrove clear-cutting can be authorized, provided that the proponent establishes "compensation measures" (DOF 2004b; recent additions to the Wildlife Law have made mangrove clear-cutting illegal, and the issue is now under debate in the Mexican Congress). In practice, compensation measures are ineffective because mangrove destruction also happens when developers fill lagoon swamps in order to gain ground for developments or to construct aquaculture ponds, resulting in a true and irreversible loss of wetland habitat (Wolanski et al. 2000).

Even when mangrove clearing is done without destroying the mudflat, the cutting of woody plants may move the mudflat ecosystem to a new steady state, from which it may not recover. Although no direct experiments have been done in the Gulf of California, experiments with black mangrove (*Avicennia germinans*) in similar mudflats in the Gulf of Mexico (López-Portillo and Ezcurra 1989 and fig. 13.4) have clearly shown that reducing mangrove cover in the mudflat results in a "catastrophic" (sensu Scheffer et al. 2001 and Scheffer and Carpenter 2003) and often irreversible shift in ecosystem states. As with the sardine fishery described above, the clearing of mangrove forests moves the lagoon ecosystem from a mature phase (the dense mangrove forest) to a disturbance phase (the aquaculture pond, or simply the barren mudflat). When the disturbed system is finally left alone, it progresses toward a new reorganization phase (the mudflat dominated by *Batis maritima* and other low-cover, salt marsh plants).

In conclusion, a number of ecosystem components within the Sea of Cortés seem to have reached an irreversible transition boundary, or a "tipping point," in the last decade.

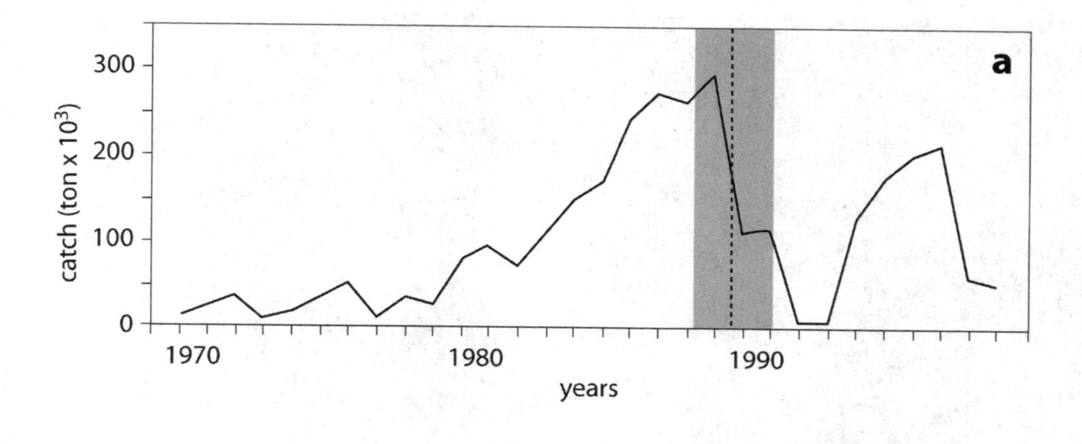

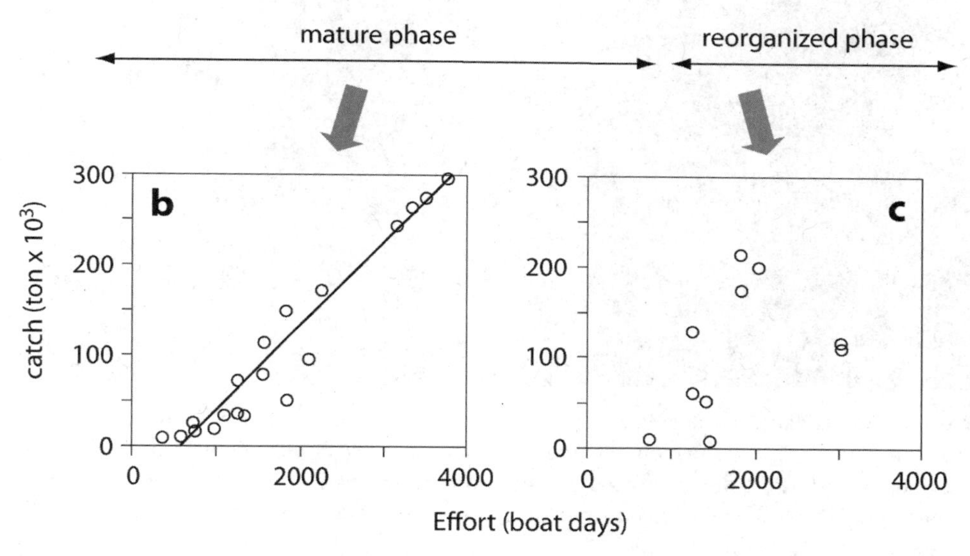

Figure 13.2 Transition between ecosystem states in the Gulf of California. (A) Total sardine catch in the Gulf of California between 1970 and 1999. (B) Linear relationship between fishing effort and catch from 1970 to 1989. (C) Nonsignificant relationship between effort and catch from 1990 to 1999. The gray area indicates the transition period. Modified with permission from Velarde et al. 2004.

Understanding which of these ecological changes are reversible and which are not, and what needs to be done to return critical ecosystem components to a functional state, is among the biggest challenges for sustainability science in the region. With these challenges in mind, we will analyze some of the most conspicuous obstacles that face environmental sustainability in the region.

The Challenges for Environmental Sustainability

While the extractive use of natural resources in the gulf seems to have reached a limit, the rapid growth of the manufacturing and services sectors is putting additional strains on the limited supply of natural resources. However, the economic shift could also open new doors for

Figure 13.3 Shrimp farms advancing on the coastal wetlands of the Yaqui River, near Ciudad Obregón, in Sonora. The occupation of mangrove swamps by new developments is noticeable, together with lines showing the dredging of the swamp's natural channels and the opening of canals that drain agricultural effluents directly into the gulf's waters. Used with permission from Google Earth 2007.

a drastic change toward sustainable resource use. The growth of these newer sectors creates an important opportunity to reduce pressure on natural resources. As a result of growing concerns regarding sustainable use of natural resources and the impacts of tourism, a few environmentally concerned business leaders have formed a group called *Noroeste Sustentable* (NOS). In short, although the gulf is undergoing extreme pressures from overexploitation in many parts, with the consequent collapse of some of its resources, it also harbors a number of successful and encouraging examples of communities that are trying to maintain their resources in a healthy and productive state, with an eye to the future.

Fisheries

With thirty-nine species listed on IUCN's Red List, it is clear that ecological degradation has already hit the gulf's biodiversity hard. Populations of five species of sea turtles have declined dramatically in the gulf (Nichols 2003; Seminoff et al. 2003). The endemic vaquita porpoise (*Phocoena sinus*) is near extinction. The

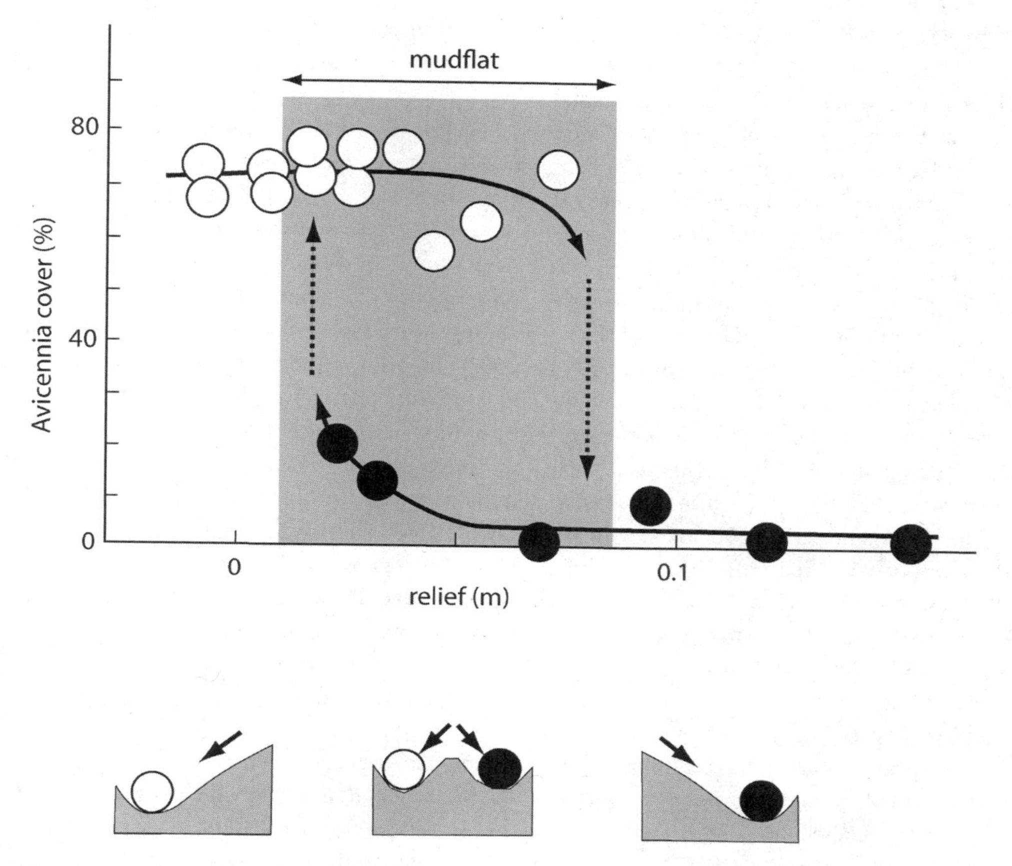

Figure 13.4 Catastrophic regime shift in mangrove ecosystems: In the lower parts of the relief gradient, frequent flooding maintains the growth of mangrove forests (open circles). In the upper hinterland, the lack of floodwater and the salinity of the soil maintain a halophyte scrub with no mangroves (black circles). In the mudflat, the system can maintain either a mangrove forest (in its mature phase) or, if cleared, a halophyte scrub (in its reorganized phase). Intermediate states are rarely seen, because once the system is transformed into a halophyte scrub, mangrove seedlings have difficulty establishing in the sun-scorched, salt-encrusted soil. Modified from López-Portillo and Ezcurra 1989.

International Committee for Vaquita Recovery estimated the current vaquita population is 150 individuals (Rojas-Bracho et al. 2006), contrasting with a previous estimate of 567 vaquitas for 1997. This new estimate takes into consideration an annual growth rate of 4% and the loss of 39 to 78 individuals per year (D'Agrosa et al. 2000). These values make the vaquita the most endangered marine cetacean in the world.

Overharvesting of fish stocks is rapidly becoming a strongly limiting factor for the success of the regional fisheries. Twenty years ago, there was a correlation between catch and effort in many of the regional fisheries: The more days the fleets operated, the more catch they brought in. Now, that correlation is largely gone, the total landings in most fisheries are chiefly independent of fishing effort, and catch per unit effort has decreased severely for many species (Sala et al. 2004; Velarde et al. 2004). Additionally, there is clear evidence that coastal food webs in the Gulf of California have been "fished down" during the last 30 years (i.e., fisheries shifted from large, long-lived species at high trophic levels to smaller, short-lived species from lower trophic levels), and the maximum length of individuals has significantly decreased in only 30 years (Sala et al. 2004).

Unfortunately, the industrial shrimp fleet is an apt example of the tragic decline of common-pool resources. Thirty years ago, the fleet was made up of approximately 700 boats, each capturing about 50 tons of shrimp per season. Today, the fleet is nearing 1,500 boats, with annual catches barely surpassing 10 tons per boat (Meltzer and Chang 2006; de la Cruz-González 2007). Despite governmental subsidies of nearly US$ 30 million each year—provided in the form of cheap fuel—many boats are facing economic collapse.

The environmental impact of this fishery is also a matter of concern. Bottom trawlers destroy some 200,000 tons of bycatch every year for a meager annual catch of 30,000 tons of shrimp (Madrid-Vera et al. 2007). While working, dragnets also impact approximately 30,000–60,000 km^2 of the seafloor (García-Caudillo et al. 2000; Brusca and Bryner 2004); some of this area is within the Upper Gulf Biosphere Reserve. The boats also collectively emit about 30,000–40,000 tons of greenhouse gases derived from the cheap subsidized fuel that keeps the already inefficient business running. Benthic habitats have been so depleted in some places that local artisanal fishermen in places such as Loreto Bay and Bahía de los Ángeles demanded the establishment of no-take zones and other marine protected areas. In often open conflict, local communities and large industrial fleets debate over the establishment of protected areas and the demand for permits that allow trawling inside some of these areas. In the gulf, conflict among sectors and the battle over particular interests have been increasingly the rule at sea (Meltzer and Chang 2006).

Fortunately, there are success stories to be told from the gulf's fisheries, and learning from them is fundamental for future conservation efforts. For example, artisanal fishermen have started to work with local researchers to understand the phenomenon of spawning aggregations (Sala et al. 2003; Erisman et al 2007), in order to identify and protect areas that are important for reproduction. As a result of pressure from local resource users, the Bay of Loreto marine park now includes two small no-take areas, and the fishermen of Bahía de Los Ángeles supported the movement to declare the area a biosphere reserve. In the San Ignacio Lagoon, fishermen have organized to preserve the environment, train community members in basic natural history knowledge, and organize whale-watching tours (Cariño et al. 2006).

The abalone and lobster cooperatives of the Pacific coast of Baja provide yet another example of long-term sustainable use (Hilborn et al. 2005). With no support from the federal government, fishermen have established strict rules for resource extraction and have developed their own enforcement program. Many generate their own electricity, run their own canneries, and finance their own schools. More than 40 years after the establishment of these communities, their productivity is still high, and their resources seem to be in fairly good shape (Chaffee et al. 2004).

But small communities and conservationists are not the only ones critical of some of the region's unsustainable modes of development. A growing number of entrepreneurs and businesspeople are also becoming committed supporters of environmental causes. Even large fishing fleets can be sustainable when their operators work in cooperation: In contrast with the decline of captures of the shrimp bottom-trawling fleet, the sardine fishery has controlled its fishing effort, and—after a past collapse—the fishery has partially recovered and is now managed sustainably (Lluch-Belda et al. 1986; Cisneros-Mata et al.1995b, 1996).

Coastal Tourism and Recreational Activities

The magnificent landscapes and the amazing density of charismatic wildlife make the Gulf of California a superb place for visitors (Tershy et al. 1999; López-Espinosa de los Monteros 2002). The first tourists in the gulf were attracted by the extraordinary catches sportfishing had to offer. Nautical tourism developed afterward, and soon the idea of connecting the region with the states of California, Oregon, and Washington in the Unites States through a series of marinas attracted developers. During President Vicente Fox's administration, this idea materialized in the form of a regional project, the *Escalera Náutica* or Nautical Stairway, with the goal of jump-starting nautical tourism in the gulf.

Over the past few years, *Escalera Náutica* has become one of the most debated projects in the region (Bowen 2004; Álvarez-Castañeda et al. 2006). Because the economy seems to be shifting from fisheries and agriculture toward the service sector (including tourism), this project seemed, in principle, a desirable path. However, experiences with unsustainable tourism in Mexico (and its sequel of failed and abandoned projects like dredged mangrove swamps and exhausted aquifers) have left a deep scar in the perception of local communities. The big challenge in the Gulf of California is to promote sustainable tourism while ensuring the preservation of the region's natural beauty and biodiversity, the very attributes that initially triggered tourism in the gulf.

ALCOSTA (*Alianza para la Sustentabilidad del Noroeste Costero Mexicano*, an alliance of several environmental organizations), among other players, has been instrumental in raising awareness regarding the *Escalera Náutica* development plan. They have spearheaded efforts to reduce the environmental impacts of the project and to make it more open to environmental conservation issues. Through public consultation, ALCOSTA developed a critical analysis of the regional environmental impact assessment for the *Escalera* project and presented it in a public hearing to federal authorities. All the general conditions proposed by ALCOSTA for the environmental authorization of the project were taken into account by the federal government in the final resolution and led President Fox to announce that the Gulf of California is a joint priority for both tourism and conservation. Hence, the *Ordenamiento Ecológico Marino* program was initiated, and a promise to enlarge the protected areas surrounding the gulf's islands to include surrounding waters was made. This occurred on World Environment Day 2004, with the signature of a coordination agreement between six ministerial secretaries and five state governments (Baja California, Baja California Sur, Nayarit, Sinaloa, and Sonora).

Recently, the state governments have adopted aggressive plans to promote tourism development by creating infrastructure. They see tourism as an opportunity to create much-needed jobs, foster economic development, and compensate for the job losses in other sectors while moving away from the growing subsidies demanded by farmers and fishermen. The question, however, remains how to promote sustainable tourism while ensuring the preservation of the region.

Case Studies of Ecosystem-Based Conservation Measures

If effective conservation in the region is to be achieved, a strategy that addresses the most critical environmental challenges needs to be developed. Such a strategy should allow for the protection of seriously endangered species, spawning aggregation areas, and endangered ecosystems such as seamounts, coastal lagoons, coral reefs, estuaries, and marine mammal habitats (Sala et al. 2002, 2003). Expansion or establishment of new protected areas is one of several approaches needed to ensure long-lasting conservation efforts (Hyun 2005); but most importantly, they require the support of the local communities and stakeholders (Aburto-Oropeza and López Sagástegui 2006).

For other areas such as coastal wetlands, the development of comprehensive plans is needed to manage and protect them effectively. The degradation of coastal wetlands is one of the gulf's most serious threats (Brusca et al. 2006), and there is little consideration for the fundamental ecosystem services they provide. Mangrove forests are being destroyed to give way to aquaculture (mostly shrimp farms) and tourism projects. Furthermore, coastal wetlands are also threatened by consumptive water use upstream and by pollution of rivers and waterways. The ecosystem services provided by estuaries and lagoons are critical for the survival of the Sea of Cortés fisheries (Aburto-Oropeza et al. 2007) and for the health of the large marine ecosystem as a whole.

In this section we offer three case studies of how ecosystem-based management (EBM) is developing in the gulf region. Each captures an ongoing initiative and thus, like the other case studies in this section of the book, provides a glimpse of both the present and possible future of marine EBM in a particular ecological and social context.

Case Study 1: The Alto Golfo Biosphere Reserve

There are myriad stories of dedicated work and heated debates around each one of the protected areas in the Gulf of California, the conflicts behind their creation, and ongoing discussions about their future resource use (Ezcurra et al. 2002). Perhaps the most emblematic of these cases is the *Alto Golfo de California y Delta del Río Colorado* in the northernmost portion of the gulf, over which discussions and debate have been, and continue to be, especially heated. Understanding the past and ongoing conflicts in the upper gulf is essential to understanding the conservation movement, including efforts to implement marine EBM, in Mexico.

The Upper Gulf of California and Colorado River Delta Biosphere Reserve is formed by part of the surrounding Sonoran Desert, the northern marine waters of the Gulf of California, and the lowermost part of the Colorado River. Its high marine biological productivity is a result of the churning of nutrients in Colorado River sediment deposits by one of the strongest tidal fluxes on the planet (Santamaría del Ángel et al. 1994; Thomson et al. 2000). This productivity makes the area very important for reproduction, nursery, and growth of many resident and migratory species (Glenn et al. 2001; Calderón-Aguilera et al. 2003; Rowell et al. 2005). Currently, the total number of marine species recorded for the reserve is 1,438, of which 11 are in danger of extinction, notably the vaquita and totoaba, both endemic to the northern part of the gulf (as discussed previously in this chapter).

The upper gulf's marine richness is reflected in its highly valuable fisheries, especially shrimp, which make this region one of the most important fishing grounds in Mexico (Brusca and Bryner 2004). Historically, the most significant economic activity for the reserve's inhabitants has been gill net and trawl fishing (McGuire and Greenberg 1993). In 1955, the Mexican fishery authority declared the region to be protected as a breeding site and nursery for birds and fish. As years passed, the region was still being subjected to an ever-growing fishing pressure. By the early 1970s, the totoaba was facing extinction due to overfishing (Lercari and Chávez 2007), forcing the federal government to decree a moratorium for totoaba harvest in the Sea of Cortés. The area was re-decreed in 1974 as a reserve zone for fisheries resource restocking. However, the depletion of the totoaba population continued catastrophically, and in 1975 the Ministry of Fisheries established a permanent ban for totoaba captures, which remains in effect today. Incidental capture of the vaquita in gill nets resulted in strong concern over its population status in the upper gulf in the 1980s, and by the early 1990s, its population was estimated to be less than five hundred (D'Agrosa et al. 2000) and it was declared endangered. At this point, Mexican federal government created the Technical Committee for the Protection of the Totoaba and the Vaquita to recommend strategies for conserving both endangered species (Rojas-Bracho et al. 2006). While some members of the committee favored immediate action to protect the upper Gulf of California from the effects of overfishing, others were of the opinion that regulating fisheries would harm the local economy. This conflict resulted in a request that research centers in Sonora develop a feasibility study for a biosphere reserve.

That study recommended establishing a marine protected area in the upper gulf (CTPTV 1993). Discussions with the local communities (Golfo de Santa Clara, Puerto Peñasco, and San Felipe, as well as the *ejidos* in the delta of the Colorado River) about the costs and benefits of the protected area took place during the first months of 1993. With the support of local businesspeople, scientists, conservationists, social leaders from the small-scale fisheries, and traditional authorities from the indigenous peoples around the Sea of Cortés, the project was presented to the Mexican government. A key element in this process was the participation of Luis Donaldo Colosio, an important politician and native of Northern Sonora, who enthusiastically supported the project.

On June 10th, 1993, the President of Mexico, Carlos Salinas de Gortari, decreed the establishment of the Biosphere Reserve of the Upper Gulf of California and Delta of the Colorado River. Many important decision makers attended the ceremony, including many cabinet members from the Mexican federal government; the governors of Sonora, Baja California, and Arizona; US Secretary of the Interior Bruce Babbitt; and the traditional governor of the Tohono O'Odham (Papago) people, whose lands extend to both sides of the Mexico–United States border. This reserve was the first marine protected area established in Mexico and included the territories of Baja California and Sonora as well as federal marine waters (INE 1995). Thus, coordination among these entities became a critically important factor to fulfill the protected area's objectives.

In 1993, the upper gulf was undergoing a socioeconomic crisis, which led to the acceptance of the protected area as a temporary solution (McGuire and Greenberg 1993). After a few years, the initial enthusiasm waned. The administration of protected areas in Mexico was at that time very small, and most of the reserves existed only on paper. There was little governmental field experience and few resources for conservation and management of protected areas, and PROFEPA, the federal authority in charge of environmental enforcement, had just been created and did not have the capacity to operate in remote areas. It would seem now that neither the federal government nor the local communities were prepared for the long-term commitment that the establishment of this reserve required. However, the creation of the National Commission for Natural Protected Areas (CONANP) in 1993 gave the reserve core funding for field operations, a director, active field staff, and a management plan.

Critical obstacles to achieving the reserve's objectives have been poor intergovernmental coordination and conflict among sectors (particularly fisheries and agriculture), poor institutional capacity, and a lack of political will to enforce the law. Thus, illegal fishing inside the reserve has grown, resulting in increased vaquita mortality. Recognizing the desperate need to protect the species, in September 2005 the Mexican government added an additional 155,500 ha protected marine polygon to the upper gulf biosphere reserve as a refuge area for the vaquita. This additional protected area covers 70% of the mammal's core habitat, a geographic range that was not well identified at the time of the establishment of the upper gulf biosphere reserve. Unfortunately, and despite the added protection, the vaquita population seems to be still in critical decline (Rojas-Bracho et al. 2006), among other species that reproduce and live in this unique and fragile area. Movement toward a more ecosystem-based approach would benefit this imperiled mammal as well as the many other species—including humans—who call this region home. But the current outlook for such a shift in perspective is uncertain.

Case Study 2: Community-Managed No-Take Marine Reserves

Community-managed no-take marine reserves have been playing a growing role in the Gulf of California. Although their establishment has been organized locally, their cumulative effect is becoming regionally significant. Two types of no-take marine reserves are present in the gulf: those inside federal marine protected areas (MPAs) and those defined by restrictions in fishing permits. Currently there are eleven MPAs established in the gulf, of which only eight include no-take zones ("core-zones" in Mexican legislation) that cover 93,125 ha, or 1.9% and 4.8% of the total area of the gulf and of the gulf's MPAs, respectively (table 13.1). The levels of full protection vary from 35% in Cabo Pulmo to a negligibly low proportion (0.08%) in Loreto Bay.

Moreover, two fishing cooperatives are in the process of obtaining permits that would facilitate sustainable fisheries and ecosystem-based management in the gulf. These two cooperatives are unique in their characteristics: The Peñasco cooperative (*Union de Buzos de Puerto Peñasco*) won the National Conservation Award in 2003 (Cudney-Bueno 2004), while the Loreto Bay cooperative (*Mujeres del Golfo*) is constituted by eight fisherwomen. The Peñasco cooperative would fish rock scallop (*Spondylus calcifer*) around Puerto Peñasco, outside the southern limit of the upper gulf biosphere reserve, while the Loreto Bay cooperative would collect four aquarium fish species of rocky reefs (*Chromis limbaughi*, *Holocanthus passer*, *Pomacanthus zonipectus*, and *Opistognathus rosenblatti*) in the southern area of the Loreto Bay national park. These cooperatives propose, in addition to the use of the traditional fishing quotas, the implementation of no-take zones as an experimental management instrument. In Peñasco, fishers are closing 800 ha, which represents 20% of the unique coquina rocky reefs of the upper Gulf of California (Cudney-Bueno 2004, 2007). Similarly, the Loreto fisherwomen have closed 30% (24 ha) of the area in five (79 ha) of the thirteen traditional fishing sites (483 ha) for aquarium species.

There is a third type of closure that can be considered as a no-take zone: where transit is prohibited for security reasons. The thermoelectric power plant in Puerto Libertad and the penal complex of the Islas Marías Archipelago are closed to all boat traffic. The latter was decreed a biosphere reserve in 2005. Although the sites have not been monitored, fishers describe high catch in areas close to the islands. For example, sport divers near the power plant describe large aggregations of the gulf grouper (*Mycteroperca jordani*), a formerly abundant species (Sáenz-Arroyo et al. 2005a, b), in the power plant area. And in the case of the Islas Marías Archipelago, many fishers operate illegally in the vicinity of the islands since the high catch warrant taking the risk of being detained.

Comunidad y Biodiversidad (COBI), a nongovernmental organization (NGO) that seeks the conservation of marine and coastal biodiversity through participatory approaches, has promoted community-based no-take zones

Table 13.1. Marine protected areas in the Gulf of California

	Total area (ha)	*Marine zone (ha)*	*No-take zone (ha)*	*Percentage no-take*	*Fully protected ecosystem*
Wildlife protection areas					
Cabo San Lucas	3,996	3,875	3,875	100	Rocky reefs
Biosphere Reserve					
Alto Golfo de California y Delta del Río Colorado	934,756	560,853	80,000	14.2	Sandy bottoms and mudflats
Bahía de los Ángeles, Canales de Ballenas y Salsipuedes	387,957	387,957	207	0.05	Wetlands and mangroves
El Vízcaino	2,546,790	40,451	—	—	—
Islas Marías	641,284	617,257	—	—	—
Isla San Pedro Mártir	30,165	29,876	821	2.6	Rocky reefs
National Parks					
Archipiélago de San Lorenzo	58,442	58,442	8,805	15.0	Rocky reefs
Bahía de Loreto	206,580	181,997	150	0.08	Rocky reefs
Cabo Pulmo	7,111	7,111	2,476	35.5	Coral reefs
Islas Marietas	1,383	1,311	—	—	—
Marine Zone of Isla Espíritu Santo	48,655	48,655	666	1.4	Rocky reefs
Total	4,863,123	1,933,910	93,125	4.8	

Sources: Vargas-Márquez and Escobar 2003; Mexico's *Comisión Nacional de Áreas Naturales Protegidas* online database available at www.conanp.gob.mx/anp/.

since 2000. Based on the experience gathered at a number of sites throughout the gulf, COBI has developed a framework to guide implementation of these zones, involving the fishing community at each step. This approach includes (1) communicating what a no-take marine reserve is, what its benefits are, and the commitments involved; (2) evaluating the acceptance/rejection of the community of use of marine reserves as a management tool; (3) designing a reserve network and developing a work plan to monitor the marine reserves; and (4) disseminating the monitoring results. In order to implement this framework, COBI initiated the Fishers Fund to support the technical, financial, and legal elements of community-based marine reserves (Torre et al. 2005).

Case Study 3: Project Pangas

One way to overcome the daunting task of researching and managing marine ecosystems in a more integrated fashion is to rely on the understanding of proxies indicating wider ecosystem processes and health. In the northern gulf, small-scale fisheries are being used as an operational proxy by PANGAS. Project PANGAS was born to address the complexity of this challenge and the growing need for local ecosystem-based management. In essence, the formula is relatively simple. Healthy and resilient local fisheries imply the existence of healthy coastal marine ecosystems. The project was developed with the dual purpose of (1) building a framework for research and management of small-scale fisheries as a means to address coastal marine ecosystem conservation and (2) initially applying and testing this framework in the northern Gulf of California. Hence its name, PANGAS, which stands for *Pesca Artesanal del Norte del Golfo de California: Ambiente y Sociedad* (Small-Scale Fisheries of the Northern Gulf of California: Society and the Environment). A *panga* is also the name given by many fishers in Latin America to the small boat used in this activity. A *panga*, then, represents a link between fishers and the marine ecosystem they depend on.

In the gulf, over a hundred species are harvested on a regular basis by the small-scale fishery fleet, which is distributed in twenty communities and at least twenty seasonal fishing camps. These species cover all trophic guilds, from top predators such as sharks and groupers, to bottom filter feeders. They are also representative of all environments (i.e., pelagic waters, benthic rocky and sandy areas, rhodolith beds). Small-scale fisheries provide approximately 50% of the world's total fish catch (Berkes et al. 2001), and this figure is likely an underestimate, as much of the worldwide catches are never officially registered. These fisheries also have an intrinsic holistic research value, providing a means to address wider ecosystem functions and health. Although it is small scale, relative to capital investment and production per boat, once added up, the cumulative outcome of this activity is anything but small. Perhaps small-scale fisheries' most important characteristic, and certainly the most relevant for the purpose of this book, is their comprehensiveness. They are not directly or technologically coupled to the extraction of a particular species. Rather, in any given coastal ecosystem where they operate, small-scale fisheries tend to target a slew of species representative of entire ecosystems.

Despite the various conservation efforts, the northern gulf is in a state of conflict between fisheries management and marine conservation, fueled by the growth of coastal human populations and marine resource use. In response to national fisheries policies and conflicts within Mexico's land-based economies, small-scale fisheries have grown considerably in the past two decades and are predicted to continue growing. Today, over three thousand *pangas* operate in the northern gulf (Cudney-Bueno 2000); this represents more than double the size of the small-scale fishing fleet 20 years ago. Considering that each *panga* is capable of carrying 1–2 metric tons, collectively the impact on the marine environment is substantial. Additionally, the *panga* fishery can easily shift as targeted species are depleted or market demands change. With this increase in effort and the flexibility of fleets, the northern gulf has also seen a rapid evolution of institutional change, the development of numerous territorial conflicts over access to fishery resources, and a downtrend in the production of most targeted species.

BUILDING LONG-TERM INSTITUTIONAL RESILIENCE THROUGH PARTICIPATORY RESEARCH AND TRAINING

The basic premise of the project is that the performance of fisheries is the result of interacting biophysical and social processes. PANGAS's research questions aim at obtaining a better understanding of these interactions. Because the northern gulf's small-scale fisheries not only cover a wide range of fishing gear, methods, and species, but also take place in a variety of environments under various social/

institutional settings, a detailed ecosystem-based analysis of all of the region's fisheries would be impossible. PANGAS's research operates at two levels. The first involves a general characterization of small-scale fisheries, primarily in terms of (1) identification and characterization of key habitats, (2) reproductive sites/times of targeted species, (3) spatial–temporal distribution of fishing activities (including gear/methods used and species targeted), and (4) existent governance structure. The second level encompasses in-depth and integrative research of small-scale fishery management units that are representative of the wider spectrum of the region's coastal marine ecosystem. PANGAS uses key commercial species as representative proxies of the region's four main types of fisheries: (1) rocky habitat fishing/commercial diving and line fisheries, (2) sand bottom fishing/longline, (3) sand bottom fishing/traps, and (4) open water fishing/gill nets.

PANGAS also emphasizes the importance of strengthening human capital for long-term research and monitoring. This is carried out via training and participation of fishers in underwater surveying, graduate degrees in interdisciplinary programs, and short training courses.

TRANSLATING RESEARCH INTO MANAGEMENT OUTCOMES

Ecosystem-based research would be futile if not translated to management outcomes. This translation demands an understanding of the main management issues and a vision for improvement that is representative of various interested parties. Interviews are conducted at various levels of governance (from fishers to local, state, and federal government officials) in order to obtain a broad understanding of the decision-making process affecting the region. These interviews also provide an understanding of the prevalent management concerns and recommendations for improvement. Parallel to this research, PANGAS is developing partnerships with key stakeholders from the fishing industry as well as with governing bodies affecting the Gulf of California. Thus, by the time management recommendations are made, they will address, as much as possible, the various concerns and recommendations of various groups.

Given Mexico's current administration and legal structure affecting marine resource use, PANGAS's management guidelines will likely fall into two main categories: (1) a combination of regional management plans for key targeted species (under the jurisdiction of the National Fisheries Commission) and/or (2) establishment of marine protected areas (endorsed by the Secretariat of the Environment). Regardless of the specific recommendation, PANGAS will first present their results through appropriate local governance structures. Depending on the regional scope of the recommendations, these institutions may be local community-based fishing groups or associations, cooperatives, federations of cooperatives, and regional councils. These initiatives will then be conveyed to the government via existing governing bodies, such as councils, the advisory group of the National Fisheries Commission, and the House of Representatives, and/or through direct meetings with leading government officials. Finally, assuming implementation of the recommendations, they will be followed by long-term monitoring that in turn contributes to an adaptive management framework that is used to assess progress and revise recommendations accordingly.

Toward a Regional Ecosystem-Based Conservation Agenda

The case studies above lead to the question, How are these local efforts contributing to ecosystem-based management of the gulf ecosystem as a whole? We see one of the primary means to be through inspiration of a regional vision, which in turn encourages more local-level action. Regional cooperation among non-governmental institutions striving to develop an ecosystem-based vision of the gulf has been attempted in the past. In December 1997, a group of scientists and conservationists convened the Coalition for the Sustainability of the Gulf of California and, after 3 years, produced a comprehensive map defining conservation priorities in the region. The conservation-priority maps produced by the coalition were critically important inputs in the governmental land use and ocean use plans for the Sea of Cortés and the surrounding coasts, and they also became the basis for other regional planning exercises (Enríquez-Andrade et al. 2005; Aburto-Oropeza and López Sagástegui 2006, and references therein).

One of the most noticeable results of this collaborative effort between regional NGOs and research groups was the presentation of a regional agenda at the Defying Ocean's End meeting in Los Cabos in 2003, where seven specific objectives were articulated in order to advance sustainability efforts in the gulf on the basis of a large-scale, ecosystem-based approach (Carvajal et al. 2004):

1. *Improve the management of regional marine and coastal protected areas.* Although impressive progress was attained in the 1990s by the Mexican government in the funding and management of its protected natural areas, many of them still subsist as "paper parks," with inadequate funding and little effective management. If the regional protected areas are to be effective in their conservation goals, they must improve in their level of funding, equipment, and staffing. In part as a result of this initiative, the Mexican Commission for Protected Natural Areas has increased its budget substantially since then for marine and coastal protected areas in the gulf.
2. *Enlarge the system of marine and coastal protected areas.* Although some marine protected areas have been created in the gulf, these cover less than 4% of the gulf's marine area. If effective conservation in the region is to be achieved, a significant increase in the marine protected areas must be obtained, reaching at least 15% of the gulf's surface. This would allow the protection of spawning aggregation areas and critically endangered ecosystems such as seamounts, coastal lagoons, coral and rocky reefs, estuaries, and marine mammal habitats (Sala et al. 2002). The regional agenda led to the creation, in 2007, of the Bahía de los Ángeles Biosphere Reserve and of the Espíritu Santo Marine Park.
3. *Develop a comprehensive plan to manage and protect priority coastal wetlands.* The degradation of coastal wetlands is one of the gulf's most serious threats. With little consideration to the ecosystem services they provide, mangrove forests are being cut for the development of aquaculture (mostly shrimp farms) and tourism projects. Furthermore, coastal wetlands in general are threatened by consumptive water use upstream and by pollution of rivers and waterways. The ecosystem services provided by estuaries and lagoons are critical for the survival of the Sea of Cortés fisheries and for the health of the large marine ecosystem as a whole. Thanks to this demand, modifications were introduced into the Mexican Wildlife Law to prevent mangrove clear-cutting, and discussions are currently ongoing in the Mexican Congress on this very important issue.
4. *Reduce the shrimp-trawling fleet and improve its fishing technology.* Many of the

strongest issues of unsustainabilty in the gulf stem from the destructive effect and the economic inefficiency of the current shrimp bottom-trawling fleet. The only alternative that will solve this growing problem is to reduce the fleet by at least 50% through a legal buyout. If effective legal means are put in place to ensure that no new fishing permits will be issued in the future—and hence that the fleet will not grow again to unsustainable levels—this will allow the negotiation of effective enforcement of the existing no-take zones, and of the introduction of better fishing gear with more-efficient excluder devices.

5. *Develop a regional plan regulating the use of land, coasts, and waters.*The main instrument in the Mexican legislation to regulate the use of space within environmental guidelines is the *Ordenamiento Ecológico*, or Ecological Planning of the Territory, which demands full and comprehensive hearings and negotiations with local governments, local businesses, and nongovernmental organizations. Because of its complexity, effective territorial planning has been difficult to achieve in the Sea of Cortés, and it is now one of the most urgent objectives to reach, with the full participation of civil society and local conservation alliances. In 2005, the Secretariat of the Environment started the process of the *Ordenamiento Ecológico* for the gulf, which was published in its general guidelines in late 2006.
6. *Reorient regional tourism toward low-impact, environmentally sustainable resource uses.* The *Escalera Náutica*, now renamed *Proyecto Mar de Cortés*, has become one of the most debated projects in the region. Most environmentalists agree that a shift in the economy from unsustainable fisheries and water-intensive agriculture into the services sector (including tourism) seems a desirable move, but the question remains of how to make the new players in the regional economy sustainable and compatible with resource conservation. Regional cooperation among nongovernmental agencies has proved critically important for this purpose. ALCOSTA, a regional alliance of NGOs, was capable of bringing a voice of alarm and concern into the *Escalera Náutica* development project, thereby transforming the initiative forever by making it much more open to environmental conservation issues.
7. *Articulate a common regional development vision.* The last point of the agenda, the development of a regional vision, is possibly the most crucial aspect in the gulf's conservation agenda. As a group, conservationists desperately need to transcend the image of negative activists and move toward a joint way of seeing the region that will enable proposals of new and sustainable modes of development, rather than attacking unsustainable alternatives. Regional conservation will be successful if, in collaboration with local business and political leaders, a regional development vision based on the long-term protection of the gulf and its resources can be pieced together collectively, and agreed upon.

This seven-point agenda parallels many of the ecosystem-based management elements highlighted in other chapters of this book. Together with the rich technical information generated by the Coalition for the Sustainability of the Gulf of California, the agenda is now being used by a group of the gulf's key stakeholders who call themselves *Noroeste Sustentable*, or NOS. Its members believe that in order to facilitate well-planned sustainable development with a regional vision, it is important to build a common, ecosystem-based regional development vision and establish a highly motivated and committed group of leaders from business, environmental organizations, civil society, and government, working together on common regional goals for sustainable development. The aim of NOS is to assess the environmental issues that the region is facing and to determine the best way in which to address them through a regionwide agreement. This agreement needs to include clear, ambitious, and measurable long-term goals for key elements of the gulf's coupled social and ecological systems, including key species and habitats, and vital industries like fisheries and tourism, as well as a comprehensive implementation plan.

NOS's efforts have already yielded a better understanding of the context necessary to develop a successful regional agreement that integrates biodiversity conservation and economic opportunity in the region. The final regional vision must work in conjunction with several other regional plans under development and promote collaborative efforts, rather than duplicating them. The next step will be to attract business leaders and government agencies to support the agreement, and build broad-based support. If successful, this important initiative will encourage the governments of the five states surrounding the Gulf of California and the federal government to agree on a viable vision for sustainable development and to commit to its implementation.

In short, the ecosystem-based vision that was developed by the Coalition for the Sustainability of the Gulf of California has contributed to myriad other regional efforts, all the way from the governmental *Ordenamiento Ecológico Marino* to the work of many nongovernmental organizations that are trying to visualize and plan their efforts within an ecosystem-based regional perspective. But the task is not easy; it involves a complex maze of administrative agencies at the federal, state, and municipal levels, and it demands an understanding of ecological and development processes at a scale that has never been attempted in Mexico before. The ultimate success of this approach is yet to be seen, but a large-scale, ecosystem-based perspective has certainly changed the way regional problems are debated, as well as the economic discourse of many local authorities.

Conclusions

Hopefully, the increasing pace of conservation efforts in the Gulf of California will be able to reverse the environmental degradation that the region has suffered and will diminish threats to its long-term sustainability in the future. There seems to be growing awareness in the region, as never seen before, of the need to take urgent action to protect the environment. Conservation groups, research institutions, federal and state governments, conscientious businesspersons, and ecotourism operators have all been contributing to the growing appreciation of the environment and to the attendant conservation actions. The involvement of local groups and their commitment as allies in conservation efforts has possibly been the single most important element in successful conservation programs.

It is now the time to develop a vision, a social pact between sectors that may drive regional development for years to come, with ever-increasing consideration for the environment, for the gulf's natural resources, and for their sustainability. The gulf receives what remains of the discharges of the Colorado River Basin, and the survival of the upper gulf is a challenge for both Mexico and the United States. Clearly, the gulf's larger basin is part of a binational wilderness, where both Mexico and the Unitd States share the responsibility of protecting their joint natural heritage. To achieve this, both countries need to develop further and continuing efforts, to promote true collaborative work. The region is but one large continuum, with shared watersheds and estuaries, species, and natural resources. The protection of these unique environments is of the uttermost importance for our survival and well-being, today and for generations to come.

There are plenty of opportunities and creative solutions to the problems the gulf is facing today, though in the end, the solution lies in the hands of all the local actors in all sectors. We need a better understanding of the gulf's resilience to the range of perturbations it faces. We need a better understanding of the gulf as a single large marine ecosystem. But perhaps more importantly, we need a shared vision, and goals, and a common commitment to regional sustainability. If we are to conserve the amazing beauty, the remarkable biological productivity, and the magnificent biological richness of this unique place, we must find new ways to cooperate, coordinating and collaborating among ourselves. We need to change the way we work, change our behaviors, and use our extraordinary collective knowledge, creativity, abilities, and capacities to achieve common goals.

Key Messages

1. A number of the Gulf of California's ecosystem components appear to have reached a tipping point in the last decade. Understanding which of these ecological changes are reversible and which are not and what needs to be done to return critical ecosystem components to a functional state are among the biggest challenges for sustainability in the region.
2. Critical obstacles to success have included poor intergovernmental coordination, conflict among sectors, poor institutional capacity, and a lack of enforcement.
3. Conservation success depends on the involvement of local groups and their commitment as allies in conservation efforts, strengthening diverse human capital, and long-term commitments by all parties.
4. Place-based efforts are contributing to larger-scale EBM in the gulf through inspiration of a regional shared vision for long-term sustainability, which in turn encourages more local action. Large-scale perspectives can also change the context within which regional problems are tackled.

Acknowledgments

The lead author (EE) gratefully acknowledges the support of the David and Lucile Packard Foundation and the Pew Fellowship Program in Marine Conservation. JT thanks the David and Lucile Packard Foundation; Marisla; the Sandler Family Supporting Foundation; The Nature Conservancy; the Tinker Foundation, Inc.; the Walton Family Foundation; and World Wildlife Fund for enabling COBI's work on fully protected marine reserves in the Gulf of California.

References

Aburto-Oropeza, O., E. Sala, G. Paredes, A. Mendoza, and E. Ballesteros. 2007. Predictability of reef fish recruitment in a highly variable nursery habitat. *Ecology* 88:2220–28.

Aburto-Oropeza, O., and C. López Sagástegui. 2006. *Red de reservas marinas del Golfo de California: Una compilación de los esfuerzos de conservación*. Mexico, DF: Greenpeace México. http://www.greenpeace.org.mx.

Álvarez-Castañeda, S. T., P. Cortés-Calva, L. Méndez, and A. Ortega-Rubio. 2006. Development in the Sea of Cortés calls for mitigation. *BioScience* 56:825–29.

Arias, E., M. Albar, M. Becerra, A. Boone, D. Chia, J. Gao, C. Muñoz et al. 2004. *Gulf of California/Colorado River Basin*. UNEP/GIWA Regional Assessment 27. Kalmar, Sweden: University of Kalmar.

Bahre, C. J., L. Bourillon, and J. Torre. 2000. The Seri and commercial totoaba fishing 1930–1965 (Seri

Indian fishermen in the Gulf of California). *Journal of the Southwest* 42:559–75.

Beman, J. M., K. R. Arrigo, and P. A. Matson. 2005. Agricultural runoff fuels large phytoplankton blooms in vulnerable areas of the ocean. *Nature* 434:211–14.

Bergman, C. 2002. *Red delta: Fighting for life at the end of the Colorado River.* San Francisco: Fulcrum.

Berkes, F., R. Mahon, P. McConney, R. Pollnac, and R. Pomeroy. 2001. *Managing small-scale fisheries: Alternative directions and methods.* Ottawa: IDRC Books.

Bowen, T. 2004. Archaeology, biology and conservation on islands in the Gulf of California. *Environmental Conservation* 31:199–206.

Brusca, R. C., and G. C. Bryner. 2004. A case study of two Mexican biosphere reserves: The Upper Gulf of California and Colorado River Delta and the El Pinacate and Gran Desierto de Altar biosphere reserves. In *Science and politics in the international environment*, ed. N. E. Harrison and G. C. Bryner Rowman. Lanham, MD: Littlefield.

Brusca, R. C., R. Cudney-Bueno, and M. Moreno-Báez. 2006. *Gulf of California esteros and estuaries: Analysis, state of knowledge and conservation priority recommendations.* Final report to the David and Lucile Packard Foundation. Tucson: Arizona-Sonora Desert Museum.

Calderón-Aguilera, L. E., S. G. Marinone, and E. A. Aragón-Noriega. 2003. Influence of oceanographic processes on the early life stages of the blue shrimp (*Litopenaeus stylirostris*) in the upper Gulf of California. *Journal of Marine Systems* 39:117–28.

Cariño, M., A. Eritrea Gámez, J. A. Martínez de la Torre, and J. de Jesús Varela. 2006. Ecoturismo, certificación y desarrollo sustentable: La empresa Kuyimá en Baja California Sur, México. In *Las regiones sociales en el siglo XXI*, ed. S. Zermeño. México, DF: Instituto de Investigaciones Sociales, UNAM.

Cariño, M., and M. Monteforte. 1999. El primer emporio perlero sustentable del mundo: La compañía criadora de concha y perla de la baja california SA, y sus perspectivas para Baja California Sur. La Paz, Baja California Sur, México: Universidad Autónoma de Baja California Sur.

Cariño, M. 1996. Historia de las relaciones hombre/naturaleza en Baja California Sur, 1500–1940. Universidad Autónoma de Baja California Sur, La Paz, Baja California Sur, México.

Carriquiry, J. D., and A. Sánchez. 1999. Sedimentation in the Colorado River delta and upper Gulf of California after nearly a century of discharge loss. *Marine Geology* 158:125–45.

Carvajal, M. A., E. Ezcurra, and A. Robles. 2004. The Gulf of California: Natural resource concerns and the pursuit of a vision. In *Defying ocean's end. An agenda for action*, ed. L. K. Glover and S. A. Earle. Washington, DC: Island Press.

Case, T. J., M. Cody, and E. Ezcurra, ed. 2002. *A new island biogeography of the Sea of Cortés.* Oxford: Oxford University Press.

Castilla, J. C., and O. Defeo. 2001. Latin American benthic shellfisheries: Emphasis on co-management and experimental practices. *Reviews in Fish Biology and Fisheries* 11(1):1–30.

Chaffee, C., B. Phillips, D. Lluch-Belda, and A. Muhilia. 2004. *An MSC assessment of the red rock lobster fishery, Baja California, Mexico.* Final report. Emeryville, CA: Scientific Certification System.

Cicin-Sain, B., P. Bernal, V. Vandeweerd, S. Belfiore, and K. Goldstein. 2002. *A guide to oceans, coasts and islands at the World Summit on Sustainable Development.* Newark, DE: Center for the Study of Marine Policy.

Cicin-Sain, B., P. Bernal, V. Vandeweerd, S. Belfiore, and K. Goldstein. 2002. *A guide to oceans, coasts and islands at the World Summit on Sustainable Development and beyond.* Revised, Post-Johannesburg version. Dover: Center for the Study of Marine Policy, University of Delaware.

Cisneros-Mata, M. A., G. Montemayor-López, and M. O. Nevárez-Martínez. 1996. Modeling deterministic effects of age structure, density dependence, environmental forcing and fishing in the population dynamics of the Pacific sardine (*Sardinops sagax caeruleus*) stock of the Gulf of California. *California Cooperative Oceanic Fisheries Investigation Report* 37:201–8.

Cisneros-Mata, M. A., M. O. Nevárez-Martínez, and M. G. Hammann. 1995. The rise and fall of the Pacific sardine, *Sardinops sagax caeruleus* Girard, in the Gulf of Californa, Mexico. *California Cooperative Oceanic Fisheries Investigation Report* 36:136–43.

Cisneros-Mata, M. A. 2004. Sustainability in complexity: From fisheries management to conservation of species, communities, and spaces in the Sea of Cortez. In, *Proceedings of the Gulf of California Conference*, 20–22. 13–17 June 2004, Tucson, AZ.

CTPTV. 1993. *Propuesta para la declaración de reserva de la Biosfera Alto Golfo de California y Delta del Río Colorado*. Internal report, Comité Ténico para la Preservación de la Totoaba y la Vaquita, Hermosillo, Son.

Cudney-Bueno, R. 2004. Los buzos comerciales de Puerto Peñasco reciben el Premio Nacional para la Conservación de SEMARNAT. *Alto Golfo Eco-Update* 1–4 April 2004.

Cudney-Bueno, R. 2000. *Management and conservation of benthic resources harvested by small-scale Hookah divers in the northern Gulf of California, Mexico: The black murex snail fishery*. Master's thesis, School of Renewable Natural Resources, University of Arizona, Tucson.

Cudney-Bueno, R. 2007. *Marine reserves, community-based management and small-scale benthic fisheries in the Gulf of California, Mexico*. PhD diss., School of Renewable Natural Resources, University of Arizona, Tucson.

D'Agrosa, C., C. E. Lennert-Cody, and O. Vidal. 2000. Vaquita bycatch in Mexico's artisanal gillnet fisheries: Driving a small population to extinction. *Conservation Biology* 14(4):1110–19.

Dayton, P. K., S. Thrush, and F. C. Coleman. 2002. *Ecological effects of fishing in marine ecosystems of the United States*. Arlington, VA: Pew Oceans Commission.

de la Cruz-González, F. J., E. A. Aragón-Noriega, J. I. Urciaga-García, C. A. Salinas-Zavala, M. A. Cisneros-Mata, and L. F. Beltrán-Morales. 2007. Socio-economic analysis of shrimp and jumbo squid fisheries in northeastern Mexico. *Interciencia* 32:144–50.

DOF (Diario Oficial de la Federación). 2004a. Acuerdo mediante el cual se aprueba la actualización de la Carta Nacional Pesquera y su anexo. *Diario Oficial de la Federación* 15 March 2004. II:1–112; III:1–113; IV:1–85; and V:1–129.

DOF (Diario Oficial de la Federación). 2003. Acuerdo por el que se dan a conocer los límites de 188 acuíferos de los Estados Unidos Mexicanos, los resultados de los estudios realizados para determinar su disponibilidad media anual de agua y sus planos de localización. *Diario Oficial de la Federación* 31 January 2003. II:1–118. http: //www.cna.gob.mx/eCNA/.

DOF (Diario Oficial de la Federación). 2006a. Acuerdo por el que se expide el Programa de Ordenamiento Ecológico Marino del Golfo de California. *Diario Oficial de la Federación* 15 December 2006.

DOF (Diario Oficial de la Federación). 2004b. Acuerdo que adiciona la especificación 4.43 a la Norma Oficial Mexicana NOM-022-SEMARNAT-2003, que establece las especificaciones para la preservación, conservación, aprovechamiento sustentable y restauración de los humedales costeros en zonas de manglar. *Diario Oficial de la Federación* 7 May 2004.

DOF (Diario Oficial de la Federación). 2006b. Decreto por el cual se aprueba el Programa de Ordenamiento Ecológico Marino del Golfo de California. *Diario Oficial de la Federación* 29 November 2006.

Enríquez-Andrade, R., G. Anaya-Reyna, J. C. Barrera-Guevara, M. A. Carvajal-Moreno, M. E. Martínez-Delgado, J. Vaca-Rodríguez, and C. Valdés-Casillas. 2005. An analysis of critical areas for biodiversity conservation in the Gulf of California region. *Ocean and Coastal Management* 48:31–50.

Erisman, B. E., M. L. Buckhorn, and P. A. Hastings. 2007. Spawning patterns in the leopard grouper, *Mycteroperca rosacea*, in comparison with other aggregating groupers. *Marine Biology* 151:1849–61.

Ezcurra, E. 2003. Conservation and sustainable use of natural resources in Baja California. In, *Protected areas and the regional planning imperative in North America*, ed. J. G. Nelson, J. C. Day, and L. Sportza, 279–96. Calgary, Alberta: University of Calgary Press and Michigan State University Press.

Ezcurra, E., L. Bourillón, A. Cantú, M. E. Martínez, and A. Robles. 2002. Ecological conservation. In *A new island biogeography of the Sea of Cortés*, ed. T. Case, M. Cody, and E. Ezcurra, 417–44. Oxford: Oxford University Press.

Ezcurra, E., R. S. Felger, A. D. Russell, and M. Equihua. 1988. Freshwater islands in a desert sand sea: The hydrology, flora, and phytogeography of the Gran Desierto oases of northwestern Mexico. *Desert Plants* 9:1–17.

Fradkin, P. L. 1984. *A river no more: The Colorado River and the West*. Tucson: University of Arizona Press.

García-Caudillo, J. M., M. A. Cisneros-Mata, and A. Balmori-Ramírez. 2000. Performance of a bycatch reduction device in the shrimp fishery of the Gulf of California, Mexico. *Biological Conservation* 92:199–205.

Glenn, E. P., P. L. Nagler, R. C. Brusca, and O. Hinojosa-Huerta. 2006. Coastal wetlands of the northern Gulf of California: Inventory and conservation status. *Aquatic Conservation—Marine and Freshwater Ecosystems* 16:5–28.

Glenn, E. P., F. Zamora-Arroyo, P. L. Nagler, M. Briggs, W. Shaw, and K. Flessa. 2001. Ecology and conservation biology of the Colorado River delta, Mexico. *Journal of Arid Environments* 49:5–15.

Hernández, A., and W. Kempton. 2003. Changes in fisheries management in Mexico: Effects of increasing scientific input and public participation. *Ocean and Coastal Management* 46:507–26.

Hilborn, R., J. K. Parrish, and K. Litle. 2005. Fishing rights or fishing wrongs? *Reviews in Fish Biology and Fisheries* 15:191–99.

Hyun, K. 2005. Transboundary solutions to environmental problems in the Gulf of California large marine ecosystem. *Coastal Management* 33:435–45.

INE (Instituto Nacional de Ecología). 2005. *Evaluación preliminar de las tasas de pérdida de superficie de manglar en México*. México, DF: Instituto Nacional de Ecología, Dirección General de Investigación de Ordenamiento Ecológico y Conservación de los Ecosistemas; informe técnico septiembre 2005. http: //www.ine.gob.mx/.

INE (Instituto Nacional de Ecología). 1995. *Programa de manejo para la reserva de la biosfera alto Golfo de California y delta del Río Colorado*. México, DF: Instituto Nacional de Ecología, SEMARNAP.

Lavín, M. F., V. M. Godínez, and L. G. Álvarez. 1998. Inverse-estuarine features of the upper Gulf of California. *Estuarine Coastal and Shelf Science* 47:769–95.

Lavín, M. F., and S. Sánchez. 1999. On how the Colorado River affected the hydrography of the upper Gulf of California. *Continental Shelf Research* 19:1545–60.

Lercari, D., and E. A. Chávez. 2007. Possible causes related to historic stock depletion of the totoaba, Totoaba macdonaldi (Perciformes: Sciaenidae), endemic to the Gulf of California. *Fisheries Research* 86:136–42.

Lluch-Belda, D., B. F. J. Magallón, and R. A. Schartzlose. 1986. Large fluctuations in sardine fishery in the Gulf of California: Possible causes. *California Cooperative Oceanic Fisheries Investigation Report* 37:201–8.

López-Espinosa de los Monteros, R. 2002. Evaluating ecotourism in natural protected areas of La Paz Bay, Baja California Sur, México: Ecotourism or nature-based tourism? *Biodiversity and Conservation* 11:1539–50.

López, A., J. Cukier, and A. Sánchez-Crispín. 2006. Segregation of tourist space in Los Cabos, Mexico. *Tourism Geographies* 8(4):359–79.

López-Portillo, J., and E. Ezcurra. 1989. Zonation in mangrove and salt-marsh vegetation in relation to soil characteristics and species interactions at the Laguna de Mecoacán, Tabasco, Mexico. *Biotropica* 21(2):107–14.

Madrid-Vera, J., F. Amezcua, and E. Morales-Bojórquez. 2007. An assessment approach to estimate biomass of fish communities from bycatch data in a tropical shrimp-trawl fishery. *Fisheries Research* 83:81–89.

McGuire, T., and J. Greenberg, ed. 1993. *Maritime community and biosphere reserve: Crisis and response in the upper Gulf of California*. Tucson: University of Arizona Press.

Meltzer, L., and J. O. Chang. 2006. Export market influence on the development of the Pacific shrimp fishery of Sonora, Mexico. *Ocean and Coastal Management* 49:222–35.

Nichols,W. J. 2003. *Biology and conservation of the sea turtles of Baja California*. PhD diss., University of Arizona, Tucson.

Páez-Osuna, F., A. Gracia, F. Flores-Verdugo, L. P. Lyle-Fritch, R. Alonso-Rodríguez, A. Roque, and A. C. Ruiz-Fernández. 2003. Shrimp aquaculture development and the environment in the Gulf of California ecoregion. *Marine Pollution Bulletin* 46:806–15.

Páez-Osuna, F. 2001. The environmental impact of shrimp aquaculture: Causes, effects, and mitigating alternatives. *Environmental Management* 28:131–40.

Robadue, D. 2002. *An overview of governance of the Gulf of California*. Narragansett: Coastal Resources Center, University of Rhode Island.

Robles, A., and M. A. Carvajal. 2001. The sea and fishing. In *The Gulf of California. A world apart*, ed. P. Robles-Gil, E. Ezcurra, and E. Mellink, 293–300. México, DF: Agrupación Sierra Madre.

Robles, A., E. Ezcurra, and C. León. 1999. *The Sea of Cortés*. Editorial Jilguero/México Desconocido, México, DF.

Rojas-Bracho, L., R. R. Reeves, and A. Jaramillo-Legorreta. 2006. Conservation of the vaquita *Phocoena sinus*. *Mammal Review* 36:179–216.

Rowell, K., K. W. Flessa, D. L. Dettman, and M. Roman. 2005. The importance of Colorado River flow to nursery habitats of the gulf corvina (*Cynoscion othonopterus*). *Canadian Journal of Fisheries and Aquatic Sciences* 62:2874–85.

Sáenz-Arroyo, A., C. M. Roberts, J. Torre, and M. Cariño-Olvera. 2005a. Using fishers' anecdotes, naturalists' observations and grey literature to reassess marine species at risk: The case of the Gulf grouper in the Gulf of California, Mexico. *Fish and Fisheries* 6:121–33.

Sáenz-Arroyo, A., C. M. Roberts, J. Torre, M. Cariño-Olvera, and R. R. Enríquez-Andrade. 2005b. *Rapidly shifting environmental baselines among fishers of the Gulf of California. Proceedings of the Royal Society B* 272:1957–62.

Sáenz-Arroyo, A., C. M. Roberts, J. Torre, M. Cariño-Olvera, and J. P. Hawkins. 2006. The value of evidence about past abundance: Marine fauna of the Gulf of California through the eyes of sixteenth to nineteenth century travelers. *Fish and Fisheries* 7:128–46.

Sala, E., O. Aburto-Oropeza, G. Paredes, I. Parra, J. C. Barrera, and P. K. Dayton. 2002. A general model for designing networks of marine reserves. *Science* 298:1991–93.

Sala, E., O. Aburto-Oropeza, M. Reza, G. Paredes, and L.G. López-Lemus. 2004. Fishing down coastal food webs in the Gulf of California. *Fisheries* 28(3):19–25.

Sala, E., O. Aburto-Oropeza, G. Paredes, and G. Thompson. 2003. Spawning aggregations and reproductive behavior of reef fishes in the Gulf of California. *Bulletin of Marine Science* 72(1):103–21.

Santamaría del Ángel, E., S. Álvarez Borrego, and F. E. Mullerkarger. 1994. Gulf-of-California biogeographic regions based on coastal zone color scanner imagery. *Journal of Geophysical Research—Oceans* 99:7411–21.

Scheffer, M., and S. R. Carpenter. 2003. Catastrophic regime shifts in ecosystems: Linking theory to observation. *Trends in Ecology and Evolution* 18:648–56.

Scheffer, M., S. R. Carpenter, J. A. Foley, C. Folke, and B. Walker. 2001. Catastrophic shifts in ecosystems. *Nature* 413:591–96.

Seminoff, J. A., T. T. Jones, A. Reséndiz, W. J. Nichols, and M. Y. Chaloupka. 2003. Monitoring green turtles (*Chelonia mydas*) at a coastal foraging area in Baja California, Mexico: Multiple indices to describe population status. *Journal of the Marine Biological Association of the United Kingdom* 83:1355–62.

Sherman, K. 1994. Sustainability, biomass yields, and health of coastal ecosystems: An ecological perspective. *Marine Ecology Progress Series* 112:277–301.

Steinbeck, J., and E. F. Ricketts. 1941. *Sea of Cortez: A leisurely journal of travel and research*. New York: Viking Press.

Steller, D. L., R. Riosmena-Rodríguez, M. S. Foster, and C. A. Roberts. 2003. Rhodolith bed diversity in the Gulf of California: The importance of rhodolith structure and consequences of disturbance. *Aquatic Conservation: Marine and Freshwater Ecosystems* 13:S5–S20.

Sykes, G. 1937. *The Colorado delta*. Carnegie Institution of Washington, publication no. 460.

Tershy, B. R., L. Bourillon, L. Metzler, and J. Barnes. 1999. A survey of ecotourism on islands in northwestern Mexico. *Environmental Conservation* 26:212–17.

Thomson, D. A., L. T. Findley, and A. N. Kerstitch. 2000. *Reef fishes of the Sea of Cortez: The rocky-shore fishes of the Gulf of California*, rev. ed. Austin: University of Texas Press.

Torre, J., A. Sáenz-Arroyo, L. Bourillón, and M. Kleiberg. 2005. Fisher fund: An initiative to encourage community-based marine reserves. In *International Marine Protected Areas Conference*, Geelong, Australia. Extended abstracts. http://www.impacongress.org/.

Torres-Orozco, E., A. Muhlia-Melo, A. Trasvina, and S. Ortega-García. 2006. Variation in yellowfin tuna (*Thunnus albacares*) catches related to El Niño-Southern Oscillation events at the entrance to the Gulf of California. *Fisheries Bulletin* 104:197–203.

Vargas-Márquez, F., and S. Escobar, ed. 2003. Áreas naturales protegidas de México con decretos federales. México, DF: Instituto Nacional de Ecología.

Velarde, E., E. Ezcurra, M. A. Cisneros-Mata, and M. F. Lavín. 2004. Seabird ecology, El Niño anomalies, and prediction of sardine fisheries in the Gulf of California. *Ecological Applications* 14(2):607–15.

Walker, B., and D. Salt. 2006. *Resilience thinking: Sustaining ecosystems and people in a changing world.* Washington, DC: Island Press.

Whitmore, R. C., R. C. Brusca, J. León de la Luz, P. González-Zamorano, R. Mendoza-Salgado, E. S. Amador-Silva, G. Holguín et al. 2005. The ecological importance of mangroves in Baja California Sur: Conservation implications for an endangered ecosystem. In *Biodiversity, ecosystems, and conservation in northern Mexico,* ed. J. E. Cartron, G. Ceballos, and R. S. Felger. New York: Oxford University Press.

Wolanski, E., S. Spagnol, S. Thomas, K. Moore, . M. Alongi, L. Trott, and A. Davidson. 2000. Modelling and visualizing the fate of shrimp pond effluent in a mangrove-fringed tidal creek. *Estuarine Coastal and Shelf Science* 50:85–97.

Young, R. H., and J. M. Romero. 1979. Variability in the yield and composition of by-catch recovered from Gulf of California shrimping vessels. *Tropical Science* 21(4):249–64.

CHAPTER 14
Eastern Scotian Shelf, Canada

Robert O'Boyle and Tana Worcester

In 1993, a moratorium was declared on cod fishing on the Eastern Scotian Shelf; so ended one of the most lucrative fisheries in the history of Nova Scotia. Almost 15 years later, there has still been no significant recovery of the resource, and the population has been declared a species of "special concern" by the Committee on the Status of Endangered Wildlife in Canada (COSEWIC 2003). Since the closure of the fishery, there have been numerous changes to the regulatory and policy environment in Canada, including new Oceans (1997) and Species at Risk (2003) acts. These acts seek, through integrated management (IM) and ecosystem-based management (EBM) approaches, to provide a more sustainable basis for the human use of Canada's three oceans (for a discussion of the distinctions between integrated and ecosystem-based management, see Rosenberg and Sandifer, chap. 2 of this volume).

In response to the Oceans Act, the Eastern Scotian Shelf Integrated Management (ESSIM) project was initiated in 1998 to explore the requirements of both IM and EBM, including how the activities of diverse ocean industries can best be collectively managed and what ecosystem-related priorities are necessary for management to address. ESSIM has been guided by nationally articulated objectives, which include considerations of biodiversity and resilience (Jamieson et al. 2001). In this chapter, we explore the opportunities and challenges for ESSIM to conserve the Scotian Shelf ecosystem, including enhancing the resilience of this system to both current and future threats.

National Legislative Context

In Canada, the Fisheries Act (rev. 1991) is the main piece of legislation used to regulate fishing and impacts to fish habitat, while the Canadian Environmental Assessment Act (1992) is designed to protect environmental quality, for example, water, air, and sediment quality. The Species at Risk Act (2003), the most recent piece of legislation, gives the government regulatory authority over human activities that would put species at risk of extinction. Collectively, all three acts provide the federal government with significant authority to manage human activity in the marine environment.

Responding to both international treaties and concerns for the cumulative impacts of human activities on its marine ecosystems, Canada enacted the Oceans Act in 1997. The Oceans Act authorized the Minister of Fisheries and Oceans Canada to lead the development of a national oceans management strategy, guided by the principles of sustainable development, the precautionary approach, and integrated management. Canada's Ocean Strategy was published in 2002 (DFO 2002b). Together, the Oceans Act and Ocean Strategy provide the basis for a new approach to managing oceans and their resources based on the premise that oceans must be managed as a collaborative effort among all government agencies and affected stakeholders through the use of new integrated ocean management tools and approaches.

When the Oceans Act was first proclaimed, there was little understanding as to what EBM meant in practical terms. Since then, Canada's approach to EBM has evolved through a series of regional initiatives, such as ESSIM, the first to be initiated (Rutherford et al. 2005) and presently the most advanced EBM and IM oceans initiative in Canada. Canada's experience with EBM implementation demonstrates two key strengths of its Oceans Act. First, the act grants the Department of Fisheries and Oceans (DFO) a clear leadership role for EBM in the marine environment, including hiring staff to facilitate its implementation. Granting authority to a single agency reduced the potential for interagency debate and established clear agency leadership. In addition, a dedicated workforce to implement EBM has ensured progress even with limited resources. Second, and perhaps less appreciated, the Oceans Act unifies conservation requirements of the three other regulatory acts (Fisheries Act, Canadian Environmental Assessment Act, Species at Risk Act, as discussed above) and provides a basis for their consistent application.

Regional Implementation: The ESSIM Initiative

The Planning Area

The boundaries of the ESSIM planning area, as with the five other large ocean management areas (LOMAs) in Canada (fig. 14.1), have been defined based upon a combination of ecological properties and administrative realities. To date, an inshore boundary has been drawn at 12 nautical miles (nm) from the coast, with the offshore boundary at Canada's exclusive economic zone (EEZ). The western boundary has been the boundary between Northwest Atlantic Fisheries Organization (NAFO) statistical divisions 4X and 4W, which split the Scotian Shelf roughly in half, and the eastern boundary is the boundary that separates DFO's Maritimes Region from the Gulf Region and the Newfoundland and Labrador Region. These boundaries have been the subject of ongoing debate and are likely to undergo further refinement. While the western boundary makes use of the existing NAFO boundary, which is used to manage several major fisheries, it also reflects significant ecological differences between the western and eastern parts of the Scotian Shelf. The western shelf has properties similar to those of the Gulf of Maine, while the eastern shelf is influenced by outflow from the Gulf of St. Lawrence (Sinclair et al. 1997). The eastern boundary of ESSIM has been more difficult to establish. Competing rationales for its location include separation of the Scotian Shelf biological community from that of the Gulf of St. Lawrence and the administrative boundary for offshore oil and gas. Unfortunately, this may ultimately result in multiple boundaries for the ESSIM area, which would diminish the initial intent of establishing ecologically based boundaries and create challenges for integrated management across many activities.

In general, no EBM boundary is going to fully coincide with all ecological and administrative features. Whatever boundaries are ultimately selected, there will need to be consideration of highly migratory large pelagic species (e.g., tunas, swordfish, and sharks) and other issues that span larger scales than ESSIM. Different interacting ecosystem components exist at different spatial scales, and it is important to have a management system flexible enough to accommodate these. Careful consideration needs to be given to community buy-in, which may influence the likelihood of successful implementation and enforcement. In our experience, where existing boundaries have been previously established using ecological criteria, making use of these boundaries may help to minimize confusion, duplication of regulatory effort, and negative stakeholder pushback. At the same time, political and administrative forces can quickly become key drivers for establishing boundaries if their ecological basis is not clearly articulated, documented, and supported by sound science.

Finally, while Canada's five LOMAs have made significant progress toward EBM, it is not clear how the lessons learned from these initiatives will be applied to Canada's other ocean areas. Addition of new LOMAs to form a comprehensive network will be important to ensure consistent implementation of EBM in Canadian waters.

Ecological Context

The Eastern Scotian Shelf is influenced by cold, low-salinity water from the Gulf of St. Lawrence and Newfoundland Shelf (fig. 14.1). Its canyons and deep basins to the west are also influenced by the Labrador Slope Water Current and the Gulf Stream. The bottom waters of the Eastern Scotian Shelf experienced gradual cooling from the mid-1980s through the 1990s, a period which corresponded with an increase in the abundance of cold-water species, such as capelin, turbot, snow crab, and shrimp. Additional information on the physical, chemical, and ecological properties of the Eastern Scotian Shelf ecosystem can be found in Breeze et al. (2002), DFO (2003), and Zwanenburg et al. (2006).

Species of commercial interest in this region have historically been dominated by

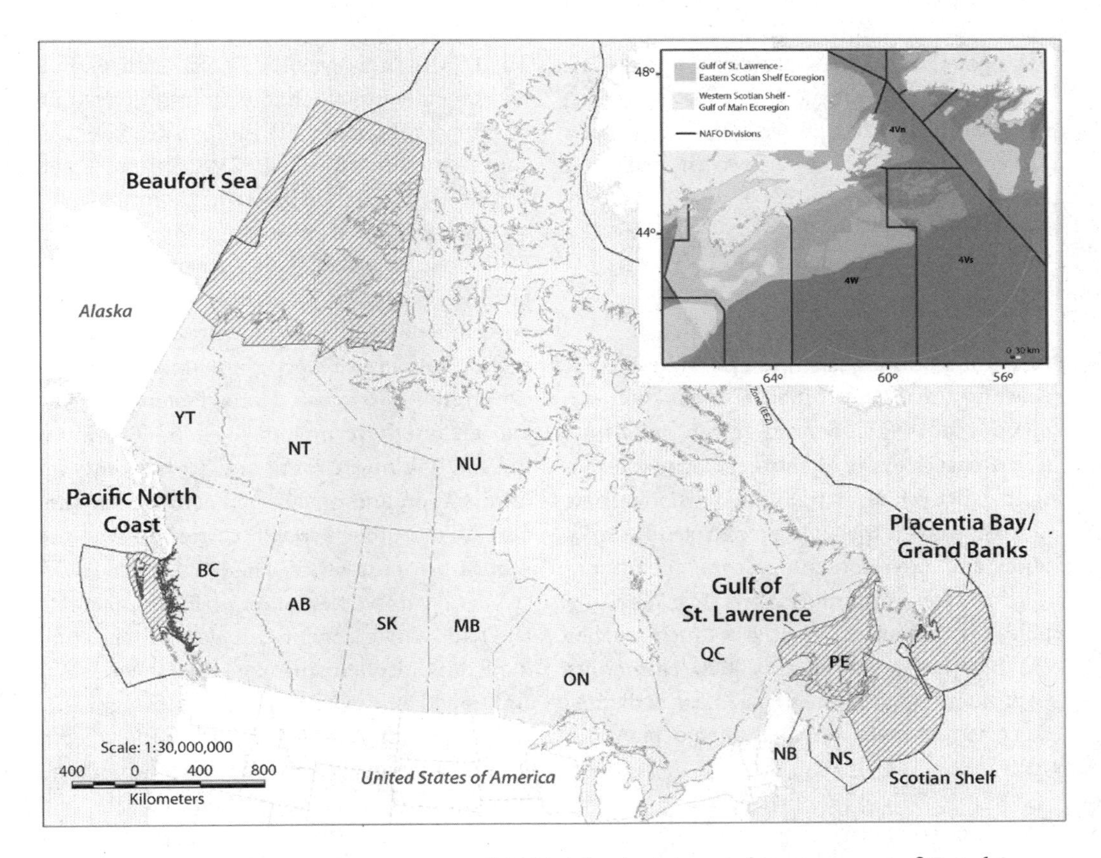

Figure 14.1 Large ocean management areas (LOMAs) for the integrated management of Canada's three oceans. Insert indicates Eastern Scotian Shelf Integrated Management (ESSIM) LOMA.

groundfish such as cod, haddock, pollock, plaice, yellowtail flounder, witch flounder, redfish, and white hake. However, most of these species suffered severe overfishing in the 1980s and are now severely depressed. Only haddock has experienced recovery in terms of abundance, although the growth rate of individuals has declined, thus reducing overall productivity. Pelagic fish species such as herring and mackerel are also harvested in the ESSIM area, although not to the same extent as groundfish. Since the collapse of the groundfish fisheries in 1993, the highest revenue species have been shrimp, snow crab, scallop, halibut, and lobster. The Eastern Scotian Shelf is also home to and visited by a wide variety of cetacean species, including the humpback whale, the fin whale, the endangered blue whale, and the endangered northern bottlenose whale. The Sable Island colony of gray seals is the largest in the world and has grown exponentially at a rate of about 12% per year over the past four decades, although the rate has slowed more recently (Trzcinski et al. 2006).

Ocean Users

Numerous human activities affect the ESSIM planning area. The initiative began with an offshore focus (i.e., beyond 12 nm), and thus the primary uses are offshore fisheries (despite significant declines in traditional fisheries, new ones have taken their place), petroleum exploration and development, marine transportation, and telecommunications (e.g., marine cables), with tourism playing a much smaller role (Coffen-Smout et al. 2001). Numerous government agencies regulate these activities, each with its own consultative and planning structures:

- Fisheries—Department of Fisheries and Oceans (DFO)
- Oil and gas—Canada–Nova Scotia Offshore Petroleum Board
- Marine transport—Transport Canada and Canadian Coast Guard
- Telecommunications—Communications Canada
- Tourism—Province of Nova Scotia

Other uses of this offshore area have included research, military exercises (by the Department of National Defense), and some limited recreational activity. Additional stakeholders include environmental agencies, coastal communities, and the general public. As ESSIM expands its purview to include the coastal zone, the diversity of interests and uses will increase substantially.

Collaborative Planning Model

The ESSIM initiative uses a collaborative, multistakeholder approach to planning, consisting of four components: (1) multistakeholder engagement through the ESSIM forum, a network of organizations, groups, and individuals who are interested in or impacted by ESSIM; (2) a smaller stakeholder advisory council to provide leadership, guidance, and advice; (3) government-level engagement consisting of both an executive decision-making Regional Committee on Ocean Management (RCOM) and an intergovernmental Federal–Provincial ESSIM Working Group; and (4) ongoing administration and operational activities in support of the other groups, carried out by the ESSIM planning office. The RCOM makes decisions on implementation of EBM efforts by DFO and Environment Canada using existing regulatory mechanisms (e.g., the Fisheries Act and Canadian Environmental Assessment Act) or by supporting the development of new regulations under the Oceans Act (e.g., through the creation of a new marine protected area).

The first ESSIM forum workshop (with about 150 people) was held in 2002 to bring communities of interest together to provide feedback and direction on the preparation of an

ESSIM plan that would outline the long-term objectives and consultative structure for IM on the Eastern Scotian Shelf (Coffen-Smout et al. 2002). A series of multi-stakeholder working groups were established to lead the process and interact with the ESSIM planning office. Overall, the ESSIM collaborative planning model has been effective in developing the ESSIM plan, which was released in 2008 (Government of Canada 2007).

Within the fisheries sector, there have also been institutional changes in response to ESSIM. When ESSIM was first initiated, there was no mechanism for the fisheries sector to have interfleet discussion. Each fishery was managed independently to meet the objectives of its integrated fisheries management plan, and cross-cutting issues were addressed on a case-by-case basis. Development of a multi-sectoral advisory structure for ESSIM stimulated a desire within the fishing community for sector-level discussion about EBM, resulting in the Scotia–Fundy Fisheries Roundtable, including representatives from all fishing fleets in the Maritimes Region, as well as some fish processors. The objectives of this group were to develop a Scotia–Fundy fishing industry perspective on marine conservation issues, address intersectoral conflicts, and further the understanding of complex ocean and fishery issues and interactions across the industry. The roundtable has been an effective forum for implementation of ecosystem-based approaches within the sector. More recently, it has also discussed the viability, productivity, and competitiveness of the fishery itself. Enhancing the capacity of the fishing industry to discuss, and hopefully resolve, socioeconomic barriers to sustainability may provide a stable basis from which to tackle conservation priorities. Indeed, resolution of these issues, including overcapacity, which is often a consequence of ill-defined access rights, is considered a prerequisite to EBM (Sinclair et al. 2002).

A Hierarchy of Objectives

In a review of five case studies of national- or multinational-level EBM implementation, Rosenberg and colleagues (chap. 16 of this volume) observe that each EBM is characterized by several key features, which include a hierarchy of goals and objectives. This hierarchy ensures that objectives defined at the highest level (linked to international, national, and state/provincial legislation and policy) are transparently linked to implementation efforts at the lowest operational level. Furthermore, having a consistent set of objectives throughout the hierarchy is important to the management and control of cumulative impacts. Certainly, pursuing different objectives in different sectors would confound management efforts to limit cumulative impacts of all sectors on identified sensitive ecosystem components (see also Rosenberg and Sandifer, chap. 2 of this volume).

The Link to National Policy: Overarching Conceptual Objectives

The overarching conceptual objectives provide overall guidance to EBM and should relate to the conservation of ecosystem structure and function. Traditionally, objectives have been defined that focus on the maintenance of the productivity of individual ecosystem components, for example, maximum sustainable yield of a fish stock. It is becoming increasingly evident that ecosystems and the human systems connected to them are "complex systems," the responses of which cannot be predicted solely by understanding the component parts (Walker and Salt 2006; Leslie and Kinzig, chap. 4 of this volume). An example of what can go wrong is provided by what happened on the Eastern Scotian Shelf. Prior to the 1980s, this ecosystem was dominated by groundfish such as cod, haddock, and pollock. As a consequence of overfishing in the 1970s and 1980s, the groundfish community was severely depleted, allowing pelagic species abundance to increase to such an extent that they now dominate the ecosystem (Frank et al. 2005). Thus, the ecosystem appears to have shifted to an alternate state, and it is not certain whether management actions could return it to a groundfish-dominated state.

Forecasting if and how certain management actions will reverse the trajectory of a given ecosystem state, such as the pelagic-dominated system described above, requires understanding the roles played by biological diversity, disturbance regimes, and interactions (Leslie and Kinzig, chap. 4 of this volume). Canada's overarching ecosystem objectives include measures of biodiversity, productivity, and habitat and thus will contribute to understanding and, ideally, bolstering the resilience of this system to perturbations (table 14.1).

Making EBM operational in a large country such as Canada, which has a number of very distinct ocean environments, requires the development of specific objectives for each LOMA based upon the overarching national objectives. The details of this process appear to differ from one case to the next, but some best practices are starting to emerge (O'Boyle et al. 2005; O'Boyle and Jamieson 2006; Rosenberg et al., chap. 16 of this volume).

Addressing Priority Regional Issues: Defining the ESSIM Conceptual Objectives

Assessments of relationships among ecosystem components have been useful in defining LOMA-relevant conceptual objectives (DFO 2005). For details of these analyses for ESSIM, see Coffen-Smout et al. 2001 (analysis of human uses), Breeze et al. 2002 (ecosystem features), Bundy 2005 (quantification of key interactions using an EcoPath model), and Zwanenburg 2006 (relationships among system components).

ECOSYSTEM OBJECTIVES

The above studies were used as background material for workshops that brought together the scientific community and other stakeholders to develop the ESSIM conceptual objectives (Government of Canada 2007). While this process was underway, DFO Science provided national criteria for the identification of ecologically and biologically significant areas. These areas are to be identified in each LOMA, based upon their ecological structure and function using the dimensions of uniqueness, aggregation, fitness consequences, resilience, and naturalness (DFO 2004). Resilience can vary, from areas where the habitat structures or species are highly sensitive, easily perturbed, and slow to recover, to areas where the habitat structures or species are relatively robust and more resistant to perturbation. In addition, criteria for ecologically significant species and community properties have been defined nationally (DFO 2006b).

The ESSIM conceptual objectives developed to date have been based upon consensus building among stakeholders, scientists, and the management community. This process, while ensuring buy-in, has resulted in very general objectives that fail to identify particular species and areas within the ESSIM area that require priority attention. The identification of specific ecologically and biologically significant areas, species, and community properties, in addition to depleted species and degraded areas within the ESSIM area, may help to provide some specificity in setting an ecological "bottom line" for monitoring and action. Possible approaches include qualitative risk assessment (Fletcher 2005) that evaluates, in qualitative terms, the risk to an ecosystem component from a defined human impact. This approach is expanded upon in Australia's ecological risk assessment (Smith et al. 2007), which consists of a first-level qualitative screening (similar to that of Fletcher 2005), followed by a second-

Table 14.1. Overarching conservation objectives used to guide Canada's ecosystem approach to management

To conserve enough components (ecosystems, species, populations, etc.) so as to maintain the natural resilience of the ecosystem

- to maintain communities within bounds of natural variability
- to maintain species within bounds of natural variability
- to maintain populations within bounds of natural variability

To conserve each component of the ecosystem so that it can play its historical role in the food web (i.e., not cause any component of the ecosystem to be altered to such an extent that it ceases to play its historical role in a higher-order component)

- to maintain primary production within historical bounds of natural variability
- to maintain trophic structure so that individual species/stages can play their historical roles in the food web
- to maintain mean generation times of populations within bounds of natural variability

To conserve the physical properties of the ecosystem

- to conserve critical landscape and bottomscape features
- to conserve water column properties

To conserve the chemical properties of the ecosystem

- to conserve water quality
- to conserve biota quality

Source: O'Boyle and Jamieson 2006.

level semiquantitative evaluation and a third-level quantitative identification of priority issues.

SUSTAINABLE HUMAN USE AND COLLABORATIVE GOVERNANCE OBJECTIVES

Sustainable human use and collaborative governance objectives for ESSIM were developed through a multistakeholder working group facilitated by DFO. Contrary to the development of the ecosystem objectives, this process was not guided by a set of overarching national objectives and did not involve DFO Science. The ESSIM planning office used a multistakeholder working group to develop an initial set of human-use elements and objectives. The objectives that were developed (table 14.2) were of a high-level nature similar to that of the healthy ecosystem objectives. The stakeholder advisory council has subsequently focused on the integration of the two ecological- and human-use objectives and has been instrumental in the continued refinement of this important part of the plan. For more information on the development and implementation of ESSIM social, economic, and institutional objectives, please see the *Eastern Scotian Shelf Integrated Ocean Management Plan* (Government of Canada 2007).

Progress to Date

Identification of Sensitive Benthic Communities

Using data on marine benthic communities, a framework to identify the areas of highest sensitivity and lowest resilience on the Scotian Shelf has been developed (fig. 14.2). Benthic communities based on their ability to sustain physiological stress and physical disturbance (Kostylev and Hannah 2007; O'Boyle and Worcester 2006) have been classified.

Table 14.2. Socioeconomic conceptual objectives in the Eastern Scotian Shelf Integrated Ocean Management Plan

Goal	*Element*	*Planning area conceptual objective*
Collaborative governance and integrated management	Integrated management	Collaborative structures and processes with adequate capacity, accessible to community members, are established.
		Appropriate legislation, policies, plans, and programs are in place.
		Legal obligations and commitments are fulfilled.
		Ocean users and regulators are compliant and accountable.
		Ocean stewardship and best practices are implemented.
		Multisectoral resource use conflict is reduced.
	Information and knowledge	Natural and social science research is responsive to knowledge needs.
		Information management and communication are effective.
		Monitoring and reporting are effective and timely.
Sustainable human use	Social and cultural well-being	Communities are sustainable.
		Sustainable ocean–community relationships are promoted and facilitated.
		Ocean area is safe, healthy, and secure.
	Economic well-being	Wealth is generated sustainably from renewable ocean resources.
		Wealth is generated sustainably from nonrenewable ocean resources.
		Wealth is generated sustainably from ocean infrastructure.
		Wealth is generated sustainably from ocean-related activities.

Source: Government of Canada 2007.

Physiological stress was determined using indicators related to the various physiological functions of an organism (fig. 14.3). For instance, water temperature influences metabolic rate, and salinity influences osmotic control. Physical disturbance was determined using information such as bottom current and grain size. Values for both physiological stress and physical disturbance were mapped for all locations on the Scotian Shelf. Based on this analysis, we predict that benthic communities with low scope for growth and low disturbance will be the least resilient to human impacts while those with high scope for growth and high disturbance will be the most resilient (O'Boyle and Worcester 2006). Although reference points for management have not been established, the relative sensitivity of these areas is starting to be considered in management decisions.

Assessment of Species at Risk and Genetic Integrity

Under the 2003 Species at Risk Act, recovery plans must be developed for endangered species The required recovery potential assessments for listed species determine the current status of the species, their recovery potential, scope for further human-induced mortality, and an evaluation of aggregate sources of mortality due to human activities. If some level of harm is allowed, this has to be allocated to each of the implicated human activities. In the ESSIM planning area, a number of species either have

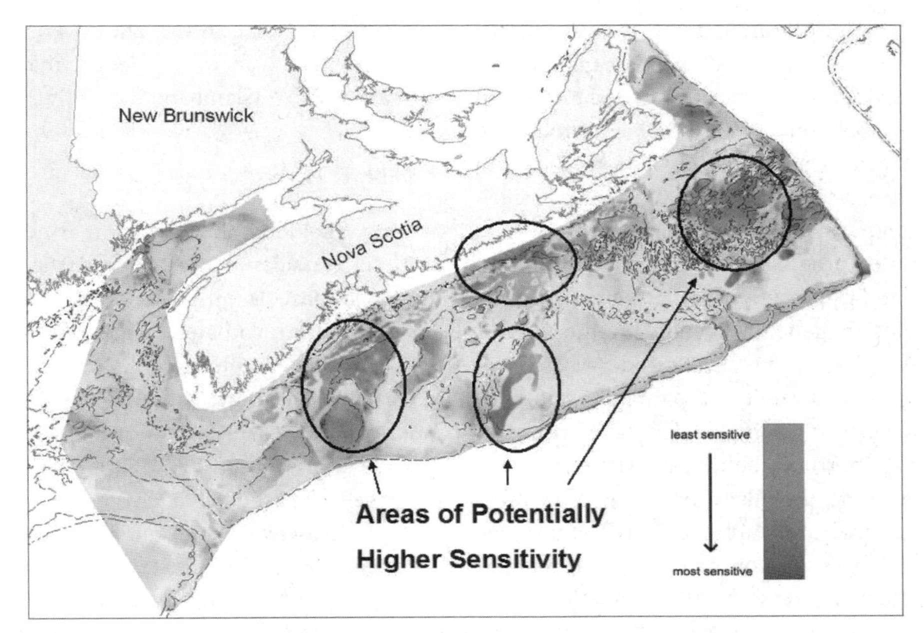

Figure 14.2 Areas of potentially high sensitivity to human impacts, using the benthic community classification illustrated in figure 14.3.

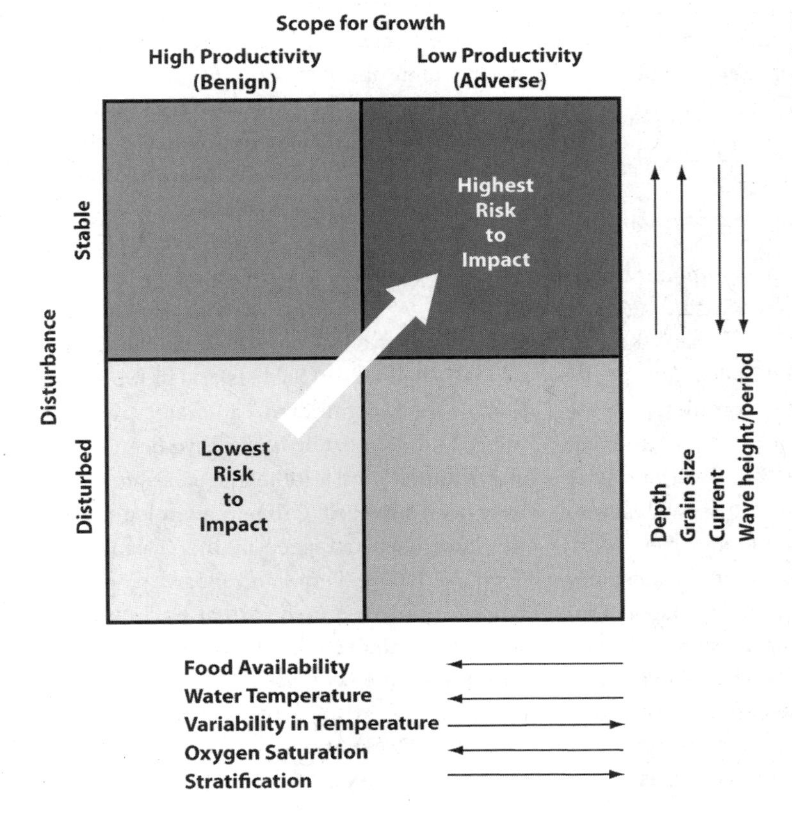

Figure 14.3 Benthic community classification system for the Eastern Scotian Shelf Integrated Management (ESSIM) initiative. From O'Boyle and Worcester 2006.

undergone or are undergoing recovery potential assessments, including porbeagle, winter skate, leatherback turtle, loggerhead turtle, and northern bottlenose whale. So far, harm has either not been permitted or been restricted to one ocean sector (e.g., fishing), making decisions on harm allocation straightforward.

Limited progress has been made on monitoring and implementing management actions to maintain genetic diversity in the ESSIM area. There are some efforts underway to evaluate genetic diversity of particular commercial species (e.g., spiny dogfish) and species at risk (e.g., northern bottlenose whale) in order to determine the likely presence of population substructure. Within fisheries, there has been some discussion of distributing fishing effort (fishing mortality) as a function of component biomass, and assessments have included the identification of distinct spawning components. Increased use of techniques for assessing genetic diversity of populations (including identification of subcomponents) and increased application of spatial analytical tools in fisheries assessments are expected to enhance efforts in this area.

Prevention of Invasive Species Introductions

Several major international shipping routes pass through the ESSIM area, including traffic between Canada and the Caribbean, the eastern seaboard of the United States, Europe, the Mediterranean, and Africa. Commercial vessel traffic through this area includes container vessels, tankers, bulk carriers, and various types of general and specialized vessels, with several government departments and agencies having responsibility for regulating marine transportation and navigation, including Transport Canada and the Canadian Coast Guard. In 2005, new Ballast Water Control and Management Regulations were proposed, and alternate ballast water exchange zones were incorporated into the proposed regulations. It has been recommended that, in the ESSIM area, ships should exchange in waters deeper than 1,000 m, west of Sable Island and the Gully.

Viability of Harvested Species

Fishing, including harvest of marine mammals, is considered to have had the largest human impact on productivity in the ESSIM area, apart from climate change. While efforts are being made to enhance the productivity of harvested species, bycatch and habitat impact monitoring present significant gaps (DFO 2002c; Gavaris et al. 2005). One species that has received particular attention is cod. Current productivity is substantially lower than historical levels (Rosenberg et al. 2005). Despite a moratorium on fishing since 1993, the stock has yet to show signs of recovery, and natural mortality rates remain high (Shelton et al. 2005). Similarly, growth rates for haddock stocks (now referred to as "pygmy haddock" by fishermen) have significantly declined (despite having high abundance) (DFO 2002a). While the underlying processes for these two cases are unclear, this may be evidence of a shift in state that could be difficult to reverse, particularly on meaningful time frames.

Impacts of Offshore Petroleum Activities

The first oil well to be drilled on the Eastern Scotian Shelf was on Sable Island in 1967. Since then, over two hundred exploratory, development, and delineation wells have been drilled, and hundreds of kilometers of seismic lines have been surveyed. Offshore petroleum activities have been managed by the Canada–Nova Scotia Offshore Petroleum Board since 1988 but have only required formal environmental assessment under the Canadian Environmental Assessment Act since 2003. The petroleum board has been an active participant of ESSIM since its inception, and even before this, the board made efforts to reflect the emerging objectives of ESSIM. Offshore chemical selection and waste treatment guidelines have been developed to minimize the impact of contaminants. A number of research programs have been funded through the petroleum industry to investigate the propagation and effects of noise in the marine environment. To date, studies in the ESSIM area have focused on effects on snow crab, cod, and northern bottlenose whales. While no reference points for noise have been established, the petroleum board has required the use of fisheries liaisons on all offshore seismic vessels to enhance communication between petroleum operators and the fishing community. Seismic operators have been asked to avoid spawning aggregations of fish and marine mammals. However, it is now the responsibility of DFO to identify where and when these areas occur.

Maintaining Habitat Integrity

One unique habitat feature within the ESSIM planning area is the Gully, an offshore deep-water canyon that is home to the endangered bottlenose whale. Before the initiation of ESSIM, there had already been calls for special protection of this area's unique features. In May 2004, the Gully MPA was the second protected area to be established under the Oceans Act. General prohibitions against disturbance, damage, destruction, or removal of any living marine organism or any part of its habitat within the MPA apply to the water column and the seabed to a depth of 15 m. Regulatory exceptions and ministerial plan approvals allow uses that do not compromise the objectives of the MPA. Search and rescue, international navigation rights, and activities related to national security and sovereignty have also been permitted in the MPA. Using a regulation under the Oceans Act appears to have been an effective way to protect this unique area; however, it was a fairly lengthy process, taking almost 6 years from the time the Gully was officially identified as an "area of interest" (December 1998) until the time the Oceans Act was enacted (May 2004).

A second measure to protect benthic habitat, taken in 2004, was the closure of a 15 km^2 area to bottom-impacting fishing gear under the Fisheries Act. Located at the mouth of the Laurentian Channel, this region is the only known occurrence of the reef-forming *Lophelia pertusa* in Canadian waters. Research survey results suggested that the reef had been damaged by fishing activity over a period of decades. The closure area was designed in consultation with representatives of active fisheries in the area (primarily redfish otter trawlers and halibut longliners), as well as the ESSIM community. The use of a fisheries closure under the Fisheries Act enabled very fast implementation of protective measures to address the primary threat; however, it is uncertain how prohibitions would be applied to nonfishing activities. In efforts to avoid piecemeal protection of coral habitat, a coral conservation plan was developed through broad consultation and was published in March 2006 (DFO 2006a). This plan provides a long-term strategy for protecting and understanding these important benthic habitats (Government of Canada 2007).

Lessons Learned

What lessons can we draw from the efforts on the Scotian Shelf for the implementation of EBM more broadly? First, EBM has certainly been facilitated by having a legislative basis for action and a clearly identified lead (DFO) as established through the Oceans Act. While progress could have been made without this legislative basis for action, no doubt it would have been much slower. There are many existing vested interests and established organizations that have power and have "evolved" to meet the demands of extant, generally sector-specific legislation, rules, and regulations. This also includes the regulators themselves, who may be committed to the status quo. Having a clear legislative basis for EBM helps to ensure that everyone understands that there is a new paradigm in which management is operating across sectors. On a related note, the Fisheries Act is currently undergoing an in-depth review. Key issues being considered include legislative changes to formally recognize the needs of EBM in the management of fisheries.

Second, it is important not to lose sight of the fact that we manage people, not ecosystems. For instance, establishing the governance boundaries has had more to do with the existing administrative units than how ecosystem components are distributed. Initial EBM efforts should build upon extant institutions and processes to illustrate the intent of EBM and develop broad-based support throughout the ocean community. There have been some interesting and positive reactions to EBM. Sectors have been encouraged to better organize themselves to respond to the challenges of EBM. Certainly, the ESSIM forum has been effective in engaging stakeholders and government agencies and developing a common vision of EBM; improved communication both within and among sectors should lead to more resilient governance structures in the future. Indeed, Hughes and colleagues (2005) point out that resilient social systems are dependent upon reliable feedback systems. These in turn depend upon communication mechanisms both within and among interacting groups, which are evolving in ESSIM. It is, however, impossible to predict the future path and timing of these developments.

Third, a hierarchical objectives structure—with overarching national objectives at the top, area-based objectives in the middle, and more-detailed sector-based operational objectives at the bottom—has been useful at a number of scales. It explicitly links internationally made commitments to regional and sector-specific management actions and can be understood by policymakers, operational managers, and stakeholders, facilitating buy-in to EBM. And it has provided a mechanism for the development of a consistent set of regionally specific objectives that work at both the planning-area and ocean-sector levels.

Fourth, experience to date has shown that the objective-setting process is as important as the product. Transparency and inclusiveness throughout the objectives' development is essential to building broad-based understanding of the goals of ESSIM. This is especially true of the integration of science into EBM, which must focus and be perceived as focusing on the priority issues. On this note, while progress is being made on the overall ESSIM plan and inclusion of ecosystem objectives in fishery management plans, a mechanism to establish priorities is still needed. For instance, regarding the socioeconomic objectives, different stakeholder groups may have different priorities that a consensual approach cannot resolve. Among ocean users (e.g., fisheries versus oil and gas), resource sharing and access are key. For NGOs and the public, how and what is impacted is critical. An objective approach based upon risk analysis is required.

Fifth, regarding management, actions are being undertaken at different spatial scales, such as the Gully MPA and coral conservation area. Certainly, a mixture of spatial management tools and best practices (gear modifications, effort controls, etc.) will be required, initially working with what exists and building from there.

Overall, good progress has been made on developing strong but flexible governance institutions with diverse venues for communication, active participation, and consensus building. Much still needs to be done to have the issues and priorities explicitly identified so that "hidden agendas" do not hold sway. EBM on the Scotian Shelf appears to be moving in the right direction.

Final Thoughts: What about Cod?

At the beginning of the chapter, we noted the collapse of the Canadian east coast cod fishery. Could ESSIM have prevented this collapse? The cause of this collapse has been debated but is generally recognized to have been overfishing (Myers et al. 1996), even though environmental conditions reduced stock productivity (DFO 2003). However, overfishing was the consequence of other problems in management (Burke et al. 1996), not least of which was overcapacity. Overcapacity, fueled by problems in resource-sharing arrangements, in turn caused enforcement problems. These issues can be and should be resolved regardless of whether or not EBM is adopted. Indeed, in 1992, under a single-stock management approach, resource-sharing arrangements were adopted within the fishery sector that largely resolved overcapacity problems and resulted in more-sustainable single-species management. It is our contention that EBM by itself would not have addressed overcapacity and thus likely would not have prevented the collapse. So the question should be, Assuming that these other issues (e.g., overcapacity) were addressed, would EBM provide the checks and balances in the management system that would lead to the avoidance of unintended consequences? We contend that EBM would lead to a more conservation-minded approach to management overall. At the stock level, management actions under an EBM approach would recognize the connections between one species and other parts of the ecosystem. At the ecosystem level, the cumulative impacts of human activities on the different various ecosystem components would be kept in view and adjustments would be made to mitigate these. Importantly, under EBM there would be a greater degree of consideration of the long-term ability of the system to provide a range of ecosystem services, rather than an attempt to maximize particular services (e.g., fishing yield) simultaneously from all relevant ecosystem components. However, it will only be through the continued efforts to design and implement EBM that we will be able to judge whether or not its full benefits will be realized.

Key Messages

1. A clear legislative basis for ecosystem-based management can greatly facilitate implementation.
2. EBM grounded in existing institutions and decision-making processes may more effectively engage diverse stakeholders.
3. A hierarchy of ecological and social objectives, including overarching objectives at the highest level and working through to operational objectives at the lowest level, is essential to EBM.
4. There is no single path to ecosystem-based management. Rather, EBM will require a mixture of management actions across different spatial and temporal scales.

Acknowledgments

The Eastern Scotian Shelf Integrated Management (ESSIM) initiative has been and continues to be the combined work of a wide variety of oceans and resource managers, scientists, and stakeholders. We would be remiss in not acknowledging the influential efforts of Joe Arbour, Tim Hall, Glen Herbert, Scott Coffen-Smout, Derek Fenton, Heather Breeze, and Melanie Hurlburt, without whom ESSIM would not be a reality. Mike Sinclair has been a strong supporter of ecosystem approaches to management and greatly influenced our ideas. Thanks to these and those countless unnamed others who have made ESSIM the success that it has been.

References

Breeze, H., D. G. Fenton, R. J. Rutherford, and M. A. Silva. 2002. *The Scotian Shelf: An ecological overview for ocean planning.* Canadian Technical Report of Fisheries and Aquatic Sciences 2393.

Bundy, A. 2005. Structure and functioning of the Eastern Scotian Shelf ecosystem before and after the collapse of groundfish stocks in the early 1990s. *Canadian Journal of Fisheries and Aquatic Sciences* 62:1453–73.

Burke, D. L, R. N. O'Boyle, P. Partington, and M. Sinclair. 1996. *Report of the second workshop on Scotia–Fundy groundfish management.* Canadian Technical Report of Fisheries and Aquatic Sciences 2100.

Coffen-Smout, S., R. G. Halliday, G. Herbert, T. Potter, and N. Witherspoon. 2001. *Ocean activities and ecosystem issues on the Eastern Scotian Shelf: An assessment of current capabilities to address ecosystem objectives.* Canadian Science Advisory Secretariat. Research document 2001/095. Ottawa: Department of Fisheries and Oceans.

Coffen-Smout, S., G. Herbert, R. J. Rutherford, and B. L. Smith, ed. 2002. Proceedings of the first Eastern Scotian Shelf Integrated Management (ESSIM) forum workshop, Halifax, Nova Scotia, 20–21 February 2002. Canadian Technical Report of Fisheries and Aquatic Sciences 2604: xiii + 63 pp.

COSEWIC (Committee on the Status of Endangered Wildlife in Canada). 2003. COSEWIC assessment and update status report on the Atlantic cod *Gadus morhua* in Canada. Ottawa: COSEWIC.

DFO (Department of Fisheries and Oceans). 2002a. *Biological considerations for the re-opening of the Eastern Scotian Shelf (4TVW) haddock fishery.* Regional Fisheries Status Report 2002/03. Ottawa: Canadian Science Advisory Secretariat.

DFO (Department of Fisheries and Oceans). 2002b. Canada's Ocean Strategy. Ottawa: DFO.

DFO (Department of Fisheries and Oceans). 2002c. *Fisheries management planning for the Canadian Eastern Georges Bank groundfish fishery.* Regional Fisheries Status Report 2002/01. Ottawa: Canadian Science Advisory Secretariat.

DFO (Department of Fisheries and Oceans). 2004. *Identification of ecologically and biologically significant areas.* Ecosystem Status Report 2004/06. Ottawa: Canadian Science Advisory Secretariat.

DFO (Department of Fisheries and Oceans). 2005. *Guidelines on evaluating ecosystem overviews and assessments: Necessary documentation.* Science Advisory Report 2005/026. Ottawa: Canadian Science Advisory Secretariat.

DFO (Department of Fisheries and Oceans). 2006a. *Coral conservation plan: Maritimes region (2006–2010).* Oceans and Coastal Management Report 2006/01. Ottawa: Canadian Science Advisory Secretariat.

DFO (Department of Fisheries and Oceans). 2006b. *Identification of ecologically significant species and community properties.* Science Advisory Report 2006/041. Ottawa: Canadian Science Advisory Secretariat.

DFO (Department of Fisheries and Oceans). 2003. *State of the Eastern Scotian Shelf ecosystem.* Ecosystem Status Report 2003/004. Ottawa: Canadian Science Advisory Secretariat.

Fletcher, W. J. 2005. The application of qualitative risk assessment methodology to prioritize issues for fisheries management. *ICES Journal of Marine Science* 62:1576–87.

Frank, K. T., B. Petrie, J. S. Choi, and W. C. Leggett. 2005. Trophic cascades in a formerly cod-dominated ecosystem. *Science* 308:1621–23.

Gavaris, S., J. M. Porter, R. L. Stephenson, G. Robert, and D. S. Pezzack. 2005. *Review of management plan conservation strategies for Canadian fisheries on Georges Bank: A test of a practical ecosystem-based framework.* ICES CM 2005/BB: 05. Copenhagen: ICES.

Government of Canada. 2007. *Eastern Scotian Shelf Integrated Ocean Management Plan. Strategic Plan.* Dartmouth, Nova Scotia: Oceans and Habitat Branch, Fisheries and Oceans Canada. http://www.dfo-mpo.gc.ca/Library/333115.pdf.

Hughes, T. P., D. R. Bellwood, C. Folke, R. S. Steneck, and J. Wilson. 2005. New paradigms for supporting the resilience of marine ecosystems. *Trends in Ecology and Evolution* 20:380–86.

Jamieson, G., R. O'Boyle, J. Arbour, D. Cobb, S. Courtenay, R. Gregory, C. Levings, J. Munro, I. Perry, and H. Vandermeulen. 2001. Proceedings of the National Workshop on Objectives and Indicators for Ecosystem-based Management. Sidney, British Columbia, 27 February–2 March 2001. Department of Fisheries and Oceans. Proceedings Series 2001/09. Ottawa: Canadian Science Advisory Secretariat.

Kostylev, V. E., and C. G. Hannah. 2007. Process-driven characterization and mapping of seabed habitats. In *Mapping the seafloor for habitat characterization*, ed. B. J. Todd and H. G. Greene, 171–84. Special Paper. St. John's, Canada: Geological Association of Canada.

Myers, R. A., J. A. Hutchings, and N. J. Barrowman. 1996. Hypotheses for the decline of cod in the North Atlantic. *Marine Ecology Progress Series* 138:293–308.

O'Boyle, R., M. Sinclair, P. Keizer, K. Lee, D. Ricard, and P. Yeats. 2005. Indicators for ecosystem-based management on the Scotian Shelf: Bridging the gap between theory and practice. *ICES Journal of Marine Science* 62:598–605.

O'Boyle, R., and G. Jamieson. 2006. Observations on the implementation of ecosystem-based management: Experiences on Canada's east and west coasts. *Fisheries Research* 79:1–12.

O'Boyle, R., and T. Worcester. 2006. *Proceedings of a Regional Advisory Process meeting on the maintenance of the diversity of ecosystem types: Phase II: Classification and characterization of Scotia–Fundy benthic communities.* Department of Fisheries and Oceans. Proceedings 2006/ 006. Ottawa: Canadian Science Advisory Secretariat.

Rosenberg, A. A., W. J. Bolster, K. E. Alexander, W. B. Leavenworth, A. B. Cooper, and M. G. McKenzie. 2005. The history of ocean resources: Modeling cod biomass using historical records. *Frontiers in Ecology and the Environment* 3:84–90.

Rutherford, R. J., G. J. Herbert, and S. S. Coffen-Smout. 2005. Integrated ocean management and the collaborative planning process: The Eastern Scotian Shelf Integrated Management (ESSIM) initiative. *Marine Policy* 29:75–83.

Shelton, P. A., A. F. Sinclair, G. A. Chouinard, R. Mohn, and D. E. Duplisea. 2005. Fishing under low productivity conditions is further delaying recovery of northwest Atlantic cod (*Gadus morhua*). *Canadian Journal of Fisheries and Aquatic Sciences* 63:235–38.

Sinclair, M., R. Arnason, J. Csirke, Z. Karnicki, J. Sigurjonsson, H. R. Skjoldal, and G. Valdimarsson. 2002. Responsible fisheries in the marine ecosystem. *Fisheries Research* 58:255–65.

Sinclair, M., R. O'Boyle, D. L. Burke, and G. Peacock. 1997. Why do some fisheries survive and others collapse? In *Developing and sustaining world fisheries resources: The state of science and management: Second World Fisheries Congress*, ed. D. A. Hancock, D. C. Smith, A. Grant, and J. P. Beumer, 23–35. Collingwood, VIC, Australia:CSIRO.

Smith, A. D. M., E. J. Fulton, A. J. Hobday, D. C. Smith, and P. Shoulder. 2007. Scientific tools to support the practical implementation of ecosystem-based fisheries management. *ICES Journal of Marine Science* 64:633–39.

Trzcinski, M. K., R. Mohn, and W. D. Bowen. 2006. Continued decline of an Atlantic cod population: How important is gray seal predation? *Ecological Applications* 16:2276–92.

Walker, B., and D. Salt. 2006. *Resilience thinking: Sustaining ecosystems and people in a changing world.* Washington, DC: Island Press.

Zwanenburg, K. C. T., A. Bundy, P. Strain, W. D. Bowen, H. Breeze, S. E. Campana, C. Hannah, E. Head, and D. Gordon. 2006. *Implications of ecosystem dynamics for the integrated management of the Eastern Scotian Shelf.* Canadian Technical Report of Fisheries and Aquatic Sciences 2652.

CHAPTER 15
Chesapeake Bay, USA

Donald F. Boesch and Erica B. Goldman

The Chesapeake Bay, the nation's largest estuary, arguably exemplifies the world's most comprehensive, sustained, and institutionalized effort to develop and apply marine ecosystem-based management (EBM). When it began in the 1980s, the Chesapeake Bay Program, an intergovernmental coalition of the US federal government and the states of the region, focused mainly on improving water quality. The program has evolved into a multifaceted set of agreements and actions affecting living resource management, habitat restoration, and human activities throughout the bay's vast watershed within a common framework.

The Chesapeake Bay stretches 332 km long, with tidal waters extending over 11,400 km^2 and 12,870 km of shoreline over the main stem of the bay and its tributaries (fig. 15.1). The Chesapeake drainage basin covers 166,000 km^2 in six states and the District of Columbia. Because the bay is relatively shallow (mean depth 6.5 m), it has relatively small volume in relation to the size of its catchment. In tandem with its modest tidal exchange, this makes the bay very susceptible to activities on land that affect inputs of freshwater, sediments, and dissolved materials to the estuary.

The Chesapeake region played a central role in the early history of the United States, and the nation's capital, Washington, D.C., sits near its center. Approximately 16 million people now live in the Chesapeake basin, with most in population centers near the tidal headwaters of estuarine tributaries, such as the Washington, Baltimore, Richmond, and Norfolk metropolitan areas. The bay includes important commercial and military ports and serves as a valuable recreational resource. Although a number of its fisheries have declined, in 2005 the bay still supported commercial fisheries worth over $113 million to harvesters (NMFS 2007). The estuary receives heavy doses of domestic and industrial waste disposal, with about five thousand discharges into the estuary or water bodies draining into it. At the same time, about 60% of the bay's watershed is forested and 25% is devoted to agriculture (Boesch and Greer 2003).

Although indigenous human populations influenced the landscape and, at least locally, affected populations of exploited marine organisms, far more significant changes in the ecosystem resulted from European settlement, mechanization, and the dramatic increase in human population (Boesch 2006). The clearing of forests for agriculture caused soil erosion that increased turbidity and sedimentation in the estuary. Agrarian expansion during the late eighteenth and early nineteenth centuries caused more sediments and nutrients, forms of nitrogen and phosphorus that the native forests had efficiently retained, to wash down into the bay, subtly altering its natural production and its food web (Kemp et al. 2005). Later in the nineteenth century, industrialization increased pollution, particularly by trace metals, and provided the mechanical means to exploit the abundant oysters, effectively strip-mining reefs once so massive that they hindered navigation by the European colonists (fig. 15.2).

The mid-1900s marked the beginning of the petrochemical period in the bay's environmental history, bringing pesticides and other manufactured organic chemicals, petroleum

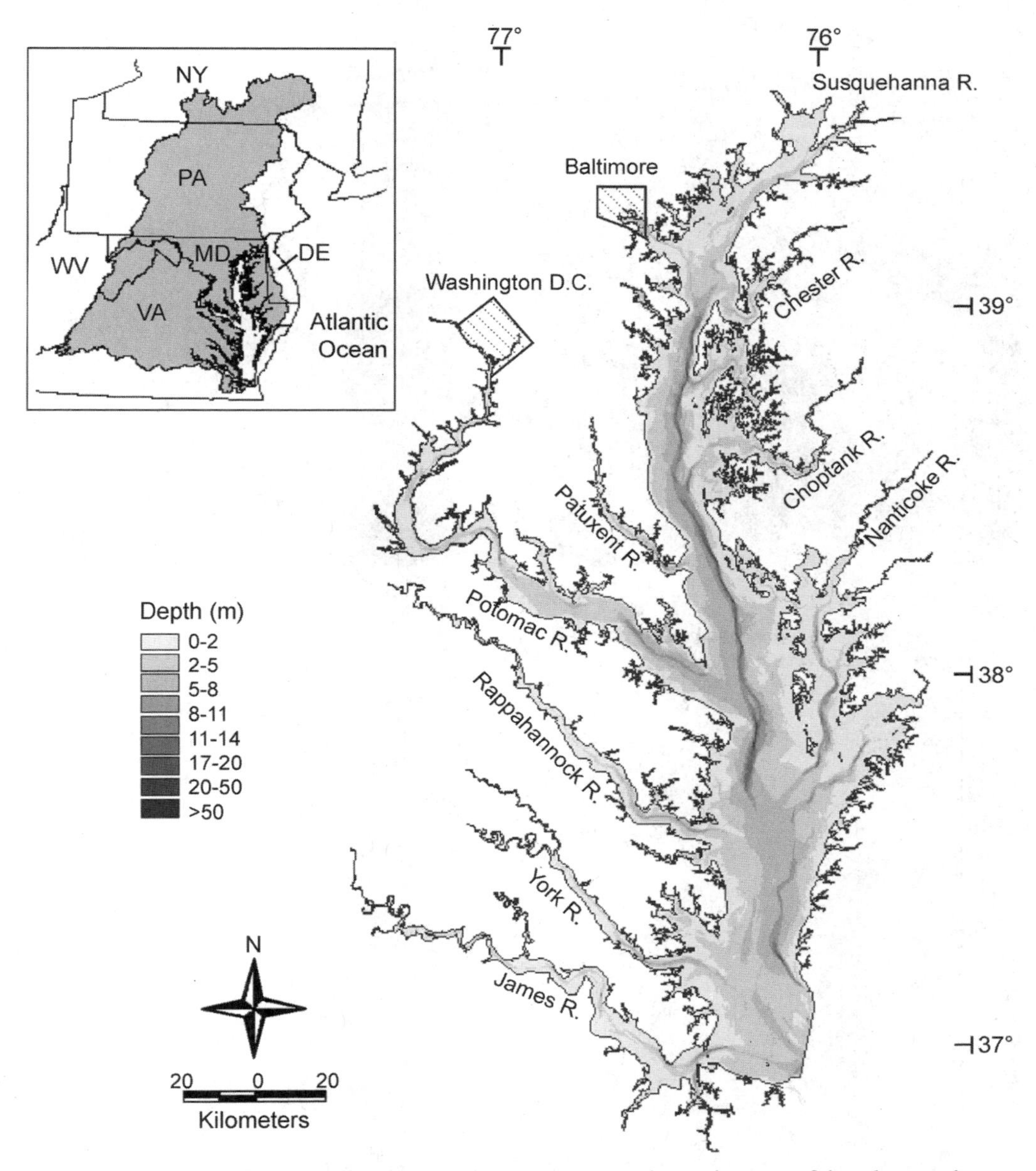

Figure 15.1 Map of the Chesapeake Bay. Despite its deep central trough, most of the Chesapeake Bay is shallow and wide, with a mean depth of 6.5 meters. The bay exists as a remnant of the drowned Susquehanna River Valley and contains numerous tidal river subestuaries. Modified with permission from Kemp et al. 2005.

by-products, and industrially produced fertilizers that allowed intensified agriculture, growing populations, and sprawling development. This exacerbated eutrophication linked to excessive nutrient loading, produced toxic hotspots, and put further pressure on already heavily exploited living resources.

During most of the twentieth century, public concern, scientific research, natural resource policies, and concerted management

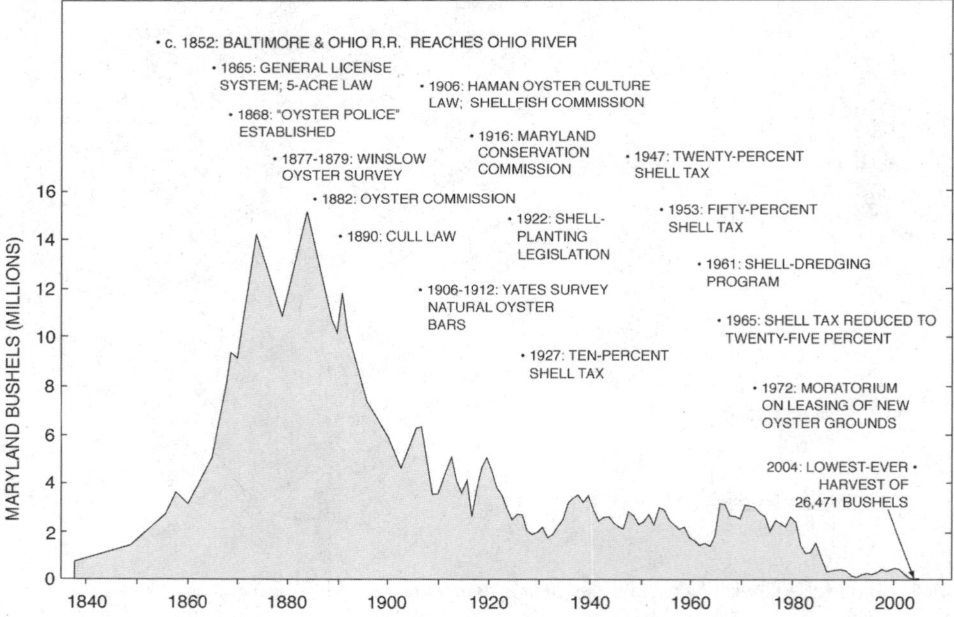

Figure 15.2 Declining oyster harvests in the Chesapeake Bay. Steel dredges and tongs virtually strip-mined the Chesapeake's historic oyster reefs in the late nineteenth and early twentieth centuries. This early-twentieth-century photograph documents the immense quantities of oysters once harvested from the bay. Note the people standing on the mounds (courtesy of Hampton History Museum, James S. Darling Oyster Packers, Hampton, circa 1910). The graph shows Maryland's plummeting oyster harvest, first after the dismantling of the virgin reefs, and then following population decline as the parasitic diseases MSX and Dermo struck remaining stocks. The major events in the history of oysters noted above failed to stop the precipitous decline. Adapted from Kennedy and Breisch 1981.

actions addressed issues separately and on relatively local scales, including fisheries exploitation, infectious waste disposal, industrial and municipal pollution, wetland loss, pesticides, channel dredging, and power plant effects (Davidson et al. 1997). But only after the devastating effects of the record floods associated with Tropical Storm Agnes in 1972 did managers, and even scientists, begin to conceive that these issues had consequences on the scale of the whole bay (Malone et al. 1993).

From the 1980s to the present, approaches to managing and restoring the Chesapeake Bay have become increasingly "ecosystemic," seeking integration, spanning spatial and temporal scales, assessing species interactions and top-down versus bottom-up controls, and transcending atmospheric, terrestrial, freshwater, and estuarine media. As a result, the Chesapeake Bay is often cited as a leading test bed, if not a model, for ecosystem-based management in various national (USCOP 2004) and international (UNEP/GPA 2006) forums. The timeline below describes the progression toward current ecosystem-based management efforts, following the bay through early efforts to periods of raised consciousness and scientific analysis (fig. 15.3).

How successful has the bay restoration effort been, and why has it not been more successful? This chapter provides an overview of the evolution of ecosystem-based management of the Chesapeake Bay and evaluates both its accomplishments and the major impediments to its implementation. We explore how the concepts of resilience in the Chesapeake ecosystem—in both its natural and human dimensions—can be applied in guiding achievable and desirable trajectories for its restoration. Our objective is not only to provide lessons learned from past efforts in this region, but also to illuminate paths forward for the restoration of this and other ecosystems.

Evolution of Ecosystem-Based Management in Chesapeake Bay

The Chesapeake Bay Program

Changes in the Chesapeake ecosystem apparent during the twentieth century petrochemical period accelerated those changes that began with European colonization, resulting in a dramatic state change from a relatively clear-water ecosystem, characterized by abundant plant growth in the shallows, to a turbid ecosystem dominated by abundant microscopic plants in the water column and stressful low-oxygen conditions during the summer (Kemp et al. 2005). This state shift largely resulted from the dramatic increase in nutrient inputs in the form of wastes from the growing human population, runoff of agricultural fertilizers and animal wastes, and atmospheric deposition of nitrogen oxides produced by fossil fuel combustion. By the mid-1980s the Chesapeake Bay was receiving about seven times more nitrogen and sixteen times more phosphorus than at the time English colonists arrived (Boynton et al. 1995). Sediment inputs to the bay, which probably peaked during nineteenth-century land clearing over most of the basin, have remained high, fed not only by soil erosion in the catchment, but also by sediments stored in streams, rivers, and reservoirs and by the retreat of tidal shorelines (Langland and Cronin 2003). To compound these effects, the drastic depletion of once prodigious oyster populations and loss of wetlands and riparian forests diminished important sinks for nutrients and sediments within both the watershed and the estuary.

The slow recovery of the ecosystem following the record floods of 1972 heightened public and political concerns about the deterioration of the bay (Malone et al. 1993; Horton 2003). The scientific community began to document the pervasive changes in the ecosystem and to identify their causes. This led to growing awareness of the bay's degradation by the public and political leaders, which in turn resulted in the

Continued Exploitation; Some Regulation (1890s-1960s)

Shared Experience & Raised Consciousness(1965-1976)

Scientific Analysis with Political Backing (1977-1983)

Implementation and Monitoring (1983-present)

Ecological Perturbations

Great oyster reefs dismantled (1890s) – mechanical harvest methods/fishing pressure

Oyster harvest one-third of peak (1930) – harvest pressure prevents reef recovery

Storm King (1933) – storm surge

Drought (1930s)

Oyster disease Dermo (1948)

Oyster disease MSX in Chesapeake Bay (1959)

Drought (1960s)

Hurricane Camille (1969) – floods

Tropical Storm Agnes (1972) – floods

Drought (1980s)

Hurricane Isabel (2003) – storm surge

1890 1900 1910 1920 1930 1940 1950 1960 1970 1980 1990 2000 2010

Year

Continued Exploitation; Some Regulation (1890-1960s)

▸**1890** – Cull Law sets minimum legal size for market oysters and required return of shells with spat and young oysters to natural oyster bars.
▸**1920s** – Shell planting legislation passes.
▸**1930s** – Reforestation programs begin to increase forested area of the watershed.
▸**1933** – Interstate conference agrees to treat the Bay as a single resource unit.
▸**1945** – Population density explodes and chemical fertilizer use expands.
▸**1948-1959** – Oyster diseases MSX and Dermo make their first appearance in the Bay.

Shared Experience and Raised Consciousness (1965-1976)

▸**1960s** – Use of chemical fertilizer grows dramatically.
▸**1967** – Chesapeake Bay Foundation established; public consciousness about the condition of the Bay starts to grow; "Save the Bay" bumper stickers appear. Media coverage of the environment deepens.
▸**1972** – Tropical Storm Agnes causes a huge influx of freshwater to the Bay, dropping the salinity of the water and washing in a pulse of nutrients and sediment.
▸**1975** – Senator Mac Mathias tours the Bay and requests that the EPA commence a study of the Bay's problems.
▸**Late 1970s** – William Warner publishes *Beautiful Swimmers*; Tom Horton publishes articles in the *Baltimore Sun*

Scientific Analysis with Political Backing (1977-1983)

▸**1977** – A six-year, $26 million EPA-funded study begins.
▸**1978** – The Chesapeake Bay Legislative Advisory Commission is created to develop a management approach.
▸**1980** – The Chesapeake Bay Commission is created, charged with developing a cooperative agreement between Maryland and Virginia to clean up the Bay.
▸**Early 1980s** – Drought conditions intensify the impact of MSX on the oyster population.

Implementation and Monitoring (1983-present)

▸**1983** – The first Chesapeake Bay Agreement is signed.
▸**1984** – The Chesapeake Bay Program, which brings together 7 states and 34 legislative initiatives, is formed. The Maryland Critical Area Law restricting development is also passed that year.
▸**1985-1987** – Another drought entrenches the parasitic diseases MSX and Dermo on the Bay's oyster population.
▸**1987** – A new Chesapeake Bay Agreement extends the 1983 compact and calls for specific benchmarks, including the 40 percent reduction in nitrogen and phosphorus by the year 2000.
▸**2000** – The Chesapeake 2000 (C2K) agreement revises the 1987 benchmarks and sets new target dates for various restoration goals.

evolution of regional management structures and restoration objectives (Hennessey 1994; Boesch et al. 2001).

A coordinated restoration effort began in 1983 "to assess and oversee the implementation of coordinated plans to improve and protect the water quality and living resources of the Chesapeake Bay estuarine systems," pulling together the three primary states in the Chesapeake watershed (Pennsylvania, Maryland, and Virginia), the District of Columbia, and the federal government to form the Chesapeake Bay Program (CBP 1999). In the years to follow, the CBP Executive Council, consisting of the governors of the three states, the D.C. mayor, the administrator of the US Environmental Protection Agency, and the chair of the Chesapeake Bay Commission (an organization of the legislatures of the three states), would endorse three major agreements and tens of directives and adoption statements related to reductions of nutrient and toxicant loadings, habitat restoration, living resource management, landscape management, and stewardship education. This institutional arrangement, including the high-level engagement of federal and state governments, has endured for over 20 years. Furthermore, the states of Delaware, New York, and West Virginia have recently joined in many CBP activities.

In 1987, the Chesapeake Bay Program entered into a landmark agreement to address the problems caused by eutrophication, which had come to be recognized as the most significant human impact on the bay. The agreement promised to reduce controllable inputs of nitrogen and phosphorus by 40% by the year 2000. As that year approached, it became clear that this target would not be reached. Furthermore, a growing understanding developed that a complex array of interrelated issues concerning environmental quality, living resources, and human activities should be addressed in a more integrated manner. The result was a new, more comprehensive agreement with a target of 2010 for achieving many of its key objectives. The Chesapeake 2000 Agreement, which includes over one hundred goals and commitments (table 15.1), comprises one of the most ambitious ecosystem management programs for any large coastal area in the world. Only the Helsinki Commission (Baltic Marine Environment Protection Commission, or HELCOM) and OSPAR (from the Oslo–Paris Convention for Protection of the Marine Environment of the North-East Atlantic) rival the Chesapeake 2000 Agreement in scope.

While the CBP may have helped to stem the further deterioration of the Chesapeake ecosystem, there are limited signs of actual improvements in the bay. For example, concentrations of a number of potentially toxic substances (some trace metals and chlorinated hydrocarbons) in sediments and organisms have declined as a result of source controls and waste treatment, but the industrialized harbors in the bay remain heavily contaminated. Other subregions also show elevated concentrations of toxicants or evidence of biological effects. Algal blooms and seasonal hypoxia in bottom

Figure 15.3 (opposite) Management eras in the Chesapeake Bay watershed. Industrialization and unsustainable extraction of resources intensified stresses on the Chesapeake Bay watershed in the latter half of the nineteenth century. By 1890, 60% of forested land had been cleared, and the destruction of historic oyster reefs was well under way. This timeline begins with early efforts toward regulation near the turn of the twentieth century and follows the Chesapeake through four distinct management eras. Superimposed are a series of perturbations, including storms, commercial oyster harvests, and oyster diseases. Tropical Storm Agnes in 1972, often cited as a sentinel ecological event, set off a major state change that moved the bay into its current eutrophied state. After Kennedy and Breisch 1981 and Constanza and Greer 1995.

Table 15.1. Goals of the Chesapeake 2000 Agreement

This agreement reaffirmed the commitments made in 1983 and 1987 by the states of Virginia, Maryland, and Pennsylvania and the District of Columbia, the Chesapeake Bay Commission, and the US Environmental Protection Agency, representing the federal government, to protect and restore the Chesapeake Bay's ecosystem. Specific objectives and actions under each goal are briefly summarized, with target dates for completing major commitments.

Living Resource Protection and Restoration

GOAL: Restore, enhance, and protect the finfish, shellfish, and other living resources and their habitats and ecological relationships to sustain all fisheries and provide for a balanced ecosystem.

Oysters: By 2010, achieve, at a minimum, a tenfold increase in native oysters.

Exotic species: By 2001, identify and rank nonnative, invasive, and terrestrial species, and by 2003, develop and implement management plans for those species deemed problematic to restoration and integrity of the ecosystem.

Fish passage and migratory and resident fish: By 2003, restore passage to more than 1,357 miles of blocked river habitat, set new goals, establish a monitoring program to assess outcomes, and revise management plans for migratory fish.

Multispecies management: By 2004, assess effects of different population levels of filter feeders; by 2005, develop ecosystem-based management plans; and by 2007, revise and implement existing plans to incorporate ecosystem approaches.

Crabs: By 2001, establish harvest targets for blue crab fishery and implement complementary state management strategies.

Vital Habitat Protection and Restoration

GOAL: Preserve, protect, and restore those habitats and natural areas vital to the survival and diversity of the living resources of the bay and its rivers.

Submerged aquatic vegetation: By 2002, revise restoration goals that expand the existing goal of protecting and restoring 114,000 acres of vegetation.

Watersheds: By 2010, develop and implement locally supported watershed management plans for two-thirds of the watershed.

Wetlands: By 2010, achieve a net gain by restoring 25,000 acres of tidal and nontidal wetlands; evaluate the potential impact of climate change, particularly with respect to wetlands.

Forests: By 2003, establish a new goal to expand the riparian forest buffer restoration goal of 2,010 miles by 2010.

Water Quality Protection and Restoration

GOAL: Achieve and maintain the water quality necessary to support the aquatic living resources of the bay and its tributaries and to protect human health.

Nutrients and sediments: By 2010, correct nutrient- and sediment-related problems in the bay and tidal tributaries sufficiently to remove them from the list of impaired waters under the Clean Water Act; by 2003, adopt new or revised water quality standards and adopt and begin implementing strategies to prevent the loss of sediment retention capabilities of lower Susquehanna River dams.

Table 15.1. *(continued)*

Water Quality Protection and Restoration (continued)

Chemical contaminants: Reduce these to levels that result in no toxic or bioaccumulative impacts on living resources; by 2010, eliminate mixing zones for persistent or bioaccumulative toxics.

Priority urban waters: Restore urban harbors; by 2010, achieve restoration goals for Anacostia River (District of Columbia).

Air pollution: By 2003, assess effects of airborne nitrogen compounds and chemical contaminants, and establish reduction goals.

Boat discharges: By 2003, establish "no discharge zones" for human wastes; by 2010, expand by 50% the availability of waste pump-out facilities.

Sound Land Use

GOAL: Develop, promote, and achieve sound land use practices that protect and restore watershed resources and water quality, maintain reduced pollutant loadings for the bay and its tributaries, and restore and preserve aquatic living resources.

Land conservation: By 2010, permanently preserve from development 20% of the land area of the watershed.

Development, redevelopment, and revitalization: By 2012, reduce the rate of harmful sprawl development by 30%; by 2005, remove state and local impediments to low-impact development; by 2010, rehabilitate and restore 1,050 brownfield sites.

Transportation: By 2002, promote coordination with land use planning to reduce dependence on automobiles; establish policies and incentives that encourage use of clean vehicle and other transportation technologies.

Public access: By 2010, expand public access points to the bay and its tributaries by 30%; by 2005, increase designated water trails by 500 miles.

Stewardship and Community Engagement

GOAL: Promote individual stewardship and assist individuals, community-based organizations, local governments, and schools to undertake initiatives to achieve the goals and commitments of this agreement.

Education and outreach: Beginning with the class of 2005, provide meaningful bay or stream experience for every student in the watershed; integrate information about the Chesapeake into school curricula and university programs.

Community engagement: Enhance small watershed and community-based actions, and incorporate local governments into policy decision making; by 2005, identify actions to address communities where poor environmental conditions have contributed to disproportionate impacts.

Government by example: By 2001, develop and use government properties consistent with goals, expand use of clean vehicle technologies, and address runoff from federal and state land.

Partnerships: Seek agreements on issues of mutual concern with Delaware, New York, and West Virginia, and establish links with community-based organizations throughout the watershed.

Source: CBP 1999.

waters occur as pervasively as ever. Sea grasses have returned in some regions but cover only a small portion of the habitat occupied in the 1950s. Oyster (*Crassostrea virginica*) populations have not recovered, because of degraded reef habitat and ravages of two parasitic pathogens, commonly referred to as MSX and Dermo.

Populations of several anadromous fishes have increased modestly as a result of the removal of barriers to upstream migration. Populations of striped bass (*Morone saxatilis*) represent perhaps the most dramatic example of recovery, increasing as a result of a 5-year moratorium on harvest followed by more precautionary management of stocks (Russell 2005). On the other hand, the very productive blue crab (*Callinectes sapidus*) fishery has danced along on the cusp of recruitment overfishing (CBSAC 2007). In the spring of 2008, the governors of Maryland and Virginia acted jointly to slash the harvest of female blue crabs by 34%, a decisive action to protect the integrity of the spawning stock (Fahrenthold 2008).

As the year 2010 fast approaches, scientists, managers, the press, and the public grow increasingly aware that the most critical goals (i.e., improvements in water quality and associated nutrient and sediment load reductions, submerged aquatic vegetation recovery, wetland restoration, recovery of stocks, and implementation of multispecies management) will not be met by the target dates. Furthermore, the CBP has been criticized by the press (Whoriskey 2004), and subsequently by the Government Accountability Office (GAO 2005), for misrepresenting progress by overemphasizing measurements of implementation processes. One major criticism focuses on the lack of alignment between projections, from a watershed model used to track presumed nutrient load reductions, and measurements of actual outcomes, such as measured loads and water quality in the estuary. Furthermore, the GAO criticized the CBP for failure to develop and apply integrated measures of outcomes that reflect the overall health of the ecosystem.

While the CBP responded by initiating annual reports (CBP 2007) that segregate measures of implementation processes (termed "restoration efforts") and measures of actual condition (termed "ecosystem health"), much remains to be done in developing integrated ecosystem indicators, refining achievable goals and target dates, and accelerating implementation of restoration activities. More-explicit consideration also should be given to some of the key principles of ecosystem-based management as articulated by the US Commission on Ocean Policy (2004): integration, sustainability, adaptation, and precaution (Boesch 2006).

Resilience of the Chesapeake Bay

Ecological Resilience

Why haven't management efforts in the bay achieved more results? Beyond the difficulties faced by the Chesapeake Bay Program, the key to the bay's slow response to restoration efforts may lie in the stability of the ecosystem's current, undesirable state, which appears resilient in its own right (Walker and Salt 2006). As described by Leslie and Kinzig (chap. 4 of this volume), resilience—the ability of a social–ecological system to undergo, absorb, and respond to change and disturbance while maintaining its functions and controls—can characterize an undesirable or degraded state as much as it can a desirable one.

Accumulated changes in the Chesapeake watershed caused the system to transform into one that is more dominated by water-column microbial processes. Loss of forest buffers, increased nutrient loading due to fertilizer use, the growing human population, decreased capacity for oyster filtration, and rapid urban development of the landscape helped create hypoxic conditions and other changes. There is a growing body of evidence that the bay has become resilient in its now-degraded state (fig. 15.4). In recent years, the extent of seasonal hypoxia in the estuary has been greater than predicted from regression with nitrate loading over a 52-year time record, suggesting that the bay may have become more susceptible to nutrient loading. While nitrate concentrations in the Susquehanna River decreased during the 1990s, the extent of hypoxia increased in both high- and low-flow years relative to the 1980s (Hagy et al. 2004). One hypothesis suggests that repeated occurrences of very high flow and nitrogen loading in the years 1993, 1994, 1996, and 1998 increased the bay's susceptibility to eutrophication and hypoxia—susceptibility that has been augmented by the effects of depleted benthic communities in the mid bay, near elimination of oyster populations from the main stem, and decline in submerged aquatic vegetation (Kemp et al. 2005).

The physical lag time between nitrogen runoff in the watershed and delivery to the estuary is a force that may slow the response of the estuary to reduction in nitrogen inputs. Nitrogen in streams that drain into the bay comes from both surface runoff and groundwater discharge. While surface water delivery correlates closely with freshwater flow, groundwater moves slowly. The average age of groundwater in the bay's tributaries is 10 years, with a range from less than 1 year to more than 50 years. The slow movement of groundwater into the bay will cause a lag time, generating a delay between the implementation of management practices to reduce nitrogen loading and the measurable improvement of water quality (Lindsey et al. 2003).

To move toward a more desirable stable state, Chesapeake Bay management efforts must first *erode* the resilience of the current state, with the aim of breaching potentially distant thresholds to access a more desirable state. Uncertainty is inherent in such a management approach because predictive capabilities cannot yet accurately forecast these anticipated threshold events. Furthermore, the bay is large and includes multiple tributaries in different states, and threshold responses are likely to vary spatially. For example, as discussed later,

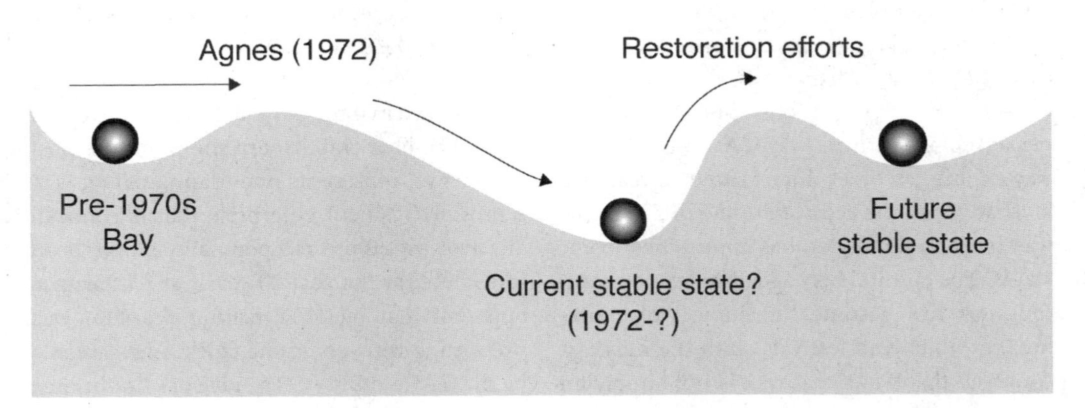

Figure 15.4 Resilience framework for the Chesapeake Bay. Resilience theory offers a conceptual framework in which to consider the changing ecological states of the Chesapeake Bay ecosystem. The erosion of buffers such as forests, wetlands, and oyster reefs diminished the resilience of the bay's precolonial ecological state. In the late twentieth century, the Chesapeake ecosystem experienced an abrupt shift to its current degraded, yet stable and resilient, state. Efforts to restore the ecosystem aim to achieve a state that provides ecosystem services similar to those provided by the bay of the past. Adapted from Gunderson 2000.

submerged vegetation has dramatically and unexpectedly recovered on the Susquehanna Flats in the uppermost bay, while hypoxia in the deep channel of the bay has proved more recalcitrant.

Social and Institutional Resilience

The social and ecological components of the Chesapeake Bay watershed are closely coupled (Costanza and Greer 1995). Ecological degradation of the estuary has forced social and institutional change, in many instances building resilience within the human dimensions of the ecosystem. As the natural resources of the bay have declined from the latter half of the twentieth century to the present, the livelihood of individuals who work on the water has been similarly jeopardized. Oysters and oystermen make a good case study in this regard. Some of the motivation for the current drive to restore the oyster population to the bay is to help bolster the bay's flagging oyster harvest and to keep the oysterman's way of life from dying out completely. But this is changing as sportfishers and waterfront-dwelling oyster gardeners emerge as important stakeholder advocates for restoring the ecological role of oysters.

The cultural image of watermen (those who earn income from harvesting the bay's living resources) is synonymous with the Chesapeake region in many ways. As this historical way of life becomes increasingly threatened by changes in the ecosystem, the image of the watermen in the region has approached iconic status. The cultural heritage of watermen communities has become the subject of museums, exhibits, and festivals, with the message conveyed that what is at risk is not simply an occupation, but an entire way of life and its associated cultural dimensions (Chambers 2006).

From an anthropological perspective, this view of Chesapeake watermen, while true in many respects, represents a distortion that links the lives of people in time and space to their dependence on particular occupations and places. Many waterfront communities, for example, reveal different sets of traditions in which mobility and adaptability are fundamental to their way of life. The necessity of combining work on the water with other occupations has historically factored into the lives of many families in these communities. Their culture is one of so-called *inherent* resilience, which enables adaptation to changing economic and environmental conditions (Chambers 2006).

The adaptability of individuals may increase social resilience as the region transforms. Increasing population density and development in coastal communities requires a continually evolving redefinition of the concept of "sense of place," one traditionally associated with historic watermen and agricultural communities (Wasserman et al. 1999). As geographically isolated watermen communities give way to retirement villages and waterfronts developed for recreation, cultural conditions may change irreversibly. Even if a recovered state of the bay—one with clear waters and abundant oysters, crabs, and fish—might rejuvenate flagging fisheries and bolster the economy of the region, the community would be culturally distinct from the 1950s or 1960s.

The Chesapeake's regional economy has also adapted in many ways to the bay's degraded state. The blue crab fishery, the estuary's most productive, represents one example of an economically resilient enterprise that is continually evolving. Blue crab populations have been in decline for the past 20 years, and it became apparent that by 1997 fishing pressure was preventing recovery of the stock. Assessments by the Bi-State Blue Crab Advisory Committee (BBCAC), a body formed by the Chesapeake Bay Commission in 1996 to explore and coordinate blue crab fishery management options, led to efforts by the three management jurisdictions to reduce fishing mortality to allow stocks to recover (BBCAC 2006). Despite these efforts, the blue crab population still hovers dangerously near the overfishing threshold (CBSAC 2007).

While the crab fishery is still the most valuable in the Chesapeake region, commercial landings are significantly lower than in years past. In addition, traditional processing practices, which require the painstaking handpicking of crabmeat from shells, no longer hold local appeal and increasingly have become jobs filled by seasonal migrant workers. New restrictions now set a quota for the number of seasonal visas that can be issued, making it increasingly difficult for processors to hire a sufficient number of pickers each season (Kobell and Guy 2005).

Meanwhile, people in the Chesapeake region and beyond still have a taste for crabmeat. Phillips Foods, one of the best-known restaurant families in the area, would historically buy only domestic product, much of it from Maryland processors. Due to the combined pressures of a fast-growing demand for fresh crabmeat year-round and a declining natural resource, Phillips Foods has looked beyond the bay for a consistent supply for the past 15 years. Much "Maryland-style crabmeat" now comes from Asia, mostly from Thailand and the Philippines. This crabmeat is not from the blue crab *Callinectes sapidus*, but from the Asian blue swimming crab, *Portunus pelagicus*, which is not as sweet but is very similar in texture. In 2005, more than 46 million pounds of crabmeat, worth nearly $300 million, entered the United States from abroad (Epstein and Desmon 2006).

Today, while local bay blue crabs still dominate the regional whole-crab market, much packaged and processed crabmeat comes from abroad. The effect of globalization on the region has helped one portion of the seafood economy, although declining stocks and stricter harvest regulations have exerted a clear toll on traditional fishers and seafood processors. One could argue that overall the economic resilience of the crab industry has increased, due in large part to the fact that the health of the industry has become decoupled from the ecological health of the bay.

Role of the Sciences in Management and Restoration

Natural Science

Scientific understanding of Chesapeake Bay processes has played an essential role in developing the goals of ecosystem-based management through the Chesapeake Bay Program and in identifying ways that these goals can be met. Governmental officials and environmental advocates alike give science high marks: Their debates principally revolve around political will and the practicality of solutions and generally not the quality of scientific evidence and interpretation (Horton 2003; Ernst 2003). Many management decisions and goals have been informed by scientific information, routinely through the active engagement of research scientists and technically trained analysts. Specific examples in which scientific understanding drove the diagnosis of impairments to the ecosystem and determined the actions needed to alleviate them include commitment to reduction of nutrient loading, application of enhanced biological nutrient removal technology, harvest restriction on striped bass and, more recently, blue crabs, and significant restrictions on the use of toxic tributyltin antifouling coatings on vessels.

The successes of environmental science in the region were no doubt aided by the fact that Virginia and Maryland have long supported applied research on the bay, helping to sustain academic research institutions with a mission and culture of service. However, adjustment of regional policy on the basis of new scientific knowledge has seldom been easy. For example, the views of scientists who initially argued that nitrogen loads to the bay must be reduced to restore water quality and ecosystem health were at the time dismissed and marginalized (Malone et al. 1993). Now reduction of nitrogen loads to the bay is the number one priority, showing the capacity for policy to catch up with robust science, when it exists.

The CBP also prides itself on the adoption of quantitative goals and use of state-of-the-art models to set goals and determine what must be done to achieve them. The centerpiece of these efforts is a series of linked models that project the amount and consequences of loads of nutrients entering the bay. These models have been applied in the inverse mode to determine the regional allocation of load reductions needed throughout the watershed to achieve the water quality requirements of important living resources (Koroncai et al. 2003). However, many other relationships among actions, goals, and environmental outcomes in the Chesapeake 2000 Agreement are, at best, integrated only through a qualitative conceptual framework. In particular, the restoration of water quality is presumed to yield significant positive benefits to fisheries production, but quantitative evidence for this relationship remains sparse (Kemp et al. 2005).

In contrast to other regions now attempting ecosystem-based management, the CBP has the benefit of a long and robust data record. The environmental history of the bay has been reconstructed using a variety of biological and geochemical markers in the sedimentary record (Cooper and Brush 1991; Colman et al. 2002; Cronin and Vann 2003; Kemp et al. 2005). A systematic monitoring program to support the CBP began in the bay in 1985 and has been sustained without interruption. While still not optimal, the monitoring program has incorporated improvements in monitoring design, new measurement methods, and continuous, automated, and synoptic observations that span temporal and regional scales. As the GAO (2005) pointed out, however, the CBP had failed to make full use of the information generated by its monitoring in the development and use of ecosystem health indicators.

Because of its historical and still-dominant focus on water pollution, the CBP perspective has been largely a bottom-up one—thinking that if the inflow of materials can be appropriately adjusted, the health of the ecosystem will improve. Of course, ecosystem conditions depend on biological interactions, including those from the top down, through which animals affect the organisms they consume or various physicochemical processes. Such top-down interactions have been increasingly incorporated into models (for example, the influence of benthic animals on geochemical fluxes and the effects of grazers and filter feeders on phytoplankton). Improved understanding of these biological interactions and the integration of top-down and bottom-up drivers is a critical requirement for managing living resources in an ecosystem context. To this end, an ecosystem-based management effort for Chesapeake Bay fisheries has recently begun (Latour et al. 2003; CBFEAP 2006).

Social Sciences and Economics

Though research on Chesapeake Bay management has emphasized the biological and physical sciences, social scientists and economists have recently begun to play an important role. Anthropological approaches, for example, have had significant impacts in the management of Chesapeake Bay, specifically as part of the analysis of the blue crab fishery and its management by the Bi-State Blue Crab Advisory Committee. That analysis had three principal components: an extensive stakeholder survey and outreach effort; habitat, predation, and economic research; and the analysis and development of sustainable harvest targets and thresholds for the Chesapeake Bay blue crab fishery (BBCAC 2006). Scientists working with the BBCAC Technical Work Group reached a consensus that for the health of the fishery it would be necessary to reduce fishing pressure on the crab stock by some 15%. This finding led to management commitments by Maryland, Virginia, and the Potomac River Fisheries Commission. But many working crabbers disagreed with the consensus, even accusing the scientists of ineptitude or outright lying (Greer 2002).

Motivated to uncover the factors driving the mistrust between scientists and watermen, Paolisso (2002, 2006) developed a cultural model for watermen's beliefs regarding the blue crab fishery. Based on intensive interviews, he identified three major beliefs held by watermen that come in direct conflict with the goals of setting targets for the blue crab fishery: First, many watermen believe, only God and nature can determine the scarcity or abundance of crabs. Second, humans cannot and need not fully understand the blue crab but should trust God's stewardship. Science, they believe, may capture generalities, but it does not have direct relevance for what watermen experience on the water day to day. Third, regulations cannot manage nature and should not attempt to change naturally occurring cycles within a fishery.

Paolisso recognized that identifying and understanding the cultural basis of disagreement between watermen and scientists are only first steps. Unfortunately, with the implementation of new harvest regulations in Maryland in 2001, significant attempts at dialogue between scientists and watermen waned. Although the cultural model for watermen's beliefs helped to predict where the different groups might come into conflict, the fishery remains a long way from any form of comanagement.

Approaches for formally incorporating public perception into decision making are, however, gaining ground. For example, one effort created a public sentiment index (PSI) for Chesapeake Bay stakeholders, which incorporated the views of interested parties such as watermen, seafood wholesalers and retailers, recreational fishers, representatives from nongovernmental organizations, scientists, and managers on a range of ecological scenarios for the state of the bay. When the PSI calculation is high, stakeholders may be willing to accept difficult policy choices and support their implementation to restore the bay ecosystem. Even though the conflicting priorities of different user groups can make consensus building difficult, the PSI framework provides one approach for engaging stakeholders in public decision making, policy formulation, and implementation (Chuenpagdee et al. 2006).

Economists and economic theory have a growing role to play in ecosystem-based management efforts in Chesapeake Bay by helping to link the state of the bay's ecology to users of its resources. Two important economic concepts, resource valuation and allowance trading, have in particular helped to shape management decisions in the region. Economic analyses in resource valuation often come into play in assigning a value of particular ecosystem services to different stakeholders (Lipton and Wellman 1995). For example, economic models can assess the impact of changes in blue crab regulations to people working in the industry and to the region as a whole (Lipton 2004). The shifting economic significance in commercial and recreational fisheries in Chesapeake Bay helps illustrate how resource valuation can be used. As the populations of commercially harvested fisheries—oysters, crabs, and striped bass—have declined, the seafood production value of the bay has diminished, reflected by lower-than-expected price increases. At the same time, however, its value as a place for recreational fishing has increased. More people live closer to the bay and spend more time on the water. In fact, fishing data collected since 1981 show an increasing trend in recreational fishing trips in Maryland and Virginia by over 150,000 trips per year. Putting numbers and figures with these different uses of the bay (e.g., recreational versus commercial) serves as an important tool for policymakers and managers (Lipton 2006).

Economists have continued to analyze assigning value to certain ecosystem services for different user groups. A study on the value of water quality to Chesapeake Bay boaters, for example, calculated the amount that boaters would be willing to pay for improvements in water quality. The study found that boaters would pay a significant amount, especially if a poor water quality condition brought potential negative impacts on human health. In addition to the other ecosystem benefits of improved water quality, this study showed that water quality does impact the enjoyment of boating and that boaters would benefit if water quality conditions were to improve (Lipton 2004).

Economists are also helping to explore allowance trading for nutrient pollution in the Chesapeake Bay. The premise behind allowance trading, already used widely in implementation of the Clean Air Act, is to create a market incentive for reducing pollutants—sulfur dioxide in the case of the Clean Air Act, and nitrogen in the case of Chesapeake Bay restoration. This approach encourages sewage treatment plants and factories to reduce pollution more than required to meet permit limits. By doing so, facilities would obtain credits they can sell to plants that have difficulty complying, thus creating a financial incentive for upgrades and other technological innovation that would decrease pollution to the bay (Stephenson and Shabman 2001). For diffuse sources of pollution, like agriculture, such an approach presents more of a challenge, especially because of the difficulty in measuring and assigning loadings. Economists do not believe this obstacle is insurmountable (Stephenson et al. 1998), but genuine caps and accurate measurements will be essential if real trading is to occur.

Cap and trade allowances for the control of point source pollution are already being implemented in Virginia and Pennsylvania. In 2005, the Virginia General Assembly passed legislation to establish the first point source, watershed-based nutrient trading system in the bay region. The Pennsylvania Department of Environmental Protection issued draft policies for the first point-source to nonpoint-source nutrient trading program that would allow credit generation for both agricultural and urban runoff control (Bancroft 2005). Maryland has taken a different approach based on requiring and financing enhanced nutrient removal (ENR) in all but very small sewage treatment plants. The Chesapeake Bay Restoration Fund, supported by a $2.50 per household per month fee, provides a dedicated revenue stream (approximately $60 million per year) for necessary ENR upgrades to publicly owned wastewater treatment systems (Blankenship 2004). The role of nutrient trading involving point and nonpoint sources under this regime is still being evaluated.

Challenges to the Implementation of Ecosystem-Based Management

Scientific

The scientific enterprise faces significant challenges in the discovery, integration, and application needed to support ecosystem-based management for restoration in Chesapeake Bay. One of the most daunting of these challenges for research comes in making the transition from ecosystem diagnosis, through prognosis, and on to treatment (i.e., management). It is much easier to draw confident conclusions about how the ecosystem changed and the likely causes of those changes than it is to make predictions about the future that cannot be supported by experimental or observational evidence. Furthermore, while the science needed to prescribe remedial actions can be based partially on diagnosis of pathology or prognosis for ecosystem integrity and recovered services, it must also extend to consider the technological and socioeconomic feasibility of treatments. At this point, environmental science lacks a similar "dedication to the cure" that drives the relatively seamless integration of diagnosis, prognosis, and treatment characteristic of medical science (Palmer et al. 2004).

Ecosystem-based management in the Chesapeake Bay remains challenged by the scientific difficulties inherent in moving from diagnosis to treatment. The problem of nutrient overenrichment offers one example. There has been extensive research on the history and driving factors of eutrophication, along with impressive technical development of prognostic models that guide ecosystem restoration goals and targets (Kemp et al. 2005). But relatively little has been invested in the development of more-effective land management practices to reduce inputs of nutrients from diffuse sources or to verify the effectiveness of those practices (Boesch et al. 2001). As a consequence, managers have identified and credited "best management practices" such as agricultural nutrient management or storm water controls, but without demonstrable outcomes supporting their efficacy (OIG 2006). Similar challenges exist for fisheries management in moving from diagnosis of a decline, through prognoses for harvest restrictions and habitat restoration, to treatment in the form of effective and enforceable regulation or habitat rehabilitation. Of course, these deficiencies are not limited to the management of Chesapeake Bay but may just be more obvious there because of the level of attention to ecosystem restoration.

Another scientific challenge in the Chesapeake region derives from the mismatch between the rate of scientific discovery and the capacity for course correction at the institutional level. There is natural resistance among managers and policymakers to deviate from objectives and processes that were difficult to commit to in the first place. Yet, scientific research continues to develop new understanding and, if true to its tradition, should constantly question conventional wisdom. Active press coverage in the Chesapeake region provides opportunities for surprising or heretical findings to enter the public domain quickly. Sometimes this might serve as an impetus for policy change; at other times it has resulted in confusion and loss of momentum in executing existing policies. A highly visible example of this challenge was the response of scientific and management communities to presumed outbreaks of the toxic dinoflagellate *Pfiesteria piscicida* in 1997 and the controversies that still linger regarding the possible link between *Pfiesteria* and neurological deficits experienced by a small group of individuals that came in contact with the bloom (Belousek 2004). While some would suggest that the steps taken to protect public health and control nonpoint source pollution were an overreaction to speculative and unvetted results, others would suggest that these were appropriate, no-regrets actions (Fincham 2007).

The Chesapeake Bay Program has a scientific and technical infrastructure, including an active scientific and technical advisory committee. It also has a very responsive and responsible communication outlet in the form of the Bay Journal (http://www.bayjournal.com/). Together these two instruments can modulate the interpretation and transfer of information. Still, substantial improvements in interpretive and translational processes will be required to respond to the increasingly interdisciplinary issues encountered in ecosystem-based management.

Another challenge stems from the fact that some in the scientific community fundamentally disagree about how the Chesapeake Bay functions as an ecological system, specifically the degree to which it is primarily physically and chemically driven, from the bottom up, versus controlled by higher trophic levels. The idea that eutrophication is caused mainly by bottom-up forcing and physical controls—through which nutrient-loading levels affect primary production and thereby light limitation and oxygen depletion—drives much of the current management focus. It has long been suggested that the loss of regulating services provided by wetlands and filter-feeding oysters are also contributing factors (Newell 2004). In fact, such feedbacks are included in water quality–ecosystem models strategically used by the Chesapeake Bay Program. On the other hand, recent emphases on the preeminence of the large reductions of harvested species (Jackson et al. 2001), depletion or extinction of keystone species (Lotze et al. 2006), and loss of biodiversity (Worm et al. 2006) as the causes of the collapse of coastal ecosystems or declines in their ecosystem services may overstate the importance of top-down controls. Such perspectives underemphasize the dramatic changes in many coastal ecosystems during the late twentieth century due to changes in the delivery of water and sediment to coastal ecosystems, as well as substantial physical alterations. Clearly both bottom-up and top-down forces affect the Chesapeake Bay. An integrated understanding of these forces will be a major scientific challenge for effective ecosystem-based management, particularly with regard to fisheries.

Finally, there is the difficulty of understanding and communicating scientific uncertainty. The CBP has depended heavily on a single array of deterministic models for strategic forecasting. These models depend on specified initial conditions that are not subject to random fluctuations. CBP has used model output that simulates the current time period (so-called nowcasts) as measures of restoration progress. This approach provides a false sense of certainty in outcomes and status. Monitoring results, on the other hand, are highly variable, with signals obscured by noise caused by environmental variability and stochastic behavior. Not only should there be more realistic portrayals of the uncertainties in model projections, but modeling and monitoring should also be more effectively integrated. An adaptive management framework, which moves iteratively between modeling and monitoring, can use each to provide information about the success of management actions (Boesch 2006).

Institutional

Despite the development of the Chesapeake 2000 Agreement, a remarkably comprehensive ecosystem-based management plan supported by strong public and political commitment, substantial institutional impediments still hinder its execution. Management of such a large watershed is unwieldy, requiring communication among federal and state governments, agencies with different responsibilities, and the states and the District of Columbia, and also between state and local governments (which is particularly important with regard to land use).

In addition, the scale of many management decisions often does not reflect the scale of ecosystem responses, reflecting a classic scale mismatch (Lee 1993; Crowder et al. 2006). While a few policies can be implemented on the scale of the whole watershed, for example, atmospheric emissions regulated under the Clean Air Act, most are executed at subregional or even local scales. They may or may not produce local responses and benefits, and their effects on the whole Chesapeake ecosystem are more difficult to assess. The Chesapeake Bay Program addresses this scale mismatch through what are called "Tributary Strategies." These assign nutrient load reductions to subunits of the watershed and engage local governments and stakeholders to find the means to accomplish those reductions. Tributary teams assembled to develop and implement the strategies have been effective in building local understanding and setting goals. They have been less effective in implementing the measures included in the strategies, mainly because they lack the authority or resources to do so.

The Chesapeake Bay Program possesses most of the necessary ingredients for adaptive management in the form of a stable governance structure, science-based planning informed by state-of-the art models, and an extensive and sustained monitoring program (Boesch 2006). However, as mentioned above, these have not been effectively integrated in a way that relentlessly tests assumptions and improves practices within a structured adaptive management program. Weak links between implementation, monitoring, and policy formation impair adaptive management in this region. A case in point: The Chesapeake Bay has more stream and river restoration projects per river mile than any place in the nation, but project records show that only 5% of these restoration efforts include any kind of performance monitoring (Hassett et al. 2005). Although baywide monitoring data do inform policy direction, political accountability for the success of those policies often lags behind their implementation. Adaptive management is not just a scientific or technical task; it must be embraced throughout the management structure (Lee 1999; NRC 2004). Most of the other large ecosystem restoration programs in the United States, such as those for the Florida Everglades and the Grand Canyon, specifically include adaptive management structures and procedures in their organization (Gunderson et al. 2008). It was only in a recent report to Congress on strengthening management, coordination, and accountability that an adaptive management framework was explicitly embraced for the Chesapeake Bay Program (CBP 2008).

Management institutions must also contend with the multiple, cumulative impacts of stakeholder activities. Particularly important for the Chesapeake Bay ecosystem are the effects of rampant urban sprawl. Development alters hydrology and the delivery of sediments, nutrients, and toxicants. It drives the addition of transportation infrastructure and consumption of resources in myriad ways that influence the estuarine ecosystem. Despite the emergence of "smart growth" policies and practices within the region, such land conversion threatens to reverse gains made in reducing pollutant impacts. If not better managed, development poses one of the most important challenges to ecosystem-based management of the Chesapeake watershed in the twenty-first century.

Societal

Each year the Chesapeake Bay Foundation, a prominent advocacy organization in the region, issues a "State of the Bay" report card (CBF 2006). It scores twelve pollution, habitat, and fisheries indicators, each on a scale of 0 to 100, with 100 representing the pre–European settlement condition. These are averaged to develop an overall health index for the bay. Since 2000, the index has never been higher than 29, corresponding to a letter grade of D, according to CBF's grading scale. Media outlets, from local newspapers to *The Baltimore Sun* to *National Geographic* magazine regularly report on CBF's State of the Bay report card. The message presented is easy to grasp. The public hears the take-home loud and clear—the bay is in bad shape and it is not getting much better.

In contrast to simple messages on ecosystem condition, the concept of ecosystem-based management presents a challenge of perception for the Chesapeake Bay. Its very premise is one of scientific complexity—that efforts to restore the bay need to focus not just on averages but also on how the different components of the ecosystem relate to each other. This premise contains inherent nonlinearities and layers of interactions in which the whole cannot be explained simply by the sum of its parts. Messages of caveats and complexity are less appealing than a single letter grade on a report card.

But what if such report card scoring could be made for each of the bay's major geographic regions? Evaluating the bay on this scale aligns better with the scale on which management actions for restoration can be executed and is more consistent with the tributary strategy approach. Communities could take ownership of their pieces of the problem and work to make them better, to improve their grades. Such a geographically explicit, integrated assessment was produced for 2006 and 2007. Fifteen tributaries in the bay were given numerical scores and letter grades based on six different indicators (three water quality, three biotic) (EcoCheck 2008). For example, the upper bay earned a high grade of C+, while the Patapsco and Back rivers each earned an F, illustrating the importance of local conditions. The media response to this new report card was no less dramatic, with headlines such as "Bay report is bleak" and "Bay is still hurting" dominating regional news following the report card's release. But the implicit message comes through differently—while the overall picture may be bleak, regional improvements can be made.

The idea that the bay's slow response to restoration efforts might be rooted in complex changes that have occurred over time is harder for the public to grasp than a linear one-to-one correlation of effort invested to improvement achieved. Furthermore, the infrastructure for management relies on complex models, which assimilate large quantities of information. How models work and the limitations of their predictions are difficult concepts to understand.

At the crux of the problem with the perception of progress in bay restoration is that the scientific enterprise is comfortable with uncertainty, while society and political institutions are not. Science moves forward by the iterative process of rejecting hypotheses and reshaping theories as new data are collected and new experiments completed. The scientific process is rooted in this forward, albeit discontinuous, momentum, and it is to be expected that the "best available science" may not be so even a year down the road. Political institutions and public understanding do not adapt in the same time frame. Politicians often want definitive answers from scientists when, in fact, political decisions tend to be made based as much or more on values than established facts (Sarewitz 2004).

Issues surrounding the restoration of Chesapeake Bay have held the political spotlight for years. The citizens of the region care deeply about the bay's murky waters and the status of oysters, blue crabs, and striped bass. In 2003, the high-level Chesapeake Bay Watershed Blue Ribbon Finance Panel (2004) was charged with developing innovative solutions for financing the multi-billion-dollar bay restoration effort. The panel called on the bay states and the federal government to make a 6-year, $15 billion investment in the creation of a regional finance authority to be charged with prioritizing and distributing restoration funds throughout the watershed. The panel made the argument that we know how to restore the bay, but we simply lack the resources to do it. While this degree of political interest in the region is powerful in many respects, the emphasis on funding may underscore a chronic mismatch between science and politics. The answer to why the bay has not yet yielded high returns, in terms of ecological improvements to dollars invested, may lie in ecosystem dynamics rather than the course of management followed over the past 25 years. But this answer may not be one that is palatable in the political arena.

Paths Forward for the Chesapeake

Management efforts in the Chesapeake region strive to move the bay from its current stable state, one that is stubbornly resilient in a degraded condition, to one that more closely resembles a bay of the past—with clearer waters, productive fisheries, healthy coastal communities, and abundant underwater grasses. The path followed by the bay as it made an abrupt shift toward the present degraded state suggests that positive feedback mechanisms may eventually accelerate restoration in a similarly nonlinear manner (fig. 15.5). For example, if light penetration improved marginally, increased benthic production would lead to uptake of nitrogen and phosphorus, to their loss to sediment sinks (denitrification and burial), and to reduced resuspension, as benthic microflora and macroflora stabilized sediments. As hypoxia in the deeper waters of the bay improved, the oxidation potential in sediments would increase, helping bind pollutants to the sediment, thereby reducing recycling of nitrogen and phosphorus back into the water column (Boesch et al. 2001; Kemp et al. 2005).

To jump-start these recovery trajectories, management efforts must erode the resilience of the bay's current degraded state. Effecting change in both the bottom-up and top-down controls could prove key. From the bottom up, reduction of nitrogen input would have a direct impact on algal biomass. From the top down, improved benthic conditions, especially increased filtration capacity of oysters or other filter feeders like mussels and clams, could reduce algal biomass by grazing. Increasing sediment accretion, especially by restoring tidal marshes, would also promote the uptake of nitrogen and phosphorus, decreasing resuspension in the water column (Kemp et al. 2005).

Recent recovery of submerged aquatic vegetation in the upper bay, particularly on the Susquehanna Flats, may reflect such positive feedback mechanisms in action. Once covered with submerged vegetation to depths of 3 meters, according to historical reports from the 1800s, the flats were devastated after Tropical Storm Agnes. But since the late 1990s, due to a combination of reduced nutrient loads and several years of drought, the flats now contain a vibrant grass bed capable of controlling its own environment. Heavy rains in June 2006, with peak flows measuring nearly half of those that followed Agnes, served as a disturbance that put these restored grasses to the test. The recovered grasses of the Susquehanna Flats proved resilient to this significant perturbation and are helping to sustain improved water quality in this region (Blankenship 2006).

In addition to abatement of nutrient and sediment loading, efforts have been undertaken to restore oyster populations in the Chesapeake Bay, both to recover filter-feeding capacity and lost habitat and to support a dying fishery and way of life. Intensive efforts to culture and plant disease-free juvenile native oysters, *Crassostrea virginica*, have been underway since the year 2000. Nearly one billion oyster spat (juveniles attached to oyster shells) have been planted at sixty sites in Maryland alone (ORP 2008). While popular notions of historically prodigious filtration of the bay by oysters are exaggerated (Pomeroy et al. 2006), targeted restoration in the tidal tributaries could reduce the effects of eutrophication (Fulford et al. 2007), particularly in conjunction with similarly targeted efforts to reduction nutrient loads.

The influence of science on ecosystem-based management in the future remains critical. If scientists could identify incipient thresholds in the system and the steps needed to cross them, more-effective outcomes could be realized, and the impetus for restoration could soar ahead. But current modeling efforts are not yet designed to forecast these threshold-type events in space and time. By the same token, if science can identify potential pitfalls, for example, ecosystem services that simply cannot be restored in the current state, management could avoid the unnecessary expenditure of effort and dollars. Success of the relationship between science and management depends on a close coupling between the two infrastructures and a capacity for adaptive change at the institutional level. Building this capacity in the Chesapeake region would require both strengthening the connection between implementation and monitoring, as discussed previously, and tightening the connection between monitoring and policy formation.

Toward that end, a potential breakthrough is the BayStat initiative, an accountability process

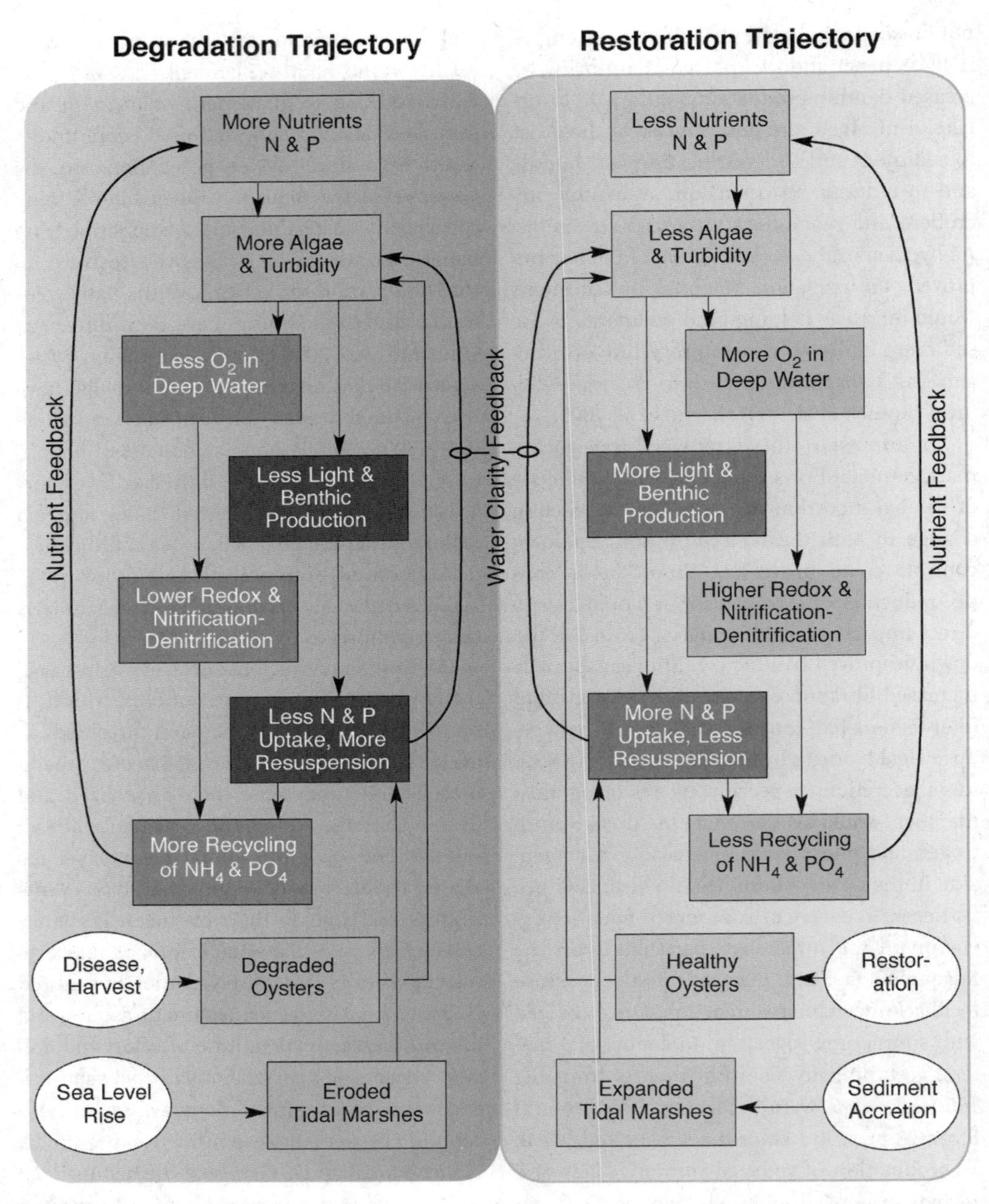

Figure 15.5 Theoretical trajectories for the degradation and restoration of Chesapeake Bay. External stressors such as oyster diseases, intense fishing pressure, and sea level rise can augment a feedback loop in which additions of nitrogen and phosphorus lead to eutrophic conditions. But in a similar manner, restoration efforts could jump-start parallel pathways that would improve water clarity. In this diagram, added nutrients affect algal biomass (light gray boxes) directly. Turbid conditions in turn influence bottom-water oxygen and nutrient recycling (medium gray boxes) and water clarity and benthic primary production (dark gray boxes). The feedback processes follow inverse pathways in the two trajectories of degradation and restoration. Adapted with permission from Kemp et al. 2005.

for measuring and evaluating efforts to restore the Chesapeake Bay being implemented by Maryland Governor Martin O'Malley. Modeled after CityStat, an accountability-based governmental management program executed while O'Malley was mayor of Baltimore, BayStat was proposed as a means, based on relentless follow-up and assessment, to use preexisting measures of the bay's health, restoration, and protection to achieve more effective coordination and implementation of programs. The governor established BayStat by Executive Order (O'Malley 2007) in February 2007, less than one month after being sworn into office, and at this writing BayStat (http://www.baystat.maryland.gov/) is being implemented with his substantial expectations and full engagement. For example, by applying criteria for cost-effectiveness, BayStat is developing a procedure for geographic targeting of agricultural management practices such as cover crops and riparian buffers, thus improving the outcomes from public investments. This initiative, if successful, could go a long way in building the true adaptive management capacity that has been wanting in ecosystem-based management of Chesapeake Bay.

Ultimately, if restoration efforts in the Chesapeake achieve their goal, a resilient ecosystem in a less-degraded state may function in a manner similar to one of the bay's previous states, but it will not be identical. While water quality may improve, certain losses the ecosystem has sustained are likely permanent. Historic oyster reefs will never be rebuilt to their previous levels. Watermen will not fully regain their former cultural and economic role. Underwater grasses will not likely completely regain their former dominance. Continued development and population growth in the region will continue to challenge the trajectory of restoration efforts. Global climate change—increasing temperatures, changing river flows, and accelerating sea level rise—will inject new uncertainties into future ecosystem states and thus effective management strategies (STAC 2008).

The Chesapeake Bay Program is currently working to revise the goals set for 2010 and to establish two-year milestones for ensuring steady progress. Ecosystem-based approaches to fisheries management are also being integrated into the Chesapeake Bay Program's institutional management framework, and new projections for population growth and climate change will be incorporated into future modeling scenarios. The future state of the bay and the success or failure of ecosystem-based management will depend on a confluence of political will, integrative thinking, and institutional flexibility, as well as a capacity to respond and adapt to external drivers.

Key Messages

1. The Chesapeake Bay is one of the world's most comprehensive, sustained, and institutionalized efforts to develop and apply marine ecosystem-based management. More than 30 years' experience indicates that successful EBM requires close connections among the science, policy, and management domains and that a number of scientific and institutional challenges to EBM still exist.
2. Scientific understanding of Chesapeake Bay—including the coupling among the social and ecological components—has played an essential role in developing the goals of EBM and in identifying ways that these goals can be met.
3. Both restoration and degradation of marine systems can occur in an abrupt, nonlinear fashion; ideally management efforts can be designed to anticipate such changes.
4. The success or failure of EBM, as well as the future state of the Chesapeake Bay, will depend on a confluence of political will, integrative thinking, and institutional flexibility.

Acknowledgments

We thank our many colleagues in the Chesapeake science and management community. Although the risk of naming any of them is to appear to overlook others, we particularly want to acknowledge Rich Batiuk, Bill Dennison, Fran Flanigan, Chuck Fox, Jack Greer, John Griffin, Verna Harrison, Tom Horton, Ed Houde, Michael Kemp, Jon Kramer, Jeff Lape, Dave Nemazie, Roger Newell, Margaret Palmer, Sandy Rodgers, Jessica Smits, and Ann Swanson. We also thank the reviewers and editors who helped us sharpen our thoughts and expressions.

References

Bancroft, D. 2005. PA nutrient trading plan deserves looking into. *Bay Journal* 15(7).

BBCAC (Bi-State Blue Crab Advisory Committee). 2006. *Blue crab 2005: Status of the Chesapeake population and its fisheries.* Annapolis, MD: Chesapeake Bay Commission. http://www.mdsg.umd.edu/crabs/bbcac/CBC-CRAB-05.pdf.

Belousek, D. W. 2004. Scientific consensus and public policy: The case of *Pfiesteria*. *Journal of Science, Philosophy & Law* 4:1–33.

Blankenship, K. 2004. "Flush tax" to fund MD nutrient reduction programs. *Bay Journal* 14(3).

Blankenship, K. 2006. Underwater grasses at the tipping point? *Bay Journal* 16(6).

Boesch, D. F., R. B. Brinsfield, and R. E. Magnien. 2001. Chesapeake Bay eutrophication: Scientific understanding, ecosystem restoration and challenges for agriculture. *Journal of Environmental Quality* 30:303–20.

Boesch, D. F., and J. Greer, ed. 2003. *Chesapeake futures: Choices for the twenty-first century.* Edgewater, MD: Chesapeake Research Consortium.

Boesch, D. F. 2006. Scientific requirements for ecosystem-based management in the restoration of Chesapeake Bay and coastal Louisiana. *Ecological Engineering* 26:6–26.

Boynton, W. R., J. H. Garber, R. Summers, and W. M. Kemp. 1995. Inputs, transformations, and transport of nitrogen and phosphorus in Chesapeake Bay and selected tributaries. *Estuaries and Coasts* 18(1):285–314.

CBF (Chesapeake Bay Foundation). 2006. *2006 State of the bay.* Annapolis, MD: Chesapeake Bay Foundation.

CBFEAP (Chesapeake Bay Fisheries Ecosystem Advisory Panel). 2006. *Fisheries ecosystem planning for Chesapeake Bay.* Trends in Fisheries Science and Management 3. Bethesda, MD: American Fisheries Society.

CBP (Chesapeake Bay Program). 1999. *Chesapeake 2000: A watershed partnership.* Annapolis, MD: US Environmental Protection Agency. www.chesapeakebay.net/agreement.htm.

CBP (Chesapeake Bay Program). 2007. *Chesapeake Bay 2006: Health and restoration assessment.* EPA 903R-07001 and EPA 903R-07002. Annapolis, MD: US Environmental Protection Agency. http://www.chesapeakebay.net/assess/index.htm.

CBP (Chesapeake Bay Program). 2008. *Strengthening the management, coordination, and accountability of the Chesapeake Bay Program.* CBP/TRS-292-08. Annapolis, MD: US Environmental Protection Agency. http://cap.chesapeakenet/rtc.htm.

CBSAC (Chesapeake Bay Stock Assessment Committee). 2007. *2007 Chesapeake Bay blue crab advisory report.* Annapolis, MD: National Oceanic and Atmospheric Administration. http://chesapeakebay.noaa.gov/docs/2007bluecrabadvisoryreport.pdf.

CBWBRFP (Chesapeake Bay Watershed Blue Ribbon Finance Panel). 2004. *Saving a national treasure: Financing the cleanup of the Chesapeake Bay.* Annapolis, MD: Chesapeake Bay Program. http://www.chesapeakebay.net/pubs/blueribbon/index.cfm.

Chambers, E. 2006. *Heritage matters: Heritage, culture, history, and Chesapeake Bay.* Chesapeake Perspectives monograph series. College Park: Maryland Sea Grant College.

Chuenpagdee, R., L. Liguori, D. Preikshot, and D. Pauly. 2006. A public sentiment index for ecosystem management of Chesapeake Bay. *Ecosystems* 9:463–73.

Colman, S. M., P. C. Baucom, J. F. Bratton, T. M. Cronin, J. P. McGeehin, D. Willard, A. R. Zimmerman, and P. R. Vogt. 2002. Radiocarbon dating, chronologic framework, and changes in accumulation rates of Holocene estuarine sediments from Chesapeake Bay. *Quaternary Research* 57:58–70.

Cooper, S. R., and G. S. Brush. 1991. Long-term history of Chesapeake Bay anoxia. *Science* 254(5034):992–96.

Costanza, R., and J. Greer. 1995. The Chesapeake Bay and its watershed: A model for sustainable ecosystem management? In *Barriers and bridges to the renewal of ecosystems and institutions*, ed. L. Gunderson, C. S. Holling, and S. S. Light. New York: Columbia University Press.

Cronin, T. M., and C. D. Vann. 2003. The sedimentary record of climatic and anthropogenic influence on the Patuxent estuary and Chesapeake Bay ecosystems. *Estuaries* 26:196–209.

Crowder, L. B., G. Osherenko, O. R. Young, S. Airamé, E. A. Norse, N. Baron, J. C. Day et al. 2006. Resolving mismatches in US ocean governance. *Science* 313:617–18.

Davidson, S. G., J. G. Merwin Jr., J. Capper, G. Power, and F. R. Shivers Jr. 1997. *Chesapeake waters: Four centuries of controversy, concern and legislation*. Centreville, MD: Tidewater.

EcoCheck. 2008. *Chesapeake Bay report card: A geographically detailed and integrated assessment of Chesapeake Bay habitat health*. Cambridge, MD: University of Maryland Center for Environmental Science. http://www.eco-check.org/reportcard/chesapeake/2007/.

Ernst, H. 2003. *Chesapeake Bay blues: Science, politics, and the struggle to save the Chesapeake Bay*. Lanham, MD: Rowman & Littlefield.

Epstein, G., and S. Desmon. 2006. Crab factory. *Baltimore Sun* 30 April 2006.

Fahrenthold, D. A. 2008. Alarm over blue crab decline. *Washington Post* 16 April 2008.

Fincham, M. 2007. Whatever happened to *Pfiesteria*. *Chesapeake Quarterly* 6:1.

Fulford, R. S., D. L. Breitburg, R. I. E. Newell, W. M. Kemp, and M. Luckenbach. 2007. Effects of oyster population restoration strategies on phytoplankton biomass in Chesapeake Bay: A flexible modeling approach. *Marine Ecology Progress Series* 336:43–61.

GAO (Government Accountability Office). 2005. *Chesapeake Bay program: Improved strategies are needed to better access, report, and manage restoration progress*. GAO 06-96. Washington, DC: Government Accountability Office.

Greer, J. 2002. Following those who follow the water. *Chesapeake Quarterly* 2(3):1–15.

Gunderson, L. H. 2000. Ecological resilience in theory and application. *Annual Review of Ecology and Systematics* 31:425–39.

Gunderson, L. H., G. Peterson, and C. S. Holling. 2008. Practicing adaptive management in complex social–ecological systems. In *Complexity theory for a sustainable future*, ed. J. Norbert and G. Cumming. New York: Columbia University Press.

Hagy, J. D., W. R. Boynton, C. W. Keefe, and K. V. Wood. 2004. Hypoxia in Chesapeake Bay, 1950–2001: Long-term change in relation to nutrient loading and river flow. *Estuaries* 27:634–58.

Hassett, B., M. Palmer, E. Bernhardt, S. Smith, J. Carr, and D. Hart. 2005. Restoring watersheds project by project: Trends in Chesapeake Bay tributary restoration. *Frontiers in Ecology and the Environment* 3:259–67.

Hennessey, T. M. 1994. Governance and adaptive management for estuarine ecosystems: The case of Chesapeake Bay. *Coastal Management* 22:119–45.

Horton, T. 2003. *Turning the tide: Saving the Chesapeake Bay*. Washington, DC: Island Press.

Jackson, J. B. C., M. X. Kirby, W. H. Berger, K. A. Bjorndal, L. W. Botsford, B. J. Bourque, R. H. Bradbury et al. 2001. Historical overfishing and the recent collapse of coastal ecosystems. *Science* 293:629–38.

Kemp, W. M., W. R. Boynton, J. E. Adolf, D. F. Boesch, W. C. Boicourt, G. Brush, J. C. Cornwell et al. 2005. Eutrophication of Chesapeake Bay: Historical trends and ecological interactions. *Marine Ecology Progress Series* 303:1–29.

Kennedy, V. S., and L. L. Breisch. 1981. *Maryland's oysters: Research and management*. College Park, MD: Maryland Sea Grant.

Kobell, R., and C. Guy. 2005. Work-visa limit snags shore employers. *Baltimore Sun* 24 January 2005.

Koroncai, R., L. Linker, J. Sweeney, and R. Batiuk. 2003. *Setting and allocating the Chesapeake Bay nutrient and sediment loads: The collaborative process, technical tools, and innovative approaches*. Annapolis, MD: US Environmental Protection Agency, Chesapeake Bay Program Office.

Langland, M., and T. Cronin, ed. 2003. *A summary report of sediment processes in Chesapeake Bay and watershed*. Water Resources Investigations Report 03-4123. New Cumberland, PA: US Geological Survey.

Latour, R. J., M. J. Brush, and C. F. Bonzek. 2003. Toward ecosystem-based fisheries management:

Strategies for multispecies modeling and associated data requirements. *Fisheries* 28:10–22.

Lee, K. N. 1999. Appraising adaptive management. *Conservation Ecology* 3(2):3.

Lee, K. N. 1993. Greed, scale mismatch and learning. *Ecological Applications* 3:560–64.

Lindsey, B. D., S. W. Phillips, C. A. Donnelly, G. K. Speiran, L. N. Plummer, J. K. Böhlke, M. J. Focazio, W. C. Burton, and E. Busenberg. 2003. *Residence times and nitrate transport in ground water discharging to streams in the Chesapeake Bay watershed.* Water-Resources Investigations Report 03-4035. Washington, DC: US Geological Survey.

Lipton, D. W., and K. Wellman. 1995. *Economic valuation of natural resources: A handbook for coastal resource policymakers.* Coastal Ocean Program Decision Analysis Series no. 5. Silver Spring, MD: National Oceanic and Atmospheric Administration.

Lipton, D.W. 2006. The changing values of an ecosystem. In *Focal Points: Chesapeake Bay Commission annual report,* 29–33. Annapolis, MD: Chesapeake Bay Commission.

Lipton, D. W. 2004. The value of improved water quality to Chesapeake Bay boaters.*Marine Resource Economics* 19:265–70.

Lotze, H. K., H. S. Lenihan, B. J. Bourque, R. H. Bradbury, R. G. Cooke, M. C. Kay, S. M. Kidwell, M. X. Kirby, C. H. Peterson, and J. B. C. Jackson. 2006. Depletion, degradation, and recovery potential of estuaries and coastal seas. *Science* 312:1806–9.

Malone, T. C., W. Boynton, T. Horton, and C. Stevenson. 1993. Nutrient loading to surface waters: Chesapeake Bay case study. In *Keeping pace with science and engineering,* ed. M. F. Uman, 8–38. Washington, DC: National Academies Press.

Newell, R. I. E. 2004. Ecosystem influences of natural and cultivated populations of suspension-feeding bivalve molluscs: A review. *Journal of Shellfish Research* 23:51–61.

NMFS (National Marine Fisheries Service). 2007. *2005 US landings by distance from shore.* NMFS. http://www.st.nmfs.gov/st1/commercial/landings/ds_8850_bystate.html.

NRC (National Research Council). 2004. *Adaptive management for water resources project planning.* Washington, DC: National Academies Press.

OIG (Office of the Inspector General). 2006. *Evaluation report: Saving the Chesapeake Bay watershed requires better coordination of environmental and agricultural resources.* EPA OIG report no. 2007-P-00004, USDA OIG report no. 50601-10-Hq. Washington, DC: Office of the Inspector General, US Environmental Protection Agency, and US Department of Agriculture.

O'Malley, M. 2007. *BayStat.* Executive Order 01.01.2007.02. Annapolis, MD: Office of the Governor. http://www.gov.state.md.us/executiveorders/01.07.02Baystat.pdf.

ORP (Oyster Recovery Partnership). 2008. http://www.oysterrecovery.org/.

Palmer, M. A., E. Bernhardt, E. Chornesky, S. Collins, A. Dobson, C. Duke, B. Gold et al. 2004. Ecology for a crowded planet. *Science* 304:1251–52.

Paolisso, M. 2002. Blue crabs and controversy on Chesapeake Bay: A cultural model for understanding watermen's reasoning about blue crab management. *Human Organization* 61:226–39.

Paolisso, M. 2006. *Chesapeake environmentalism: Rethinking culture to strengthen restoration and resource management.* Chesapeake Perspectives. College Park, MD: Maryland Sea Grant College.

Pomeroy, L. R., C. F. D'Elia, and L. C. Schaffner. 2006. Limits to top-down control of phytoplankton by oysters in Chesapeake Bay. *Marine Ecology Progress Series* 325:301–09.

Russell, D. 2005. *Striper wars: An American fish story.* Washington, DC: Island Press.

Sarewitz, D. 2004. How science makes environmental controversies worse. *Environmental Science and Policy* 7:385–403.

STAC (Scientific and Technical Advisory Committee). 2008. *Climate change and the Chesapeake Bay: State-of-the-science review and recommendations.* Edgewater, MD: Chesapeake Research Consortium. http://www.chesapeake.org/stac/Pubs/climchangereport.pdf.

Stephenson, K., and L. Shabman. 2001. The trouble with implementing TMDLs. *Regulation* 24:28–32.

Stephenson, K., P. Norris, and L. Shabman. 1998. Watershed-based effluent trading: The non-point source challenge. *Contemporary Economic Policy* 26:412–21.

UNEP/GPA (United Nations Environment Programme/Global Programme for Action for the Protection of the Marine Environment from Land-based Activities). 2006. *Ecosystem-based management: Markers for assessing progress.* The Hague: UNEP/GPA.

USCOP (US Commission on Ocean Policy). 2004. *Sustaining our oceans: A public resource—a public trust. Final report to the President and Congress.* Washington, DC: USCOP.

Walker, B., and D. Salt. 2006. *Resilience thinking: Sustaining ecosystems and people in a changing world.* Washington, DC: Island Press.

Wasserman, D., M. Womersley, and S. Gottlieb. 1999. Can a sense of place be preserved. In *Philosophies of place,* ed. A. Light and J. M. Smith. Lanham, MD: Rowman & Littlefield.

Whoriskey, P. 2004. Bay pollution progress overstated. *Washington Post* 18 July 2004, p. A.01.

Worm, B., E. B. Barbier, N. Beaumont, J. E. Duffy, C. Folke, B. S. Halpern, J. B. C. Jackson et al. 2006. Impacts of biodiversity loss on ocean ecosystem services. *Science* 314:787–90.

CHAPTER 16
Lessons from National-Level Implementation Across the World

Andrew A. Rosenberg, Marjorie L. Mooney-Seus, Ilse Kiessling, Charlotte B. Mogensen, Robert O'Boyle, and Jonathan Peacey

Ecosystem-based management (EBM) recognizes that human and ecological communities are interdependent and interact with their physical environment to form distinct ecological units or ecosystems. Human actions and their consequences impact the functioning and resilience of coastal and marine ecosystems and the services they provide (see McLeod and Leslie, chap. 1 of this volume; Shackeroff et al., chap. 3 of this volume). So, ideally, EBM should integrate management of human activities and impacts across existing political and jurisdictional boundaries.

Here we consider the development and implementation of ecosystem-based systems of ocean management by five governments: Australia, Canada, the European Union (EU), New Zealand, and the United States (USA). While all five case studies are consistent in their recognition of the need to coordinate management of human uses of ocean resources, the approaches to implementing EBM vary considerably. Australia, Canada, and the EU have adopted top-down integrated management legislation or, in the case of the EU obligatory policy, dictated agency coordination of sector management from the highest governmental level (i.e., multinational or national). Others are advancing EBM through existing national sector-management legislation or regional and localized efforts while integrated management legislation is being developed.

Variations in the approach to implementing EBM are, to some extent, related to the structures of the respective governments. The top-down approaches underway in Canada, Australia, and New Zealand may be facilitated by their parliamentary systems. In the United States, under a weaker federal system, individual states are taking the lead in advancing EBM. With authority divided between Member States and the Union, EBM in the EU presents different challenges: Overarching policy is then implemented by Member States or regions.

Despite the nuances of governmental and institutional structures, EBM implementation requires a strategic approach to be effective. This entails adopting an overarching mandate to advance EBM in an integrated fashion with clearly defined regulatory authority and consistent objectives, standards, and monitoring protocols. The knowledge and experiences of localized efforts must be incorporated into management decisions. It is imperative to establish a clearly articulated strategy for stakeholder outreach and involvement at all stages of the process.

This chapter explores the policy context and coordination efforts for advancing EBM in these five case studies. The challenges of EBM implementation are identified for each example. Then, we present a framework for effective EBM implementation, strong political support, a systematic and coordinated approach, and clear management objectives focused on ecosystem services.

Australia

Policy Context

Policy direction has strongly influenced EBM development in Australia. The principal EBM framework for the marine environment is Australia's Oceans Policy (AOP). The AOP was introduced in December 1998 and incorporated a commitment to implement "an integrated and ecosystem-based oceans planning and management system for all Australia's marine jurisdictions" (Commonwealth of Australia 1998). The AOP defines integrated, ecosystem-based oceans management as an approach that recognizes the structure and function of ecosystems and the interdependence of human uses and ecosystem health. EBM is described as a management approach that recognizes ecological, social, and cultural objectives for an ecosystem but makes ecological sustainability the primary goal of management. Australia subscribes to the widely accepted definition of sustainable development published in the Bruntland report (1987): "development that meets the needs of the present without compromising the ability of future generations to meet their own needs." The AOP is a considerable shift in approach to marine planning and management in Australia from a sector-based system (i.e., focused on a single resource or activity) with fragmented management arrangements to a system that integrates the diverse elements of the ocean comprehensively and recognizes the combined interactions of human activities and the value of a precautionary approach to management (Commonwealth of Australia 1998).

The AOP defines a range of institutional arrangements to support the implementation of the policy. One of these arrangements—regional marine planning—was established under the AOP as the key tool to achieve integrated and ecosystem-based oceans planning and management (DEH 2006). Only one plan was completed under the regional marine planning process—the South East Regional Marine Plan, which was released in May 2004. The South East Plan brings together key information, identifies high-level objectives for the region, and sets out a detailed action plan for the pursuit of those objectives. It identifies lead agencies and time frames for delivery, as well as actions that range across sectors. However, the South East Plan does not address indicators of ecosystem health or sustainability or issues arising from cumulative impacts, nor does it reconcile conflicts between competing users, deliver coordination across the entire government, or integrate management actions in the pursuit of resource allocation across sector-specific interests.

Six years after it was introduced, a review of the implementation of AOP was conducted in response to concerns about slower-than-anticipated progress in implementing regional marine planning and the failure of the process to secure cross-sectoral engagement and support within government agencies (TFG International 2002). Based on the findings of this review, the Australian Government launched a new model of planning in 2005, known as marine bioregional planning, under the Commonwealth Environment Protection and Biodiversity Conservation Act 1999 (EPBC Act). Using existing arrangements under the EPBC Act, marine bioregional planning is intended to drive nationally consistent decision making on marine biodiversity conservation with the objective of securing long-term protection of Australia's marine environment.

The EPBC Act is Australia's most comprehensive environmental legislation, and it provides a strong foundation for implementing EBM across Australia's ocean jurisdiction. The commonwealth marine environment (i.e.,

3–200 nautical miles offshore) is identified as a "matter of national environmental significance" under the EPBC Act. Any action that will have, or is likely to have, a significant impact on the commonwealth marine environment requires assessment and approval under this law. Bringing marine bioregional planning under the EPBC Act has given new impetus to the implementation of Australia's Oceans Policy by streamlining the marine planning process and providing greater guidance about marine environment conservation priorities.

Mode of Coordination

Section 176 of the EPBC Act specifically outlines what a bioregional plan will include, namely, the following:

- Descriptions of the biodiversity, economic, social, and heritage values of a planning region
- The objectives of the plan relating to biodiversity and other values
- How the community will be involved in planning
- Mechanisms for monitoring and reviewing the plan over time

The marine bioregional planning process also includes nested planning for the identification and establishment of marine protected areas (MPAs) in commonwealth-managed waters. EBM principles guide other sections of Australia's EPBC Act, as well, such as the provisions for strategic assessment of marine activities as they relate to the protection of environment and heritage objectives. Commonwealth and state export fisheries are assessed under the EPBC Act to ensure that they are managed in an ecosystem-based fashion. However, the significance of marine bioregional planning under the EPBC Act is that it provides the statutory link between the application of EBM and decisions made by the Minister for the Environment and Water Resources about export fisheries and relevant marine matters. While not bound to specific actions, the minister must consider marine bioregional plans when making decisions under the EPBC Act, and thus the plans can significantly influence activities in and adjacent to the commonwealth marine area.

Marine bioregional plans will include a detailed ecological profile of the planning region (for example, see DEWR 2007) and objectives, strategies, and actions for ecological conservation and protection in commonwealth waters. Marine bioregional planning is centered on the protection of conservation values based on available knowledge of biodiversity, and ecosystem processes and plans provide a strategic framework for the government to inform, coordinate, and prioritize its marine conservation programs (e.g., development of marine protected areas, species recovery plans, and assessment of sustainable fisheries). As such, marine bioregional planning is intended to inform marine managers and industries about conservation issues and priorities in each marine bioregion, assist in understanding the impacts of actions on the commonwealth marine environment, and determine the circumstances under which actions can take place (DEWHA 2003).

In Australia, EBM is implemented through the marine bioregional planning process on the basis of planning regions defined both by ecological attributes (e.g., ocean currents, geology, and biota) and pragmatic management considerations. Besides the completed South East Regional Marine Plan, bioregional marine plans are being developed in four other regions: the north, northwest, southwest, and central east. All of the plans, including a national network of marine protected areas, are expected to be complete by 2012.

Marine bioregional plans are intended to have a life of between 5 and 10 years. Cumulative risk assessment is an important part of the process, and monitoring and review arrangements, including a set of indicators by which to monitor the health of the bioregion, will be significant components of the plans. Monitoring efforts also are intended to link with national State of the Marine Environment Report arrangements and include indicators relevant to determining marine environmental health on both a regional and national scale.

Challenges

Australia now has a legislative context for EBM through the marine bioregional planning process under the EPBC Act, and the potential for clarification of administrative relationships between ecosystem protection and sector management is now improved. A significant challenge to the implementation of EBM in Australia has been a lack of involvement in, and ownership of, EBM planning and management processes by sector-specific Australian Government agencies and industries that are likely to have significant roles in implementation or to be affected by it. Furthermore, although the AOP is the principal framework for EBM within Australia's marine environment, it is a commonwealth policy that is yet to be explicitly endorsed by Australia's states and Northern Territory. The AOP provides little guidance on the resolution of tensions between competing sectoral interests, maintenance of existing sectoral and jurisdictional management arrangements, and effective, ongoing implementation of cross-sectoral management measures (Westcott 2000). State/Northern Territory jurisdiction over other matters that affect the management of the oceans (e.g., watershed management) further complicates achievement of effective ecosystem-based, integrated oceans planning and management.

Canada

Policy Context

Strong political support for EBM and integrated planning and management in Canada enabled the country to develop legislation first and then national policy. At least part of the motivation for advancing EBM nationally was due to Canada's efforts to comply with relevant international laws. Canada's Oceans Act (COA), enacted in 1997, paved the way for a national Oceans Strategy based on the principles of sustainable development, and a commitment to err on the side of caution (i.e., to take a precautionary approach) when scientific information is lacking or incomplete (DFO 2002).

Mode of Coordination

Under the auspices of the COA, Canada's maritime boundaries are defined in accordance with the provisions of the United Nations Convention on the Law of the Sea (UNCLOS). Canada established a national system of regional planning areas, referred to as large ocean management areas (LOMAs). The LOMA boundaries are based on more than ecological considerations. Administrative boundaries, human community structure, and ecological characteristics also are taken into account. So far, the national network of LOMAs consists of Placentia Bay and Grand Banks, Eastern Scotian Shelf, Gulf of St. Lawrence, Beaufort Sea, and Pacific North Coast.

LOMAs are in varying stages of the planning process. The first phase of the plan is to establish a foundation for achieving the long-term objectives of the COA and Canada's Oceans Strategy (DFO 2005a); later phases will focus on implementation. O'Boyle and Worcester (chap. 14 of this volume) discuss the experiences and lessons learned in the Eastern Scotian Shelf Integrated Management (ESSIM) LOMA.

All LOMAs are following national guidelines and using a common hierarchy of planning tools. National guidelines stipulate that the ecosystem must first be described, followed by the identification of the key human stressors. An Ecosystem Overview and Assessment

Report has been developed for each LOMA to describe the structure and functioning of the ecosystem, the human activities (e.g., fishing) and resulting stressors (e.g., dragging), and the impacted ecosystem components (e.g., benthic community) (DFO 2005b). Within each of the five LOMAs, marine managers and stakeholders have begun drafting conceptual objectives, and some areas have begun to explore regional and sector-specific operational objectives. Priority ecosystem objectives are under development and will be prioritized after identification of ecologically significant and/or degraded ecological areas and species (DFO 2007a). Operational objectives explore the LOMA ecosystem objectives in detail and identify the most suitable indicator (measurable quantity) and reference point (level of the indicator) to achieve the objectives. It has been suggested that the suite of operational objectives at the LOMA level be the basis for determination of overall ecosystem health (O'Boyle et al. 2005). Operational objectives also are defined to control impacts of various sectors (e.g., fishing, oil and gas exploration). These controls are linked back to the LOMA-level operational objectives, thus leading to the control of cumulative impacts across ocean sectors. Many of these sector-based operational objectives already exist but are not part of the overall EBM framework.

The Scotian Shelf LOMA provides a good example of how stakeholder involvement can enhance the planning process, particularly to help reconcile current operational objectives with EBM. Input from ocean communities was gathered through directed meetings with relevant organizations, groups, and individuals, as well as through a large, multistakeholder discussion during the second Eastern Scotian Shelf Integrated Management (ESSIM) forum workshop in 2003. Based on these interchanges, the ESSIM forum secretariat plans to work collaboratively with interested parties, both regionally and nationally, over the longer term to develop the plan for eventual implementation (ESSIM 2003).

In Canada, EBM relies on guidelines, regulations, and best practices developed under acts and policies other than the COA to achieve its ends. It is expected that management actions will consist of a suite of regulatory tools similar to that currently used (e.g., quotas, gear specifications). However, greater consideration is being given to both area restrictions (e.g., marine protected areas) and codes of practice to better manage sector-specific activities (e.g., fishing, oil and gas development). Identified ecologically and biologically significant areas will aid in the identification of areas requiring special management consideration.

Challenges

Similarly to Australia, an important challenge for Canada is the need to engage provincial and municipal governments in the implementation of EBM in the coastal zone. The goal is to extend the LOMAs inshore eventually, although the initial efforts have been farther offshore where there are fewer jurisdictional issues (e.g., on the Scotian Shelf). Also, while suites of ecosystem objectives are being developed for each LOMA, there is a need to determine precisely how the scientific criteria on ecologically and biologically significant areas and species, degraded areas, and depleted species—which are intended to focus attention on the key conservation issues—will inform the conservation priorities. In addition, a structured prioritization process with broader stakeholder input using defined overarching social and economic objectives is needed to inform LOMA objectives. Experience with the fishing sector on the Scotian Shelf indicates that stakeholder engagement is most effective when it is sector-based and begins with dialogue on current practices that are EBM compliant. This discussion still needs to be expanded to include the other sectors. Finally, predicting ecosystem responses to a particular management action is difficult—hence the appeal of the resilience framework described by Leslie and Kinzig (chap. 4 of this volume). With competing stakeholder interests, the possibility of advocacy for a particular action not supported by science must be countered with effective peer review and advisory processes.

European Union

Policy Context

The EU started developing a Marine Thematic Strategy for the Protection and Conservation of the European Marine Environment (the EU Marine Strategy) in 2002 (CEC 2002). The strategy stems from the Sixth Environmental Action Programme (6EAP) and is based on the recognition that existing measures at Community and national levels are inadequate to deal with the threats to the EU marine environment. To address this shortcoming, a Marine Strategy Directive was put forward as a new Community-level instrument for marine conservation.

The key emphasis of the Marine Strategy and proposed directive is the connection of science, policy, and management in Europe's regional seas (e.g., Baltic, North, and Mediterranean seas). The strategy intends to promote sustainable use of the seas and to conserve marine ecosystems by 2021 and requires Member States to cooperate wherever possible at the level of existing regional seas conventions. The strategy aims to consolidate and integrate the patchwork of policies, legislation, programs, and action plans at the national, regional, EU, and international levels in place to control and reduce pressures and impacts on the marine environment. This provides a basis for advancing EBM, inter alia, in EU legislation, policies, and programs under different commission directorate generals; the Convention for the Protection of the Marine Environment of the North-East Atlantic; the Barcelona Convention (Mediterranean); and the Helsinki Commission (HELCOM).

The European Commission also is developing a governance framework for integrated policymaking (e.g., the development of an integrated surveillance network for European waters and maritime spatial planning, and integrated coastal zone management to improve compatibility of different marine activities). The Commission also plans to create an internal program to enhance coordination between the sector-specific policy initiatives related to maritime affairs. Moreover, an action plan, which was released in 2007 with broad-based stakeholder input, lists a range of concrete initiatives that will be taken as first steps toward a more consistent, integrated EU maritime policy (CEC 2007). A number of stakeholders, including some 279 Member States, came out in support of the policy, and some implementation has begun (e.g., by the Netherlands, Germany, O SPAR, HELCOM, and Denmark) (EC 2006).

Mode of Coordination

The Marine Strategy is primarily focused on the protection of the regional seas bordered by EU countries, but its objectives and principles have broader application to adjacent seas. It also contains specific objectives in order to reduce the EU's footprint in other parts of the world, most notably on the high seas. This allows the EU to help meet its obligations for conservation and protection of the marine environment under various international and regional agreements to which it is party. The strategy captures the essence of some key features of EBM, such as integrated and adaptive management guided by clear objectives, identification of appropriate environmental indicators, and monitoring of long-term impacts.

Once planning boundaries were defined

based on biogeographic and oceanographic criteria, overarching conceptual objectives and corresponding action steps were set to guide member countries in developing their own conceptual and operational objectives with corresponding timelines for their achievement. Each coastal Member State must develop, in coordination with other Member States and other countries with which it shares a given marine region, an implementation plan containing an assessment of the pressures and threats impacting the marine environment, along with their and associated costs. These plans also will contain monitoring and assessment programs. The coastal states must do the following:

- Assess the current environmental status of their waters within a suggested 4 years after entry into force of the Marine Strategy Directive
- Determine what constitutes good environmental status in their waters within a suggested 4 years after entry into force of the directive
- Establish a series of environmental targets within a suggested 5 years after entry into force of the directive
- Implement a monitoring program within a suggested 6 years after entry into force of the directive
- Develop and operationalize a program of measures by 2016 and 2018, respectively

Challenges

Issues related to the conservation and sustainable use of marine ecosystems and biodiversity in the EU are influenced by a number of Community policy sectors. While conservation of the marine environment has traditionally been addressed as a part of EU environmental policy, the management of EU fisheries falls exclusively under the Community's Common Fisheries Policy. Therefore, it is important that appropriate mechanisms, both at Community and Member States levels, are in place or will be implemented to assure a smooth interplay between the marine and fisheries policy sectors.

In order to capture the specific needs and challenges faced by Europe's different marine regions, it is only logical that specific targets are defined at the regional rather than the EU level. The Marine Strategy Directive does not provide criteria for defining and judging "good environmental status." Given that this is the target of all actions under the directive, the lack of criteria can be seen as one of the main weaknesses of the proposed directive. Environmental indicators need to be developed to explore the progress in maintaining and recovering biodiversity and fish stocks, reducing synthetic substances in the marine environment, and preventing eutrophication.

Shared jurisdictional boundaries pose the greatest challenge to the EU. At the multinational level, the EU Water Framework jurisdiction extends out 6 miles; the EU, whose authority extends throughout the entire exclusive economic zone, or EEZ (3–200 nautical miles offshore), regulates fisheries. The EU has the sole right to initiate legislation, and Member States must abide by EU regulations. Member States also carry an obligation to protect the environment under UNCLOS. In addition, there's an EU commitment to protect the marine environment in 6EAP, which presents a framework for environmental policy for the EU for the period 2002–2012. However, this does not suggest that Member States have a collective responsibility to protect the marine environment. However, the European Parliament's initial reaction was in favor of regional implementation (like the EU), but there was some limited resistance to the notion of collective responsibility. The issue of individual Member State versus collective responsibility is indeed an important one. The political challenge is whether Member States should develop individual marine strategies or regional strategies.

Another current issue being explored by the European Parliament and the Commission of the European Communities is whether MPAs should be mandatory via legislation or through nonbinding, less formal political agreements. The Commission maintains that it is unnecessary to make MPAs compulsory, because the obligation already exists under EU law (i.e., the EC Habitats Directive, Natura 2000, already calls for protected areas because it is designed to protect the most seriously threatened habitats and species across Europe).

New Zealand

Policy Context

New Zealand does not have a national EBM policy or legislation, but there is strong political leadership for most of the elements of EBM. This is demonstrated in the development of the Fisheries Act, the Coastal Policy Statement of the Resource Management Act (RMA), the Biodiversity Strategy (NZ 2000), and the Marine Protected Areas Policy (DCMF 2005). Such leadership is apparent in the quick decisions taken to reduce allowable catches of depleted fish stocks and to close fisheries when populations of protected species are threatened, in the substantial network of no-take marine reserves and marine protected areas throughout the country, and in the decision to develop an Oceans Policy. Motivation for this political leadership stems in part from New Zealand's desire to satisfy its obligations under international agreements.

A coordinated approach to oceans management was first discussed in New Zealand national policy in 2000. However, an Oceans Policy was not completed, due to factors including disagreement between the government and Maori (New Zealand indigenous people) over ownership of nearshore and seabed areas. These issues were addressed by new legislation in 2004. Development of the Oceans Policy recommenced in 2005 and is expected to focus initially on improving management of environmental effects of human activities in the EEZ. Efforts to promote EBM continue to occur through the actions of existing national sectoral agencies as well as at the local level.

Mode of Coordination

New Zealand has advanced EBM through existing public laws and infrastructure. The Fisheries Act 1996 (NZP 2007) is the major sectoral legislative instrument related to EBM. The purpose of the Fisheries Act is "to provide for the utilization of fisheries resources while ensuring sustainability," and it requires adverse effects of fishing to be avoided, remedied, or mitigated. Effect is defined in a way that includes past, present, and future effects and cumulative effects. Environmental principles for maintaining species viability and biological diversity, and information principles designed to ensure a precautionary approach, are also part of the Fisheries Act. Under the act, catch limits for most harvested fish species are required to be set to allow stocks to be maintained at or above the biomass that produces the maximum sustainable yield, modified as necessary by the interdependence of stocks. Taken as a whole, the act gives priority to maintaining biological sustainability; beyond this point it requires biological and utilization considerations to be balanced. Under Fisheries Act regulations, bycatch limits and bycatch mitigation devices are required in areas where protected species are abundant. In 2001, under Fisheries Act regulations, areas covering approximately 115,000 km^2 of seamount-like features were closed to trawling to protect benthic communities (Alder and Wood 2004). More recently, the government implemented a fishing industry proposal to close an additional 1.1 million km^2 (31%) of New Zealand's exclusive economic zone to bottom trawling and dredging, to protect benthic communities (MOF 2006). The Ministry of Fisheries (MOF) also released the Strategy for Managing the Environmental Effects of Fishing, which included proposed standards to define acceptable limits of environmental effects of fishing on different parts of the aquatic environment (MOF 2005).

The Biodiversity Strategy, released in 2000, applies across the land, territorial sea (0–12 nautical miles from the coast), and exclusive economic zone (12–200 nautical miles) and was developed to address New Zealand's obligations under the Convention on Biological Diversity. The strategy does not use the term "ecosystem-based management" but sets out a series of goals to maintain biodiversity—including at an ecosystem level—and identifies goals for coastal and marine ecosystems. It is implemented through a range of conservation-related and sectoral legislative instruments and, within the territorial sea, through the RMA. The RMA is New Zealand's primary cross-sectoral planning instrument, providing for interjurisdictional management. It specifies the activities that are prohibited in the coastal marine area unless expressly authorized. The RMA also requires the Department of Conservation, a central government agency, to prepare a coastal policy statement containing national priorities for the preservation of the natural character of the coastal environment, implementation of New Zealand's relevant international obligations, and matters to be included in regional coastal plans in regard to the preservation of the natural character of the coastal environment. Like the Fisheries Act, the RMA seeks to balance ecological and utilization considerations while ensuring biological sustainability. Under the RMA, assessments by regional councils of proposed activities affecting natural resources take into consideration ecosystem objectives established by the regional council. Long-term planning is inherent in the RMA, which recognizes the needs of future generations and the need for the sustainable management of natural and physical resources.

In the absence of an overarching oceans policy, various initiatives were undertaken to improve management integration. One example is passage of legislative amendments by which the effects of proposed new aquaculture operations on existing fishing operations are assessed. Another example is the development of the Marine Protected Areas Policy, developed to meet an objective of the Biodiversity Strategy, under which Marine Reserves Act and Fisheries Act tools will be used in a more coordinated manner to help ensure appropriate protection of marine habitats (DCMF 2005). The Biodiversity Strategy adopts a target of 10% of marine habitats to be protected. The Marine Protected Areas Policy builds on this target by requiring a comprehensive and representative network of marine habitats and ecosystems to be protected. An example at the local level where EBM principles were applied with some success is the Fiordland Marine Conservation Strategy. This project has provided a multi-interest forum where those involved in the environment and fisheries management of the fiords and surrounding coast have worked together across agency and sector boundaries (Guardians of Fiordland's Fisheries and Marine Environment 2003).

Challenges

A major challenge for implementing EBM is that different administrative boundaries are used by the regional councils, the Ministry of Fisheries (e.g., quota management areas and fisheries management areas), and the Department of Conservation. None of these boundaries coincide with *iwi* (Maori tribe) boundaries, which are becoming increasingly important in managing activities, especially in nearshore oceans environments. Another challenge is that the major sectoral legislative instruments for controlling mining and oil extraction do not contain specific requirements relating to EBM, and there is no formal cross-sector integration of management measures in the EEZ.

While the different statutes provide for aspects of EBM to be applied to many elements of New Zealand's oceans environment, there are important gaps in the current legislation, including the following:

- Lack of overall direction for management of ocean resources and ecosystems
- Lack of an explicit requirement to apply a precautionary approach
- Inadequate consideration of the effects of different activities
- Poor integration between management processes and activities established under different acts

The RMA falls short of providing a single oceans planning framework in two important ways. First, it does not address the effects of fishing on the marine environment, which are addressed under the Fisheries Act. The Fisheries Act and RMA require that decision makers under each act take into account plans developed under the other. However, in practice, the linkages are generally weak. Second, the RMA applies only to the territorial sea (within 12 nautical miles) and not to the EEZ. Over time, the Oceans Policy is expected to address these challenges.

United States

Policy Context

Until recently there was limited political leadership for implementing EBM at the national level in the United States. As a result, the approach to ocean governance has been fragmented and inefficient. US ocean governance consists of overlapping and conflicting laws, regulations, and management bodies at state and federal levels, with most efforts focused primarily on the resolution of issues on a sector-by-sector basis (Cicin-Sain and Ehler 2002; USCOP 2004; Searles Jones and Ganey, chap. 10 of this volume). The movement toward development of a US national policy was stimulated largely by the work of a federal task force, the US Commission on Ocean Policy (USCOP 2004), and by nongovernmental efforts of the Pew Oceans Commission (POC 2003).

The sixteen-member US Commission on Ocean Policy was established in 2000 under the Oceans Act and consisted of a group of experts in various marine fields, who were charged with developing a coherent, comprehensive, and long-range national policy for exploration, protection, and use of ocean and coastal resources. The commission developed a set of recommendations for EBM implementation that included overarching conceptual objectives. To improve scientific understanding, the commission also urged that "regional ecosystem assessments" be conducted to provide a vehicle for comprehensively and periodically analyzing the status of an ocean region, establish baselines for ocean ecosystem health, and describe existing or potential impacts from human activities (USCOP 2004). In response to the recommendations of these commissions, the federal government developed a US Ocean Action Plan (US Executive Branch 2004), established the Cabinet Committee on Ocean Policy, initiated legislative changes in the primary fisheries law for the United States, enacted legislation to enhance efforts to clean up and prevent marine debris, adopted the UN resolution to stop destructive fishing practices on the high seas, and established the Northwestern Hawaiian Islands Marine National Monument—the largest single area of conservation in the nation's history and the largest protected marine area in the world (COP 2007).

To some degree, international policy also influenced the advancement of EBM in the United States, even though it has yet to ratify either UNCLOS or the Convention on Biological Diversity. This is because the United States has ratified the UN Fish Stocks Agreement; is implementing the FAO Code of Conduct for Responsible Fisheries; endorses Principle 7 of the Global Compact, which advocates taking a precautionary approach to business decisions; and endorses Agenda 21, a comprehensive plan of action to be taken by organizations of the UN system, governments, and major groups in every area in which humans impact the environment. While not as powerful as UNCLOS or the Convention on Biological Diversity, these nonbinding agreements still wield significant influence in promoting the adoption of EBM within domestic policy and laws and likely helped to influence at least some US state and federal regulators in their actions.

Mode of Coordination

If pending legislation is passed, then plans are to, inter alia, establish two high-level governmental positions—a national administrator and a presidential oceans policy advisor—as well as a committee on oceans policy and a council of advisors on oceans policy. In addition, certain regions will be designated for EBM and a regional partnership will be established for each region. There also will be continued strengthening of regional ocean-observing systems under IOOS (HR 21, 2007). In the meantime, the United States has begun to coordinate management of ocean use through existing sector-specific federal laws. In addition, some US states are enacting ocean management legislation for state waters.

Despite the current lack of a national legislative mandate for EBM, since the late 1990s some revisions made to existing national legislation reflect some important efforts to conserve marine ecosystems. These include modifying fisheries legislation to address bycatch and habitat protection; requiring consultation between agencies with regard to habitat impacts, endangered species concerns, and overall environmental impacts, including cumulative impacts under marine mammal protection, endangered species protection, and national environmental policy legislation; and coordinating efforts for coastal zone management and between state and federal governments through the Coastal Zone Management Act (CZMA) of 1972. Some have argued that the CZMA probably has the greatest potential for becoming the legislative basis for EBM (Dunnigan 2005). Searles Jones and Ganey (chap. 10 of this volume) discuss changes to several pieces of US legislation in greater detail.

Should overarching federal legislation be passed in the future, both the US Commission on Ocean Policy and the Pew Oceans Commission suggested that a good starting point would be to define planning boundaries by refining existing regional fishery management council boundaries. Another option would be to consider past regional planning efforts by the US Environmental Protection Agency (EPA) and NOAA. EPA identified some critical coastal areas through its large-scale ecosystem-based protection efforts, which focused on regions spanning 100,000 km^2 each (USEPA 1995). These efforts were ecosystem-based in that they sought to consider social, economic, and ecological concerns surrounding a large, complex, highly beneficial, and irreplaceable ecosystem. Several of these key coastal programs are still being supported by EPA (e.g., in the Great Lakes, Puget Sound, Gulf of Mexico, Chesapeake Bay, and Gulf of Maine).

In the mid-1980s, Dr. Kenneth Sherman of NOAA's National Marine Fisheries Service and Dr. Lewis Alexander of the University of Rhode Island pioneered the concept of large marine ecosystems (LMEs). These areas typically include expansive ocean areas, generally greater than 200,000 km^2 (77,220 square miles), functioning as ecosystems with distinct bathymetry, hydrography, and biological productivity. NOAA's jurisdiction includes all or part of eleven

LMEs. These include the East Bering Sea, Gulf of Alaska, California Current, Gulf of California, Gulf of Mexico, Southeast US Continental Shelf, Northeast US Continental Shelf, Scotian Shelf, Newfoundland–Labrador Shelf, Insular Pacific–Hawaiian waters, and the Caribbean Sea.

Further action is being taken at the state level to embrace EBM. For instance, the governor of California developed an action plan to reinvigorate efforts to put into practice measures identified under the California Ocean Protection Act. The plan sets forth four planning area ecosystem objectives and thirteen actions (operational/use objectives) designed to reinvigorate ocean and coastal management in California (COPC 2006). In addition, comprehensive ocean management legislation was passed or is pending in several US states, including New York, Oregon, Washington, Massachusetts, New Jersey, and Florida (USJOCI 2007).

Challenges

National political leadership and a clear mandate are needed to adopt an ecosystem-based approach to management in the United States. Despite the blueprint provided by the work of the national commissions, progress has been slow. Without national integrated management and EBM legislation, dedicated funding for EBM is lacking. Generally, federal and state agencies operate autonomously to manage their respective sectors, even though some have adopted broader ecosystem objectives and a few voluntary arrangements to promote better coordination of activities. Still apparent are the inherent struggles to coordinate management efforts across jurisdictional boundaries between state agencies (i.e., jurisdiction generally extends from 0–3 nautical miles) and federal agencies (i.e., generally 3–200 nautical miles), as well as between the United States and neighboring countries (i.e., Canada, Mexico, Cuba, and the Bahamas).

Interactions between sectors are only weakly addressed in the current system, as are cumulative impacts of human activities, and there is no mechanism for evaluating and resolving trade-offs in ecosystem services. Without the means to address these key principles of EBM, the existing sector-by-sector management approach cannot really evolve into an ecosystem-based approach.

The United States is the only case study country that has not acceded to UNCLOS or the Convention on Biodiversity. This hinders progress toward a national EBM effort, because meeting international obligations can be a powerful driver of change in national policy as well as a means of improving international efforts to implement an ecosystem-based approach.

A Framework for EBM Implementation

Based on the analysis of existing policy and practice in the five case study regions, EBM is characterized by several key features: policy direction and authority, a hierarchy of goals and objectives, defined regions, and explicit ecosystem monitoring and management review processes.

Policy Direction and Authority

Strong political leadership is essential to EBM, associated in many cases with legislative change and mandates, clear ecosystem conservation goals, and long-term funding of implementation, monitoring, and enforcement measures. Political support also underpins the establishment of institutions with EBM mandates and/or direction to existing agencies to implement an EBM approach within their decision-making and regulatory processes. Ideally, there also is impetus to better integrate management across regulatory authorities. Through a more coordinated, complementary approach, gaps in management can be identified and remedied, and movement can be made toward "true" ecosystem-based management, embodying the principles described in detail in this volume.

A Hierarchy of Management Goals and Objectives Relevant to the Protection of Ecosystem Values

A hierarchy of planning and management objectives consists of *overarching conceptual goals* at the highest level possible (e.g., national or multinational) that link international directives with national/regional EBM objectives (see table 16.1). These are long-term goals (e.g., more than 10 years) but, under an adaptive management process, may be modified over time, as the implementation of management activities influences regional objectives. These lay the foundation for *regional-level objectives* that include conceptual objectives, operational objectives, and finally, sector-specific operational objectives. *Conceptual objectives* are medium-term (e.g., 5–10 years) and multisectoral and adapt the overarching conceptual goals to regional needs, issues, and desires so they are applicable to a specific "ecosystem" management area. At this level, ecosystem services and their relative values are identified, pressures and/or risks are identified, and time frames for implementing measures to protect ecosystem values are determined. Importantly, existing governance arrangements should be defined at this level, including sectoral and cross-sectoral arrangements and gaps in management responsibility. *Operational objectives* provide specific direction to activities within a specific management area. They relate to multisectoral management actions and the management of cumulative impacts of human activities on the ecosystem. They can be refined, as new data and information become available, perhaps through the development of monitoring arrangements with stakeholders. However, if data and information are insufficient, proxies or recent trends can be used to guide management decisions in the short term. Finally, *sector-specific operational objectives* are consistent with and linked to EBM goals and objectives. New operational objectives may need to be developed, if they do not already exist, and existing sector management may have to be expanded or redirected to be more ecosystem-focused (e.g., in cases where single-species management is occurring and only action on the target species is managed). Essential to the success of sector management is the effectiveness of monitoring and enforcement of new and existing management measures. For more discussion of this hierarchy, see also O'Boyle and Worcester (chap. 14 of this volume).

Defined Ecosystem Planning and Management Regions

A key component of EBM is identification of planning regions relevant to the conservation and protection of ecosystem functioning. An ecosystem planning and management region provides the basis for governance and a focus for coordination of multisectoral activities toward ecosystem objectives. While ecosystem planning and management regions ideally are defined on ecological features and characteristics, the changing dynamics of marine ecosystems and the political realities of existing regulatory boundaries require taking into account political, social, economic, and administrative considerations as well as ecological issues when defining EBM boundaries.

Ecosystem Monitoring and Management Review Processes

EBM is not a short-term project. It is a dynamic, long-term process, which requires regular updating and adaptation. Data and information must be continually enhanced; policies, administrative arrangements, and options for problem resolution must be assessed on an ongoing basis; and the administrative system

Table 16.1. Hierarchy of EBM goals and objectives

Stakeholder input is important across all stages.

Objectives	*Function*	*Example*
Overarching conceptual goals	High-level goals that link international/national directives with national/regional objectives	Conserve coral biodiversity.
Ecosystem planning area conceptual objectives	Linkage of overarching conceptual goals to regional ecosystem management needs and priorities	Restore coral biodiversity to pre-1980s levels.
Ecosystem planning area operational objectives	Multisectoral management and monitoring of cumulative impacts	Limit disturbed area of deep sea corals to 5% of total pre-1980s area.
Sector-specific operational objectives	Development or redirection of sectoral management tactics to ensure consistency with ecosystem management objectives	Reduce impacts on deep sea coral biodiversity by the fishing sector.
Subsector operational objectives	Application of ecosystem objectives within the activities of an individual sector	Limit area/prevent use of gear types used by a particular fishery known to impact deep sea corals.

must be robust (i.e., able to adapt to changing circumstances as well as ecosystem condition and concerns). Such learning and adaptation requires sustained monitoring and evaluation of trends in the condition and use of the ecosystems in question, as well as the effectiveness of governance responses, to periodically refine the design and operation of the EBM program over time.

Discussion

Experience in Australia, Canada, the EU, New Zealand, and the United States indicates that EBM is not characterized by a rigid set of rules and that there are many factors that affect the nature and scope of EBM initiatives. Differences in history, governance, and patterns of resource use within each case study region, for example, may be seen to have significantly influenced EBM efforts and strategies. Yet, while each case study region presents a unique experience in the development and operation of EBM, key elements of EBM are present in management approaches (table 16.2), and a number of common lessons may be identified. Each of the case studies addresses some key elements of EBM (e.g., maintaining structure and function of the ecosystem, managing human impacts on the ecosystem, considering links between living and nonliving resources, considering long time scales and cumulative impacts, having defined processes for stakeholder input and involvement in decision-making process, and having

Table 16.2. EBM elements evident in the case studies' core international agreements and conventions

	Maintains structure and function of ecosystem	*Manages human impacts on ecosystem*
Australia (national legislation and policy)	Australia's Oceans Policy,[1] Environment Protection and Biodiversity Conservation Act 1999, national representative system of marine protected areas	Australia's Oceans Policy, Environment Protection and Biodiversity Conservation Act 1999, national representative system of marine protected areas
Canada (national legislation and policy)	Oceans Strategy[2]	Ocean Action Plan[3]
European Union (multinational policy)	EU Marine Strategy	EU Marine Strategy[7]
New Zealand (national sector-specific legislation and policy)	Biodiversity Strategy, Wildlife Act (WA), Marine Mammals Protection Act (MMPA), Fisheries Act (FA)	Strategy for Managing the Environmental Effects of Fishing, Resource Management Act (RMA), MMPA, Marine Reserves Act (MRA), WA
United States (state policy and law and national sector-specific legislation and policy)	US Commission on Ocean Policy (USCOP),[12] Pew Oceans Commission, Marine Mammal Protection Act (MMPA), National Marine Sanctuaries Act (NMSA), Endangered Species Act (ESA), Magnuson–Stevens Fishery Conservation and Management Act (MSFCMA), Clean Water Act (CWA), Coastal Zone Management Act (CZMA), Marine Plastic Pollution Research and Control Act (MPPRCA)	USCOP, Pew Oceans Commission, MMPA, NMSA, ESA, MSFCMA, CWA, CZMA, MPPRCA
Convention on Biological Diversity (binding)	Articles 8d; 9c, d; 10a, b	
FAO Code of Conduct for Responsible Fisheries (voluntary)	Articles 2 i; 6.1; 6.2; 6.3; 6.6; 6.7; 7.2.2.d; 7.2.2.e; 7.2.3; 7.6.3; 7.6.4; 7.6.9; 8.4.3; 8.4.6; 8.4.7; 8.4.8; 8.11; 12	Articles 6.7; 7.2.2.f
UN Convention on the Law of the Sea (UNCLOS) (binding)	Articles 61.2; 61.3; 61.4; 64; 65; 66; 67; 68; 119; 120; 150b	Articles 145; 194; 195; 196; 204, Sections 5, 11

[1]Commonwealth of Australia 1998.

[2]Uses different terminology but in essence says the same thing: "maintaining health and integrity." DFO 2002, p. 12.

[3]DFO 2005a, p. 16. [4]DFO 2002, p. 13. [5]DFO 2002, p. 4. [6]One of three founding principles in the Oceans Act.

[7]CEC 2002, p. 2–3. [8]CEC 2002, p. 3. [9]CEC 2002. [10]CEC 2002, p. 4.

[11]CEC 2002, p. 20. [12]USCOP 2004.

onsiders links between living and onliving resources	*Cumulative and Long term*	*Stakeholder Involvement*	*Precautionary Approach*
ustralia's Oceans Policy, Envi- nment Protection and Biodi- rsity Conservation Act 1999, tional representative system of arine protected areas	Australia's Oceans Policy, Environment Protection and Biodiversity Conservation Act 1999, national representative system of marine protected areas	Australia's Oceans Policy, Environment Protection and Biodiversity Conservation Act 1999, national representative system of marine protected areas	Australia's Oceans Policy, Environment Protection and Biodiversity Conservation Act 1999, national representative system of marine protected areas
ceans Strategy[4]	Oceans Strategy[5]	Oceans Act	Oceans Act[6]
J Marine Strategy[8]	EU Marine Strategy[9]	EU Marine Strategy[10]	EU Marine Strategy[11]
odiversity Strategy, RMA	FA, RMA	FA, RMA	FA, MRA
ational Environmental Policy t (NEPA), national standard 9 MSFCMA	California's Ocean Action Plan, USCOP, Pew Oceans Commission	NOAA Ecosystem Principles Advisory Panel, USCOP, NOAA Fisheries pilot projects, CWA, MSFCMA	USCOP, Pew Oceans Commission
		Articles 8j, e; 10d, e	Article 14a, b implied
ticles 2 a, g; 6.8; 6.9; 7.2.2.g; .2; 8.7; 8.9.1.d; 8.9.1.e; 10	7.3.3 long-term management objectives	Articles 6.4; 6.13; 6.15; 6.16; 7.1.6; 7.19; 8.5.1	Articles 6.5; 7.4.3; 7.5
ticles 56.1.a; 67; 145		Article 118	Article 206

precautionary approach language) in multinational (i.e., EU) or national frameworks, either through overarching national ocean law and/or policy or through national sector-specific laws. The problem is that these elements are not always effectively implemented.

EBM depends upon the willingness of participants to make it happen. There are few explicit or short-term incentives for EBM, and thus it requires political credibility and legitimacy to make it work. EBM has international political legitimacy through such mechanisms as UNCLOS, Agenda 21, and the Convention on Biological Diversity. While there is growing commitment to EBM at a multinational and/or national level worldwide, as demonstrated by a growing number of policy and legislative tools in the case study areas, in some cases the lack of a legal basis for EBM has inhibited its implementation. This is not often due to a lack of sector-specific legislation, but rather due to a lack of appropriate legislation designed to deal with cross-sectoral conservation and management issues. In other cases, the completion of an EBM system or plan is perceived as an endpoint rather than a beginning, at which time political, community, and financial commitment to ongoing management has lapsed.

Since EBM is a process that has the capacity to challenge the status quo, policy leadership and direction are arguably even more necessary than they are for conventional management approaches. Participatory decision making is one source of legitimacy within EBM. EBM is also dependent on the extent to which pertinent and affected sectoral interests are involved in decision-making processes. In New Zealand, Australia, Canada, and the United States, the involvement of indigenous peoples will be imperative in implementing new national policy providing direction to EBM. Multisectoral planning and management is a key factor in EBM, and the exclusion of certain sectoral interests (though potentially affording less scope for conflict) has the potential to weaken the ultimate management outcomes. Thus there must be a participatory, transparent, and iterative process for stakeholder involvement even though the method for input can vary depending on specific institutional arrangements to build support for EBM.

The extent of autonomy of subnational levels of government and the nature of their power with respect to the national government are significant factors in determining how EBM has developed in case study areas. The legal jurisdiction of subnational governments has likely influenced whether EBM has advanced at the international, national, subnational, or local levels or through an approach involving multiple levels. For example, while there is increasing state support for adoption of EBM legislation in US state waters, the federal government lags behind in its efforts in national waters. In Canada's Scotian Shelf project, provinces were engaged in discussions about offshore EBM, but the jurisdictional issues have yet to be resolved in inshore areas. The latter also is the case in Australia. The EU faces the challenge of coordinating efforts of its many Member States with abutting jurisdictions and overlapping ecosystems. Nevertheless, it is anticipated that the EU Marine Strategy will provide a legal basis for Member State implementation efforts.

In such large and administratively complex governance units as Australia, Canada, the United States, and the EU, federalism allows a diversity that can respond to broadly different issues and circumstances. However, the fragmentation of federal division of powers presents challenges where harmonization of policy and effort between and within levels of government (i.e., horizontal and vertical coordination) is necessary. Planning frameworks, policy developments, and regional programs have tended to evolve in a number of the case study regions in a national policy vacuum, and links between initiatives are only slowly beginning to be defined. With a commitment to marine environmental legal and/or policy frameworks involving coordination at a national scale, the potential foundation for implementation of EBM can be stronger than ever.

Experience suggests that when EBM is implemented in concert with integrated management, coordinated management of regional sector activities and clear lines of management accountability may be more effectively enacted and enforced. Integrated management is a process that recognizes that planning and management need to be integrated across geographic and temporal scales, as well as spheres of government, to meet multisectoral objectives. Within the EBM context, integrated management can provide the framework within which multinational, federal, state, and/or local agencies communicate and coordinate their efforts to protect ecosystems.

EBM requires a dynamic, comprehensive approach toward maintaining ecosystem services. Each case study area has begun to identify ecological goals and objectives. However, most of the case studies have made limited progress in defining social and economic goals and objectives and in understanding how these relate to ecological goals and objectives. Furthermore, since EBM is not an end within itself, it becomes imperative to define not only goals and objectives but also corresponding performance standards in order to determine the effectiveness of the management system employed.

Environmental policy is fraught with uncertainty about ecological, economic, physical, and/or technological conditions, and the efficacy of particular management actions may themselves be uncertain. This means that in order for EBM to be effective, there has to be opportunity for modification and ongoing review in light of changing knowledge, circumstances, and system characteristics (see Leslie and Kinzig, chap. 4 of this volume). Where an adaptive approach is employed, EBM has the capacity to respond to management lessons at multiple levels as well as monitor changes in state and understanding of ecosystem functioning to revise management objectives accordingly.

An important factor for the future of EBM will be assessment of the benefits of the approach versus the administrative costs of its establishment. In implementing EBM, governments inevitably have to consider the costs of making it work. Evidence to date suggests that the costs of EBM are associated with short-term administration and planning and longer-term political costs of changes in power distribution (e.g., state and federal relations or national and international management body relations). However, the longer-term socioeconomic and conservation benefits warrant commitment to the principles and practice of EBM.

Key Messages

1. Ecosystem-based management is underway at national and multinational scales worldwide: Australia, Canada, the European Union, New Zealand, and the United States face similar challenges in the management of some of the largest marine jurisdictions in the world, and they are all developing marine policy and management programs based on ecosystem-based management principles to address them.
2. Challenges to EBM implementation include fragmented institutional arrangements, entrenched interests, and unresolved jurisdictional complexity.
3. Our recommendations for future implementation of EBM include the following:
   - Articulating clear goals for EBM planning and implementation that link local, regional, national, and international scales and concerns
   - Developing a clear political directive and long-term support for EBM

- Coordinating actions of regulatory agencies
- Enabling a participatory, proactive, and transparent learning process that enables adaptation to changing knowledge, environmental conditions, and improved technologies

Acknowledgments

The authors of this chapter wish to thank the David and Lucile Packard Foundation for their support of this collaborative effort. The lead author dedicates his work to the memory of Lenore E. Rosenberg, who passed away during the workshop at which this chapter came to fruition.

References

Alder, J., and L. Wood. 2004. Managing and protecting seamounts ecosystems. In *Seamounts biodiversity and fisheries*, ed. T. Morato and D. Pauly. University of British Columbia, Fisheries Centre Research Reports 12(5). http://www.fisheries.ubc.ca/publications/reports/fcrr.php.

Bruntland, G., ed. 1987. *Our common future: The world commission on environment and development.* Oxford: Oxford University Press.

CEC (Commission of the European Communities). 2007. *An integrated maritime policy for the European Union: Impact assessment summary.* Commission staff working document. Accompanying document to the communication to the commission to the European Parliament, the Council, the European Economic and Social Committee and the Committee of the Regions. Brussels 10.10.2007. SEC(2007) 1280. http://ec.europa.eu/maritimeaffairs/sectoral_dgenv_en.html.

CEC (Commission of the European Communities). 2002. *Towards a strategy to conserve and protect the marine environment.* Communication from the commission to the Council and the European Parliament. Brussels 2.10.2002. COM(2002) 539 final. http://europa.eu/scadplus/leg/en/lvb/l28129.htm.

Cicin-Sain, B., and C. Ehler. 2002. *Improving regional ocean governance in the United States.* Proceedings of the 9 December 2002, Washington, DC, workshop sponsored by US National Oceanic and Atmospheric Administration, Office of Ocean and Coastal Resource Management; Center for the Study of Marine Policy, University of Delaware; US Environmental Protection Agency, Ocean and Coastal Protection Division; and Coastal States Organization. http://www.ocean.udel.edu/cmp/pdf/RegionalProceedings.pdf.

Commonwealth of Australia. 1998. *Australia's Oceans Policy,* vol. 1. Canberra. ISBN 0 642 54592 8. http://www.environment.gov.au/coasts/oceans-policy/publications/policy-v1.html.

COPC (California Ocean Protection Council). 2006. *A vision for our ocean and coast: Five year strategic plan.* Sacramento: COPC. www.resources.ca.gov/copc/docs/OPC_Strategic_Plan_2006.pdf.

DCMF (Department of Conservation and Ministry of Fisheries). 2005. *Marine Protected Areas Policy and implementation plan.* Wellington, NZ. http://www.biodiversity.govt.nz/seas/biodiversity/protected/mpa_policy.html.

DEH (Department of Environment and Heritage). 2006. *Marine bioregional plans: Fact sheet.* National Oceans Office. Canberra: Commonwealth of Australia.

DEWHA (Department of Environment, Water, Heritage and the Arts). 2003. *Oceans policy: Principles and process.* National Oceans Office. Canberra: Commonwealth of Australia. ISBN 1 877043 257.

DEWR (Department of the Environment and Water Resources). 2007. *The south-west bioregional marine plan. Bioregional profile. A description of the ecosystems, conservation values and uses of the south-west marine region.* Canberra: Commonwealth of Australia.

DFO (Department of Fisheries and Oceans). 2005a. *Canada's oceans action plan: For present and future generations.* Ottawa: Canadian Science Advisory Secretariat. K1A 0E6. www.dfo-mpo.gc.ca.

DFO (Department of Fisheries and Oceans). 2002. *Canada's ocean strategy.* Ottawa: Canadian Science Advisory Secretariat.

DFO (Department of Fisheries and Oceans). 2007a. *Guidance document on identifying conservation priorities and phrasing conservation objectives for large ocean management areas.* Science Advisory Report

2007/010. Ottawa: Canadian Science Advisory Secretariat.

DFO (Department of Fisheries and Oceans). 2005b. *Guidelines on evaluating ecosystem overviews and assessments: Necessary documentation.* Science Advisory Report 2005/026. Ottawa: Canadian Science Advisory Secretariat.

Dunnigan, J. 2005. *NOAA's ecosystem approach and plans.* Presentation to Marine Fisheries Advisory Committee 11 January 2005.

EC (European Commission). 2006. *Consultation on maritime policy.* Maritime Affairs. http://ec.europa.eu/maritimeaffairs/policy_en.html.

ESSIM. 2003. *The Eastern Scotian Shelf Integrated Management (ESSIM) initiative: A strategic planning framework for the Eastern Scotian Shelf ocean management plan.* A discussion paper prepared for the ESSIM forum. January 2003. Oceans and Coastal Management Division. Maritimes Region. Ottawa: Fisheries and Oceans Canada.

GFFME (Guardians of Fiordland's Fisheries and Marine Environment, Inc.). 2003. *Fiordland marine conservation strategy.* Te Kaupapa Atawhai o Te Moana o Atawhenua.

HR (House of Representatives). 2007. 21: Oceans Conservation, Education, and National Strategy for the Twenty-first Century Act. Introduced to the 110th Congress on 4 January 2007 by Representative Sam Farr. http://www.govtrack.us/congress/bill.xpd?bill=h110-21.

MOF (Ministry of Fisheries). 2006. World's largest EEZ conservation measure proposed. Press release. Wellington, NZ: MOF. http://www.fish.govt.nz/en-nz/Press/Press+Releases+2006/February+2006/Worlds+largest+EEZ+marine+conservation+measure+proposed.htm?wbc_purpose=Basic%26WBCMODE February 14.

MOF (Ministry of Fisheries). 2005. Strategy for managing the environmental effects of fishing. Wellington, NZ: MOF. http://www.fish.govt.nz/NR/rdonlyres/B341907D-D284-4E3D-BC19-9628D3301902/1505/SMEEFPapa2.pdf.

NZ (New Zealand). 2000. *The New Zealand biodiversity strategy.* February. ISBN O 478 21919 9. http://www.biodiversity.govt.nz/pdfs/picture/nzbs-whole.pdf.

NZP (New Zealand Parliament). 2007. Fisheries Act 1996 Amendment Bill. Wellington, NZ. http://www.parliament.nz/en-NZ/PB/Legislation/Bills/d/9/4/00DBHOH_BILL7780_1-Fisheries-Act-1996-Amendment-Bill.htm.

O'Boyle, R., M. Sinclair, P. Keizer, K. Lee, D. Ricard, and P. Yeats. 2005. Indicators for ecosystem-based management on the Scotian Shelf: Bridging the gap between theory and practice. *ICES Journal of Marine Science* 62:598–605.

POC (Pew Oceans Commission). 2003. *America's living ocean: Charting a course for sea change. A report to the nation.* Washington, DC: Pew Trusts.

TFG International. 2002. *Review of the implementation of oceans policy: Final report.* 25 October 2002.

USCOP (US Commission on Ocean Policy). 2004. *An ocean blueprint for the twenty-first century.* Final report. Washington, DC: USCOP. ISBN 0 9759462 0 X.

US Executive Branch. 2004. *US Ocean Action Plan. The Bush Administration's response to the US Commission on Ocean Policy.* Washington, DC. http://ocean.ceq.gov/actionplan.pdf.

USEPA (US Environmental Protection Agency). 1995. *A phase I inventory of current EPA efforts to protect ecosystems.* EPA841-S-95-001. Office of Water (4503F). Washington, DC: USEPA.

USJOCI (US Joint Ocean Commission Initiative). 2007. *Regional and state ocean activities summary.* Meridian Institute. Last updated 7 March 2007. http://www.jointoceancommission.org/index.html.

Westcott, G. 2000. The development and initial implementation of Australia's integrated and comprehensive oceans policy. *Ocean & Coastal Management* 43:853–78.

CHAPTER 17
State of Practice

Karen L. McLeod and Heather M. Leslie

As the preceding case studies illustrate, there is no one path to ecosystem-based management (EBM). Rather, EBM is a shift in mindset to considering whole ecosystems and the many services that they deliver, the effects of multiple activities on those systems and their services, and the implications of decision making for the health of ecosystems and human well-being. Thus, EBM relies not on prescription, but on adapting a set of approaches to a particular social, ecological, and historical context. In this chapter we will evaluate the current state of EBM practice through the lenses of four themes: scale, stakeholders, integration of science and management, and success.

Embracing a Multitude of Scales

One of the main challenges for implementing EBM is designing strategies to maintain or restore target ecosystem states and services at a multitude of geographic scales. The efforts described in this volume vary in geographic scale from a small estuary (Morro Bay, at < 10 km^2) to an entire large marine ecosystem (Gulf of California, at 375,000 km^2) to national- and multinational-level implementation, and they vary in age from the nascent effort in Morro Bay to over three decades of activities in Chesapeake Bay. New visions for sustainability and associated changes in governance are emerging through multinational and national policies (e.g., Australia, Canada, and the European Union), regional initiatives (Chesapeake Bay and Puget Sound), and local efforts by scientists, user groups, or nongovernmental organizations (e.g., Morro Bay and the upper Gulf of California). A crucial lesson that emerges from these cases is that no matter what the geographic focus of the effort, it is critical to consider drivers of change operating at scales both smaller and larger than the focal scale. For example, practitioners in Puget Sound recognize that the sound is part of the larger Georgia Basin ecosystem that extends well into British Columbia. Thus, effective restoration of the sound will require coordination with other states and Canada to manage or accommodate climate change impacts, to consider the effects of transportation and trade policies, and to evaluate harvest regulations for widely ranging species such as salmon, herring, and marine birds. At the same time, effective restoration will also require thinking small and recognizing the roles of distinct subregions, as is the case for strategies to recover salmon (Ruckelshaus et al. 2002).

At large geographic scales, overarching policy directives for EBM at the multinational (European Union), national (Australia and Canada), and regional (Puget Sound) levels aim to ensure consistent decision making. However, these directives are not sufficient in and of themselves; local- and regional-scale planning and implementation are also critical. For example, in Australia, marine bioregional planning (initially under the Oceans Policy, but now under the Environmental Protection and Biodiversity Conservation Act) was established as the key tool to achieve EBM. Similarly, Canada's approach under its Oceans Act hinges on regional planning areas, and the European Union's Marine Strategy relies on implementation by Member States. Despite explicit attempts to engage stakeholders at smaller scales, top-down directives may suffer from a lack of involvement in and ownership of planning and management by specific agencies, user groups, or local communities.

In the absence of an overarching mandate providing a shared vision, goals, and common commitment to sustainability, large-scale change can be incredibly difficult. For example, in the Gulf of California, numerous groups have outlined the need to understand and manage the region as a single large marine ecosystem, but few changes have taken place on the ground at this scale beyond the articulation of conservation priorities. That being said, place-based efforts in the gulf are planting the seeds for larger-scale EBM through inspiration of a regional shared vision for long-term sustainability, which in turn, encourages more local action. Similar opportunities to knit together local-level efforts are evident in the Gulf of Maine and Gulf of Mexico and have borne fruit on the US west coast. Local-level efforts can also foster higher-level engagement, as is the case for the West Coast Governors' Agreement on Ocean Health, which is looking to a set of six pilot projects (one of which is Morro Bay) to inform a larger, future vision to be articulated in new coastwide policies.

Key paths forward include (1) better understanding which drivers to manage at which scales and how drivers interact across scales, (2) building and nurturing smaller-scale efforts that engage local communities under the auspices of larger-scale policies such as Canada's Oceans Act, and (3) effectively connecting local-scale initiatives such as Morro Bay to regional and large marine ecosystem scales.

Broadening Constituencies

A key aspect of EBM is broadening the spectrum of stakeholders from single interests (e.g., those with a financial stake in the harvest of a particular fishery) to the multitude of perspectives that cut across sectors, jurisdictions, economic interests and classes, and races and ethnicities. This raises fundamental questions. How do we meaningfully engage a greater diversity of interests and perspectives? How do we ensure that they work together toward common goals, rather than at cross-purposes? When they have different subgoals, how do we weight one aim, need, or value against another? For example, how do we ensure social equity when evaluating trade-offs among services? Although most people share the goal of wanting a healthy ocean, healthy means different things to different people, and people differ in their willingness to make necessary trade-offs and sacrifices, particularly when it means accepting short-term "losses" in favor of long-term sustainability.

In some cases, the primary motivator for engagement of diverse interests is an overarching mandate, such as in Canada with the Oceans Act. Under the act, multiple stakeholders have engaged in the process to develop a hierarchy of goals and objectives (including ecosystem-level objectives for each large ocean management area, or LOMA) through a network of organizations, groups, and individuals who are interested in or impacted by the initiative within each LOMA. A subset of these stakeholders contributes to a smaller advisory council that provides leadership, guidance, and direct interactions with the government. Notably, Canada's regional planning began with an offshore, rather than an inshore, focus, thus limiting the range of players at the table. As the initiative expands its purview to include the coastal zone, the diversity of interests and uses will increase substantially, and along with these, the challenges of the task at hand. In Puget Sound, the state-appointed Puget Sound Partnership worked with their diverse members (representatives of public and private institutions, tribes, and citizen groups), a scientific working group, and the general public to articulate a set of ecosystem-wide goals and initial actions for achieving the governor's vision for a healthy sound.

In the absence of a mandate, the impetus for meaningfully engaging people can be the leadership provided by a local coalition, such as the San Luis Obispo Science and Ecosystem Alliance (SLOSEA) in Morro Bay. SLOSEA brings together scientists, resource managers, and others who live and work in the ecosystem to facilitate interagency communication and management coordination under a collective vision. In the Gulf of California, civic groups like COBI, *Noroeste Sustentable,* and PANGAS are working with stakeholders to articulate a vision for their local ecosystems. In the case of PANGAS, scientists are engaging with resource users and managers in the northern gulf to further involve them in long-term research and monitoring and create robust partnerships among representatives of the fishing industry and various decision-making bodies and to improve management.

One key to not only engaging a broad spectrum of stakeholders, but also carefully balancing their diverse needs and values is to design decision making such that the trade-offs inevitable in comprehensive management are transparent. Although there will be some win-win situations (e.g., ecosystem protection through no-take marine reserves that also serve to enhance surrounding fisheries, Gell and Roberts 2003), there will certainly be costs associated with management actions taken under EBM. This is not to say that over the long term there will necessarily be more costs or trade-offs under EBM, but rather that an integrated EBM approach seeks to fully account for potential and realized trade-offs among competing uses and nonuses, rather than making such decisions on a more ad hoc basis (see also Rosenberg and Sandifer, chap. 2 of this volume). There is an urgent need for the development and testing of tools for evaluating trade-offs among different human activities and ecosystem services, as indicated by the case studies in this section and synthesis of other EBM efforts across the globe (McLeod et al. in prep.; Tallis et al. 2008).

Strengthening Connections Between Science and Management

Ecosystem-based management hinges on integration, bridging historical gaps across management sectors, among scientific disciplines, and among science, management, and policy. The roles of science for decision making are rich and varied, but chief among them is to provide information about the likely consequences of different policy options (Lubchenco 1998). Importantly, the breadth of ecological and social processes that are considered may affect the range of management strategies employed or the expected policy outcomes. For example, in Chesapeake Bay, historical attention to the role of bottom-up processes biased the focus of restoration efforts toward water quality, and there is now a need to shift to a more comprehensive view that incorporates both bottom-up and top-down perspectives. As illustrated by the Puget Sound case study, a more holistic view of the ecosystem (specifically, the potential effects of salmon hatcheries) produced a richer range of policy options and outcomes than any of the simpler views. Thus, science can convey the suite of ecologically plausible possibilities. Notably, the ultimate drivers of change in marine systems will be social and political in nature. Thus, while science plays a critical role in informing decision making, the existence of robust scientific data and even strong connections between that science and management does not guarantee that policies or their implementation will indeed reflect the latest science.

A key area of science in support of EBM is evaluating how various human activities interact to affect the delivery of ecosystem services.

In places such as Morro Bay, water quality, sedimentation, eelgrass distribution, and social factors affect oyster production. In Chesapeake Bay, there is growing awareness of the potential interactions among inputs of nutrients, toxicants, and sediment and losses of potential sinks for these inputs (oysters, wetlands, and riparian forests). Yet, to our knowledge, there is not a case in the marine realm in which there is a robust assessment of the combined effects of different factors on the delivery of ecosystem services (although new efforts in Puget Sound are designed to do exactly this). What is known about how various activities interact? Would we characterize interactions as additive, multiplicative (more than the sum of their parts), or antagonistic (less than the sum of their parts)? In a meta-analysis of experimental studies to date, Crain and colleagues (2008) showed that while interactions between pairs of stressors tended to be equally multiplicative, additive, or antagonistic, when a third stressor was introduced, in most cases the type of interaction changed, and generally this change was in a "worse" direction (i.e., shifted from antagonistic to additive or from additive to multiplicative). A global assessment of the cumulative threats of different human activities to marine ecosystems relied on an additive model because there is insufficient information on how different activities interact, particularly outside of experimental settings (Halpern et al. 2008). Clearly this is a rich area for further research.

In order for science to inform decision making, the diversity of perspectives within the scientific community must be effectively integrated and synthesized (see also Rosenberg and Sandifer, chap. 2 of this volume). In Puget Sound, this occurred through the efforts of a team of scientists from state, federal, tribal, academic, NGO, and private institutions to summarize what is known about the Puget Sound ecosystem, key threats to its functioning, and major gaps in scientific understanding (Sound Science 2007). Ruckelshaus and colleagues (chap. 12 of this volume) note that the process of developing the state of science for the region (i.e., *Sound Science*) initiated fruitful discussions among members of the diverse community of scientists in the region, many of whom tend to work on small pieces of the overall ecosystem. Such relationship building is critical not only to the development of meaningful syntheses of the science, but also for catalyzing productive collaborations among scientists from the cross-section of disciplines that will ultimately be essential to successful EBM.

In addition to scientific teams that inform state-mandated actions such as the Puget Sound Partnership, local-scale efforts can facilitate synthesis and enable scientists and managers to directly interact. For example, in Morro Bay, SLOSEA provides local managers with data to inform decision making and catalyzes regular and sustained interactions among diverse members of the group (scientists, managers, and other stakeholders). In the Gulf of California, research in support of more sustainable management of small-scale fisheries (which is considered to be a proxy for overall ecosystem health) through PANGAS incorporates interviews with decision makers at a variety of levels. Such interviews provide an understanding of the prevalent management concerns and recommendations for improvement from multiple perspectives. Thus, research and subsequent recommendations for policy change made by PANGAS will address, as much as possible, the concerns and recommendations of a variety of groups.

Moreover, there are substantial needs to better integrate perspectives from the natural and social sciences in order to understand linkages between the ecological and social domains, predict system dynamics, and effectively design management strategies. At present, the social sciences are significantly underrepresented in coastal and ocean management (with the notable exception of economics). Thus, a key path forward is to increase communication and transdisciplinary exchange so as to better understand feedbacks between the social and ecological components of marine ecosystems and to integrate knowledge from the social sciences and humanities with the biogeophysical scientific view that tends to dominate in marine management decisions.

Assessing Success

The final, and perhaps most critical, theme we will explore with respect to the state of practice of EBM is how to assess success. How can we measure the success of existing initiatives, many of which are still in the planning stages? Presently, the state of practice provides us with more questions than answers. Chief among these (as stated eloquently by Boesch and Goldman in chap. 15 of this volume) is how do we make the shift "from ecosystem diagnosis, through prognosis, and on to treatment (i.e., management)"? Ecologists and others are very skilled at defining the problem but have only recently developed scenarios and other tools that contribute to solutions (e.g., Peterson et al. 2003; MA 2005a; Carpenter et al. 2006; McClanahan et al. 2008).

Ultimately, EBM efforts aim to achieve the long-term health of coastal and ocean ecosystems *and* human well-being. Thus, measuring success will require being able to define "health" and "well-being" and to measure them (or indicators thereof) across a range of ecosystem types and geographic scales. In Puget Sound, Governor Gregoire laid out a clear vision for restoring the sound by 2020. This includes "a thriving natural system, with clean marine and freshwaters, healthy and abundant native species, natural shorelines and places for public enjoyment, and a vibrant economy that prospers in harmony with a healthy Sound" (PSP 2006). The present challenge for the region is to figure out what the measurable ecological, social, and institutional outcomes might be to achieve this vision and, just as important, what management measures will help achieve those outcomes. What are the appropriate indicators of health for human well-being, species, habitats, water flows, and water quality? From a scientific perspective, can we generate quantitative targets to indicate "how much" of each indicator is needed in order to ensure a healthy condition?

On a related note, measurable changes in ecosystem health are rarely linked to changes in economic activity, quality of life, or other aspects of well-being. Thus, even if we can show that some measure of ecosystem quality has improved, we are often unable to show that such improvements have led to changes in how people use, enjoy, or benefit economically from these systems. Consequently, in Morro Bay, ongoing efforts are aimed at collecting and monitoring economic indicator data and linking changes in these metrics to changes in the ecosystem. From a larger perspective, the Millennium Ecosystem Assessment championed a new focus on the connections between ecosystem services, human welfare, and economic systems (MA 2005b). While this view is gaining traction, much work remains to show when, if, and how ecosystem service projects can benefit both people and nature (Goldman et al. 2008; Tallis et al. 2008).

Chesapeake Bay, the most "mature" case examined in this volume, is a prime illustration of the many challenges associated with trying to measure success. The Chesapeake 2000 Agreement includes over a hundred goals and commitments and thus is one of the most ambitious EBM programs for any large coastal area in the world. However, while management efforts to date may have lessened further deterioration of the bay, there are limited ecological signs of success and inadequate social and economic data from which to evaluate nonecological outcomes. The number of potentially toxic substances has declined, but industrialized harbors remain heavily contaminated.

Algal blooms and seasonal hypoxia continue to be pervasive. Oyster populations have not recovered. Sea grasses have rebounded in some regions of the bay but still cover only a small proportion of what they once did. In addition, the Chesapeake Bay Program has been criticized for poorly differentiating between measures of implementation process and measures of actual condition. These phenomena are by no means unique to the Chesapeake, and they present challenges for any EBM implementation effort.

The nature of success may be patchy in time and space, including substantial lag times in ecosystem responses to management or restoration efforts. For example, while the overall state of Chesapeake Bay may be dismal (none of the thirteen health indicators of the 2007 "State of the Bay" improved, and those for phosphorus, blue crabs, and water quality actually fell), restoration efforts tend to be on the scale of individual tributaries and may foreshadow eventual improvements in the system at larger spatial scales. Thus, evaluating success over a multitude of scales is critical. It is also crucial to recognize the possibility of a state shift, as whether or not a system has crossed a threshold could bear heavily on the success of conservation or restoration efforts. If the system has already crossed a threshold, it may be much harder (or in the worst case scenario, impossible) to reverse changes that have already occurred. On the other hand, a focus on indicators of resilience and a more holistic perspective on how and why change is occurring may enable management to bolster elements that could reverse an undesirable situation or prevent a system from crossing a threshold in the first place.

Evaluation of management effectiveness developed in other domains, including biodiversity conservation, marine protected areas, and coastal zone management, can help inform assessment of success of ecosystem-based management efforts in coastal and ocean areas (Margoluis and Salafsky 1998; Olsen 2003; Pomeroy et al. 2004). While these other domains have much in common with marine EBM, frameworks for evaluating progress and success must be tailored to the considerably broader and deeper goals of EBM as well as the varied contexts in which initiatives are being developed. Clearly an interdisciplinary research approach will be vital—neither the biophysical nor the social sciences alone will be sufficient to investigate the factors that contribute to the success or failure of EBM efforts in varied social, institutional, and ecological contexts (for related discussions, see Orbach 1995 and Christie 2004).

Ushering in a New Era

The current state of practice provides plentiful guidance regarding the planning aspects of EBM: policy changes in support of new approaches, how to engage diverse interests at various scales, or how to effectively connect science to management. However, the leading edges of EBM lie in the nuts and bolts of implementation. How do we more effectively work across scales or across sectors? How do we manage human activities in the coastal and marine realm in light of large-scale drivers such as climate change? How do we assess progress toward ecosystem-based management and ensure that our social and ecological systems are sufficiently resilient to make course corrections as needed?

The implementation of EBM in any context requires a shift in mindset from managing one thing in many places (sectoral management) to managing many things in one place (area-based management). Key aspects of this shift include embracing change in our own behavior, societal expectations, and management; working to enhance both ecosystem and human well-being; and balancing use and protection. Moving forward with the implementation of EBM means recognizing the full spectrum of possible services from any given system, and the feedbacks and linkages between the social and ecological components in that system. Transparent decision-making frameworks for evaluating trade-offs among services or sectors, and institutions that can operate across multiple scales in order to respond more rapidly to changing and often surprising conditions, also are needed. We anticipate that the syntheses of EBM case studies in 10 or 20 years' time will yield an abundance of studies from which to choose and that EBM may well be indistinguishable from "business as usual."

Acknowledgments

We thank Sarah Lester and John Meyer for their comments on an earlier version of this chapter, as well as the contributors of this volume and the many other colleagues with whom we have interacted regarding these ideas.

References

Carpenter, S. R., E. M. Bennett, and G. D. Peterson. 2006. Scenarios for ecosystem services: An overview. *Ecology and Society* 11. http://www.ecologyandsociety.org/vol11/iss1/art29.

Christie, P. 2004. MPAs as biological successes and social failures in Southeast Asia. In *Aquatic protected areas as fisheries management tools: Design, use, and evaluation of these fully protected areas*, ed. J. B. Shipley, 155–64. Bethesda, MD: American Fisheries Society.

Crain, C. M., K. Kroeker, and B. Halpern. 2008. Interactive and cumulative effects of multiple human stressors in marine systems. *Ecology Letters* 11:1304–15.

Crowder, L. B., G. Osherenko, O. R. Young, S. Airamé, E. A. Norse, N. Baron, J. Day et al. 2006. Resolving mismatches in US ocean governance. *Science* 313:617–18.

Gell, F. R., and C. M. Roberts. 2003. Benefits beyond boundaries: The fishery effects of marine reserves. *Trends in Ecology and Evolution* 18:448–55.

Goldman, R. L., H. Tallis, P. Kareiva, and G. C. Daily. 2008. Field evidence that ecosystem service projects support biodiversity and diversify options. *Proceedings of the National Academy of Sciences* 105:9445–48.

Halpern, B. S., S. Walbridge, K. A. Selkoe, C. V. Kappel, F. Micheli, C. D'Agrosa, J. F. Bruno et al. 2008. A global map of human impacts on marine ecosystems. *Science* 319:948–52.

Lubchenco, J. 1998. Entering the century of the environment: A new social contract for science. *Science* 279:491–97.

MA (Millennium Ecosystem Assessment). 2005a. *Ecosystems and human well-being: Scenarios*. Washington, DC: Island Press.

MA (Millennium Ecosystem Assesssment). 2005b. *Ecosystems and human well-being: Synthesis*. Washington, DC: Island Press.

Margoluis, R., and N. Salafsky. 1998. *Measures of success: Designing, managing and monitoring conservation and development projects*. Washington, DC: Island Press.

McClanahan, T. R., J. E. Cinner, J. Maina, N. A. J. Graham, T. M. Daw, S. M. Stead, A. Wamukota et al. 2008. Conservation action in a changing climate. *Conservation Letters* 1:53–59.

McLeod, K. L., K. Heiman, M. Caldwell, J. Day, J. Lubchenco, S. Murawski, M. Peloso, A. A. Rosenberg, J. Sanchirico, and S. Yaffee. In prep. The art of the possible: Marine ecosystem-based management. *Conservation Letters*.

Olsen, S. B. 2003. Frameworks and indicators for assessing progress in integrated coastal management initiatives. *Ocean and Coastal Management* 46:347–61.

Orbach, M. 1995. Social scientific contributions to coastal policy making. In *Improving interactions between coastal science and policy: Proceedings of the California Symposium*, 49–59. Washington, DC: National Academies Press.

Peterson, G. D., G. S. Cumming, and S. R. Carpenter. 2003. Scenario planning: A tool for conservation in an uncertain world. *Conservation Biology* 17:358–66.

Pomeroy, R. S., J. E. Parks, and L. M. Watson. 2004. *How is your MPA doing? A guidebook of natural and social indicators for evaluating marine protected area management effectiveness*. Cambridge, UK:

International Union for Conservation of Nature (IUCN).

PSP (Puget Sound Partnership). 2006. *Sound health, sound future: Protecting and restoring Puget Sound.* Olympia, WA. http://www.psp.wa.gov/.

Ruckelshaus, M. H., P. Levin, J. B. Johnson, and P. M. Kareiva. 2002. The Pacific salmon wars: What science brings to the challenge of recovering species. *Annual Review of Ecology and Systematics* 33:665–706.

Sound Science. 2007. *Sound Science: Synthesizing ecological and socioeconomic information about the Puget Sound ecosystem.* M. Ruckelshaus and M. McClure, coordinators. Seattle, WA: National Oceanic and Atmospheric Administration (NMFS), Northwest Fisheries Science Center.

Tallis, H., P. Kareiva, M. Marvier, and A. Chang. 2008. An ecosystem services framework to support both practical conservation and economic development. *Proceedings of the National Academy of Sciences* 105:9457–64.

PART 5
Looking Ahead

CHAPTER 18
Toward a New Ethic for the Oceans

Kathleen Dean Moore and Roly Russell

An angry fisherman stands in a crowded room, shouting at a fisheries biologist whose PowerPoint presentation shows a dead zone off the Oregon coast. Farther north, lawyers representing the interests of corporations, sport fishers, and First Nations people debate long into the night. Meanwhile, ice shelves are calving off Antarctica, oil derricks are pumping off the Florida coast, albatross are dying off in the Bering Sea, a trawling family is auctioning off its boat, and a garbage scow is off-loading into the ocean. Worldwide, an estimated 25% of major fish stocks are overexploited, depleted, or recovering from depletion (FAO 2006).

In lab meetings, conferences, and publications, scientists and managers are calling for new ways to think about the management of ocean resources. Some are calling as well for a new ethical framework that will shape new management approaches. The Pew Oceans Commission states that "a change in values—not only what we value, but *how we value*—is essential to protecting and restoring our oceans and coasts" (POC 2003, emphasis added).

Marine ecosystem-based management (EBM) can be seen as part of new directions in thought about how best to manage human behavior in ocean ecosystems. Although the definition of EBM is contested, we understand it here to mean "an integrated approach to management that considers the entire ecosystem [with a goal of maintaining] an ecosystem in a healthy, productive and resilient condition so that it can provide the services humans want and need" (McLeod et al. 2005; McLeod and Leslie, chap. 1 of this volume). Setting itself in contrast to approaches that focus on managing target species under many individual species-specific plans, EBM focuses on protecting and enhancing the well-being of the ecosystem as a whole (POC 2003). It rests on the assumption that a resilient, thriving ecosystem will provide services now and—at least as important—as much and as good in the future. Some consider EBM to be sufficiently different from single-species management as to constitute a Kuhnian paradigm shift (Kuhn 1970; Yaffee 1999; but see Stanley 1995).

Our goal here is to provide some insights into the moral and ethical context of EBM. That ethical landscape is like a Darwinian entangled bank (Darwin 1964), grown wild and complex after years of interaction—the elaborately constructed theories, the full-throated songs of imagination and conscience, the wisdom of those who thrive in their places, and, under the damp earth, worms of absolutism, intractable conflict, and the murmurs of marine ecosystem collapse. Can we make any sense of it?

In this chapter, we shall address two sets of questions:

1. Where are we? Where, on conceptual and historical maps of the ethical landscape, are the current discussions about marine management playing out? What assumptions are currently being made about how moral decisions should be made? What moral questions do people ask in the context of EBM, and why are those questions so difficult to answer?
2. Where do we go next? Can we begin to imagine a new marine ethic that is robust enough to support the paradigm shift in thinking that EBM requires? Does

the evolution of ecological management practices call for a coevolution of ethics?

The Ethical Landscape of Ecosystem-Based Marine Management

A Conceptual Map

First, a quick sketch of Western ethics: A conceptual map of the landscape of ethical inquiry can be drawn as a line divided into three parts, corresponding to the three steps involved in a moral action (Rachels 2006). Whether it's restoring salmon on the Klamath River, setting a start date for Bering Sea crabbers, establishing marine protected areas along the California coast, or protecting Chesapeake Bay oysters—any decision to choose one management plan over another is not a singular event, but a series of events over time. If one were to draw a simple timeline of a generic policy decision, it would begin with an *intention* that results in a *process* (an *act* or *decision* or *policy*), which itself has *consequences* into the future. Each of these three stages of the timeline raises a different set of moral questions. In reverse order—

1. *Consequence—questions of value*
   The enterprise of managing human intervention in the oceans raises questions of goals, and thus of *values*, of what we judge to be good or bad. What are we managing the ocean *for*? That is, what desired states are we trying to achieve? What values are we trying to maximize, and whose, and to what ultimate end? What *is* a thriving ecosystem? What is human well-being? Questions of value raise corresponding metaethical questions: When goals compete, how do we decide among them? Does anything have value in itself, or is all value instrumental and contingent? What sorts of arguments can we bring to bear on questions of ultimate good?
2. *Process—questions of rightness and justice*
   Marine management decisions also raise questions about the nature of the process (the action taken, the policy chosen, or the decision made)—in and of itself, regardless of its consequences. They thus raise questions of what we judge to be *right and wrong, just and unjust*. Is the action or process just? Was it fairly agreed upon? Does it violate or respect human rights or the rights of other beings or entities? And the larger questions: What is a just allocation of resources? Do we have responsibilities to provide for the needs of future generations? Ought we to consider the interests of species other than human, or of ecosystems themselves? These questions are importantly different from questions about outcomes; a process itself can be wrong or unjust, even if its outcomes happen to be overwhelmingly positive. And of course, the opposite can be true: A morally justified act or process can have disastrous results.
3. *Intention—questions of virtue*
   Finally, managing human interventions in oceans raises also questions we don't often ponder when so many of our decisions respond to crises and the focus is on what we should *do*, rather than who we should *be*. These are questions regarding human character and intention, and thus of *virtue*. What can we ask of people in relation to the ocean, what sort of generosity or wisdom? Is self-interest an acceptable premise for management plans, or should humans aspire to generosity, a kind of sharing, a larger notion of individual thriving, a sense of personal identity based on relation to place or to the other? What can we learn from the wisdom of ocean communities and their traditions? What does it really mean to be fully human, to thrive as a human being within the context of complex, interdependent biotic and abiotic communities?

Western ethics has developed three major categories of moral theories, corresponding to the three parts of the moral timeline (see box 18.1).

1. *Consequentialist moral theories* weigh the morality of processes by the effect they have or are likely to have in the future. The right plan is the one that is most likely to result in the maximum possible good, which is to say, the plan that maximizes what one most values. In the Western world, the dominant consequentialist moral theory is utilitarianism. It posits human happiness as the justifying value or goal of a practice; the right policy is thus the one that is most likely to increase the sum total of human well-being.

   Within this moral context, discussion of the goals or desired states of marine management policies becomes significant. If we are to judge a practice by its outcomes, it is paramount to be clear about what it is that society values most, or what results it wants to maximize. It is important, too, to be clear about how an affected community's goals might be different from those of another affected community, from society in general, or from the goals of marine scientists. The theories raise questions of goal definition as well, which accounts for the importance of what are often called "indicators," signs by which progress toward the maximization of values such as ecosystem health or human well-being can be measured (Kaufman et al., chap. 7 of this volume).
2. *Deontological moral theories* focus attention not on the intention or results, but on the nature of the act (process, practice, decision) itself. The unfortunate label comes from the Greek word *deon,* which means duty (Wolf 2003). On this theory, the achievement of desired states doesn't justify a process; a process is judged right or wrong according to its characteristics. Is it just? Was it fairly chosen? This accounts in part for the emphasis on fair process in decisions about ecosystem-based management. Are all the players at the table? Are they heard? Are all interests accounted for? Are all rights respected?
3. *Virtue ethics* focus on the origin of the act in the character of the actors, their intentions or motivations, rather than on the nature of the act itself or its consequences. A right act, a justifiable process, is one that embodies the virtues of a good person or community of good will—often such qualities as humility, respect, caring, empathy, reciprocity, and sharing. Arguments for the conservation of long-standing traditions of fishing or indigenous communities may be seen as virtue-based. So may discussions of what it means to be a good citizen of the environment (Leopold 1949; Wolf 2003). Other than this, we see little discussion of virtue in debates about marine ecosystem-based management, perhaps not a surprise in contemporary Western culture. Rather, humans are usually assumed to be self-interested and even shortsighted, and policies (taxes, markets, and stakeholder meetings, for example) are often designed in hopes that they might create a context in which self-serving actions can result in the common good—making a virtue ethic irrelevant.

These moral theories are constantly at play as people or committees struggle to choose the best policies and make the best decisions about marine ecosystem management. Whether policymakers are aware of these underlying moral theories or not, they shape the questions asked, the answers admitted as relevant, the arguments seriously considered.

A Historical Map

Where is marine management located now on the map of moral theory, how did it get there, and where might it go next? In other words, what historical story can be told about

Box 18.1. Three Moral Theories and the Questions They Ask of Marine Ecosystem-Based Management Policies

Moral theory	*Locus of moral value*	*Basis of moral judgment*	*Questions for ecosystem-based management policies*
Consequentialist ethics	Consequences	Which policy is most likely to maximize the value to be achieved and least likely to diminish it?	What are the central goals or desired end states of the ecosystem-based management policy? How can the desired states be defined, and how can their achievement be measured? Is the policy conducive to clearly defined indicators or "targets" for ecosystem properties?
Deontological ethics	Act/policy/process	What are the characteristics of the process?	Is the process just? Was it fairly arrived at? Does it appropriately respect the interests and rights of the entities affected?
Virtue ethics	Character, intention	What aspects of character and integrity does the policy embody?	Does the policy embody the appropriate virtues, such as humility, respect, caring, empathy, reciprocity, and sharing?

the changing ethical context and moral assumptions of decision making in marine management?

First, we note that most moral reasoning in EBM is profoundly consequentialist; policies are judged almost solely by their outcomes. Even the superficially simple question, What are we managing *for*? presupposes that results justify processes. This is exemplified in attempts to derive lists of goals for ecosystem management (e.g., Slocombe 1998; Parris and Kates 2003; also see the stated objectives/goals for case studies in this volume in tables 12.1, 14.1, and 15.1) or calls to create such goals (e.g., USCOP 2004). It is evident as well in arguments for modeling. For example, Walters and Martell (2004) argue that fisheries "management is a process of making choices. There is no way to make choices without making at least some predictions about the comparative outcomes of choices, and these predictions cannot be made without some sort of 'model' for how the world works." Of course, there *are* ways to justify choices without predicting outcomes, but not if a choice can only be justified by its results.

Predictions and lists of desired states are important *just because* they are premises in consequentialist arguments: A policy is the right one if it maximizes achievement of the stated goals, wrong if it fails to do so. This is an old and venerable approach for ocean management, rooted most likely in prior ocean management paradigms that focused on single-species or single-issue management. Indeed, this moral framework applies to much of our global behavior, reaching well beyond the bounds of ocean management.

Second, most EBM decisions still rest on anthropocentric moral reasoning (Dallmeyer 2003), although this is an area of important and accelerating change. As people have been managing the oceans for a variety of single species, they have, on a larger scale, been trying to manage the world for a particular single species—*Homo sapiens,* whose well-being is assumed to be the summum bonum, the ultimate good. We refer again to this book's definition of EBM: "The goal of ecosystem-based management is to maintain an ecosystem in a healthy, productive and resilient condition so that it can provide the services humans want and need" (McLeod et al. 2005; McLeod and Leslie, chap. 1 of this volume). Ecosystem-based management adopts an *anthropocentric* perspective to the extent that it grants humans a special ethical standing and adopts human well-being as the goal of management decisions, as this definition and others seem to do (e.g., NOAA 2005; UNEP 2006).

In this respect, current "mission statements" for EBM are still rooted in Gifford Pinchot's famous resource conservation ethic, in which management practices should aim for "the greatest good for the greatest number [of people] for the longest time" (Callicott 1992; Pinchot 1947). Some trace the idea back millennia to the doctrine of dominion or final causes (Ehrenfeld 1981). Based on Genesis, this doctrine asserts that the natural world was created for the benefit of humankind, for humans to dominate and/or steward (White 1957; Stanley 1995). Merchant (1990) argues that Christian notions of dominion were magnified by the theories of the Enlightenment, in which the primary goal of knowledge was to control the natural world for human ends.

EBM may offer a new—some may claim revolutionary—approach to management decisions. But to the extent that it justifies processes by their ultimate benefit to humans, it is working within the utilitarian framework that has served resource managers for as long as they have been in business.

But utilitarianism has changed over the years. Indeed, it has undergone an evolution of sorts, and an interesting one. Early utilitarian policymaking focused narrowly and directly on human well-being as the justificatory goal of fisheries policies. Direct links to human thriving trumped all other values; marine ecosystems were thought to have value to the extent that they directly promoted human interests—fish to eat, seas to sail, jobs to provide for families. If sea lions eat the same fish we eat, we ought to cull sea lions.

Over time, utilitarianism got smarter, incorporating ecological knowledge about the myriad ways that human well-being depends inextricably on the "health" of the ecosystems that support our lives. As it turns out, for example, acknowledging the complexities of marine food webs may mean that culling sea lions is more likely to *decrease* the quantity of fish available for our own harvests, because of indirect relationships between sea lions and "our" fish that involve other species (Yodzis 2001). Under this complexity-allowing framework, efforts to safeguard the health of the ecosystem are good because they safeguard our own health. For example, the US Commission on Ocean Policy (USCOP 2004) stated, "While humans have always depended on particularly valued marine species for food, medicine, and other useful products, there has been a tendency to ignore species that do not have a clear, recognizable impact on society. However, it is now understood that every species makes some contribution to the structure and function of its ecosystem."

At the turn of the twenty-first century, scientists began explicitly to name and categorize the goods and services that ecosystems directly and indirectly provide to humans (Daily 1997; MA 2005). In that naming, they called attention to what had been too much overlooked in the scientific literature—the instrumental value of ecosystems and their importance as means to the end of (and even necessary conditions for) human thriving. Some of the values are direct: Provisioning services and cultural services give us what we need and/or want—salmon steaks, something to trade for money, vitamins, SCUBA diving vacations in the Bahamas. Others of the values are indirect: Supporting services and regulating services provide the necessary conditions for other goods and services more obviously connected to the possibility of human thriving—predictable weather and ocean currents, cubic miles of krill, eelgrass nurseries, or mangroves to hold coastal shorelines in place and temper tsunamis (UNEP 2006).

There was to be yet another expansion in the concept of the human interests that can be served by marine ecosystems, to include spiritual values that are more difficult to describe and measure. This expanded notion of utility, Kellert says (2003), is "rooted in human dependence on nature and the marine world not just for bodily comfort and physical health, but also for creative capacity, intellectual prowess, emotional relations, spiritual connection and more." Thus a utilitarian ethic can value, for example, a sense of ourselves as belonging to a maritime people or a place, shades of blue to make the heart ache, *Moby Dick*, *The Old Man and the Sea*, and the comfort that a dying Rachel Carson found in the repeated movements of wave and tide (Carson 1998).

Most important, in EBM, sustainability over time (that is, delivery of goods and services in perpetuity) is taken to be an essential component of ecosystem values (see WWF 2006; NRC 2006). This represents an important expansion of the scale of moral concern, a recognition that our moral responsibilities toward human well-being extend beyond this generation into the future. If the measure of our policies is that they increase the well-being of human beings (see MA 2005), then our moral obligations extend beyond our immediate self-interest to the interests of generations of people not yet born or even imagined (e.g., UN 2000). Philosophers have begun to give careful thought to what principles might guide policies toward future generations (see box 18.2).

So the utilitarian moral theory at the foundations of EBM has come a long way from its origins in dominion and narrow humanism, as our understanding of human self-interest has become more sophisticated and expansive. On this account, human well-being is still the highest value. We are just considerably better informed and more imaginative about what is in our long-term interest, understanding that in order to take care of human interests, we will have to take care of ecosystems as well. This is an argument, we parenthetically note, with the same form as old arguments advocating good care of slaves.

As the realm of valued outcomes of management decisions has widened, scientific knowledge supporting management decisions has changed, creating a concurrent change in the policy approaches. Policies are now beginning to explicitly recognize perspectives of interconnectedness, surprise, precautionary principles, and resilience theory (e.g,. Guerry 2005; UNEP 2006; Francis et al. 2007). For example, the model emphasized in this volume focuses mainly on a precautionary approach to management, in which complexity and nonlinearities in the behavior of the ecosystem necessitate preparation for and expectation of surprises in the system (POC 2003). As Tudela and Short (2005) wrote, "The central issue here is that *recognizing the complex nature of exploited marine ecosystems should lead to reducing our expectations of science to make accurate quantitative predictions.*" The model recognizes also the importance of managing for *resilience* in ecosystems, that is, the extent to which ecosystems can absorb disturbance and continue to function without slowly degrading or unexpectedly flipping into alternate states (Francis et al. 2007; Leslie and Kinzig, chap. 4 of this volume).

There was to be another expansion in consequentialist moral theory—a potentially revolutionary change. Say one grants that a policy is morally justified if it maximizes the ultimate value or maximizes other values that are instrumental to that end. That doesn't mean that we have to agree that humans and human happiness are the only things good in themselves, that is, good as ends. Since the 1949 publication of Aldo Leopold's *A Sand County Almanac* and Richard Sylvan's article "Is There a Need for a New, an Environmental, Ethic?" in 1973, Western ethics has been invited to imagine that the thriving of individual animals, species, a diversity of species, or an ecosystem itself might have intrinsic value. That is, in addition to their instrumental value as means to human ends, these sorts of entities might have value as ends in themselves.

Thus our marine management policies might be judged according to whether they advance the well-being not only of humans and what serves humans, but of the ecosystems in their own right (Grumbine 1993). No longer an anthropocentric theory, this is a biocentric

Box 18.2. What Guidelines Might Direct Our Policies toward Future Generations?

Philosophers propose that the following principles might guide our policies that affect future generations (Partridge 2001):

1. First, Do No Harm

Harm to future persons is likely to result from disruptions in fundamental biotic and ecosystemic conditions. It is easier to prevent harm to future generations than to actively promote their happiness. Thus, this commonsense principle emphasizes mitigating evil to future generations, rather than choosing and promoting their good.

2. Leave as Much and as Good for Others

While we might not be able to leave for the future exactly the resource that we deplete, we should leave the opportunity to satisfy the need that the depleted resource satisfied. This entails that we use recycling technologies and renewable resources and that we preserve natural ecosystems.

3. Preserve the Options

While we cannot predict the solutions to scarcity, we owe future generations a full range of opportunities to find and implement their own solutions. This entails humble and careful preservation of the components of systems, even if we do not now understand their importance.

4. Anticipate and Prevent

Because it is much easier and cheaper to foresee and prevent a future harm than to remedy it, we have a "duty of anticipation" to conduct the studies that will predict the impact of policies on the future.

5. Justly Forbear

Once we anticipate a possible harm from a planned course of action, we should justly forbear from that action, even if that means forgoing immediate advantage for the sake of future generations.

or ecocentric moral theory, an important expansion of the sphere of our moral concern (Leopold 1949). Accordingly, Callicott (1992) reformulates Leopold to say, "Human economic activities are right when they tend to co-create healthy biotic communities, they are wrong when they tend otherwise."

So where is EBM now on the map of moral theories? Many EBM projects seem to presuppose a consequentialist ethic with two, perhaps competing, notions of the good, working to maximize both ecological integrity and the goods and services that promote human well-being. The course charted by Washington State's Governor Christine Gregoire for the Puget Sound Partnership may be an example: "Puget Sound forever will be a thriving natural system, with clean marine and freshwaters, healthy and abundant native species, natural shorelines and places for public enjoyment, and a vibrant economy that prospers in productive harmony with a healthy Sound" (PSP 2006; see Ruckelshaus et al., chap. 12 of this volume). How these goals can be reconciled is the primary challenge of EBM.

Moral Quandaries, Briefly Mapped

This is approximately the moral landscape in which EBM operates as this book goes to press. It is not, as managers will attest, an easy place to work. By its very nature, a consequentialist moral theory presents certain problems and complexities, and these are the very problems that predictably bedevil those who would make management decisions (box 18.3). These are the quandaries that create the shouting fishermen, the table-pounding lawyers, the empty nets, the despairing families selling their trawlers while sport fishers ship home boxes of frozen fish. What are the moral difficulties and ensuing decision constraints imposed by EBM's ethical framework, and what (if any) guidance does the framework provide to escape these moral conflicts?

1. CONFLICTING PRIMARY GOALS

When there is disagreement about the summum bonum, the ultimate good, the desired state that a management plan should be designed to achieve, consequentialism offers no help. Should ecosystem-based management maximize human thriving or ecosystem thriving? In theory, there is no conflict, since humans are part of the ecosystem. But in practice, this is exactly the point where discussion often breaks down. Goal setting in ocean management often reaches an impasse when it comes to defining and prioritizing primary goals (Leslie and McLeod 2007). However, considering the high likelihood of irremediable differences among competing visions of the primary goal, consequentialism provides no guidance for prioritizing or comparing them (Wolf 2003).

2. CONFLICTING SECONDARY GOALS

Even if we could all agree on the ultimate goal of planning for the future of the ocean, we are still left with the difficulty that there will be conflicts in choosing a path to arrive at this desired state. These differences establish conflicts in secondary goals, and there is nothing in a consequentialist moral theory that tells us how to navigate the differences. For example, if we agree that human well-being is the ultimate goal, how do we decide whether to approach this state by primarily conserving the ecosystem for future human societies, with a secondary concern for conserving the way of life and jobs in fishing communities, versus the opposite ranking? And what is human well-being? Potential options for defining the good life are diverse, from a superconsuming, individualist, hypergrowth, high-profit, now-rather-than-later worldview to a small-scale, modest, communitarian, ecocentric worldview, and most points between these extremes.

Or, suppose that we agree on the ultimate ends of ecosystem-based management; do we carry out that management through a systems-science type engineering approach or via a complexity-inclusive precautionary principle approach (e.g., Tudela and Short 2005)? The goal-oriented moral framework alone provides us no clear direction to address these problems.

In September of 2000, a large group of world leaders agreed upon a set of target-based and time-constrained global goals that became known as the Millennium Development Goals (UN 2000). This was an important step toward a global-scale agreement on fundamental goals: We will unite to reduce poverty, hunger, and disease, while increasing literacy, gender equity, and environmental sustainability. But a problem rapidly arises: How do we compare these goals? As the Marine and Coastal Ecosystems and Human Wellbeing review rightly states, trade-offs in reaching these goals are inevitable (UNEP 2006). Yet *how* do we logically compare these goals in order to arrive at such a compromise? Mangroves in Southeast Asia and Latin America are being destroyed and replaced with aquaculture ponds for shrimp; as a result, coastal children are going hungry, but North Americans are enjoying inexpensive shrimp. How can these values be reconciled or even balanced?

Box 18.3. Some Difficulties of Making Policy Decisions within the Context of a Narrowly Consequentialist Paradigm

1. Conflicting primary goals
The consequentialist framework does not provide any clear direction regarding how to prioritize conflicting fundamental goals of what we are aiming to achieve.

2. Conflicting secondary goals
If we agree on two important goals yet realize one incurs a necessary debt in another, how do we proceed? A consequentialist framework does not offer advice in this context.

3. Competing temporal and geographical (spatial, political) scales
What directions do we take when confronted with the complexities of balancing current needs with those of future generations, for example? Consequentialism may imply that we take no actions that will limit future options, but how then do we operationalize this policy in a world of uncertainty and surprises without halting necessary behaviors?

4. Cheating the outcomes
Is it acceptable to shortcut to a desired state if the actions involved in this trajectory are dangerous, inhumane, or malicious? Consequentialism might imply that these may all be acceptable in the race to achieve a particular outcome. Are they?

5. Inconsistency with an ecological worldview
If ecological science teaches that humans are ecological entities in the same way that other species are, then it becomes difficult to argue within a consequentialist framework that the system should be managed primarily for human utilitarian purposes.

3. COMPETING SCALES

A further issue is common to all types of ecological management: Conflicts among goals focused on different spatial, temporal, and political scales are difficult to resolve. How are we to decide whether to maximize short-term goals (usually at considerable cost to the future and disregard of the past) or long-term sustainability (usually at considerable cost to the present)? On what geographical or political scale do we make decisions, understanding that benefits in one place often impose costs in another, raising issues of environmental justice? Indeed, actions taken at one location and at one scale may propagate, attenuate, or amplify through a complex and connected ecosystem (Yodzis 2001; Guichard and Peterson, chap. 5 of this volume; Leslie and Kinzig, chap. 4 of this volume). How do we make decisions in this landscape of different, and often conflicting, temporal, spatial, and political scales? Once again, this is a question for which our consequentialist model does not provide any clear direction.

4. "CHEATING" THE OUTCOMES

Justifying a practice by whether it achieves a given end invites people to "cheat the outcomes" by satisfying the "measurables" without achieving a sustainable outcome: "Tell me the minimum I have to do to comply or to satisfy the appearance of compliance." Other deleterious consequences of the practice are outweighed by the value of the end product, on the principle that the end justifies the means. For example, around the globe, unutilized bycatch in commercial fisheries between 1992 and 2001 is estimated at 8% of total harvested catch (Kelleher 2005). Thus, even if we accept that the global goals for total fish catch may be appropriate and sustainable, attempts to meet this target as easily and as economically as possible are inducing behavior that seems dangerous for the system—even while the stated goals are being met tidily. On a more banal level, the temptation to cheat the outcomes can lead to academic discussions and managerial battles about the numbers in the indicators, rather than the health of the system.

5. INCONSISTENCY WITH AN ECOLOGICAL WORLDVIEW

We save the most important issue for last: If one accepts an *ecological* worldview that does not grant special status to humans, it is difficult to justify a *moral* worldview that privileges human well-being over the well-being of other species. We believe the conceptual inconsistency between an anthropocentric moral framework and an ecological worldview is a defeating problem.

One who holds an ecologist's view of the world has, ipso facto, a view of the place of humans in that world. Although human cultural and social systems may present unique ecological connections, humans are one part, among many parts, of a complex and only partially understood network of interdependencies and interactions with the rest of the biotic, cultural, and geochemical worlds. This is no food-web pyramid, with humans at the pinnacle; rather, humans occupy but one rather unremarkable intersection in a complex and multidimensional web (fig. 18.1). On the other hand, an anthropocentric worldview puts humans at the apex of creation, "especially created and uniquely valuable" (Callicott 1992). Thus it is frankly inconsistent for ecologists to make anthropocentric moral judgments about ecosystem-based management plans on the basis of an ecological worldview. It is an inconsistency that requires one to presuppose ecological holism at the metaphysical level, and human exceptionalism at the ethical level (Nelson and Vucetich 2004). Are humans the pinnacle of creation, their interests the ultimate good of the universe, or are they not? No, ecology seems to tell us. Yes, utilitarianism must insist (Callicott 1992).

This effort to hold incompatibles together might be why we continue to find moral decision making in this context so hard. At any rate, there is a baseline moral imperative for consistency: Our ethical theories should not require us to hold two incompatible views about the place of humans in the natural order. We're telling each other two different stories here, and they don't match up. If we're going to have a paradigm shift, we don't want to get caught up in the cogs, and this may be exactly what is happening. The constraints on decision making imposed by a utilitarian framework have the potential to hobble our attempts at long-term agreement about how to behave properly in our ocean environment.

So while consequentialist reasoning can help us measure the relative worth of acts, decisions, plans, and policies by their relation to desired outcomes, it cannot in itself help us decide what those desired outcomes should be. Moreover, it has no way to speak to issues of virtue and environmental justice. The principle of the "greatest good for the greatest number" has not brought out the best in us. Nor has it been kind to oppressed people (Kliskey et al., chap. 9 of this volume).

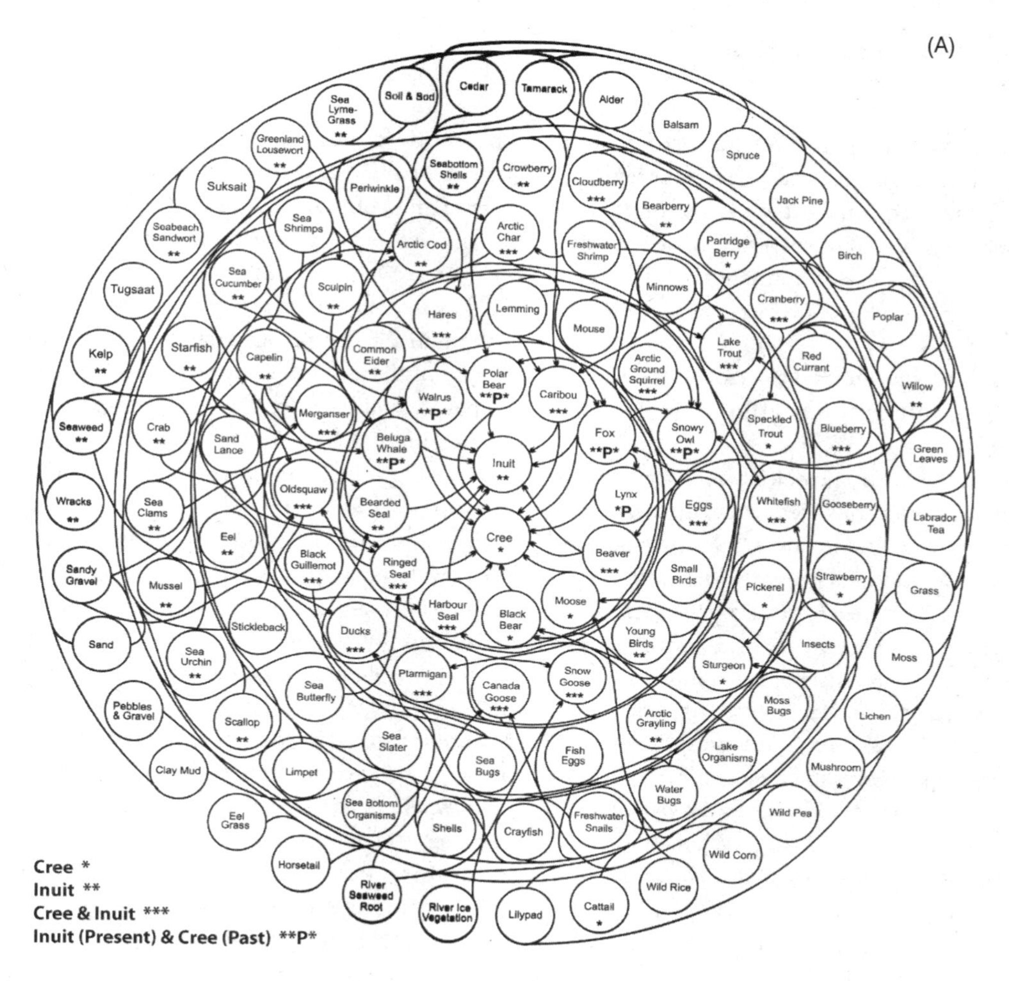

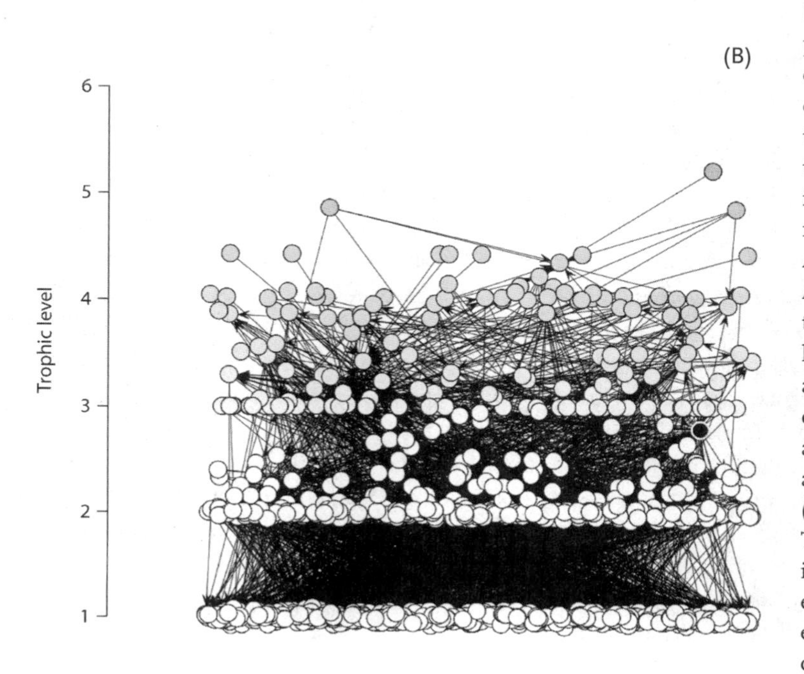

Figure 18.1 How worldviews alter our perceptions of where humans belong in ecosystems. (A) "Traditional" perception of humans in the center of a complex, interconnected food or resource web. Here the ecosystem uses of First Nations people in Northern Canada (Cree and Inuit) are represented, with humans in the center. (B) A food web from Sanak Island—an Alaskan Aleutian island—where the vertical positions in the web are determined by trophic level (i.e., the number of steps that an animal is removed from plants). The white circles represent plants; on the first level above them are herbivores, then omnivores, and progressively higher-level carnivores (with darker gray as trophic level increases). The black circle indicates where humans fall in this particular ecosystem as calculated by ecological trophic level. A from McDonald et al. 1997. B from Wood et al., unpublished data.

Thus we should not be surprised to find ourselves often at an impasse, without a shared moral understanding or the tools of moral reasoning that would justify the choice of one management policy over another.

Toward a New Ecosystem-Based Marine Management Ethic

We have tried to show that questions about marine ecosystem-based management are usually asked within an artificially and unnecessarily confined moral space (fig. 18.2). That space privileges anthropocentric, consequentialist moral theory and pays relatively little attention to virtue ethics and deontological ethics. That is, when considering the moral questions raised by marine ecosystem-based management, policymakers generally bypass questions of virtue and duty and move quickly to questions of value, asking, What are the human goals of management policies, and how can they best be reached? Wolf (2003) calls this narrow focus "unacceptably impoverished." Given the difficulties raised or left unresolved by the narrow focus, a new approach to EBM ethics is in order. A truly ecosystem-based management, to the extent that it amounts to something of a revolution in thought about oceans, might be expected to require a similar paradigm shift in ethical theory.

A new approach might begin by complicating the texture of the moral reasoning in marine ecosystem-based management. Just as there are many ways of knowing (although scientists not surprisingly privilege the scientific method), there are many ways of deciding what is right (although Western thinkers not surprisingly privilege utilitarian moral thought [see Kliskey et al., chap. 9 of this volume]). What increase in creativity might result if policymakers were to reason within the framework of a virtue ethic, for example, asking not, What do we want from the natural resource? but rather, What do we want from ourselves? Should we learn to share? Should we develop new concepts of the commons? Can there be human thriving without ecosystem thriving? How do we fully realize our humanity? How deeply should we care about future generations?

Similarly, a deontological approach, focusing not on the outcomes, but on the nature of the policy itself, might make more room for consideration of issues of equity and justice. Thus, it might help policymakers ward off the disregard of human rights and the rights of nonhuman beings that sometimes result from a narrow or overly eager pursuit of desired states. Ecosystem-based management offers an opportunity to think about the interests of all beings. A good example is the seventh and final objective for sustainability in the Gulf of California defined at the Defying Ocean's End meeting (see Ezcurra et al., chap. 13 of this volume), where it was agreed to "articulate a common regional development vision" and focus on positive policy actions rather than the more typical approach of negating proposed options.

Even within the consequentialist framework, it may be time to take quite seriously the ecological view of humankind as part of a web of life, rather than king of the ash heap. Exploring the Gulf of California, John Steinbeck came to this realization:

> Man is related to the whole thing, related inextricably to all reality, known and unknowable. This is a simple thing to say, but a profound feeling of it made a Jesus, a St. Augustine, a Roger Bacon, a Charles Darwin, an Einstein. Each of them in his own tempo and with his own voice discovered and reaffirmed with astonishment the knowledge that all things are one thing and that one thing is all things—a plankton, a shimmering phosphorescence on the sea and the spinning planets and an expanding universe, all bound together by the elastic string of time (Steinbeck 1941, quoted in Kellert 2003).

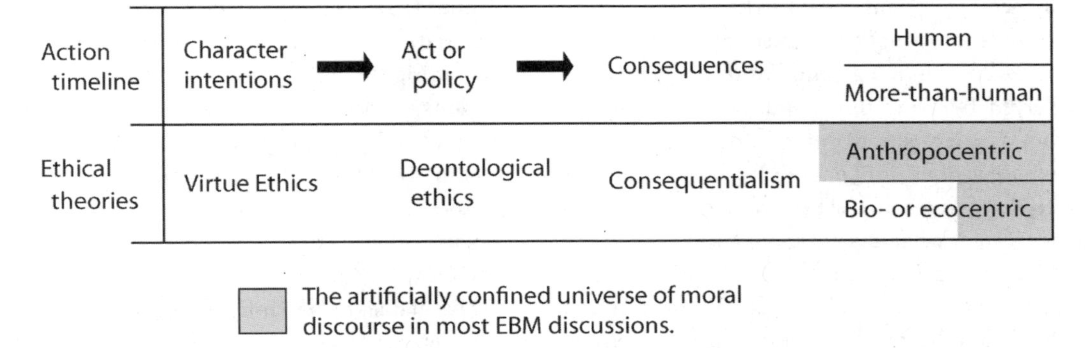

Figure 18.2 The universe of moral discourse.

If this is so, then logical consistency—and, one might add, moral integrity—require an ecocentric moral view, a framework for policies that take the well-being of the ecosystem (or biodiversity) to be an end in itself. Once people truly accept their moral connection with the whole, the door is open to EBM policies that diminish human harm to the ecosystem and strive for balance and shared thriving (Kellert 2003).

We note that the moral significance of the ecological worldview is captured in a new moral theory with roots in the social ethic of care advanced by Nel Noddings and others over the last few decades (Noddings 2003). Its central insight is that moral obligations do not grow from calculations of advantage or consideration of principles, or even a calculation of the greatest good for the greatest number. Moral obligations grow from relationships. A mother takes responsibility to care for her child because she is the child's mother, not from any abstract sense of duty. In much the same way, our duties to others grow from our relationships with them—as family members, as teachers, as community members, as citizens, or simply as people we interact with. Our experience of caring and being cared for is the basis of a social ethic.

Aldo Leopold agreed: "All ethics so far evolved rest upon a single premise: that the individual is a member of a community of interdependent parts" (Leopold 1949). Moreover, evolutionary and ecological science reminds us that we are in relationship not just with people, but with the places that sustain us and the beings that inhabit those places. As we expand our sense of kinship, so also our sense of obligation expands beyond the boundaries of our immediate relations to the larger world. It leads us to acknowledge an expanded duty of care, a duty to consider the interests of ecosystems, what makes them thrive.

Our purpose is not to propose a new ethical foundation for marine ecosystem-based management. Rather it is to show that new ethical foundations are possible, even necessary, and to suggest that ecosystem-based management might offer the opportunity to think more carefully and creatively about the principles that can guide us through "ecological situations so new or intricate, or involving such deferred reactions, that the path of social expedience is not discernible"—which is, as it turns out, Aldo Leopold's (1949) definition of an ethic.

"From so simple a beginning endless forms most beautiful and wonderful have been, and are being, evolved," Darwin wrote (1964, p. 490). We hope the same for a new EBM ethic: that from a single-dimensioned morality might evolve a complicated, multifaceted ethic that is a closer match to the heaving, heaped-up astonishments of the marine world. And from an ecological worldview might emerge a more complete sense of the fullness of what it means to be human, not as bundles of interests to be served by every thing and policy that exists, but as members of interrelated communities whose actions, for better or for worse, affect everything they touch. Then we might have a better idea of what it is we should seek.

Devising marine ecosystem-based management plans for specific areas of concern will be a complicated and arduous process involving our best understanding of ecosystems and of the conditions under which ecosystems are most likely to thrive; careful consideration of the social, cultural, and spiritual contexts for management decisions; strong decision-making institutions and strong institutions for scientific and moral education; expertise in suitable technologies; and a deep sense of moral responsibility for the future. We cannot, of course, predict what those plans might be. But the moral landscape upon which the decisions will be made suggests a set of constraints on what is morally defensible. We do not know the necessary and sufficient conditions for good policy, but we can suggest some of the necessary conditions.

A morally justifiable marine ecosystem-based management policy will do the following:

1. Rest on the ecological and moral premise, following Aldo Leopold, that every human "individual is a member of a community of interdependent parts . . . to include soils, waters, plants, and animals," not a ruler or conqueror or owner. That is, a good plan will understand humans as part of the ecosystems that support their lives. If not this, then what has ecology been trying to teach us?
2. Consider and weigh the interests of all the entities involved, human and nonhuman, understanding that the entities that support human lives are not only "natural resources," but beings with morally significant vital needs of their own.
3. Honor the precautionary principle, whereby in cases of scientific uncertainty regarding the ramifications of particular actions, priority will be given to environmental and human health without having to wait until the reality and seriousness of those adverse effects become fully documented scientifically (adapted from the COC 2000).
4. Be equitable, respecting the rights of all people, not imposing undeserved or arbitrary costs on some, to the undeserved or arbitrary benefit of others. A policy should honor basic human rights to life and to the basic conditions for meaningful life. Insofar as the single-minded pursuit of self-interest often incurs inequitable social and environmental cost, it will avoid excess in favor of moderation.
5. Promote the development of the most generous in people, rather than take advantage of the most self-interested. Ecosystem-based management policies will be an invitation to (re)consider how membership in thriving bio-cultural communities is an essential element of what it means to thrive as a human being.
6. Respect the full panoply of values inherent in marine organisms and ecosystems—not just their monetary values, but also their intrinsic values and their often inchoate, perhaps immeasurable or unfathomable, aesthetic and spiritual values.
7. Honor the present generation's obligation to the future, to leave a world that is at least as resilient and full of possibility as the world that was left to us.
8. Be pluralistic, cross-disciplinary, and cross-cultural, inviting a robust practical discourse about the whole universe of bio-cultural values and opening the discourse to considerations of virtue and duty, as well as outcomes (Norton 2003).
9. Be enriched by an epistemological pluralism, honoring many ways of knowing, respecting the wisdom of people who have

lived in a place for a very long time. A policy should speak in many tongues—including the languages of natural and social science; languages of the unpredictable and local as well as the generalizable and universal; and the literary, historical, religious, and spiritual languages of astonishment and celebration.

10. Incorporate management for resilient, flexible, and adaptive *dynamics* as goals rather than management only for particular *states*, thus expecting the inevitable surprises, admitting the unknowable, acknowledging ecological complexities, and leaving room for future social and ecological evolution.

We believe that the ongoing evolution of human understanding of our moral duties, especially our obligations to nonhuman species, to ecological systems, and to generations unborn, mandates the adoption of an ecosystem-based management approach to oceans. Moreover, the evolution of new management approaches based on ecological principles and scaled to ecosystems can awaken an expanded sense of moral responsibility. What if we thought of ourselves as citizens of the biotic world—not *only* as the nineteenth-century "economic man," but as members of interrelated communities whose actions, for better or for worse, and to greater or lesser extent, affect everything they touch? We believe that the opportunity exists for a coevolution of ecological ethics and ecological management practices, each of them requiring of the other a renewed commitment to the thriving of the bio-cultural whole.

References

Callicott, J. B. 1992. Principal traditions in American environmental ethics: A survey of moral values for framing an American ocean policy. *Ocean and Coastal Management* 17:299–325.

Carson, R. 1998. *The sense of wonder*. New York: HarperCollins.

COC (Commission of the European Communities). 2000. *Communication from the commission: On the precautionary principle*. Brussels: COC.

Daily, G. C., ed. 1997. *Nature's services: Societal dependence on natural ecosystems*. Washington, DC: Island Press.

Dallmeyer, D. G. 2003. *Values at sea: Ethics for the marine environment*. Athens, GA: University of Georgia Press.

Darwin, C. (1964 [1859]). *On the origin of species: A facsimile of the first edition*. Cambridge, MA: Harvard University Press.

Ehrenfeld, D. W. 1981. *The arrogance of humanism*. New York: Oxford University Press.

FAO (Food and Agriculture Organization of the United Nations). 2006. *The state of world fisheries and aquaculture*. Rome: FAO.

Francis, R. C., M. A. Hixon, M. E. Clarke, S. A. Murawski, and S. Ralston. 2007. Ten commandments for ecosystem-based fisheries scientists. *Fisheries* 32(5):217–33.

Grumbine, R. E. 1993. What is ecosystem management? *Conservation Biology* 8(1):27–38.

Guerry, A. 2005. Icarus and Daedalus: Conceptual and tactical lessons for marine ecosystem-based management. *Frontiers in Ecology and the Environment* 3(4):202–11.

Kelleher, K. 2005. Discards in the world's marine fisheries: An update. FAO Fisheries Technical Paper no. 470. Rome: Food and Agriculture Organization (FAO).

Kellert, S. 2003. Human values, ethics, and the marine environment. In *Values at sea: Ethics for the marine environment*, ed. D. Dallmeyer. Athens: University of Georgia Press.

Kuhn, T. S. 1970. *The structure of scientific revolutions*, 2nd ed. Chicago: University of Chicago Press.

Leopold, A. 1949. *A Sand County almanac*. New York: Oxford University Press.

Leslie, H. M., and K. L. McLeod. 2007. Confronting the challenges of implementing marine ecosystem-based management. *Frontiers in Ecology and the Environment* 5(10):540–48.

MA (Millennium Ecosystem Assesssment). 2005. *Ecosystems and human well-being: Synthesis*. Washington, DC: Island Press.

McDonald, M., L. Arragutainaq, and Z. Novalinga, ed. 1997. *Voices from the Bay: Traditional ecological knowledge of Inuit and Cree in the Hudson Bay Bioregion*. Ottawa: Canadian Arctic

Resources Committee and Municipality of Sanikiluaq.

McLeod, K. L., J. Lubchenco, S. R. Palumbi, and A. A. Rosenberg. 2005. *Scientific consensus statement on marine ecosystem-based management*. The Communication Partnership for Science and the Sea (COMPASS). Signed by 221 academic scientists and policy experts with relevant expertise. http://www.compassonline.org/pdf_files/EBM_Consensus_Statement_v12.pdf.

Merchant, C. 1990. *The death of nature*. New York: HarperOne.

Nelson, M. P., and J. Vucetich. 2004. O holismo na ética ambiental (Holism in environmental ethics). In *Éticas e políticas ambientais (Environmental ethics and environmental policies)*. Lisbon, Portugal: Centro de Filosofia da Universidade de Lisboa.

NOAA (National Oceanic and Atmospheric Administration). 2005. *New priorities for the twenty-first century: NOAA's strategic plan*. Washington, DC: NOAA.

Noddings, N. 2003. *Caring: A feminine approach to ethics and moral education*. Berkeley: University of California Press.

Norton, B. 2003. Marine environmental ethics: Where we might start. In *Values at sea: Ethics for the marine environment*, ed. D. G. Dallmeyer. Athens: University of Georgia Press.

NRC (National Research Council). 2006. *Dynamic changes in marine ecosystems: Fishing, food webs, and future options*. Washington, DC: National Academies Press.

Parris, T. M., and R. W. Kates. 2003. Characterizing a sustainability transition: Goals, targets, trends, and driving forces. *Proceedings of the National Academy of Sciences* 100(14):8068–73.

Partridge, E. 2001. Future generations. In *A companion to environmental philosophy*, ed. D. Jamieson. Oxford, UK: Blackwell.

Pinchot, G. 1947. *Breaking new ground*. New York: Harcourt Brace.

POC (Pew Oceans Commission). 2003. *America's living ocean: Charting a course for sea change. A report to the nation*. Washington, DC: Pew Trusts.

PSP (Puget Sound Partnership). 2006. *Sound health, sound future: Protecting and restoring Puget Sound*. Olympia, WA. http://www.psp.wa.gov/.

Rachels, J. 2006. *The elements of moral philosophy*. New York: McGraw-Hill.

Slocombe, D. S. 1998. Defining goals and criteria for ecosystem-based management. *Environmental Management* 22(4):483–93.

Stanley, T. R. 1995. Ecosystem management and the arrogance of humanism. *Conservation Biology* 9(2):255–62.

Steinbeck, J. 1941. *Log from the Sea of Cortez*. Mamaroneck, NY: Paul P. Appel.

Sylvan, R. 1973. Is there a need for a new, an environmental ethic? In *Proceedings of the fifteenth world congress of philosophy*, 205–10. Sofia, Bulgaria: Sofia Press.

Tudela, S., and K. Short. 2005. Paradigm shifts, gaps, inertia, and political agendas in ecosystem-based fisheries management. *Marine Ecology Progress Series* 300:282–86.

UN (United Nations). 2000. 55th Session, article 60.

UNEP (United Nations Environment Programme). 2006. *Marine and coastal ecosystems and human wellbeing: A synthesis report based on the findings of the Millennium Ecosystem Assessment*. Nairobi: UNEP.

USCOP (US Commission on Ocean Policy). 2004. *An ocean blueprint for the twenty-first century*. Final report. Washington, DC: USCOP. ISBN 0 9759462 0 X.

Walters, C. J., and S. J. D. Martell. 2004. *Fisheries ecology and management*. Princeton, NJ: Princeton University Press.

White, L. 1957. The historical roots of our ecologic crisis. *Science* 155:1203–7.

Wolf, C. 2003. Environmental ethics and marine ecosystems: From a "land ethic" to a "sea ethic." In *Values at sea: Ethics for the marine environment*, ed. D. G. Dallmayer. Athens: University of Georgia Press.

Wood, S., R. Russell, D. Lund, N. Huntly, and H. D. G. Maschner. Unpublished data. Sanak Biocomplexity Project.

WWF (World Wildlife Fund). 2006. *Living planet report*. Gland, Switzerland: WWF.

Yaffee, S. 1999. Three faces of ecosystem management. *Conservation Biology* 13(4):713–25.

Yodzis, P. 2001. Must top predators be culled for the sake of fisheries? *Trends in Ecology and Evolution* 16(2):78–84.

CHAPTER 19
Ways Forward

Karen L. McLeod and Heather M. Leslie

Connections between ourselves and the coasts and oceans in which we work and play ensure our very existence on this planet and enable us to thrive. We depend on coasts and oceans for a multitude of ecosystem services, ranging from those obvious to even our young children (food and places to frolic) to those that are more subtle but no less important (water filtration and shoreline stabilization). The continued delivery of these services is at risk, and we cannot cast blame on a single cause. It is not exclusively fishing, population growth, coastal development, or runoff from agricultural practices that is driving the degradation of coastal and ocean ecosystems. Instead, it is the synergies among a host of policies, activities, and decisions made by individuals, governments, and multinational conglomerates over centuries, resulting in no corner of the oceans left untouched (Jackson et al. 2001; Lotze et al. 2006; Halpern et al. 2008).

Most marine scientists, practitioners, and decision makers appreciate and are aware of this complexity. Many also see the need for a management perspective that considers ecosystems as an interconnected whole. However, we have not yet embraced such thinking as business as usual, nor is it reflected in most current management or policies. How can we move from management systems designed to consider each sector of human activity in isolation to management systems that embrace the inherent connections among activities and explicitly deal with trade-offs among activities and ecosystem services? We will begin with a summary of key concepts integral to ecosystem-based management (EBM) that have emerged from this volume, followed by a discussion of future research needs, and conclude with suggested strategies for moving forward with implementation over the near-term and longer time horizons.

Key Concepts

1. Embrace a Common Goal

Enabling connections between ourselves, coasts, and oceans will require a means for sectors to work toward a common, overarching goal (Leslie and McLeod 2007). While sustaining the long-term ability of systems to deliver a range of ecosystem services may be the core goal of EBM for the oceans (Rosenberg and McLeod 2005), this concept is too generic to be readily applied on the ground. Each sector of human activity (fisheries, coastal development, tourism, etc.) must consider how it affects ecosystem structure, functioning, and key processes, and ultimately all sectors must collectively work toward the common goal of maintaining key components of ecosystem health. This sharply contrasts with current management practices that tend to focus on the short-term provision of single services and under which individual sectors are often working at cross-purposes (Rosenberg and Sandifer, chap. 2 of this volume).

EBM requires not only this type of collective action on the part of multiple sectors of human activity, but also broadening of the spectrum of perspectives that have a stake and voice in management. In the case of Puget Sound, the state-appointed Puget Sound Partnership worked with its diverse members (representatives of public and private institutions, tribes, and citizen groups), a scientific working group, and the general public to articulate a set of ecosystem-wide goals that aligned with the governor's vision for a healthy sound (Ruckelshaus et al., chap. 12 of this volume). In other cases, such as Morro Bay, a local coalition, the San Luis Obispo Science and Ecosystem Alliance (SLOSEA), brought together scientists, managers, and others who live and work in that system to develop a collective vision (Wendt et al., chap. 11 of this volume). Similarly, in the Gulf of California, civic groups such as COBI, *Noroeste Sustentable,* and PANGAS are working with stakeholders to articulate a vision for their local ecosystems (Ezcurra et al., chap. 13 of this volume).

Although most people share the goal of a healthy ocean, healthy means different things to different people. People also differ in their willingness to make necessary trade-offs and sacrifices, particularly when it means accepting short-term "losses" in favor of long-term sustainability. A crucial next step is to better define ecosystem health for coasts and oceans and to connect ecosystem health to human well-being in specific places. This will allow societies to better define common goals for our coasts and oceans.

2. Maintain Options for the Future

A critical lesson from resilience science for EBM is the need to maintain options for the future in the face of uncertainty; abrupt, potentially irreversible shifts in ecosystem state; and inevitable change (Leslie and Kinzig, chap. 4 of this volume). Improving our capability to understand and predict thresholds between states, the drivers that contribute to these shifts, and the anticipated responses of individuals and societies will allow us to better forecast—and manage for—future changes. Local and traditional ecological knowledge can provide a longer-term and richer knowledge of a place, offering a critical context within which to understand and respond to change (Kliskey et al., chap. 9 of this volume). In the meantime, we must proceed with caution and a healthy dose of humility and not allow concern over short-term costs to blind us from the critical importance of taking the long view and preventing potentially irreversible ecosystem damage. The most catastrophic changes in ecosystems and their services identified in the Millennium Ecosystem Assessment involved nonlinear or abrupt shifts from a "desirable" state to a state that is less desirable, such as the shift from coral to algal dominance on tropical coral reefs (Carpenter et al. 2006). Thus, the precautionary approach is to presume that alternate ecosystem states (including undesirable states) exist until data indicate otherwise and to manage with the expectation of surprise. The numerous and often overwhelming effects of climate change on the oceans are another case in point: We currently lack strong predictive power regarding specific impacts, and yet we must proceed with the design and implementation of options for both mitigation (e.g., reducing carbon emissions) and adaptation (e.g., improved responses to coastal flooding).

Maintaining options for the future will require management institutions and the infrastructure that supports them (both social and scientific) to adapt to changing conditions, learn from the past, build resilience against surprise, and accept an uncertain future. Managing with an eye toward long-term sustainability may be one of the most challenging aspects of EBM, as it will certainly entail short-term costs, such as creating the necessary scientific infrastructure or realigning a sectoral management system to ensure that ultimately all sectors are working toward a common goal (see Rosenberg and Sandifer, chap. 2 of this volume). One key aspect of taking a long-term view is that we must preserve the components of systems, even if we do not currently understand or appreciate their importance (Partridge 2001, as cited in Moore and Russell, chap. 18 of this volume). For individuals and many societies, the recognition that our moral responsibilities extend beyond this generation is certainly not new. However, this ethic is not currently reflected in ocean policies writ large, particularly in the Western world.

3. Engage a Diversity of Approaches and Perspectives

Humans value ocean ecosystems in countless ways, and a key aspect of EBM is embracing and engaging a diversity of perspectives and approaches. In the words of Moore and Russell (chap. 18 of this volume), EBM "will speak in many tongues—including the languages of natural and social science; languages of the unpredictable and local as well as the generalizable and universal; and the literary, historical, religious, and spiritual languages of astonishment and celebration." Ecosystem-based management for the oceans will require small-scale, iterative change (evolution), as well as overarching systemic change (revolution). Thus, the question is how to proceed now with evolutionary change while laying the groundwork for revolution when the time is right, rather than one versus the other.

One aspect of this diversity is the need to employ strategies that emphasize both protection and use. At present, for example, we often discuss biodiversity conservation strategies (e.g., no-take marine reserves) and multiple-use strategies (e.g., ocean zoning) as mutually exclusive approaches. Instead, we must consider this entire continuum of approaches and evaluate the relative costs and benefits of various strategies. The emphasis in EBM must be on transparent decision making that acknowledges trade-offs among options, as further developed later in this chapter.

Strategies must also be employed across a diversity of scales, including everything from local, site-based efforts to the scales of large marine ecosystems (LMEs). Thus, instead of asking the question, What is the correct scale for EBM decision making? we should instead be asking, What decisions are best made at my particular scale of decision making, and how do I account for the connections between my focal scale and both smaller and larger scales? The present challenge is to better understand how to effectively scale the knowledge, strategies, and tactics from smaller-scale pilots, such as the effort in Morro Bay, California (chap. 11 of this volume), to larger scales, such as the entire California Current LME. Specifically, how do actions taken at one location propagate through complex and highly connected social–ecological systems?

Another important aspect of diversity is effectively incorporating a multiplicity of perspectives, particularly when framing a common goal (a crucial step in the EBM discussion, above). Interdisciplinary teams of scientists, practitioners, and other stakeholders are best engaged at the outset to take full advantage of all relevant approaches and sources of knowledge (including local and indigenous knowledge, Kliskey et al., chap. 9 of this volume) in the design and implementation of EBM. Increasing communication and transdisciplinary exchange will allow us to better understand feedbacks between the social and ecological components of marine ecosystems and create the broad base of public and political support needed to sustain EBM.

Future Research Needs

One of the key messages of this book is that we have the scientific knowledge, institutional frameworks, and on-the-ground experience to move forward with marine EBM now. However, as EBM matures, information and science needs will grow and require us to further develop science support for EBM (see box 19.1).

Research is particularly needed to better understand the linkages between social and ecological systems, an inherently interdisciplinary endeavor. Rosenberg and Sandifer (chap. 2 of this volume) argue that the research enterprise for EBM is just as fragmented as the management realm. The case studies in part 4 also highlight these research challenges, whether we look at the Chesapeake Bay or the Eastern Scotian Shelf. How do we bring natural and social scientists together and enable them to develop and answer EBM-relevant questions of common interest? There is no single answer to this question, but regardless of the specific path forward, incentives need to be in place to facilitate interdisciplinary research. Moreover, natural and social scientists must better integrate their fields to address the questions that will arise as EBM is implemented, to develop effective partnerships, and to be able to convey the outcomes of such work to policymakers and the public.

Strategies for Implementation

The leading edges of EBM lie in the nuts and bolts of implementation. Practitioners may embrace the underlying principles of EBM, yet grapple with how to implement it, resulting in a sizable gap between concept and practice (Arkema et al. 2006). Below, we describe five strategies that can be employed to implement EBM, progressing from those that can be implemented right now to those that will require longer time frames.

1. Create Incentives for Stewardship and Collaboration

A focus on the local scale is fundamental to understanding how people relate to the sea. After all, this is the scale at which people engage with and experience the world. We must also remind ourselves that it is the cumulative effects of an unfathomable number of individual decisions that have led us to our current predicament, and just as importantly, it will be people who act as "cooperative, recuperative, restorative agents" of change (Shackeroff et al., chap. 3 of this volume). Thus, moving forward with EBM must entail building and nurturing a local constituency, developing trust and understanding among this constituency, and creating incentives for stewardship and collaboration. Management approaches that better reflect a diversity of perspectives are most likely to succeed over the long term.

Strategies to effectively engage local communities and municipalities are evident in the six small-scale EBM pilot projects underway along the US west coast. As a case in point, in Morro Bay, California, a local coalition (the San Luis Obispo Science and Ecosystem Alliance, or SLOSEA) of scientists, resource managers, and other stakeholders who live and work in the ecosystem are working to improve information sharing, issue identification, collaboration, and management within their estuary and the surrounding area. In the absence of any mandate for agencies and other actors to work across jurisdictional boundaries, SLOSEA facilitates interagency communication and management coordination and gleans knowledge about the ecosystem from scientists, fishermen, resource managers, businesspeople, and environmentalists. Such collaboration is possible thanks to a collective vision (i.e., a common goal) that guides the continuing evolution and development of SLOSEA's activities and to a strong place-based identity that enables diverse players to expand their perspective beyond their own immediate concerns (Wendt et al., chap. 11 of this volume). Importantly, as reiterated many times throughout this volume, EBM cannot proceed only at this small scale, as drivers of change operate over a multitude of scales. Thus, the ultimate focus must be on fostering an environment of collaboration and stewardship across many scales, from those of local communities to state and federal agencies to collaborations for protection and management

Box 19.1. Outstanding Research Questions Related to Marine EBM

1. **What can we learn from how humans have interacted with coastal and marine environments in the past?**

   Historical ecology and paleoecology offer powerful approaches to understanding past interactions of humans with coastal and marine environments. See work by Jackson and colleagues (2001), Rosenberg and colleagues (2005), and Salomon and colleagues (2007) for examples of how contemporary data can be combined with historical data to help us understand how social systems and ecological conditions have changed through time. Such data provide a critical context for decision making.

2. **How do we translate knowledge developed at the local scale to broader geographic scales?**

   There is often a mismatch between the relatively local scales at which ecological and social data are collected and the broader geographic scales at which management must operate. How do we resolve the challenges of integrating across a range of spatial and temporal scales? Can we develop frameworks that adequately capture the key elements of both social and ecological systems to better understand the range of factors affecting the flow of ecosystem services?

3. **What are the cumulative impacts of human activities on ecosystem health and human well-being?**

   Indicators to track ecosystem health and human well-being are needed, both to understand how coupled systems respond to individual management actions and to evaluate progress toward EBM goals more generally. While there are many candidate indicators, assessing progress (How are our actions affecting ecosystem health and human well-being?) requires a core set that can be feasibly measured across different ecosystem types, geographic scales, and sociopolitical contexts. Further, we need a more explicit understanding of the expected responses of ecosystem components to combined, rather than individual, threats.

4. **How can trade-offs among ecosystem services and sectors be more systematically assessed?**

   Given the inherent trade-offs among ecosystem services and human activities in the coastal and marine realms, a more mechanistic understanding of the biogeophysical, social, and economic factors influencing the flow of ecosystem services is needed. Ultimately, such understanding must underpin the development of decision-making tools to explicitly assess trade-offs and evaluate the likely outcomes of particular management strategies.

5. **How do we evaluate the success of ecosystem-based management efforts?**

   Documentation of lessons learned from EBM projects to date—such as what is captured in the chapters in part 4 of this volume—and synthesis of these lessons across EBM and similar efforts is vital in order to elucidate what contributes to success. Ultimately, we are managing people and people's interactions with coastal and marine environments, rather than the environment itself. Thus, the core indicators of management success need to incorporate information on the social, institutional, and economic dimensions of the systems of interest, as well as ecological factors.

across much larger regional scales, such as that envisioned in the West Coast Governors' Agreement on Ocean Health in the United States (see http://westcoastoceans.gov/).

2. Build on Small Successes and Within-Sector Reform

To many, the implementation of EBM is daunting. After all, social, political, and institutional change does not come easily. For policymakers, balancing the varied interests of constituents while ensuring stewardship of public trust resources is challenging. Managers must reconcile the mismatch between ecological processes and jurisdictional authority in order to meet multiple, potentially conflicting, objectives. Synthesis, translation, and application of scientific knowledge for EBM are new areas for many scientists and require additional institutional and financial support.

It is important to keep in mind that there are many paths to EBM, many of which will start small, building on small successes and within-sector reform. In New York's Great South Bay, for example, representatives from state government, the private sector, and nongovernmental institutions have come together to develop an innovative place-based strategy focused on restoration, research and education, and sustainable aquaculture in order to restore the ecological functioning and ecosystem services of the bay. Alternatively, governmental mandates can play a significant role, as in the Chesapeake Bay, where restoration of sea grass beds, oyster reefs, and other bay habitats and changes in farming, waste management, and development practices are helping to shift that system to a more desirable one characterized by abundant fisheries, clean water, and vibrant habitats (Boesch and Goldman, chap. 15 of this volume).

From a scientific perspective, while numerous sectors have begun to use ecosystem principles in decision making, fisheries have arguably made the most progress toward applying an ecosystem approach. Ecosystem considerations such as bycatch or the effects of fishing on trophic interactions or habitats are increasingly incorporated in fisheries decision making across numerous scales (Murawski 2007). For example, in the waters surrounding Antarctica, the Convention on the Conservation of Antarctic Marine Living Resources (CCAMLR) has institutionalized ecosystem principles to account for key predator–prey relationships, such as those involving krill, since the early 1980s (Constable et al. 2000; Ruckelshaus et al. 2008). In the United States, the reauthorization of the Magnuson–Stevens Fishery Conservation and Management Act (2007) contains expanded authorities to consider the ecosystem impacts of fishing, and fisheries management in places such as the Bering Sea–Aleutian Islands are based on complex multispecies models (Ruckelshaus et al. 2008). Importantly, ecosystem-based fisheries management (or an ecosystem approach applied to any individual sector), while a step in the right direction, is only a partial solution, as (1) it does not fully account for the impacts of fisheries on nontarget species, (2) it cannot regulate other activities that affect fished species, such as pollution or habitat loss, and (3) its goal is to maximize the production of a single service, as opposed to multiple services (Rosenberg and McLeod 2005).

From a legal perspective, numerous provisions in state and federal laws either permit or require consideration of ecosystem principles, and work is currently underway by the Environmental Law Institute to specify existing opportunities within state and federal laws in various regions of the United States (see http://www.eli.org/Program_Areas/ocean_ebm.cfm). For example, in the United States, prudent use of the National Environmental Policy Act (NEPA) or similar statutes, which require that all federal agencies consider the likely consequences of their actions on the environment before they take them, could provide a useful bridge from conventional management to new decision-making approaches that explicitly consider both indirect and cumulative effects. However, the present challenge is that NEPA does not mandate taking a particular course of action (Searles Jones and Ganey, chap. 10 of this volume).

From a management perspective, implementation of EBM must include expanded monitoring and evaluation to assess how key components of an ecosystem are changing over time and ultimately how those changes affect the flow of services. For example, how is the amount of salt marsh area changing over time, and how does that affect shoreline protection and estuarine water quality? While conventional indicators of ecological, physical, or social patterns remain important, monitoring programs must increasingly incorporate information on key processes, such as those that sustain biodiversity (e.g., recruitment or dispersal) or affect the production of ecosystem services (e.g., rates of nutrient cycling). An emphasis on process as opposed to pattern allows us to better understand why things are changing (and potentially reverse those changes), rather than to simply measure changes. Ideally, some indicators will also provide early warning signals of potential changes in a system where abrupt, nonlinear responses to policy and management actions are more likely (Leslie and Kinzig, chap. 4 of this volume). Importantly, expanded monitoring does not necessarily mean measuring more indicators, but instead suggests a need to hone the lists of possible metrics (of which there are hundreds) to an essential core set that will elucidate both how and why systems are changing (see Kaufman et al., chap. 7, and Ruckelshaus et al., chap. 12 of this volume). Ultimately, all EBM implementation efforts must include robust plans to manage adaptively, acknowledging that information about coastal and ocean ecosystems and the values that people place on them will necessarily change over time.

This discussion should not be taken to suggest that EBM can proceed exclusively in an iterative, evolutionary fashion. For one, we may run out of time to make the large-scale course corrections needed to deal with daunting challenges such as ocean acidification (Orr et al. 2005) or the loss of polar sea ice (Serreze et al. 2007). From a legal perspective, the statutory tools that do exist to implement aspects of EBM are "scattered throughout various statutes, applied only to certain agencies or to certain resources or only under certain circumstances" (Searles Jones and Ganey, chap. 10 of this volume), suggesting the need for a more overarching legal mandate. Finally, as emphasized by numerous parties, incremental gains, while absolutely necessary, are not sufficient to tackle the scope of the challenges facing our coasts and oceans today (POC 2003; USCOP 2004; Crowder et al. 2006; Leslie and McLeod 2007).

3. *Employ Area-Based Management Tools*

The implementation of EBM in any context requires a shift in mindset from managing one thing in many places (sectoral management) to managing many things in one place (area-based management). As discussed above, EBM is about acknowledging the simultaneous need to conserve, protect, and use our coasts and oceans in a manner that maintains long-term ecosystem health and human well-being. While we have employed area-based management tools for decades, within any given location, different areas have been established for different purposes by agencies with different authorities, usually to achieve single-sector goals (e.g., fisheries or the protection of marine mammals; see fig. 2.2 showing the Gulf of Maine). Thus, the areas do not complement or leverage one another, and the result is an overlapping set of requirements that are effective in providing some ecosystem protections but lack a holistic focus on the suite of activities affecting a particular place and the range of services we seek from it (Rosenberg and Sandifer, chap. 2 of this volume).

Ideally, at relatively large scales, EBM for the oceans will include comprehensive spatial planning and zoning. Such zoning would define areas in which compatible activities could occur, separate incompatible activities, allow for multiple levels of protection, and establish a process for monitoring, reviewing, and adapting zoning designations as necessary (Crowder et al. 2006). Zoning would also establish rules for the siting of emerging ocean uses such as offshore aquaculture, wind farms, wave energy, and liquefied natural gas terminals, thus ensuring greater regulatory certainty for both current and future users. By employing the full spectrum of levels of protection, ranging from no-take marine reserves through multiple-use areas and specific-use set-asides, comprehensive zoning accommodates the need to simultaneously consider use and protection. Marine spatial planning in support of ocean zoning could simplify the rules, facilitate coordination among agencies, reduce conflicts among user groups, and determine if sets of activities should be prohibited or allowed within different areas in a system of contiguous zones, much as occurs on land (Crowder et al. 2006; Young et al. 2007; Sivas and Caldwell 2008; Rosenberg and Sandifer, chap. 2 of this volume). The result will be a more coherent system than our current piecemeal approach.

To date, the most prominent example of comprehensive zoning is Australia's Great Barrier Reef (GBR) Marine Park. The GBR's scheme includes seven different types of zones, ranging from no-go zones (called preservation zones) to no-take marine reserves (a third of the GBR area) to multiple-use areas in which numerous different types of activities are allowed. The emphasis of this zoning scheme is to provide a spectrum of zones with different objectives, rather than solely allowing or prohibiting certain activities in specific zones. The objectives of a particular zone determine which activities are allowed within that zone without a permit and which activities may occur with a permit. Other countries, including Belgium, China, Germany, the Netherlands, and the United Kingdom, have implemented or are experimenting with marine spatial planning (Crowder et al. 2006 and references therein), and Massachusetts just passed legislation to facilitate marine spatial planning and zoning within its state waters.

4. Make Trade-offs Explicit

Integrated, multi-objective management seeks to fully account for potential and realized trade-offs among different ecosystem services (which often correspond to objectives). Attempts to optimize a single service often lead to reductions or losses of other services (Holling and Meffe 1996). Such trade-offs occur over space and time, among competing uses and nonuses, and between the present and future provision of services (Rodríguez et al. 2006; see also Rosenberg and Sandifer, chap. 2 of this volume). Thus, an important aspect of EBM implementation will be to create mechanisms that allow explicit and transparent decision making with respect to these trade-offs, rather than having trade-offs occur as unintended consequences of various actions or making such decisions on an ad hoc basis.

One option for assessing trade-offs is to assign values to services, through either market or nonmarket valuation techniques (de Groot et al. 2002; Farber et al. 2002; Chee 2004; Barbier 2007; Barbier, chap. 8 of this volume). Valuation is useful for quantifying and analyzing gains and losses and determining who wins and who loses when making particular decisions (Barbier, chap. 8 of this volume). Valuation is also complex and the results controversial, particularly for services that are not easily (or should not be) monetized. Moreover, simply assigning value (particularly for an entire system) does not always provide a basis for concrete decision making. Instead, we must ultimately understand how a change in ecosystem functioning results in a change in a service, in other words, employ the production function (or ecological endpoints) approach (Wainger and Boyd, chap. 6 of this volume).

How can we design strategies to ensure that ecosystems can continue to provide the range of ecosystem services society expects and needs? Novel frameworks under development for assessing trade-offs among ecosystem services will allow managers to visualize these trade-offs, reveal suboptimal management options, and help inform if and how much of a trade-off exists among a particular set of services (Nelson et al. 2008; Lester et al. in review). On the US west coast, this approach is being used to illustrate the trade-off between biodiversity preservation in no-take marine reserves versus fishery revenue and the trade-offs among wave energy production, crab fishery revenue, and coastal real estate value (Lester et al. in review). The biggest stumbling block to applying such frameworks to real-life situations is a lack of models that can predict the delivery of ecosystem services both now and in the future. However, such models are being developed. The Natural Capital Project has developed a tool, Integrated Valuation of Ecosystem Services and Tradeoffs (InVEST), to model and map the delivery, distribution, and economic value of services, under both current and future management scenarios (see http://www.naturalcapitalproject.org/InVEST.html). This tool has not yet been applied to coastal or marine settings, but these applications are underway as this volume is going to print. Another related tool under development that can foster a better understanding of trade-offs among services is the MIMES (Multiscale Integrated Models of Ecosystem Services) suite of dynamic ecological economic models aimed at integrating our understanding of ecosystem functioning, ecosystem services, and human well-being across a range of spatial scales (see http://www.uvm.edu/giee/mimes/).

5. Develop a New Mandate

The current system of ocean governance in most nations around the world is largely a historical artifact that isolates decisions within individual sectors. It reflects a conspicuous mismatch between what we currently know about marine ecosystems and our management of activities that affect these systems. While we can certainly make great strides within the constraints of current policies, ultimately, the needed leadership and direction to heal divisions across a fragmented management system will require new national or multinational mandates (POC 2003; USCOP 2004; Rosenberg et al., chap. 16 of this volume). Canada passed its Oceans Act granting authority to a single agency in 1997, establishing clear leadership and unifying conservation requirements of three other regulatory acts. Australia passed its own Oceans Policy in 1998 and Environment Protection and Biodiversity Conservation Act in 1999, and together these provide strong foundations for the implementation of EBM for its oceans. Yet, in the United States (and numerous other places), no single agency is responsible for ensuring that the collective effects of all of our activities (be they aimed at conservation or exploitation) leave us with an intact, functioning ocean that will continue to meet society's wants and needs.

Importantly, while the presence of an overarching national or multinational policy cannot ensure success, it can certainly facilitate improved management. Many of the overarching (and therefore cross-agency) priorities necessary for implementation of EBM fall into jurisdictional gray areas where no agency takes responsibility for those priorities. A mandate can help drive agencies to allocate scarce time and money resources to address those priorities for which they currently do not have clear responsibility. Moreover, such a mandate is often necessary to effectively foster collaborative and integrated decision making and comprehensively address cross-sectoral cumulative impacts, interactions among activities, and trade-offs among services. A mandate should also include decision rules and clear authority to resolve conflicts and coordinated, inclusive planning and decision-making structures that are scaled to reflect ecosystem, rather than political, boundaries (Searles Jones and Ganey, chap. 10 of this volume).

Concluding Thoughts

Ecosystem-based management, in both principle and practice, is intended to streamline management. It no longer ignores the connections among management actions or the cumulative nature of impacts, and thus it should result in more-efficient decision making. Notably, our vision for EBM is not something extra for managers or other practitioners to do on top of their current "day jobs." Rather, it is a different way of doing business that will require new resources in terms of both personnel and funding. Integration and transparent negotiations around trade-offs are "fundamentally different from tinkering around the edges of the existing management process" (Rosenberg and Sandifer, chap. 2 of this volume).

Considerable effort is currently being directed at the planning stages of EBM for the oceans. The time has come to move more quickly and efficiently from planning to implementation. This will require overcoming our own fear of change. It is our hope that the windows of opportunity to embrace and empower such change will emerge soon. It may also require acknowledging, as Moore and Russell do in chap. 18 of this volume, that we are but one species and "members of interrelated communities whose actions, for better or for worse, affect everything they touch" in a highly dynamic world full of surprises at every turn.

Acknowledgments

We thank all of the contributors of this volume for joining us on the journey that created this book. We also are grateful to J. Lubchenco, T. Baldwin, and our families for their foresight and perseverance throughout the project and to the David and Lucile Packard Foundation for support.

References

Arkema, K., S. C. Abramson, and B. M. Dewsbury. 2006. Marine ecosystem-based management: From characterization to implementation. *Frontiers in Ecology and the Environment* 10:525–32.

Barbier, E. B. 2007. Valuing ecosystem services as productive inputs. *Economic Policy* 22(49):177–229.

Carpenter, S. R., R. DeFries, T. Dietz, H. A. Mooney, S. Polasky, W. V. Reid, and R. J. Scholes. 2006. Millennium Ecosystem Assessment: Research needs. *Science* 314:257–58.

Chee, Y. E. 2004. An ecological perspective on the valuation of ecosystem services. *Biological Conservation* 120:549–65.

Constable, A. J., W. K. de la Mare, D. J. Agnew, I. Everson, and D. Miller. 2000. Managing fisheries to conserve the Antarctic marine ecosystem: Practical implementation of the Convention on the Conservation of Antarctic Marine Living Resources (CCAMLR). *ICES Journal of Marine Science* 57:778–91.

Crowder, L. B., G. Osherenko, O. R. Young, S. Airamé, E. A. Norse, N. Baron, J. Day et al. 2006. Resolving mismatches in US ocean governance. *Science* 313:617–18.

de Groot, R. S., M. A. Wilson, and R. M. J. Boumans. 2002. A typology for the classification, description and valuation of ecosystem functions, goods and services. *Ecological Economics* 41:393–408.

Farber, S. C., R. Costanza, and M. A. Wilson. 2002. Economic and ecological concepts for valuing ecosystem services. *Ecological Economics* 41:375–92.

Halpern, B. S., S. Walbridge, K. A. Selkoe, C. V. Kappel, F. Micheli, C. D'Agrosa, J. F. Bruno et al. 2008. A global map of human impacts on marine ecosystems. *Science* 319:948–52.

Holling, C. S., and G. K. Meffe. 1996. Command and control and the pathology of natural resource management. *Conservation Biology* 10:328–37.

Jackson, J. B. C., M. X. Kirby, W. H. Berger, K. A. Bjorndal, L. W. Botsford, B. J. Bourque, R. H. Bradbury et al. 2001. Historical overfishing and the recent collapse of coastal ecosystems. *Science* 293:629–38.

Leslie, H. M., and K. L. McLeod. 2007. Confronting the challenges of implementing marine ecosystem-based management. *Frontiers in Ecology and the Environment* 5(10):540–48.

Lester, S. E., C. Costello, J. A. Barth, S. D. Gaines, and B. S. Halpern. In review. Ecosystem service trade-off analysis. *Ecological Applications.*

Lotze, H. K., H. S. Lenihan, B. J. Bourque, R. H. Bradbury, R. G. Cooke, M. C. Kay, S. M. Kidwell, M. X. Kirby, C. H. Peterson, and J. B. C. Jackson. 2006. Depletion, degradation, and recovery potential of estuaries and coastal seas. *Science* 312:1806–9.

Murawski, S. A. 2007. Ten myths concerning ecosystem approaches to living marine resource management. *Marine Policy* 31:381–690.

Nelson, E., S. Polasky, D. J. Lewis, A. J. Plantinga, E. Lonsdorf, D. White, D. Bael, and J. J. Lawler. 2008. Efficiency of incentives to jointly increase carbon sequestration and species conservation on a landscape. *Proceedings of the National Academy of Sciences* 105:9471–76.

Orr, J. C., V. J. Fabry, O. Aumont, L. Bopp, S. C. Coney, R. A. Feely, A. Gnanadesikan et al. 2005. Anthropogenic ocean acidification over the twenty-first century and its impact on calcifying organisms. *Nature* 437:681–86.

Partridge, E. 2001. Future generations. In *A companion to environmental philosophy*, ed. D. Jamieson. Oxford, UK: Blackwell.

POC (Pew Oceans Commission). 2003. *America's living ocean: Charting a course for sea change. A report to the nation.* Washington, DC: Pew Trusts.

Rodríguez, J. P., T. D. Beard Jr., E. M. Bennett, G. S. Cumming, S. Cork, J. Agard, A. P. Dobson, and G. D. Peterson. 2006. Trade-offs across space, time, and ecosystem services. *Ecology and Society* 11(1):28. http://www.ecologyandsociety.org/vol11/iss1/art28/.

Rosenberg, A. A., and K. L. McLeod. 2005. Implementing ecosystem-based approaches to management for the conservation of ecosystem services. In Politics and socio-economics of ecosystem-based management of marine resources, ed. H. I. Browman and K. I. Stergiou. *Marine Ecology Progress Series* 300:270–74.

Rosenberg, A. A., W. J. Bolster, K. E. Alexander, W. B. Leavenworth, A. B. Cooper, and M. G. McKenzie. 2005. The history of ocean resources: Modeling cod biomass using historical records. *Frontiers in Ecology and Environment* 3:84–90.

Ruckelshaus, M., T. Klinger, N. Knowlton, D. P. DeMaster. 2008. Marine ecosystem-based management in practice: Scientific and governance challenges. *BioScience* 58:53–63.

Salomon, A. K., N. M. Tanape Sr., and H. P. Huntington. 2007. Serial depletion of marine invertebrates leads to the decline of a strongly interacting grazer. *Ecological Applications* 17:1752–70.

Serreze, M. C., M. M. Holland, and J. Stroeve. 2007. Perspectives on the Arctic's shrinking sea-ice cover. *Science* 315:1533–36.

Sivas, D. A., and M. R. Caldwell. 2008. A new vision for California ocean governance: Comprehensive ecosystem-based marine zoning. *Stanford Environmental Law Journal* 27:209–70.

USCOP (US Commission on Ocean Policy). 2004. *An ocean blueprint for the twenty-first century.* Final report. Washington, DC: USCOP. ISBN 0 9759462 0 X.

Young, O. R., G. Osherenko, J. Ekstrom, L. B. Crowder, J. Ogden, J. A. Wilson, J. C. Day et al. 2007. Solving the crisis in ocean governance: Place-based management of marine ecosystems. *Environment* 49:20–32.

About the Editors of *Ecosystem-Based Management for the Oceans*

Karen L. McLeod, a marine ecologist, is the Director of Science for the Communication Partnership for Science and the Sea (COMPASS), based at Oregon State University. Born in a small town in eastern Pennsylvania not too far from the Atlantic, a fascination with oceans led her to the Turks and Caicos Islands where she first encountered the astounding beauty and fragility of coral reefs and learned firsthand about the challenges inherent in marine conservation. After receiving a BA from Franklin and Marshall College and an M.S. from the University of South Florida, she headed west to pursue a PhD at Oregon State University, with field seasons studying predator-prey interactions among coral reef fishes in the Bahamas. As a doctoral student, she became increasingly concerned about the gap between what we know about oceans and the management and policies that govern our interactions with oceans, and committed herself to working to the interface between science and policy. Her current research interests include ecosystem services, the coupling between human and natural systems, and more broadly the scientific underpinnings of management and policy in marine systems. She has been actively involved in numerous projects and publications related to ecosystem-based management (EBM), most notably the development of a scientific consensus statement (signed by over 200 experts in 2005) that defined marine EBM and its key underlying principles. Dr. McLeod's work has appeared in *BioScience*, *Ecology*, *Frontiers in Ecology and the Environment*, and *Science*, among others, and this is her first book. She currently resides in Corvallis, Oregon where she and her husband, three-year old daughter, dog, and two cats enjoy their proximity to mountains, old growth forests, and the sea.

Heather M. Leslie, a marine conservation scientist, is the Peggy and Henry D. Sharpe Assistant Professor of Environmental Studies and Biology at Brown University. In her research, teaching and policy work, Dr. Leslie seeks to better understand the drivers of ecological and social dynamics in marine systems, and to more effectively integrate science into marine policy and management. Born in southern California and raised in New England, Dr. Leslie has always been close to the sea. She has worked on coastal science and conservation issues throughout the US as well as in Mexico and New Zealand. Dr. Leslie received an AB in Biology from Harvard University and a PhD in Zoology from Oregon State University, and conducted postdoctoral research at Princeton University. Her research interests include coastal marine ecology, design and evaluation of marine conservation strategies, and human-environment linkages in coastal areas. Her work has appeared in the *Proceedings of the National Academy of Sciences*, *Ecology*, *Conservation Biology*, and *Frontiers in Ecology and the Environment* and has been covered by *The New York Times*, the BBC, *Science Daily*, and the Environmental News Service. This is her first book. She lives in Providence, Rhode Island, close to Narragansett Bay, with her husband and 3-year-old son.

Contributor Biographies

Octavio Aburto-Oropeza is a marine biologist and a professor at the Universidad Autónoma de Baja California Sur. His research has focused on commercially exploited reef fish and marine protected areas management in the Gulf of California. He leads a research group that studies the importance of mangroves for the local fisheries and is part of a regional project that deals with the conservation of marine top predators and spawning aggregations. Currently, Octavio is a PhD student at the Scripps Institution of Oceanography.

Lilian (Naia) Alessa is an Associate Professor of Biological Sciences at the University of Alaska, Anchorage. She heads the Resilience and Adaptive Management Group at UAA and has served on the board of the Arctic Research Consortium of the United States. She currently conducts extensive research on human adaptation to climate change, funded by the National Science Foundation, including International Polar Year projects such as the Indigenous Arctic Observing Network. Canadian-born and raised, Alessa holds a PhD in cell biology from the University of British Columbia and has extensive training in cognitive psychology. Her studies of cellular organization greatly inform her current approaches to social ecological complexity.

Edward B. Barbier is the John S Bugas Professor of Economics, Department of Economics and Finance, University of Wyoming. Professor Barbier has over 25 years experience as an environmental and resource economist, working on natural resource and development issues as well as the interface between economics and ecology. His applied work has focused on land degradation, wildlife management, trade and natural resources, coastal and wetland use, tropical deforestation, biological invasions, and biodiversity loss.

Brad Barr received a BS from the University of Maine and an MS from the University of Massachusetts and is completing a PhD at the University of Alaska. He is a Senior Policy Advisor with NOAA's National Marine Sanctuary Program. He serves on the Boards of Directors of the George Wright Society, Coastal Zone Canada Association, and Science and Management of Protected Areas Association, and he is a member of the IUCN World Commission on Protected Areas. His current research focus is ocean wilderness.

Donald F. Boesch is Professor and President at the University of Maryland Center for Environmental Science. A biological oceanographer, he has conducted research along the US East and Gulf Coasts and in Australia and the East China Sea. He has been particularly engaged in scientific assessments related to the coastal ocean environment and large-scale ecosystem restoration, notably for the Chesapeake Bay, the Mississippi River Delta, the Greater Everglades, and the Baltic Sea. He was a scientific advisor to both the Pew Oceans Commission and the US Commission on Ocean Policy, both of which emphasized the need for ecosystem-based management.

James W. Boyd is an economist dedicated to improved conservation and environmental protection. He is currently a Senior Fellow at Resources for the Future. He has held visiting professorships at Stanford University and Washington University in St Louis. He received his PhD in Applied Microeconomics from the Wharton School and has served on National Academy of Science and other advisory panels, including most recently the US Environmental Protection Agency's Committee on Valuing Ecological Systems and Services.

Larry B. Crowder is Stephen Toth Professor of Marine Biology and Director of the Center for Marine Conservation in the Nicholas School of the Environment and Earth Sciences at Duke University. Crowder's experience lies in marine ecology and fisheries. His research centers on population and community ecology. His applied interests include protected species/fisheries conflicts, especially bycatch in fishing gear. Recently his focus has also included marine conservation biology and policy as well as international affairs.

Richard Cudney-Bueno is an Associate Program Officer in the Conservation and Science Program of The David and Lucile Packard Foundation. He holds an MS and PhD in Natural Resources and Cultural Anthropology from the University of Arizona and has worked on science, management, and conservation in the Gulf of California for 15 years. He is co-founder of the PANGAS Project, a multi-institutional and bi-national initiative for ecosystem-based research and management of coastal fisheries. Richard is also an Adjunct Professor at the School of Natural Resources of the University of Arizona, a Research Associate of the Institute of Marine Sciences at the University of California-Santa Cruz, a Senior Fellow of the Environmental Leadership Program, and member of the Advisory Board of Advanced Conservation Strategies.

Marìa de los Ángeles Carvajal is from Sonora, Mexico. She received her MSc in Marine Science at the ITESM Guaymas Campus, in Mexico. She served as Executive Director for the Gulf of California region with Conservation International (CI) for 16 years, during which she played a major role in implementing a regional conservation strategy and establishing six protected areas. She is a founder and member of EcoCostas, a Latin America Network; a founder and member of ALCOSTA (a Mexican Alliance) and a 2006 Donella Meadows Fellow. Currently, she is developing a local conservation organization, SuMar–Voces por la Naturaleza.

Timothy Essington is an Associate Professor at the School of Aquatic and Fishery Sciences, University of Washington. He studies the role of fish and fisheries in marine food webs in a range of ecological settings, including the central Pacific, the Baltic Sea, and Puget Sound.

Exequiel Ezcurra has devoted his career to the study of northwestern Mexico. He has published more than 150 research papers and books, including 3 books and an award-winning film on the Gulf of California. He was honored with a Conservation Biology Award and a Pew Fellowship in Marine Conservation, and was President of Mexico's National Institute of Ecology and Provost of the San Diego Natural History Museum. Currently, he is Director of the University of California Institute for Mexico and the United States (UC MEXUS) and Professor of Ecology at the University of California, Riverside.

Steve Ganey is Director of Regional Fisheries Initiatives for The Pew Charitable Trusts. Previously, he spent five years directing the Regional Marine Conservation Project for a consortium of philanthropic foundations, where he provided strategic direction for U.S. marine conservation advocacy. He is experienced in ocean and fisheries conservation and management and has served as staff for several organizations, including the Pew Oceans Commission and Alaska Marine Conservation Council.

Erica B. Goldman is a science writer for the Maryland Sea Grant College, part of the University of Maryland Center for Environmental Science. She holds a doctorate in biology from the University of Washington and has studied firsthand some of the world's great marine ecosystems--including Antarctica, Chesapeake Bay, Puget Sound, Lake Baikal in Siberia, and deep-sea hydrothermal vents in the eastern Pacific. She's committed to bridging the gap between science and policy to help preserve the marine environment for future generations.

Anne D. Guerry became an ecologist after seeing the light in the forests of New England. She received her BA from Yale University, MS in Wildlife Ecology from the University of Maine, and PhD in Zoology from Oregon State University. She is currently Lead Scientist of the Marine Initiative of the Nature Capital Project, a joint venture among Stanford's Woods Institute for the Environment, The Nature Conservancy, and World Wildlife Fund; her work focuses on modeling and mapping ecosystem services provided by marine and coastal systems. Anne aims to do good ecological science in the service of society and to still have some time to see the light in the forests with her family.

Frédéric Guichard is an Associate Professor of Biology at McGill University in Canada. He received his PhD from Laval University in Quebec in marine ecology and spent 2 years as a postdoctoral fellow at Princeton University in theoretical ecology before starting his current position. His research interests are focused on the importance of space for understanding of ecological dynamics. His research projects involve theoretical and field studies of larval transport, and its implications for demographic fluctuations, species coexistence, speciation, and ecosystem processes.

Elliott L. Hazen is a postdoctoral researcher at the Duke University Marine Laboratory. His interests lie within fisheries ecology, more specifically, understanding the link between food webs and oceanography in marine ecosystems. Although most of his research focuses on predator–prey ecology, he is also involved with projects to understand the effects of anthropogenic change on marine ecosystems.

Leah Bunce Karrer is a marine conservation scientist with Conservation International's Center for Applied Biodiversity Science, where she directs the Marine Management Area Science Program. She is one of the few, but increasingly popular, marine social scientists in the world. She earned her doctorate from Duke University, advised by Michael Orbach, and her bachelor's from University of Pennsylvania in marine biology. She lives in Washington, D.C., with her husband, who is an increasingly sustainable urban developer.

Les Kaufman is a marine ecologist from Boston University and Conservation International. He received his doctorate from Johns Hopkins University in 1980 under Jeremy Jackson and did his postdoctoral work at Harvard with Karel Liem. His principal area of expertise is the evolutionary biology of coral reef and tropical great lakes fishes, but his interests in natural science range widely from rain forest trees to planetary nebulae. Les lives in Boston with his wife Jackie, who is a research psychologist, and college student son, Justin, who is into human–computer interfaces.

Ilse Kiessling is a Research Fellow with the School of Environmental Research, Charles Darwin University, Australia. She holds a PhD in marine policy and an honours degree in Antarctic science from the University of Tasmania, Australia. Dr. Kiessling has worked with the Australian federal Department of the Environment and Water Resources since 2000 and has extensive experience from the Antarctic to the tropics in marine conservation policy and management.

Ann P. Kinzig is an Associate Professor in the School of Life Sciences at Arizona State University. She received her PhD in Energy and Resources from Berkeley (1994). Before arriving at ASU, she was a postdoctoral researcher and lecturer at Princeton University. Dr. Kinzig's research interests focus broadly on ecosystem services, conservation–development interactions, and the resilience of natural resource systems. She is an Aldo Leopold Leadership Fellow and was selected as the first AAAS Roger Revelle Fellow in Global Stewardship in 1998.

Andrew (Anaru) Kliskey comes from Aoatearoa/New Zealand. He received BS, MA, and PhD degrees from the University of Otago, NZ. He was a postdoctoral researcher at the University of British Columbia, BC, and a Lecturer and then Senior Lecturer in Geography and GIS at the University of Canterbury, NZ. He is an Associate Professor of Biology and Geography & Environmental Studies and a member of the Resilience and Adaptive Management Group at the University of Alaska, Anchorage.

Phil Levin is a research biologist with NOAA Fisheries in Seattle. He is a community ecologist and conservation biologist specializing in fishes in a variety of marine, estuarine, and freshwater habitats. His focus is on complex interactions among ecological processes that ultimately produce ecological patterns. He is currently focused on developing new ecological approaches to solving fisheries and conservation problems.

Jane Lubchenco is Distinguished Professor of Zoology and Valley Professor of Marine Biology at Oregon State University. Her PhD is from Harvard University. A marine biologist and environmental scientist, she co-founded the Leopold Leadership Program, the Communication Partnership for Science and the Sea (COMPASS), the Partnership for Interdisciplinary Studies of Coastal Oceans (PISCO) and Climate Central – organizations devoted to advancing or communicating scientific knowledge to public audiences. She served on the Pew Oceans Commission.

Don Maruska is strategic advisor and facilitator for the San Luis Obispo Science and Ecosystem Alliance. In his consulting practice, he guides groups to solve issues through the process described in his book How Great Decisions Get Made (AMACOM 2004). His 35 years of experience include being CEO of three Silicon Valley companies and recipient of a national Innovation Award in 1988. Don earned his MBA and JD from Stanford University and his BA from Harvard University.

Charlotte B. Mogensen is Head of Section at the Ministry of Food, Agriculture and Fisheries, Denmark, working on international fisheries. Ms. Mogensen worked as Fisheries Policy Officer at WWF European Policy Office, Brussels, advising the WWF network on aspects of the Common Fisheries Policy. She was also a Marine Species Adviser for the Joint Nature Conservation Committee, the UK governmental agency advising on recovery and precautionary management of fish stocks, including sharks and deep-sea fishes. Ms. Mogensen is currently the Regional Vice Chair for the Northeast Atlantic region of the IUCN Shark Specialist Group.

Marjorie L. Mooney-Seus is co-owner and founder of Fort Hill Associates LLC, a management consulting company. Prior to 2000, Ms. Mooney-Seus was manager of the New England Aquarium Conservation Department. She has a Master's Degree in Environmental Policy from Tufts University and graduated cum laude with a BA in Communications from the University of Massachusetts, Amherst.

Kathleen Dean Moore is Distinguished Professor of Philosophy and University Writer Laureate at Oregon State University, where she teaches environmental ethics and the philosophy of nature. Her most recent books are *The Pine Island Paradox, Holdfast: At Home in the Natural World,* and *Riverwalking: Reflections on Moving Water,* explorations of our cultural, spiritual, and moral connections to the natural world.

Robert O'Boyle joined Canada's Department of Fisheries and Oceans at the Bedford Institute of Oceanography in 1977 as an assessment scientist. During 1987–1995, he headed the research programs supporting management of the region's marine resources, and since 1996, he has led the peer review process for science advice. He was Associate Director of Science during 2000–2007 and is now an emeritus scientist at BIO, pursuing research on an ecosystem approach to management.

Jonathan Peacey is currently National Manager, Fisheries Operations with the New Zealand Ministry of Fisheries. Previous roles include Acting Science Manager of the NZ Seafood Industry Council, Fisheries Director of the London-based Marine Stewardship Council, and Chief Policy Analyst with the NZ Ministry of Fisheries. He has also worked as a consultant for a range of fisheries clients. He has a Master's Degree in Resource Management and a BSc (honours) in Zoology from University of Canterbury.

Linwood Pendleton is a Senior Fellow, Director of Economic Research at The Ocean Foundation. He is an expert in marine and coastal economics and policy. Before joining The Ocean Foundation, Dr. Pendleton was a tenured Associate Professor of Environmental Science and Engineering at UCLA and retains an adjunct position there. Dr. Pendleton has an advanced degree in Ecology from Princeton and in Public Policy from Harvard and a doctorate from Yale's School of Forestry and Environmental Studies.

Charles "Pete" Peterson is a marine ecologist and conservation oceanographer at the University of North Carolina in Chapel Hill. He received his bachelor's degree from Princeton and his doctorate from UCSB under Joe Connell. His career has melded academic research with policy development and implementation through 25 years of participation on fisheries and environmental management commissions in North Carolina. He lives with his family on one of the "Southern Banks" of North Carolina, where he advises his sons that land ownership is a single-generation lease from Nature rather than an inheritable legacy.

Garry Peterson is Canada Research Chair and Assistant Professor jointly appointed in Geography and the School of Environment at McGill University, in Montreal, Canada. He is a board member of the Resilience Alliance, an international network of researchers exploring the dynamics of coupled social–ecological systems. He is also currently a researcher at the Stockholm Resilience Centre at Stockholm University in Sweden.

Andrew A. Rosenberg is a Professor in the Institute for the Study of Earth, Oceans, and Space at the University of New Hampshire. He received his PhD in Biology from Dalhousie University in 1984. Dr. Rosenberg was a member of the US Commission on Ocean Policy and is a Senior Scientist with Communication Partnership for Science and the Sea (COMPASS). He was formerly Deputy Director of NOAA Fisheries and has worked extensively in marine resource science and management.

Mary Ruckelshaus is a research biologist with NOAA Fisheries in Seattle. She is a population biologist who uses basic ecological and evolutionary principles to design recovery strategies for imperiled species and ecosystems. Currently she is the chief scientist for the Puget Sound Partnership, a public-private entity charged with restoring the Puget Sound ecosystem in Washington State.

Roly Russell is an ecologist who now spends his time using history to try to understand the complexities of what leads societies to behave in sustainable ways. He just finished a postdoctoral appointment at Columbia University in New York, engaged in research on adaptive and sustainable behavior in social–ecological systems, both ancient and contemporary. He is now living in British Columbia with his family, working on understanding the history of political, social, and ecological change in the Great Plains.

Paul A. Sandifer is Senior Scientist for Coastal Ecology for NOAA's National Ocean Service. Previously, he spent 31 years with the South Carolina Department of Natural Resources, where he served as Director of Marine Resources and as agency director. He is experienced in fisheries and natural resources management, aquaculture, and coastal research and has served on numerous boards and advisory bodies, including the US Commission on Ocean Policy and as Chair of the Atlantic States Marine Fisheries Commission.

Janis Searles Jones is the Vice President for Legal Affairs with the Ocean Conservancy. Janis is a lawyer and activist with broad experience in ocean conservation, public lands, fisheries, and other environmental issues. Janis formerly acted as Senior Regional Counsel and Policy Advisor for the Pacific Regional Office of Oceana and as a staff attorney for the Alaska office of Earthjustice.

Janna M. Shackeroff works in policy and strategic initiatives for NOAA's Papah naumoku kea Marine National Monument. Dr. Shackeroff contributed to this volume while a PhD candidate at the Duke Marine Laboratory in the Nicholas School of the Environment and Earth Sciences, and while a Morris K. Udall Foundation Dissertation Fellow in US Environmental Policy and Conflict Resolution. She specializes in interdisciplinary approaches to ocean research, management, and conservation through research in the historical ecology and human dimensions of coastal areas. Her interests lie in the study of human–marine environmental relationships and working with local communities.

Jorge Torre is the Executive Director of Comunidad y Biodiversidad, A.C. Jorge arrived at the Gulf of California in 1988, and since then he has been involved in marine conservation projects, including studies on endangered species, inventories and monitoring programs, fisheries impacts evaluations, and ecological traditional knowledge. In recent years, he has focused on the application of management effectiveness indicators and the design and evaluation of community-based marine reserves. He received a PhD from the University of Arizona.

Lisa A. Wainger has over 15 years of experience developing integrated ecological and economic analysis tools for government agencies, nonprofits, and businesses that are used to communicate changes in ecological conditions in terms of socioeconomic impacts. Dr. Wainger is currently a Research Associate Professor at the University of Maryland Center for Environmental Science. She received her PhD from the University of Maryland, College Park, in an interdisciplinary program that combined landscape ecology and environmental economics.

Dean E. Wendt is founding director of the San Luis Obispo Science and Ecosystem Alliance. His expertise is in marine biology and marine conservation. He is an Associate Professor at the Center for Coastal Marine Sciences, Cal Poly, San Luis Obispo. He served 2 years on the Master Plan Science Advisory Team as part of the implementation of the California Marine Life Protection Act. Professor Wendt earned his AM and PhD from Harvard University.

Tana Worcester started working at the Bedford Institute of Oceanography as an environmental assessment biologist after receiving her Master of Marine Management degree from Dalhousie University. Since 2003, she has been a science advisor on a variety of oceans and habitat-related issues and has been closely involved with the issues and implementation of the Eastern Scotian Shelf Integrated Management Initiative. She is currently the Coordinator of the Centre for Science Advice in the Maritimes Region of Canada's Department of Fisheries and Oceans.

Index

Notes: Italicized page numbers indicate boxes, tables, or figures. The following abbreviations have been used throughout the index: EBM for ecosystem-based management, TK for traditional ecological knowledge, and LK for local knowledge.

Island Press | Board of Directors